T0258548

IET HISTORY OF TECHNOLOGY SERIES 22

Series Editor: Dr B. Bowers

Television

an international history
of the formative years

Other volumes in this series:

Television
an international history
of the formative years

R.W. Burns

The Institution of Engineering and Technology in association with The Science Museum, London

Published by The Institution of Engineering and Technology, London, United Kingdom

First edition © 1998 The Institution of Electrical Engineers
Reprint with new cover © 2007 The Institution of Engineering and Technology

First published 1998
Reprinted 2007

The Institution of Engineering and Technology
Michael Faraday House
Six Hills Way, Stevenage
Herts, SG1 2AY, United Kingdom

www.theiet.org

British Library Cataloguing in Publication Data
A catalogue record for this product is available from the British Library

ISBN (10 digit) 0 85296 914 7
ISBN (13 digit) 978-0-85296-914-4

Typeset in the UK by Blackpool Typesetting Services Ltd
Printed in the UK by Bookcraft, Bath
Reprinted in the UK by Lightning Source UK Ltd, Milton Keynes

Contents

Appendices

Bibliography

Index

Acknowledgments

For the reasons given in the preface, this book has been based, in the main, on written, primary source documents. The existence of well-staffed archive centres and libraries with well-referenced holdings of such materials is, of course, a *sine qua non* of any written history which has as its objectives accuracy, balance and impartiality. Fortunately for the author of this book, a great deal of published and unpublished documentation exists.

Of the many organisations that have assisted me in my researches, special mention must be made of Radio Rentals Ltd who generously provided me with a travelling scholarship to enable me to undertake some of my research work in the US. Mr S L McCrearie, of this company, was most helpful in arranging for me to visit the Bell Telephone Laboratories at Murray Hill and the Radio Corporation of America at Princeton. I wish to express my gratitude to him and to Radio Rentals Ltd for their kindness and maganimity.

It was a joy to research at Bell Telephone Laboratories, and at the laboratories of the Radio Corporation of America, and to experience their hospitality, generosity and friendliness. Mr G Schindler and Miss R Stumm of BTL, and Mr A Pinsky and Mr Russinoff of RCA, were always accommodating and ever tolerant of the many day-by-day requests which I put forward and I am most appreciative of their efforts. In addition, Mrs E Romano and Miss M Lyons of the American Telephone and Telegraph Company and Mrs Stepno of RCA were very obliging in giving me photocopies and photographs.

In Washington it was a pleasure to meet and obtain the expert assistance of Miss K Fagen, Mr J Dixon and Mr J Linthicum, Federal Communications Commission; Mr J Finiston and Mr Hess, National Archives; Mr E Schamel, National Record Centre (Suitland); and Dr E Sivowitch, Smithsonian Institution. I am much indebted to them and to the staff at both the Library of Congress and the US Patent Office (Crystal City) who made accessible various research materials.

Similarly in New York I received valuable assistance from Mr W Howard, National Broadcasting Company; Miss C Heinz, Broadcast Pioneers Library; and Mr J Poteat, Television Information Office.

More recently, Miss E H Fladger, of the Schaffer library, Schenectady, and Dr S Hochheiser, of AT&T Achives, Warren, have kindly sent me pre-war memoranda, reports and photographs relating to the television activities of the General Electric Company and the Bell Telephone Laboratories, respectively.

In Germany, Dipl. Ing B Rall, of AEG's Forschungsinstitut, Ulm, and Dr O Blumtritt, of the Deutsches Museum, Munich, were very helpful in endeavouring to locate pre-war photographs of German television equipment.

Many persons, libraries and archive centres in the United Kingdom have supplied me with primary source materials, photocopies and photographs. I am especially grateful to the late Mr J A Lodge, Thorn-EMI; Mrs R Edge, EMI Music Archives; Mr P Leggatt, formerly of the BBC; Mrs J Kavanagh, BBC Written Archives Centre; Mrs J Farrugia, Post Office Archives and Records Centre; the late Mrs B Hance, and Mr R Rodwell, Marconi Historical Archives; Mr C J Somers, formerly of Ferranti Ltd; and Mr R M Herbert, formerly of Baird Television Ltd.

Since a vast amount of my time and endeavours were spent in libraries and archive centres I should like to express my gratitude for the courtesy and support of the staffs of the libraries and archive centres mentioned above, and also of the library and records staff of the Institution of Electrical Engineers, the Science Museum, the National Information Reference Service, the British Library, the British Newspaper Library, the British Lending Library, the Public Record Office, the Nottinghamshire County Council, the University of Nottingham and Nottingham Trent University.

Many quotations from written primary and secondary sources have been included in this book. I am most grateful to the following persons for giving me permission to use extracts from their organisations' publications and/or private papers: the editors of *The Daily Telegraph, Daily Mail, The Birmingham Post and Mail, Folkstone Herald, Evening Telegraph and Post, Hastings and St Leonards Observer, Sunday Express, The Times, The Guardian, Glasgow Herald, The New York Times, Daily Express, Evening Standard, Electronics and Wireless World, Nature*, the journals of the Royal Television Society, *The Engineer*, the journals of the Institute of Physics and the Institute of Electrical and Electronics Engineers; Sir Clifford Cornford, Mr B C Owers, Mr R Price, Mr O P Sutton and Mr G C Arnold.

Especially warm thanks are due to those persons who gave me—in all cases most willingly—permission on behalf of their organisations to use photographs and drawings. To the following I am deeply indebted: Mr S L McCrearie, formerly service director of Radio Rentals Ltd; Ms R Pearce of Granada Communications; Mr G D Speake, formerly deputy director of research, GEC Research (Marconi Research Centre) Mr W P Lucas, formerly administration manager, Thorn EMI Central Research Laboratories; the late Mr J A Lodge, formerly archivist of Thorn EMI Central Research Laboratories; Mr D P Leggatt, formerly chief engineer, external relations, BBC; the controller of Her Majesty's Stationary Office; Mr B C Owers, secretary, The Rank Organisation; Mr A Kemp, editor of the *Glasgow Herald*; Mr V Yates, company archivist and historian, Selfridges; Mrs E Romano of the Photo Centre, AT&T; Mr A Pinsky, of Scientific Information Services, RCA; and Mr R M Herbert and Professor D Shoenberg.

Preface

Until 1873 no discovery had been made of a physical phenomenon which would enable moving images of scenes and objects to be transmitted electrically from one place to another. This position changed in 1873 when the photoconductive property of selenium was discovered. Its discovery stimulated much interest in 'seeing by electricity' and numerous proposals were advanced from 1878 by which it was hoped that television would be realised. However, for more than 45 years success eluded the early inventors. Then, on 26 January 1926, Baird demonstrated for the first time anywhere a rudimentary form of television. Progress was now rapid and many inventors, scientists and engineers in several countries (but principally in the UK, the USA, Germany, France, the USSR and Japan) engaged in the development of a system that could have public entertainment value.

Just ten years after Baird's achievement work commenced on the London television station. The world's first regular, public, high-definition television broadcasting system was inaugurated in the UK on 2 November 1936. Soon afterwards similar high-definition systems were established in the USA, France, Germany, the USSR and Japan. Unfortunately, the onset of hostilities in 1939 curtailed the spread of television, worldwide, and from 1940 to 1945 very little progress was accomplished in the domestic television field.

The objective of this book is to present a balanced history of world television from 1878 to 1940 based, in the main, on written, primary source documents, i.e. on private letters, memoranda, committee minutes and reports, and on published articles, papers, patents, editorials, reports and letters in the technical and non-technical press. The book considers the factors—technical, financial and general—which led to the setting-up of the world's earliest public high-definition television broadcasting services.

Since the unpublished and published literature on television is vast, some constraints have had to be imposed on the scope of this book. It does not deal with the pre-war history of colour television or the pre-war history of cinema television except in some isolated instances. As a consequence, the work and achievements of the Scophony company have not been considered.

Furthermore, because the number of pre-war television patents runs into many hundreds some judicious selection has had to be effected. A description of every television patent would not only be extremely tedious but also unnecessary. Many of these patents were either naive or simplistic in concept, or were minor variants of other, more basic, patents. A great number of patents were never utilised commercially. For those that had some worth, or which are important historically because they highlight trends in development, an attempt has been made to present groups of patents in a synoptic, and thereby a meaningful, way (see Tables 6.1, 7.1, 15.1 and Appendix 2).

A contextual approach has been adopted for this history to stress the influence of one discipline upon another and to illustrate the point that sometimes progress in a given field can only be attained when discoveries in related fields have been demonstrated. Thus, Chapters 1 (Images and Society) and 4 (Persistence of vision and moving images) seek to confirm and to explain the very appreciable public interest, over many centuries, in the formation and display of images by any means, and why it was that a few years after Willoughby Smith's letter of 1873 which announced the photoconductive property of selenium, much attention was given by respected scientists and inventors to 'seeing by electricity'. Chapters 2 (Images by wire, picture telegraphy) and 7 (Developments of importance to television) describe some of the inventions and fundamental scientific and engineering advances which impacted on the progress of television. In addition, by this approach the general nature of the mechanics of development can be delineated and compared, e.g., by reference to the histories of cinematography and television.

In writing any history it is of course essential that the events which are portrayed are interpreted from a contemporaneous position. Considerable endeavour has been taken to present the views of scientists and engineers, newspaper reporters and editors, cartoonists and others so that the evolution of television is seen from the perspective of the times and not from the standpoint of a later generation. Accordingly, the unfolding saga has been profusely illustrated by contemporary quotations.

Table 23.2 shows that only four television standards were in operation, worldwide, in 1955. These were based essentially on the 1930s work of EMI and RCA. Since some controversy still seems to exist over whether EMI's system owed much to that of RCA, an examination of the available evidence (some of which has not been previously published) has been undertaken in an attempt to resolve the issue. This evidence is given in Chapter 19 and indicates that the EMI all-electronic television investigations were independent of similar investigations being advanced by RCA. Further new (unpublished) material evinces the fact that Baird Television Ltd, in c. 1930, approached first RCA and then HMV (which became part of EMI) in order to form a commercial association.

Much care and effort have been spent to ensure that an impartial and accurate history has been written. More than 1300 references, to both primary and secondary sources, have been detailed, and the text has been illustrated with many diagrams, photographs and tables of data.

Abbreviations

a.c.	alternating current
AT&T	American Telephone and Telegraph Company
BBC	British Broadcasting Corporation
BIT	Baird International Television Ltd
BTDC	Baird Television Development Company
BTL	Baird Television Ltd
BTL	Bell Telephone Laboratories
CBS	Columbia Broadcasting System
CDC	Compagnie des Compteurs
c.r.o.	cathode ray oscilloscope
c.r.t.	cathode ray tube
d.c.	direct current
DRP	Deutsches Reichspost
EMI	Electric and Musical Industries Ltd
ERPI	Electrical Research Products Inc
FCC	Federal Communications Commission
FRC	Federal Radio Commission
GE	General Electric (US)
GPO	General Post Office (UK)
HF	High frequency
HMV	His Master's Voice
ICAN	International Congress on Air Navigation
M-EMI	Marconi-EMI Television Company Ltd
MWT	Marconi Wireless Telegraph Company Ltd
NBC	National Broadcasting Company
NHK	Nippon Hoso Kyokai
NTSC	National Television Standards Committee
p.e.	photoelectric
POED	Post Office Engineering Department
PTT	Postes, Télégraphes et Téléphones

RCA	Radio Corporation of America
RMA	Radio Manufacturers Association (UK)
RMA	Radio Manufacturers Association (US)
RPZ	Reichspost Zentralamt
RRG	Reich Rundfunk Gesellschaft
RTS	Royal Television Society
TAC	Television Advisory Committee
TSC	Technical Subcommittee (of the TAC)
UHF	ultra high frequency
VHF	very high frequency
WTB	Wireless Telegraphy Board

Part I

The era of speculation
1877 to c. 1922

Chapter 1

Images and society (c. 16th century to c. mid-19th century)

Television was privately demonstrated, in a rudimentary way, on 26 January 1926, by John Logie Baird, a Scottish electrical engineer. On that day he showed his crude, flickering, half-tone televised images to a group of approximately 40 scientists from the Royal Institution, London[1]. No other person, anywhere, had previously reproduced such images.

Baird was one of a small number of lone inventors who, during the first half of the 1920–30 decade, worked determinedly to devise a system of 'seeing by electricity'. Other experimenters included C F Jenkins of the US, D von Mihaly, a Hungarian working in Germany, and E Belin of France.

By any criterion Baird's achievement was an outstanding one. For nearly 50 years from 1878 scientists, engineers and inventors in the UK, the US, France, Germany, Russia and elsewhere had sought to find a solution to the problem of transmitting, electronically, moving images from one place to another[2]. Their objective had been to implement a method which would enable vision signals to be sent to a distant observer in a way similar to that which had permitted sound signals to be transmitted. In 1877 Alexander Graham Bell had invented the telephone which allowed 'hearing by electricity', or telephony, to be readily introduced, and soon afterwards proposals were advanced for systems which it was thought would enable 'seeing by electricity', or television, to be realised.

However, the problem of distant vision was of an altogether different order of complexity compared with that of telephony and success eluded the many persons, from laymen to university professors, who engaged in propounding and advancing ideas for its solution (see Table 1.1).

The essential ingredient which made this activity possible was the discovery of the photoconductive properties of selenium by Willoughby Smith and his assistant May in 1873[3]. Prior to 1873, no usable effect had been demonstrated which related changes of light flux to changes of an electrical quantity—a necessary requirement in any system of television. From this fact it may be concluded that the history of television dates from Willoughby Smith's 1873 letter to the editor of *The Journal of*

Table 1.1   Dates (and names of inventors) of some distant vision proposals for the period 1878–1924

1878 de Paiva	1904 von Jaworsky	1920 Kakourine
1879 Perosino	and Frankenstein	1920 Egerton
1879 Senlecq	1904 Ribble	1921 Whiston
1880 Carey	1906 Lux	1921 Schoultz
1880 Ayrton and Perry	1906 Rignoux	1922 Belin
1880 Middleton	1906 Diecjmann and Glage	1922 Valensi
1880 Sawyer	1907 Rosing	1922 Jenkins
1880 Le Blanc	1908 Campbell Swinton	1922 Rtcheoloff
1881 Senlecq	1908 Adamian	1923 Baird
1881 Bidwell	1908 Anderson and	1923 Hammond
	Anderson	
1882 Lucas	1908 Sellers	1923 Zworykin
1884 Nipkow	1909 Ruhmer	1923 Gardner and Hineline
1889 Weiller	1910 Schmierer	1923 Nisco
1890 Sutton	1910 Ekstrom	1923 Western Electric
1893 Pontois	1910 Hoglund	1923 Stephenson and
		Walton
1894 Majorana	1911 Rosing	1923 Robb and Martin
1894 Jenkins	1911 Campbell Swinton	1924 Sequin and Sequin
1895 Nystrom	1911 Rosing	1924 Blake and Spooner
1897 Szczepanik	1914 Lavington Hart	1924 Alexanderson
1898 Vol'fke	1915 Voulgre	1924 Hoxie
1898 Dussaud	1915 Dauvillier	1924 Takayanagi
1899 Bolumordivinov	1917 Nicolson	1924 Apollinar and Zeitlin
1902 Coblyn	1919 von Mihaly	1924 McCreary
1902 von Bronk	1919 Sandell	1924 d'Albe
1903 Belin and Belin	1920 Baden Powell	1924 Dieckmann

the Society of Telegraph Engineers, but such a view would ignore all the work which had been undertaken from 1843 on picture telegraphy—the process by which images of documents and the like can be sent electrically from one place to another—and which had an influence on early distant vision schemes.

In picture telegraphy and television, scanning and synchronising are fundamental operations. Whether the object, whose image is to be transmitted, is of a two-dimensional form as in picture telegraphy, or of a three-dimensional nature as in television, it is crucial in practice for it to be analysed, point by point by a scanning beam, and for the received image to be synthesised by another scanning beam synchronised to that at the transmitter. An important difference between picture telegraphy and television is the speed of scanning. In television engineering an image must be scanned in a fraction of a second—typically 1/50 of a second—whereas in picture telegraphy practice the duration of a scan can be several minutes.

The principles of picture telegraphy and television, although not commercially adopted until the 20th century, were slowly evolved during the 19th century[4], a century in which the display of images, by various means, fascinated all who saw them. Photography, cinematography, seeing by electricity and picture telegraphy all had their beginnings during this time. The first three of these are concerned with the capture of an image of a static or moving scene or object and, with picture telegraphy, the subsequent display of that image to one or more persons. There is a symbiotic relationship between the mediums of picture telegraphy and television, and still and cine photography, and the individual media forms progressed somewhat contemporaneously with their associated forms (see Table 1.2)[5].

During the early formative stage of television the knowledge and experience which had been gained from experiments on picture telegraphy was of beneficial use. Later, in the formative stage, when practical television systems were being

Figure 1.1   *Public interest in shadow theatre. The earliest televised objects/images were displayed as silhouettes, i.e. without half tones*

Reproduced by permission of the Science Museum

Table 1.2   The mediums of picture telegraphy and television, and still and cine photography progressed somewhat contemporaneously

Date	Photography/cinematography Picture telegraphy/television
1898	first image on paper/leather (Wedgewood)
1814	sensitivity of iodine to light demonstrated (Davy)
1826	first permanaent photographic image (Niepce)
1835	silver chloride process described (Fox Talbot)
1837	printing electric telegraphs (Davy, Morse)
1839	Daguerrotype method devised (Niepce and Daguerrotype)
1843	first patent on picture telelgraphy (Bain)
1848	first picture telegraphy apparatus constructed and demonstrated between Brighton and London (Bakewell)
1851	wet collodion process introduced (Archer)
1862	first public picture telegraphy service, between Lyon and Paris; later abandoned (Caselli)
1869	public picture telegraph service between Paris and Lyon, service soon abandoned (Meyer)
1871	siver bromide plates with gelatine coating as binding agent introduced (Maddox)
1872	picture telegraphy experiments between Paris and Marseilles (D'Arlincourt)
1873	photoconductive property of selenium described (Smith)
1878	first ideas advanced for 'seeing be electricity', or television, (see Table 1.1)
1878	series of successive pictures of moving objects obtained (Muybridge) (see Table 4.2)
1879	zoogyroscope, a sequence photograph projector demonstrated (Muybridge)
1885	sensitised paper film rolls available (Eastman)
1889	celluloid film introduced (Eastman)
1889	various cameras and projectors for exposing and projecting film suggested/demonstrated (see Table 4.3)
1893	work on phototelegraphy (Amstutz)
1896	cinematography firmly established (see Table 4.3)
1907	Paris–London phototelegraphy service inaugurated but short lived (Korn)
1926	first public demonstration of television (Baird)

introduced, it was found desirable to base certain aspects of their design on distinct factors which had influenced cinematography. Persistence of vision, flicker, the number of lines necessary for good definition[6], the production techniques of television programmes—lighting, make-up, scene composition and stage scenery—were matters which were partly or wholly determined from the practice of motion picture production.

This symbiosis, and the successes which were being achieved with one of the visual communication methods, probably had the effect of stimulating those who were eager to further the other forms. Although for nearly 50 years, from 1878 to 1926, the early distant vision workers and proponents lacked real success, there was never a lengthy period that was devoid of some television activity.

An important determinant in the advancement of the new media was the public interest. The public, particularly in the 19th century, had a great curiosity for exhibitions, spectacles, and shows of all genres, from those which were based on the display of human deformities and characteristics to those which were of an informative, educative or stimulating nature. This enthusiasm also prevailed in the 20th century and when television schemes were being publicly demonstrated in the 1920s much interest was evinced, Figure 1.2. When, at the British Association meeting held in Leeds in 1927, Baird showed some flickering, ghostly, low-definition, peep show like images of a person being televised by his noctovisor the exhibit had such a popular appeal that the police had to be called out to regulate the queues[7]. And when in 1930 he displayed his large screen 30-line televised images at the London Coliseum the public interest was so great the house full notice had to be brought out. This favourable public support was a feature, too, of the exhibitions of the system in Berlin, Paris and Stockholm[8].

For centuries peep shows, magic lantern shows, theatre and pageants had attracted the purveyors of magic and those for whom personal power and mastery were prime objects[9]. The earliest arrangers of two-dimensional moving images were magicians. Pliny the Elder relates that the god Hercules would regularly show himself, gigantic in stature, among the vapours of the fire kindled in his temple at Tyre; Aesculapius often displayed himself to his worshippers at Agrigento, and the temple at Enginium, also in Sicily, was so celebrated for apparitions of the two divinities, Hera and Aphrodite, that the shrine became a place of pilgrimage. Iamblichus tells us that it was the priests, who were also magicians, who were responsible for these appearances, and that they were always accompanied by smoke and vapour[10].

These illusions were created by means of concave metal mirrors, mostly fabricated from silver, and it was upon their surfaces that the magicians, and later the wandering entertainers of the Middle Ages known as the tregetours, depended for their most spectacular effects. Skill was needed to create illusions by the use of concave mirrors alone and probably only simple-minded or superstitious audiences could be duped. But with the invention of the magic lantern in the 17th century the illusionists and sorcerers had an instrument which much enhanced their powers of producing impressive phantasms.

The magic lantern was invented in c. 1640 by Anthanasius Kircher, a Jesuit priest and a scientist of some repute, and was described by him in his book 'Ars Magna Lucis et Umbrae' of 1645. His pictures, Figure 1.3, were painted on long strips of glass, every part of which was opaque except the figures, and his screen was placed between the spectators and the lantern. This could move towards or away from the screen so that, by skilful manipulation of his apparatus, Father Kircher could make his figure images appear at one moment as big as giants, the next as small as dwarfs. They would advance, retreat, dissolve into seeming nothingness and then reappear in quite different shapes[11].

The phantasmagoria, Figure 1.4, which functioned in a similar way to Kircher's lantern, was named by Philipsthal in 1802 when he exhibited it with enormous

OLYMPIA — 1928

Injured one: "I WILL see that Televisor, even if they kill me."

Figure 1.2   The crude images produced by Baird's televisor attracted much public attention

Reproduced by permission of the Science Museum Source: *Television and Short Wave World* September 1928

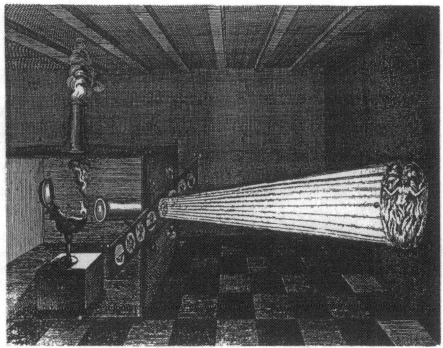

Figure 1.3   *Depiction of the magic lantern invented in c. 1640 by Athanasius Kircher and described by him in his book 'Ars Magna Lucis et Umbrae'*

Reproduced by permission of the Science Museum

general interest in London and Edinburgh. Sir David Brewster, a noted scientist, has left a description of Philipsthal's technique[12]:

'The small theatre of exhibition was lighted only by one hanging lamp, the flame of which was drawn up into an opaque chimney or shade when the performance began. In this semi-obscurity the curtain rose and displayed a cave with skeletons and other terrific figures in relief upon its walls. The flickering light was then drawn up beneath its shroud, and the spectators, left in total darkness, found themselves in the midst of thunder and lightning. A thin transparent screen had, unknown to the spectators, been let down after the disappearance of the light, and upon it the flashes of lightning and all the subsequent appearances were represented. The thunder and lightning were followed by the ghosts, skeletons and known individuals whose eyes and mouths were made to move by the shifting of combined slides. After the first figure had been exhibited for a short time, it began to grow less and less, as if removed to a great distance, and at last vanished in a small cloud of light. Out of this same cloud another figure began to appear and gradually grew larger and larger and approached the spectator until it attained its perfect development.'

Figure 1.4   An example of phantasmagoria

Reproduced by permission of BFI Stills, Posters and Designs

One writer has said: 'The aerial images projected by magicians with the aid of concave mirrors and the phantoms flung upon the screen by the lantern were the distant ancestors of the fantastic, irrational, magical aspects of the film.'[13] The same comment applies to the images now seen on television screens. Another means, the peep shows, and the panoramas and dioramas which were developed from them, offered their audiences two-dimensional images which were the precursors of modern newsreels and travelogues as seen on television.

Peep shows and panoramas have a relationship with the camera obscura. In the latter an actual scene is reflected into a darkened chamber, but in the former painted or contrived scenes are used. The principle of generating images in a camera obscura or dark box was known to Alhazen (c. 1000), Roger Bacon (c. 1267) and others, and the camera obscura itself had been described by Leonardo da Vinci in his notebooks (which, however, had remained unpublished until 1797). G B della Porta, the Italian natural philosopher, gave an account of such a device in his Magia Naturalis (1588). Porta's camera obscura initially consisted of a darkened room with a small hole in one wall so that, when the sun shone on objects outside the room their images were projected onto the inner wall opposite the hole. Following Porta's suggestions, artists began using the cameras to aid them in composing their pictures and to interpret three-dimensional scenes, by recording them as two-dimensional likenesses, on paper or canvas.

A number of improvements to the camera obscura were made in the 16th century and included in Porta's greatly enlarged second edition of his Magia Naturalis (1589). Geronimo Cardano in 1550 proposed the addition of a biconvex lens to obtain a brighter image and in 1568 Daniello Barbaro suggested the use of a diaphragm to secure sharper images. Igrazio Denti in 1573 corrected the reversed image by means of a concave mirror behind the lens and Friedrich Risner (d. 1580) described in his works (published in 1606) methods of enlarging and of reducing the image. He suggested using a portable box rather than a room.

During the 17th century several writers, including Robert Boyle (1669) and Robert Hooke (1679), made reference to portable forms of camera obscura and by the beginning of the 19th century the camera had long been in use.

The camera obscura can show only the scenes and incidents in its immediate vicinity; the peep show, however, by use of mirrors and sometimes of lenses, can give glimpses of distant, ancient or legendary vistas. Alberti is supposed to have made a number of combination clocks and peep shows[14]. His peep show, framed and artificially lighted, consisted of modelled groups of figures placed against a painted background; they were seen not directly but by reflection in a mirror fixed inside the lid of the box[15].

In the 17th century the makers of peep shows preferred scenes from everyday life to those taken from antiquity. A peep show, in the National Gallery of London, painted only on the sides and bottom of a box, shows the interior of a Dutch house but such was the ingenuity of the artist (Hoogstraaten) that a remarkable illusion of the three-dimensional reality is created. Tiles, two chairs and a dog are painted partly on the sides and partly on the bottom of the box, and a table is painted wholly on the floor. Several inclined mirrors and an application of the science of perspective enables the effect to be produced.

Lenses were fitted to the eyepieces of the peep shows in the 17th century, both to enlarge the images and to enhance their apparent three-dimensional presentation.

Peep shows continued to be constructed throughout the 18th and 19th centuries: Charles Dickens in 'Our mutual friend' describes a peep show at a village fair. The peep show was made in various sizes, Figure 1.5, from those suitable as toys to those which would allow a score or more of people to view the scene at the same time. Sometimes melodramas would be shown—in which the peep show operator would pull the pictures up and down in rapid succession—but also subjects of a patriotic nature, for example, the Coronation of George IV, Queen Victoria's visit to the city of London and the Battle of Waterloo, would be displayed (see also Figure 1.6).

The public's fascination with perspective and the optical effects which could be achieved by subtle and dramatic lighting led to the transformation of the peep show into the great popular entertainment, the panorama and the diorama of the 19th century. It was the painter Philip de Loutherbourg who, while experimenting with receding scenes lit from behind, conceived the novel idea of changing the peep show into a picture house which 'like the camera obscura, would hold the spectators as well as the moving images and would exhibit scenes marvellously animated by means of light and so cunningly depicted that they could be viewed directly with as great an illusion of plasticity as if seen through a lens'.

Loutherbourg's eidophusikon had its first performance on 26 February 1781. The performance started with a vision of dawn over London seen from Greenwich Hill. 'The distant city lay shrouded in mist, which gradually lifted as the sun grew

Figure 1.5   A peep show of 1848

Reproduced by permission of the Science Museum

stronger. Cattle loomed up in the foreground grazing in Greenwich Park, while as the day advanced more and more shipping moved up and down the Thames. This scene was followed by a musical interval, the music being specially chosen to harmonise with the picture, and next, in quick succession, scenes of rural life, a stupendous sunset, a fight between peasants and wolves in the Swiss Alps and, finally, the English Fleet advancing to the relief of Gibraltar'.[16]

Loutherbourg later sold his eidophusikon to a Mr Chapman, who had been associated with its management, and it spent many years on provincial tours before it was destroyed by fire at the beginning of the last century. The eidophusikon stimulated the invention of the panorama, and all the other 'oramas', which featured so prominently in the social life of the Regency and Victorian periods. According to one author 'there were almost as many houses showing entertainments of this kind in London [of 1840] as there are cinemas today [1960]'[17]. Of the subjects that were shown mention can be made of the 'Moving diorama of the polar expedition, being a series of views representing the progress of His Majesty's ships the Hecla and Envy in their endeavours to discover a north-west passage from the Atlantic to the Pacific Ocean', at the Covent Garden Theatre; and the 'Grand moving picture of a voyage to the Isle of Wight including a visit to Cowes Regatta', at the Drury Lane Theatre. These dioramas were featured during the intervals of the plays which were being

Figure 1.6 *A 'peep show' of c. 1925 J L Baird is viewing an image produced by his receiving*
equipment

Reproduced by permission of Radio Rentals Ltd

performed, in a way similar to the additional attractions which were a characteristic
of the early cinemas. It is interesting to note that Baird displayed his early large-
screen television images at the Coliseum, when the intervals were in progress.

The enthusiasm for these dioramas spread all over Europe and the US and inventors devised many related forms of entertainment such as the betaniorama, the
cyclorama, the giorama, the pleorama, the kalorama, the kineorama, the octorama,
the physiorama, the typorama, the udorama and the uranorama.

Of the disparate 'oramas', the invention of the panorama is generally attributed
to Robert Barker, an Edinburgh painter. On 19 June 1787 he patented 'an entire
new contrivance or apparatus called by him 'La nature a coup d'oeil' for the
purpose of displaying views of nature at large, by oil painting, fresco, water colours,
crayons or other modes of painting or drawing'. Within two years Barker had
exhibited his first canvas in Edinburgh, Glasgow and London, had obtained financial support from a joint stock company and had received the approbation of Sir
Joshua Reynolds for his apparatus—'I find I was in error in supposing your invention could never succeed, for the present exhibition proves it is capable of producing effects and representing nature in a manner far superior to the limited scale of
pictures in general.' Soon, a more wieldly and distinctive name was coined: one of
Barker's friends suggested 'panorama' (from the Greek for all-embracing view) and
from 1791 the word entered the common vocabulary of the public.

Figure 1.7   A cross-sectional view of a panorama theatre. The painted murals and viewing platforms are clearly shown

Reproduced by permission of BFI Stills, Posters and Designs

From London the exhibition of panoramas spread across the European continent to Paris, Berlin and as far away as St Petersburg, and the panorama vogue caused considerable enthusiasm and competition[18].

One historical picture of the 'Taking of Seringapatam'—in the Mysore war of 1799—was seen by the general public at Somerset House where they 'poured in by hundreds and thousands for even a transient gaze—for such a sight was altogether as marvellous as it was novel. You carried it home, and did nothing but think of it, talk of it and dream of it'. A contemporary account of the 200 feet long painting which evoked such excitement was given by T F Dibdin in his 'Reminiscences of a literary life'[19].

> 'The learned were amazed, and the unlearned were enraptured. I can never forget its first impression upon my own mind. It was as if a thing dropped down from the clouds—all fire, energy, intelligence and animation. You looked a second time, the figures moved, and were commingled in hot and bloody fight. You saw the flash of the cannon, the glitter of the bayonet, the gleam of the falchion. You longed to be leaping from crag to crag with Sir David Baird, who is hallooing his men on to victory! Then again, you seemed to be listening to the groans of the wounded and the dying . . . '

Panoramas became the newsreels of the Napoleonic era. In the United Kingdom the roll of panoramic subjects reflected the public interest in current affairs and the exploits of Nelson, Wellington—the battles of the Nile, Copenhagen, Trafalgar, Corunna, Salamanaca, Waterloo . . .

Elsewhere, in Dublin, Rome, Berlin, Naples and other places, some of the early panoramas were devoted to subjects of a purely topographical or educational interest. The poet John Ruskin asserted[20] that panoramas at their best had displayed 'an attention to truth, and a splendour and care in the execution' which made them 'very truly a school both in physical geography and art'. Whatever their artistic worth, panoramas informed adults and children alike in a manner that was pleasant and beneficial. The paintings added a pictorial dimension to the news items of the daily papers, which were not illustrated and which remained in this state for several decades of the 19th century. Thus, like the present day figures at Madame Tussaud's, London, panoramas added a visual portrayal of the written description.

An especially successful and financially lucrative panorama was that of the Coronation of George IV in July 1821. H A Barker's painting grossed £10 000—at a time when a labourer in London worked for £0.75 per week[21]—and doubtless aided Barker to retire at the early age of 48[22].

The educational benefit of the pictures was augmented by the provision of sixpence (2.5p) booklets which could be bought at the exhibitions. These served not only as souvenirs or mementos of the events but also guided the viewer with outline descriptions of the scenes being observed by summarising their history, geography and topicality.

Of especial interest to the British, with their worldwide empire, were the subjects of geography and travel. For the low wage-earning members of the general public tourism was an unaffordable luxury; for others, although travel was perhaps delightful to contemplate, there were hazards and discomforts to be faced in practice. These were graphically delineated by an anonymous writer to *Blackwood's Magazine*, in an article, published in 1824, in praise of panoramas[23].

'What cost a couple of hundred pounds and half a year, half a century ago, now costs a shilling and a quarter of an hour. Throwing out of the old account the innumerable miseries of travel, the insolence of public functionaries, the roguery of innkeepers, the visitation of banditti, charged to the muzzle with sabre, pistol, and scapulary, and the rascality of the custom-house officers who plunder passport in hand, the indescribable désagréments of Italian cookery and the insufferable annoyances of that epitome of abomination, an Italian bed . . .

'Now the affair is settled in a summary manner. The mountain or the sea, the classic vale or the ancient city, is transported to us on the wings of the wind . . . Constantinople with its bearded and turbaned multitudes . . . Switzerland, with its lakes covered with sunset . . . and now Pompeii, reposing in its slumber of two thousand years, in the very busy Strand [of London] . . . '

In an age which lacked commercial travel operators—Thomas Cook's pioneering conducted continental tours did not start until the mid-1850s—and which also lacked illustrated newspapers and magazines, the 'oramas' were media forms which, within their limitations, offered a role that was to be vastly improved upon by television, *viz.* to entertain, to inform and to educate the public. Effectively, the panorama and its variants made the public receptive to the value of pictorial imagery of reported topics.

Compared with the eidophusikon the early panoramas had two shortcomings, namely their lack of motion and variety. As one anonymous writer put it in 1815: 'Painting is one of the attractive arts cultivated by the ingenuity of man; but in order to complete the pleasure to be derived from it, it's necessary that motion should be imparted to the sublime scenery it copies'.[24] The second limitation was related to the absence of movement. For the price of admission, only one, or at most two, scenes were seen, but at the pantomimes a rapid succession of pictures could be observed by means of Loutherbourg's apparatus.

Moving panoramas were introduced at the beginning of the 19th century and much ingenuity was displayed by the exhibitors of these panoramic vistas. A contemporary account (1820) of the delights to be seen in Covent Garden's 'Harlequin and Friar Bacon, or, The Brazen Head' mentions that a cut-out vessel was drawn across the stage while, behind it, a painted scene rolled in the opposite direction to represent the crossing from Holyhead Bay to Dublin. 'Twilight darkens and still the packet sweeps along, and still remote vessels pass her; the steam boat is seen smoking on its way; the moon rises, throws its rays upon the water and with midnight is gone; sky brightens, and morning shews the mountains round the bay of Dublin . . . this whole scene received great applause'.[25]

Panoramas reached the peak of their dominance in the London theatre area in the 1850s; thereafter a decline set in, although there were at least a dozen such entertainments in the capital in 1870. This fall coincided with the emergence in the 1840s of illustrated journals and magazines. The engraved representations of current events could be produced much faster than those provided by the most rapid painter, and when photographs of news items began to appear, in the same decade, the fidelity of the scenes of the panoramas were compared with those of the new visual medium. 'Even the Daguerrotype pictures—Nature's own transcript of herself—leave us something to desire; how much then must the most cunning of mortals [painters] fall short when dealing with the evanescent changes of light and shadow!'[26] A photograph at that time was not a perfect image of an actual event, but it surpassed in faithfulness anything which could be represented by a panorama. And the science and technology of photography were just in the early formative stages of their development. Photography could be an art form but also it could be a formidable documentary instrument.

The introduction of photography in 1839 was an essential step towards the ultimate achievement of photography of motion, or cinematography[27]. At approximately the same date the earliest devices—the thaumotrope and the phenakistoscope—which demonstrated persistence of vision on which cinematography and television are based, were being offered to the public[28].

Photography and cinematography had a common drawback; they were not instantaneous in portraying distant events and objects. It was necessary for the film or prints to be transported from the scene of a photographic assignment to the printing works of a publishing house. And in the 19th century such journeys could take many days or several weeks. A similar constraint prior to c. 1790 had prevented the rapid dispatch of messages from one place to another.

For millennia the speed of communications had remained that of the fastest runner or the swiftest horse. The Romans considered the creation of some form of public post absolutely crucial for serving the military and administrative needs of their government. Good communications was a *sine qua non* for good organisation. Under the Republic officials in the provinces had their own tabellari and messengers who had been provided with a diploma by the emperor or a provincial governor could requisition carriages from the towns that lay on or near the road along which they passed. Even so, one historian has estimated that these messengers could only cover, on average, about 50 miles a day, so that Constantinople could be reached from Rome in 24 days, and Alexandria in 54.

These times were drastically reduced in France during the Napoleonic era. Then, an extensive network of semaphoric stations, based on the work of Claude Chappe, enabled signals to be received in Paris from Lille in two minutes, from Calais in four minutes 55 seconds, from Strasbourg in five minutes two seconds and from Brest in six minutes 50 seconds. By 1852 France had a system of 556 semaphore stations stretching over a total distance of 4800 km[29].

A substantial increase in the rate of information transmission came with the work of Soemmering, Schilling, Cooke and Wheatstone, Gauss and Weber, and Morse in the early 19th century on electric telegraphy[30]. Their experiments and the stimulus provided by the growth of the railways led to electric telegraphs which, at the end of the 1830–1840 decade, could transmit messages hundreds of miles almost instantaneously. Moreover, some of the telegraph receivers printed the received signals on paper. And if marks, in the form of dots and dashes, could be received then surely means could be suggested for analysing a picture into a series of picture elements which, after conversion to an electrical quantity, could be sent to a distant point where they could be reconstituted into a synthesised image of the original.

In the 1840s, more or less contemporaneously with the rise of photography and the nascent interest in persistence of vision, the first steps towards the electrical transmission of printed type and illustrations were taken. Consequently, attention must now be turned to a brief consideration of these steps and of the relationship between picture telegraphy and television.

References

1 BURNS, R.W.: 'British television, the formative years' (Peter Peregrinus Ltd., London, 1986) Chapter 2, pp. 47–72
2 BURNS, R.W.: 'Seeing by electricity', *IEE Proc. A*, Jan. 1986, **133**, (1), pp. 27–37
3 SMITH, W.: Letter to Latimer Clark, *Journal of the Society of Telegraph Engineers*, 1873, **2**, pp. 31–33
4 BURNS, R.W.: 'The electric telegraph and the development of picture telegraphy' *in* 'History of electrical engineering', IEE conference publication, 1988, pp. 80–84

5 VIVIE, J.: 'Historique et développement de la technique cinematographique' (BPI, Paris, c. 1944), pp. 43–45
6 Ref.1, Chapter 16, pp. 367–382
7 Ref.1, p. 59
8 Ref.1, pp. 158–162
9 Zglinicki, F. von.: 'Der weg das film' (Rembrandt-Verlag, Berlin, 1956)
10 Cook, O.: 'Movement in two dimensions' (Hutchinson, London, 1963) Chapter 1
11 J.A.Cl.: Article on 'Magic' *in* 'Encyclopaedia Britannica', 9th edn., **XV**, pp. 207–211
12 Ref.10, p. 20
13 Ref. 10, p. 23
14 Ref. 10, p. 24
15 ALTICK, R.D.: 'The shows of London' (The Belknap Press of Harvard University Press, Cambridge, Mass., 1978) p. 56
16 Ref. 10, pp. 29–30
17 Ref. 10, p. 31
18 OETTERMANN, S.: 'Das Panorama Die Geschichte eines Massenmediums' (Syndikat, Frankfurt, 1980)
19 Ref. 15, p. 135
20 Ref. 15, p. 174
21 PRIESTLY, H.: 'The what it cost the day before yesterday book, from 1850 to the present day' (Kenneth Mason, Hampshire, 1979)
22 Ref. 15, p. 177
23 Ref. 15, p. 181
24 Ref. 15, p. 198
25 Ref. 15, p. 199
26 Ref. 15, p. 194
27 CRE, B.: 'Muybridge and the chronophotographers' (Museum of the moving image, London, 1992)
28 THOMAS, D.B.: 'The origins of the motion picture' (HMSO, London, 1964)
29 APPLEYARD, R.: 'Pioneers of electrical communication—Claude Chappe', *Electrical Communication*, 1929–1930, **8**, pp. 63–80
30 BURNS, R.W.: 'Soemmering, Schilling, Cooke and Wheatstone and the electric telegraph' *in* 'History of electrical engineering', IEE conference publication, 1988, pp. 70–79

Chapter 2

Images by wire, picture telegraphy (1843–c. 1900)

The first proposal for transmitting facsimiles electrically from one place to another was contained in a British patent[1] dated 27 November 1843. In this Alexander Bain, a Scottish clock and instrument maker, described 'Certain improvements in producing and regulating electric currents and improvements in electric timepieces and in electric printing and signal telegraphs'. His patent was comprehensive and he put forward seven different ideas for developments in electric telegraphy: the sixth of these related to his 'improvement for taking copies of surfaces, for instance the surface of printer's types at distant places'.

Bain[2] was born in October 1810 at Houstry, in the parish of Watten in the county of Caithness. He was one of eleven children of John Bain, a crofter, and his wife Isabella Waiter.

Alexander Bain received only a very basic education. He was employed as a herdsboy during much of a year and attended school in the winter. Some time after he left school—the date is not known—he became an indentured apprentice to John Sellar, a watchmaker of Wick, but did not complete his apprenticeship.

The turning point in Bain's early life stemmed from a lecture on 'Light, heat and electricity' which he attended in January 1830 in Thurso. He must have been keen to hear the speaker for after the meeting he had to walk, in bitterly cold weather, 13 miles back to his father's cottage and then, next morning, walk eight miles to Sellar's shop in Wick.

In the 1830s in Caithness opportunities for advancement were minimal. Bain decided to travel to London to seek work and to avail himself of the educational facilities which were accessible in the capital. Arriving there in 1837 he found employment as a journeyman clockmaker in Clerkenwell and soon began to attend lectures, exhibitions and demonstrations at the Adelaide Gallery of Popular Science and at the Polytechnic Institution. By 1838 Bain had begun to contemplate how a clock could be operated from an electric battery. His ideas progressed during the next two years and in 1840 he showed a model of his electric clock to one of his colleagues. It was the first electromagnetic clock ever invented. He also devised at

*Figure 2.1   Alexander Bain (1810–1877) was the first person anywhere to propose a method
for transmitting facsimiles electrically from one place to another. 'He was not a
commercial man but his inventive powers were most wonderful. He has given the
world some invaluable inventions.'*

about this time an electromagnetic printing telegraph[3]. Later, an amalgam of his general notions on electric clocks and electric printing telegraphs led to his invention of apparatus for sending black and white images from one place to another.

Bain's many diverse inventions seem to have been soundly based and eminently practicable for the period in which they were advanced. In a report[4], published in April 1844, on one of his telegraphs *The Times* noted: 'The results have proved highly satisfactory, and established the rapidity and accuracy of communication and the simplicity of the means by which it is accomplished. Mr Bain has proved himself a most ingenious and meritorious inventor of a very novel and efficacious instrument.'

In April 1850 he demonstrated his electrochemical telegraph (in which the image producing feature was comparable to that of the facsimile apparatus) in the Elysee Palace before the President of the Republic and some notable figures of the French Government. During the exhibition 'as an instance of the extraordinary powers of the telegraph' a despatch containing 1327 letters 'was conveyed between Lille and Paris in the space of 55 seconds, being at the rate of nearly 1500 letters per minute'.

Highton[5] an early writer on electric telegraphy, mentioned in his book that, around 1850, Bain's telegraph was one of the three most commonly used in America, coming after Morse's in general use, although in rapidity of signalling it was the fastest. And Schaffner[6] who wrote 'The telegraph manual' (1859) referred to 'the many ingenious contrivances invented by Mr Bain' and said: 'He was not a commercial man but his inventive powers were most wonderful. He has given the world some invaluable inventions'.

After Bain's 1843 patent on picture telegraph apparatus was enrolled many inventors put forward various devices and systems to further the progress of this application of electrical science but, despite some attempts at commercialisation in the 1860s and the first decade of the 20th century, permanent picture telegraphy services did not commence until the 1920s. Then several schemes—the Siemens – Karolus of Germany, the Belin of France, the Marconi of the UK and the RCA of the USA, *et al.*—were introduced and facsimile transmission became a feature of modern communications.

R H Ranger was one of the engineers who participated in the design of the RCA system. His 1925 paper on 'Transmission and reception of photoradiograms' includes the following acknowledgement to the work of Bain[7]:

> 'The transmission of pictures electrically had its inception almost at the same time as straight telegraphy for in 1842 [sic] Alexander Bain, an English physicist [sic] first proposed a device to send pictures from one place to another by electric wires. His plan is so basically correct that it is only right, at the start, to show the simplicity of his plan and how, generally we are all following in his footsteps.'

Figures 2.2 and 2.3, taken from Bain's 1843 patent[1], shows the arrangement which he submitted for sending a copy of the surface of printer's type. Essentially, the oscillatory motion of the pendulum combined with the vertical controlled motion of the metal frame to cause the stylus to scan, indirectly, the surface of the type. The

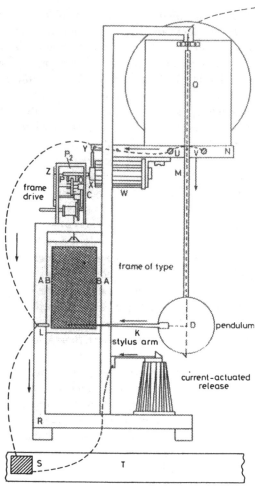

Figure 2.2   Bain's facsimile apparatus as shown in British patent 9745, dated 27 November 1843. The frame of type was scanned in two dimensions by the linear motion of the frame drive and the sinusoidal motion of the pendulum

transmitting and receiving instruments, which were similar in construction, were synchronised by arranging that the two pendulums actuated an electric circuit so that if one preceded the other by a slight amount in its swing it was held until the other had reached the same position, when both then started a new stroke. The two pendulums were thus the basic synchronisers of the system—an indispensable feature of any facsimile or television system. On each swing the frame descended by a given constant amount so that the whole surface was scanned uniformly.

At the transmitter the metal frame was filled with short insulated wires, parallel to each other and at right angles to the plane of the frame, so that they made contact with the raised surface of the metal type on one side and the moving stylus, attached to the pendulum, on the other side. Consequently, as the stylus moved across the

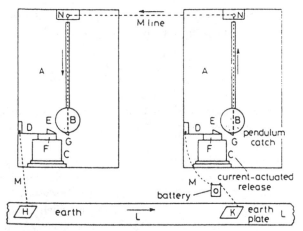

Figure 2.3   *The transmitter and receiver pendulums of Bain's facsimile apparatus were synchro-nised at the ends of every swing. The use of line synchronising pulses in modern television can be traced back to Bain's notions*

Source: British patent 9745, 27 November 1843

frame, an electric circuit containing the stylus, the frame and type was continually made and broken according to the arrangement of the type.

The receiving frame held two thicknesses of damp paper which had been previously saturated with a solution composed of equal parts of prussiate of potash and nitrate of soda. At the back of the paper there was a smooth metal plate which pressed the paper into contact with the ends of the parallel wires that filled the frame, as in the transmitting frame. By chemical action it was intended that the making and breaking of the current in the circuit should discolour the paper at the receiver to give a copy of the original surface.

This, then, was Bain's invention: it did not contain any radically new discovery nor even a new electrical principle, other than that of scanning, but was based on a sensible application of the technology which was available at the time to the solution of a new problem. His proposals represented a natural development of the science of electric telegraphy and were made apparently realistic by the advances which had occurred previously in this field. Thus, his use of electrochemical marking followed the practice which had been put forward in a patent[8] in 1838 by Edward Davy for a chemical marking telegraph. In his scheme 'three wires were to be used, and the points of the metal wires were to be caused to press, by means of the motion of mag-netic needles, upon chemically prepared fabric at the distant or receiving station'. The fabric to be used was calico or paper, and it was to be moistened with a solu-tion of hydriodate of potass and muriate of lime.

Davy described the operation in the following way:

'The motion of a needle to the right [should cause] a mark to be made on one part of the fabric, and the motion of the same needle to the left [should cause] a mark to be made on another part of the fabric; and the same for each needle attached to the respective wires. Thus the single or combined marks [can be made] to express letters, or other desired symbols.'

Although Davy's patent was bought by the old Electric Telegraph Company it was never utilised. It seems likely that Bain knew of Davy's idea, for much work on electric telegraphs was being carried out at that time. Also, because Bain had made several applications for patents prior to 1843, he was probably aware of the patent literature on this subject.

The discolouration of certain chemically treated papers was not new, for in 1800 a Mr Cruickshanks of Woolwich had noticed that the colour of litmus paper was changed by the galvanic current[9]. Cruickshanks's discovery was made while he was repeating and extending the experiments on electrolysis which had been initiated by Nicholson and Carlisle in the same year as Volta's discovery of the voltaic pile. Following this observation, much work was undertaken on electrolysis by various workers, culminating in the great work of Faraday in 1834. Davy's, Bain's and later Bakewell's use of the above stated effect really represented an extension of the employment of the electrolytic cell which Soemmering[10] had used in 1809 as a detector of electricity—albeit in a different form.

The novel concept incorporated into Bain's invention was undoubtedly the principle of automatically scanning a two-dimensional array and transmitting, automatically signals dependent upon some variable characteristic of the surface. Scanning is a vital requirement in all facsimile and, *mutatis mutandis*, television systems, but had not been proposed prior to Bain's 1843 patent. In their book 'Engineers and electrons: a century of electrical progress' J D Ryder and D G Fink state: '[Bain's] concept embodied all the geometrical and timing methods of the modern television system'.

It is rather surprising that Bain did not extend his invention to include the transmission of drawings, maps and the like. This was left to Bakewell[11] to accomplish in 1848, and as a consequence some controversy took place in 1850 as to who was actually the first to suggest the facsimile transmission of handwritten letters. In a letter[12] to *The Times* dated 17 November 1850 Bain wrote: 'My copying telegraph is capable of transmitting not only manuscripts written with all the characters of the autograph, but also of delineating at a distance any figure whatever which can be traced by drawing, stamping, etc. Thus a paper profile of a fugitive could, by its means, be transmitted in a few moments to all parts of the kingdom to which telegraphic wires extend.'

Bain's 1850 letter to *The Times* is interesting as it shows his invention had not been put into practice[13] because it required greater accuracy in the mechanism and more perfect insulation of the wire than had yet been realised. A further factor was possibly the high cost of sending telegraphic signals. Highton[14] gives the rates charged, in 1850 in England, for a telegraphed message of twenty words as follows:

London to Birmingham	112 miles	32.5p
London to Hull	200 miles	47.5p
London to Glasgow	420 miles	50.0p

Assuming the same speed of signalling for a modern facsimile transmission as for a telegraph message and a *pro rata* increase in charges, the cost of sending a page of a letter or a diagram of similar size would have been many pounds. These prices may be compared with the costs in the UK of goods and services at that time: a labourer worked for approximately 75p per week, rent for a working class family was about 27p per week, coal was 86p per ton in London and an ounce of tobacco and a gallon of beer were 1p and 5p, respectively.

The first instrument to be practically demonstrated was that constructed by F C Bakewell and patented by him on 2 December 1848[11]. In this system the message was written with a nonconducting liquid, such as varnish, on tin foil, and the tin foil then wrapped around the cylinder of the transmitting equipment. In the receiver's cylinder a paper, thoroughly moistened with a solution which was readily decomposed by an electric current, was placed. Marks were produced on the paper whenever the electric circuit, which comprised the transmitting and receiving cylinders and associated apparatus, was completed. The solution preferred by Bakewell consisted of a mixture of one third part of muriatic acid, one third part water and one third part of a saturated solution of prussiate of potass, Figure 2.4.

Both cylinders were rotated at equal, uniform speeds by means of weights and a clockwork-type mechanism, and each cylinder was traversed by a metal style which was carried in a traversing nut mounted on a lead screw. Hence, whenever the transmitter style pressed on the exposed tin foil, the circuit was closed through the moistened paper and a mark was recorded.

The preparation of the master surface was much simpler in Bakewell's apparatus than in Bain's and, furthermore, his use of rotating cylinders and associated linearly moving styles was the forerunner of many 20th century facsimile machines.

The problem of synchronising two nonmechanically linked mechanisms was to exercise the minds of inventors in the fields of still picture transmission and television for very many years. Bakewell soon encountered this difficulty for a newspaper report on his system noted[15] 'The chief difficulty with which he has had to contend lay in making the revolutions of the two cylinders correspond exactly, and this he has endeavoured to overcome by means of an electromagnetic regulator, which acts upon the receiving instrument and checks its motion so as to keep pace with the other. The machine is still in an experimental state, and evidently short of perfection: but sufficient success has been secured to render the practical result aimed at almost certain, and it is impossible to witness the delicacy and ingenuity of the process without feelings of surprise and delight.'

Neither Bain's nor Bakewell's designs were subsequently used for a regular service, but in a letter[16] to *The Times* in July 1894 Mr Armytage Bakewell observed that 40 years previously his father's copying electric telegraph had successfully transmitted autographic messages between Brighton and London. Invisible dispatches which could be rendered legible by the recipient had also been sent by this system.

Great interest was taken by the Prince Consort in Bakewell's invention and the inventor had the honour of exhibiting his instruments and of explaining their mechanical and electrical principles to His Royal Highness at Buckingham Palace.

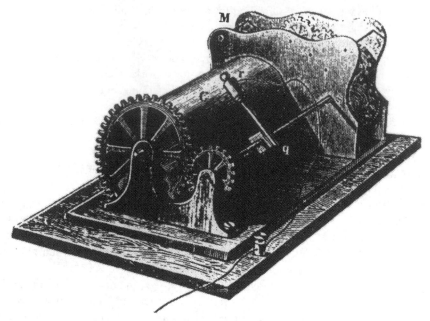

Figure 2.4   *F C Bakewell was the first person to suggest using cylindrical scanners in picture telegraphy systems. He described his scheme in a British patent dated 2 June 1849.*

Source: *Handbuch der Photographie und Teleautographie*, Leipzig 1911

The copying electric telegraph was later exhibited at the Great Exhibition of 1851 and received the highest award, the Council Medal.

The Abbé G Caselli, of Florence, in the provisional specification for his 1855 patent, gave a possible explanation[17] for the nonuse of the invention: 'The principle on which facsimile copies of messages may be produced through the medium of the electric current is well known; but it is the rapidity of transmission that is required to render this principle of practical value . . . '

Later, in the complete specification of the patent, he stated another view. 'The principal barrier to success in a machine of this nature is to obtain a perfect synchronism of motion in the machine which transfers the dispatches, and that at the opposite end of the line which receives and fixes them on the paper.'

A third reason was given by T A Dillon in his patent of 1879[18]. In this he put forward the suggestion that the original document to be copied and transmitted should be enlarged before it was scanned and that correspondingly at the receiver the copy should be reduced. 'Because of the enlarged nature of the letters, the practical difficulties which operated against the Bain, Bakewell, Caselli and Bonetti automatic copying systems are obviated,' he claimed.

Actually, the nonimplementation of either the Bain or the Bakewell schemes was probably due to a combination of all three factors mentioned above, plus an overriding limitation based on economic grounds, as events were to show in the latter half of the 1860–70 decade.

Caselli's pantographic telegraph was the first to be utilised, on a regular basis, anywhere in the world, Figure 2.5.

The first notification in England of the application of Caselli's invention was contained in the foreign intelligence column[19] of *The Times* for 22 February 1862: 'A new system of telegraph has been submitted to the [French] Emperor, to which its inventor, M Caselli, has given the name of 'pantograph'. This telegraph has been already worked at Florence and Leghorn. It transmits autograph messages and drawings with all the perfections and defects of the originals. An inhabitant of Leghorn wrote four lines from Dante and they appeared in the same handwriting at Florence. A portrait of the same poet was painted at Leghorn, and it was reproduced at Florence line for line and shade for shade. A bill of exchange was drawn in the same manner, and its authenticity admitted. The Emperor was much pleased at the trial made in his presence, and he proposes to establish it in France.'

The first dispatch was sent from Lyon to Paris on 10 February 1862[20]. Later, Le Corps Legislatif ordered the installation of the pantelegraph on the railway between these two cities and from 16 February 1863 the public was able to forward messages. In 1867 the director of telegraphs, a Monsieur de Vougy, sanctioned the setting-up of a second line on the Marseille to Lyons route and his department provided the necessary metallised paper, at the rate of 0.20Fr for each square centimetre of image transmitted. Unfortunately, the public did not appreciate the importance of the enterprise and after a few years the State abandoned the service.

A similar system[21] was employed by a French telegraph engineer named Meyer, except that he used synchronously running metal cylinders, much the same as those employed by Bakewell. Meyer's apparatus was put into operation between Paris and Lyons in 1869, but after a short period it, too, was taken out of service.

Another French engineer, d'Arlincourt, was not discouraged by the lack of success of the Caselli and Meyer projects for he carried out some experiments between Paris and Marseille in 1872 with a comparable scheme. Although it was favourably commented upon at the Vienna Exhibition of 1873, it was quickly cast aside like its predecessors[22].

Nevertheless, the pioneer thoughts and work of Bain, Bakewell, Caselli, Meyer, d'Arlincourt and others showed the way to a more satisfactory solution of the problem and possibly stimulated others to make attempts at the realisation of an efficient, reliable and economic facsimile service. Their endeavours highlighted the areas where new ideas could be applied:

1   conversion of the tonal gradations of a drawing, picture, message or photograph to an electrical signal;
2   synchronisation of the transmitting and receiving instruments;
3   conversion of the received electrical signals to produce visible impressions, corresponding faithfully with those on the original document, on a sheet of material at the receiver;
4   scanning of the transmitted and received surfaces.

Apart from improvements in synchronisation techniques, further developments in picture/facsimile transmission were dependent upon the utilisation of recent dis-

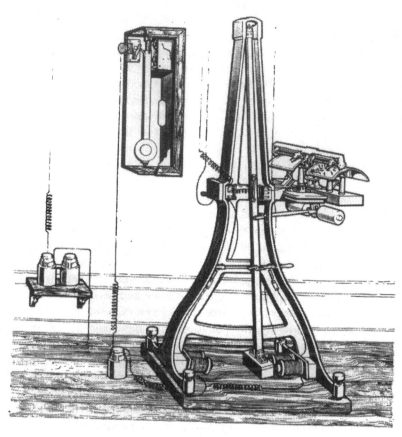

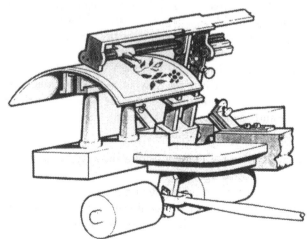

Figure 2.5   *L' Abbé Caselli in 1862 demonstrated apparatus, for transmitting facsimiles, which was based on the principles advanced in 1843 by Bain*

Reproduced by permission of the Science Museum

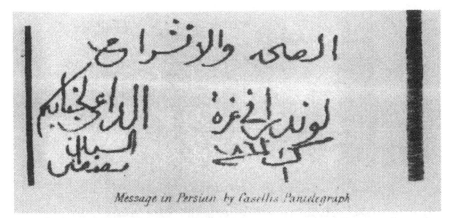

Figure 2.6   *Caselli's pantelegraph was able to transmit messages in nonRoman script*

Source: BOYER, J.M.J.: 'La transmission télégraphiques des images et des photographies', Pans, 1864

coveries which would allow intermediate tones to be transmitted. Bain's, Bakewell's and Caselli's apparatuses worked by sending pulses of current along the propagation path; the signals were essentially telegraphic rather than telephonic in character: either a signal was present or it was absent. This meant that although line drawings, diagrams and letters could be faxed from one place to another, it was not possible to reproduce electrically portraits which comprised graded tones. And yet, as noted previously, the development of photography was proceeding contemporaneously with the advancement of electric telegraphy and picture transmission.

Four years before Bain put forward his notions for taking copies of surfaces, L J M Daguerre in France and W H Fox Talbot in England had publicised the first practical techniques for creating permanent images by the agency of light. The really important factor in their work was that they had each discovered and published a way of developing a latent image so that it became visible on paper or on a plate.

The problem which had faced artists and scientists using the camera obscura during the early years of the 19th century had been how to fix the image which they had obtained by the action of light, without having to trace it onto translucent paper. Clearly, a light sensitive chemical was required which was capable of being developed and fixed. Berzelius in his 'Text book of chemistry', published in 1808, had listed more than one hundred substances which had their chemical or physical structure altered by light, and indeed the influence of light on silver nitrate had been reported by an Italian physician named Angelo Sala in 1614. Not surprisingly, several workers in Britain, Europe and America experimented with these substances in the late eighteenth and early nineteenth centuries in the hope of obtaining permanent pictures.

This activity was probably spurred on by the demand for inexpensive naturalistic pictures, particularly portraits, which existed towards the end of the eighteenth century. A simple way of reproducing pictures was by means of silhouettes, made by

Figure 2.7   Meyer's apparatus was used commercially between Paris and Lyons in 1869

Source: BOYER, J.M.J.: 'La transmission télégraphiques des images et des photographies', Paris, 1864

tracing the outline of a projected image of the face and filling it in with black paint. G L Chretien in 1786 invented a machine, the physiontrace, in which the projected image of a head was traced by a stylus, and by a pantograph arrangement an engraving tool could cut a copper plate that could be inked and printed. Aloys Senefelder invented lithography in 1798, but although it was introduced in Paris in 1802 it was not until 1813 that it became a success and a fashionable hobby.

Thomas Wedgwood, son of the famous Josiah Wedgwood, and Sir Humphry Davy achieved some fame with their use of siver nitrate and silver chloride in 1802, and Nicephone Niepce in 1822 triumphed in making a heliographic copy of an engraving on a glass plate coated with bitumen. The first permanent camera picture was taken by Niepce in 1826 using his asphalt process on a pewter plate. The

Figure 2.8   *D'Arlincourt in 1872 carried out some experiments on picture transmission between Paris and Marseille. His apparatus was not a commercial success*

Source: BOYER, J.M.J.: 'La transmission télégraphiques des images et des photographies', Paris, 1864

exposure was inordinately long, about eight hours on a bright summer's day, and hence the shadow and intermediate tone effects recorded on the plate represented a distortion of the scene viewed at a given instant of time.

Another person who was experimenting with silver salts at this same time was L J M Daguerre, a painter. In December 1839 Niepce and he formed a partnership. Subsequently, a full description of their inventions and methods was presented at a joint meeting of the Academie des Sciences and of the Academie des Beaux Arts, by Francois Arago, on 19 August 1839.

Meanwhile, Fox Talbot had been working with light sensitive substances from 1835 and had used paper coated with silver chloride. He publicised his process in 1839, after the first announcement of the daguerrotype but before the official description of it to the two Academies. Two years later he patented the calotype process which used silver iodide as a sensitive material together with silver nitrate and potassium iodide. The sensitivity of Talbot's paper was further increased by treatment with gallic acid, the sensitising properties of which had been discovered in 1837 by J B Reade.

Another important advance was made in 1851 when F S Archer introduced the wet collodion process, in which silver salts were coated on glass in a film of collodion. The plates were exposed while still wet and gave very clear glass negatives. Talbot believed that his patents covered the collodion process, but following a law suit in 1854 (Talbot versus Laroche), a favourable verdict was given to Laroche and the use of Archer's process was thenceforth free for general utilisation, as its inventor had intended. The British patent of the daguerrotype expired in 1853 and, as Talbot did not renew the calotype patents, all forms of photography could be used in Britain, without restrictions, by amateurs and professionals.

Thus, when Caselli patented his pantographic telegraph in 1855, the art of photography was well established. And yet the transmission of photographic images by electric telegraphy was not advanced until Amstutz put forward his ideas in 1893[21]. The actual development of a suitable method was made in 1907 when Korn of Munich and the Belgian inventor H Carbonelle published their independent processes.

The major difficulty which faced the early inventors in this field was the lack of a photoelectric cell. Prior to Willoughby Smith's notification[23] of 1873, the only known relationship between the production of an electrical effect and its optical cause was the disclosure which Edmund Bacquerel[24] had made in 1839. He had noted that an electric current was established in a cell containing two dissimilar liquids when exposed to light. The finding, however, was not really appropriate for the electrical transmission of images as a sensitive galvanometer was needed for the observation of the current. This meant that the obvious method of transmitting a photographic image by electrical means—the projection of a scanning beam of light through a photographic transparency onto a photoconductive cell—was not available until the arrival of the selenium cell.

A consequence of this limitation was that only photographic methods which created or were capable of creating images in relief could be considered for use in facsimile systems.

By the time that Caselli carried out his experiments the technique for producing photographic images in relief was known: for during 1852 Fox Talbot had perfected and patented a process for making printing blocks directly by photographic methods using a coating of potassium dichromate and gelatine on steel plates. He called this process photoglyphic engraving. A combination of Bain's apparatus and Fox Talbot's technique could possibly have led to an earlier realisation of the suggestion advanced by Amstutz in 1893. He employed the properties of a dichromated gelatin film for his method of facsimile transmission. In this, a reversed negative was imprinted upon a glass sheet which had been coated with a gelatin emulsion containing potassium dichromate so that the lightest parts of the negative (corresponding to the shadows of the original) passed more light to the gelatin, making it hard, and *vice versa*. A simple washing in water then dissolved the softer parts of the emulsion so as to leave an image in relief in which the thickness of the film at each point was dependent on the original intensity of illumination.

Amstutz applied the film to a rotating cylinder and scanned its surface with a stylus, the movements of which caused, by a suitable mechanism, the electric current to vary in the transmission line. This use of a stylus to scan a relief surface was first put forward by Edison in 1881 when he presented his autographic telegraph to the Exposition Internationale d'Électricité de Paris. Edison's apparatus was probably based on his phonograph of 1877. In operation, the sender of a message wrote his dispatch with a hard pencil so as to emboss the sheet of soft paper, which was then scanned by the agency of the stylus and cylinder combination.

The progress of still picture transmission proceeded in a rather erratic manner. Following Bain's and Bakewell's pioneering ideas on the subject in the 1840–50 decade, advances in the techniques of scanning and synchronisation occurred which enabled several practical schemes to be implemented in the 1865–75 period, only to lapse subsequently into disuse after a lack of commercial success. Then, in the first few years of the twentieth century, renewed interest in the subject emerged which led to equipments being developed by Korn (1907), Carbonelle (1907), Berjenneau (1907), Semat (1909), Belin (1907) and others. But again, although some of these schemes were capable of giving quite good results, their uneconomic viability consigned them to the warehouse of proven but unsuitable for the times inventions.

Undoubtedly, the progress and development of picture transmission stimulated interest in television or 'seeing by electricity' or 'distant vision' as it was known before the 1900s. Soon after the photoconductive property of selenium was discovered in 1873, several proposals for distant vision systems were made. Scientists and inventors were well aware of the basic problems to be solved and only a lack of basic hardware prevented progress from being forwarded. Essentially, similar difficulties to those which had been encountered in the transmission of still pictures had to be overcome, namely those associated with scanning, synchronisation, opticoelectrical and electroptical transducers. Not surprisingly, some of the devices and techniques which had proved successful in facsimile work were used, at least initially, during the progress of television. Devices such as selenium cells, mirror galvanometers, Kerr cells and synchronisation methods based on tuning forks and phonic wheels were all utilised in the march towards practical facsimile and television systems. In the early history of distant vision little, if any, consideration was paid to the theoretical differences between the two transmission systems: basically, distant vision was treated from an empirical viewpoint and it was not until 1908 that Shelford Bidwell and Campbell Swinton showed the enormity of the problem which had to be solved before television could become a reality.

References

1 BAIN, A.: 'Certain improvements in producing and regulating currents and improvements in electric time pieces and in electric printing and signal telegraphs'. British patent 9745, 27 November 1843
2 BURNS, R.W.: 'Alexander Bain, (1810–1877). Some aspects of his life' *in* 'History of electrical engineering', IEE conference publication, 1989
3 BAIN, A., and WRIGHT, Lt.T.: 'Application of electricity to control railway engines and carriages, mark time, give signals, and print intelligence at distant places'. British patent 9204, 21 December 1841

4 ANON.: a report, *The Times*, April 1844
5 HIGHTON, E.: 'The electric telegraph, its history and progress' (John Weale, London, 1852)
6 SCHAFFNER, T.P.: 'The telegraph manual: a complete history and description of semaphoric, electric and magnetic telegraphs of Europe, Asia, Africa and America, ancient and modern' (Pudney and Russell, New York, 1859)
7 RANGER, R.H.: 'Transmission and reception of photoradiograms', *Proc. IRE*, 1926, **14**, pp. 161–180
8 DAVY, E.: 'Telegraphs'. British patent 7719, 4 July 1838
9 FAHIE, J.J.: 'A history of the electric telegraph to the year 1837' (E and H M Spon, London, 1884)
10 BURNS, R.W.: 'Soemmering, Schilling, Cooke and Wheatstone and the electric telegraph' *in* 'History of electrical engineering', IEE conference publication 1988, pp. 70–79
11 BAKEWELL, F.C.: 'Electric telegraphs'. British patent 12 352, 2 June 1849
12 BAIN, A.: Letter to *The Times*, November 1850
13 BURNS, R.W.: 'The electric telegraph and the development of picture telegraphy' *in* 'History of electrical engineering', IEE conference publication 1988, pp. 80–84
14 Ref. 5, p. 168
15 ANON.: Report in *The Times*, November 1850
16 BAKEWELL, F.C.: Letter to *The Times*, July 1894
17 CASELLI, G., and NEWTON, A.V.: 'Electric telegraphs'. British patent 125 232, 10 November 1855
18 DILLON, T.A.: 'Transmitting messages and printed matter, etc., by electrical cables'. British patent 1347, 4 April 1879
19 ANON.: Report in *The Times*, February 1862
20 BOYER, J.M.J.: 'La transmission télégraphiques des images et des photographies' (Paris, 1864)
21 KORN, A., and GLATZEL, B.: 'Handbuch der Phototelegraphie und Telautographie' (Nemnich, Leipzig, 1911)
22 THORNE-BAKER, T.: 'The telegraphic transmission of photographs' (Constable, London, 1910)
23 SMITH, W.: Letter to Latimer Clark, *Journal of the Society of Telegraph Engineers*, 1873, **2**, p. 31
24 BECQUEREL, E.: 'Recherches sur les effets de la radiation chimique de la lumière solaire, au moyen des courants électrique', *Compte Rendu*, 30 July 1839, **9**, pp. 145–149

Chapter 3

Seeing by electricity, the earliest notions (1878–1880)

The 1873 discovery of the effects of light upon the resistance of a selenium bar is important historically, not so much for any practical value which selenium might have had for the purpose, but for the glut of schemes and proposals which were made for television systems in the years which followed.

Selenium, which belongs to the sulphur and tellurium family, is a nonmetallic element and was first discovered by Berzelius in 1817 in a red deposit found at the bottom of sulphuric acid chambers when pyrites containing selenium was used[1]. Like sulphur it exists in several modifications, being obtained as a dark red amorphous powder, as a brownish black glass mass, as red monclinic crystals or as a bluish grey, metal-like crystalline mass. In its natural state selenium is almost a nonconductor of electricity, its specific conductivity being forty thousand million times smaller than that of copper, but Knox, in 1837, found that on being annealed it became a conductor having a large resistivity compared to that of copper.

It was this property which led to its use in certain experiments by Willoughby Smith Figure 3.1, and to the discovery which he reported in a letter[2] to Mr Latimer Clark, then vice president of the Society of Telegraph Engineers.

Wharf Road
4 February 1873

My Dear Latimer Clark
Being desirous of obtaining a more suitable high resistance for use at the Shore Station in connection with my system of testing and signalling during the submersion of long submarine cables, I was induced to experiment with bars of selenium— a known metal of very high resistance. I obtained several bars, varying in length from 5 cm to 10 cm, and of a diameter from 1.0 mm to 1.5 mm. Each bar was hermetically sealed in a glass tube, and a platinum wire projected from each end for the purpose of connection.

Figure 3.1   Willoughby Smith, with his assistant J May, discovered the photoconductive property of selenium in 1873. The discovery led to the establishment of the science and technology of 'seeing by electricity', or 'distant vision'

The early experiments did not place the selenium in a very favourable light for the purpose required, for although the resistance was all that could be desired—some of the bars giving 1400 MΩ absolute—yet there was a great discrepancy in the tests, and seldom did different operators obtain the same result. While investigating the cause of such great differences in the resistance of the bars, it was found that the resistance altered materially according to the intensity of light to which they were subjected. When the bars were fixed in a box with a sliding cover, so as to exclude all light, their resistance was at its highest, and remained very constant, fulfilling all the conditions necessary to my requirements; but immediately the cover of the box was removed the conductivity increased from 15 to 100 per cent, according to the intensity of the light falling on the bar. Merely intercepting the light by passing the hand before an ordinary gas-burner, placed several feet from the bar, increased the resistance from 15 to 20 per cent. If the light be intercepted by glass of various colours, the resistance varies according to the amount of light passing through the glass.

To ensure that the temperature was in no way affecting the experiments, one of the bars was placed in a trough of water so that there was about an inch of water for the light to pass through, but the results were the same; and when a strong light from the ignition of a narrow band of magnesium was held about 9 in above the water the resistance immediately fell more than two-thirds, returning to its normal condition immediately the light was extinguished.

I am sorry that I shall not be able to attend the meeting of the Society of Telegraph Engineers tomorrow evening. If, however, you think this communication of sufficient interest, perhaps you will bring it before the meeting. I hope before the close of the session that I shall have an opportunity of bringing the subject more fully before the Society in the shape of a paper, when I shall be better able to give them full particulars of the results of the experiments which we have made during the last nine months.

I remain
Yours faithfully
Willoughby Smith

It is interesting to note that, although Berzelius had discovered selenium in 1817 and had published his 'Textbook on chemistry' in 1808, which had listed more than a hundred substances that had their chemical/physical properties influenced by light, he did not make the observation which was described in Smith's letter. Of course, Berzelius did not have access to a sensitive galvanometer in 1808, as Becquerel had in 1839, otherwise the properties of selenium might have been noted earlier than in 1873 and the progress of picture transmission, following Bain's lead, advanced with greater rapidity. The development of the galvanometer had to await Oersted's pronouncements of 1820 and the observation made by Schweigger in the same year which led to the use of his multiplier as a means of creating a magnetic effect many times larger than could be realised by using a wire with a single turn.

Smith's letter was published in the *Journal of the Society of Telegraph Engineers* and at the meeting at which it was read the chairman remarked that he thought it indicated a very interesting scientific discovery, and one about which it was probable they would hear a good deal in future. The chairman had himself witnessed the experiments, and could confirm all that Smith had stated. Selenium's sensibility to light, the chairman remarked, 'was extraordinary, that of a mere lucifer match being sufficient to affect its conductivity'.

The chairman's statement was quite prophetic: Willoughby Smith's discovery gave rise to much attention and speculation in the scientific world at the time.

Soon after the publication of the letter, a Lieutenant Sale, at the request of the editor of *Nature*, Sir Norman Lockyer, conducted some experiments on 'The action of light on the electrical resistance of selenium' and communicated his results to the Royal Society in 1873[3]. Sale found that the effect produced on exposure to light was sensibly instantaneous, but that on removing the light source the return to normal resistance was not so rapid. This time constant effect was to be a factor of much concern to inventors of distant vision systems.

Smith forwarded a further letter to the Society of Telegraph Engineers on 3 March 1876. An extract from this letter is given below because it relates to the reason for Smith's use of selenium.

While in charge of the electrical department of the laying of the cable [from] Valentia to Hearts' Content in 1866 I introduced a new system by which ship and shore could communicate freely with each other during the laying of the cable without interfering with the necessary electrical tests. To work this system it was necessary that a resistance of about one hundred megohms should be attached to the shore end of the cable. The resistance which I first employed was composed of alternate sheets of tinfoil and gelatine, and, although they answered the purpose, still the resistance was not constant enough to be satisfactory. While searching for a more suitable material the high resistance of selenium was brought to my notice, but at the same time I was informed that it was doubtful whether it would answer my purpose as it was not constant in its resistance. I obtained several specimens of selenium and instructed Mr May, my Chief Assistant at our works at Greenwich, to fit up the system we adopt on shore during the laying of cables, using selenium as the high resistance, and employ the spare members of the staff as though they were on shore duty and report to me on the subject. . . . It was while these experiments were going on that it was noticed that the deflections varied according to the intensity of light falling on the selenium. One of many of the experiments made was as follows:

Time	Selenium closed in box, gas in room not burning	Cover off box, gas not burning	Cover off box, two ordinary gas burners alight in room
1′	1483	1419	1047
2′	1483	1405	1018
3′	1483	1405	1018
	(resistances in megohms)		

In each case the temperature in the box was 71.5°C. During the laying of the 1873 and 1874 Atlantic cables, the Lisbon and Madeira, Madeira and St Vincent, St Vincent and Pernambuco, and the Australian and New Zealand cables, I have with success adopted selenium bars protected from the action of light.

The earliest published accounts of schemes for seeing by electricity which included some details of the equipment which might be used were those of Senlecq[4] (1879), Ayrton and Perry[5] (1880), Carey[6] (1880), Sawyer[7] (1880) and Le Blanc[8] (1880). Others who contributed views or suggestions were W L[9] (1882), Figuier (1877), de Paiva[10] (1878), Perosino[11] (1879), Redmond[12] (1879) and Middleton[13] (1880). A de Paiva was the first person to write a brochure on seeing by electricity. His work, 'La téléscopie électrique, basée sur l'emploi du selenium', was published in *Porto*, by A J da Silva. In addition, scanty announcements were made in the technical press of proposals by a Dr H E Licks[14] of Pennsylvania (1880) and Connelly and Mac Tighe[15] of Pittsburgh (1880).

Some of these workers claimed that they had given thought to the problem of distant vision for a number of years before they felt confident to commit their ideas in writing. Ayrton and Perry's plan was suggested to them in 1877 and more immediately by a cartoon, Figure 3.3, in the magazine Punch; Sawyer wrote that the principles 'and even the apparatus for rendering visible objects at a distance through a single wire' were described in the fall of 1877 to a Mr James G Smith, a former superintendent of the Atlantic and Pacific Telegraph Company; Sencleq claimed that his apparatus was invented in the early part of 1877 and some writers have stated that Carey's first idea was mooted in 1875 although no published evidence seems to exist to substantiate this statement.

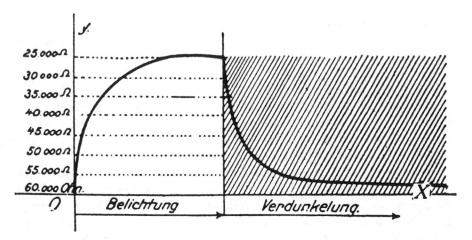

Figure 3.2   *Graph showing the change of resistance of selenium with change of light flux. This time constant effect was a matter of concern to experimenters for more than 50 years*

Reproduced by Permission of Krayn Verlag
Source: Das elektrische Fernsehen und das Telehor, Krayn Verlag, Berlin, 1923, p. 21

EDISON'S TELEPHONOSCOPE (TRANSMITS LIGHT AS WELL AS SOUND).

Figure 3.3   Punch's Almanac for 1879 was the first magazine to illustrate, by means of a cartoon, a possible future application of 'seeing by electricity'

Both the initial proposals of Carey and Senleq were for the reproduction of still pictures rather than of moving scenes, and so not surprisingly the known techniques of picture telegraphy were adopted to some extent.

Carey suggested using in one of his receiving instruments chemically prepared paper while Senleq's first receiver was to employ a tracing point of blacklead or pencil for 'drawing very finely'. In addition, his transmitter was to consist of an 'ordinary camera obscura containing at the focus an unpolished glass and any system of autographic telegraphic transmission'.

Willoughby Smith's discovery enabled an opticoelectrical transducer to be devised for a distant vision transmitter, but no progress had taken place on the design of a realistic electro-optical transducer for incorporation into a receiver. The only known chemical/physical properties which had been utilised in picture telegraphic systems at that time were the marking of chemically treated paper by an electric current and the marking of a sheet of plain paper by a pencil-electromagnet arrangement. Consequently, when the early workers appreciated that such reception methods would not suffice for moving picture reproduction, they had to consider what effects could be employed. Unfortunately, there were very few that were suitable for the purpose.

The carbon arc had been spectacularly demonstrated by Sir Humphry Davy at a Royal Institution lecture in 1808 and was later used in lighthouses and elsewhere for lighting, but in the absence of electronic amplifiers this form of illumination was of no importance to the inventors of distant vision in the 1870s — although many years later, in 1931, Baird demonstrated the feasibility of having a modulated arc as the light source in his low-definition television system.

Nevertheless, the 1870–80 decade was one of considerable activity in electrical science and on 18 December 1878 Joseph Swan showed an incandescent carbon filament lamp in operation at a meeting of the Newcastle Chemical Society. Ten months later, on 19 October 1879, Edison's lamp with a carbonised sewing thread filament was successfully exhibited. These events may have influenced Senlecq and Carey, for both of these inventors advocated utilising incandescent platinum filaments in their receiving instruments. Sawyer thought that a spark produced by two platinum wires connected to an induction coil would solve the problem. The observation that a platinum wire glowed when an electric current passed through it was first noted by Staite and Moleyns in 1859 and, although this property was never subsequently the basis for a source of lighting in a television receiver, Baird did construct a honeycomb mosaic of lamps for a large-screen televisor which he demonstrated to a cinema audience in 1930.

Ayrton and Perry made an original contribution in their 1880 paper to *Nature* when they put forward the notion that the then recently discovered Kerr effect could feature in a 'seeing by electricity' receiver. Notwithstanding some criticism from a Mr J E H Gordon, a week later, the Kerr effect was to be employed in such receivers until 1936. Ayrton and Perry's alternative proposal for showing an image was not to vary the brightness of a lamp directly but to control its output flux indirectly by means of a variable shutter operated by the received current. In this way, a very high luminosity lamp could be utilised with a disc attached to a modified galvanometer. Here, again, developments which had occurred earlier during the advancement of electric telegraphy were incorporated in the design of the shutter mechanism.

One of the fundamental questions that had to be answered by Ayrton, Perry, Carey, Senlecq, Sawyer and others concerned the methods which had to be adopted in transmitting the varying electrical signals to the receiver from the selenium cell or cells. Two basic schemes were possible. First, in those transmitting systems which were based on a mosaic of small cells (Figures 3.4 and 3.5) each individual cell could be connected by a conductor to a separate receiving element of a corresponding receiver mosaic. In this manner the problem of synchronisation would vanish, but the construction of the necessary transmission line would pose immense difficulties. Carey, and Ayrton and Perry, advanced such a solution. It is possible that they were influenced either by the working of the human eye and optic nerve or by the early developments which had taken place in electric telegraphy. During the latter half of the 18th century and first quarter of the 19th century several inventors, CM (1753), Lesage (1774), Linget (1782), Reiser (1794) and Soemmering (1809) had described inventions for signalling in which the number of conductors between the sending and receiving equipments was equal to the number of letters of the alphabet, and the numbers of sending and receiving elements were each made equal to this number. Ayrton and Perry's and Carey's distant vision ideas were thus analogous to these telegraphic systems. The second possibility was to employ some form of scanning procedure at the transmitter and receiver, as had been adopted in picture telegraphy by Bain, Bakewell and Caselli. Senlecq was the first to propose this

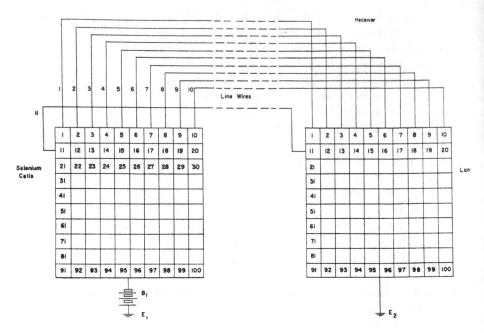

Figure 3.4   *The use of a mosaic of light-sensitive cells at the transmitter, and a mosaic of light-emissive elements at the receiver, eliminated the need for synchronising means between the transmitter and receiver*

Source: *Journal of the SMPE*, **63**, November 1954, p. 231

method and later much thought was given to scanning by Le Blanc, Nipkow, Weiller, Rosing and others.

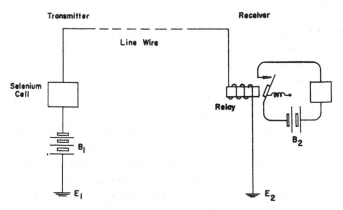

Figure 3.5   *In one form of Figure 3.4, the transmitter elements would be selenium cells, and the receiver elements might be lamps operated, via relays, by the transmitted currents*

Source: *Journal of the SMPE*, **63**, November 1954, p. 231

It is perhaps not surprising that many notions and inventions were put forward which were impracticable at the time of their advancement, either because a great deal of effort would have been required to perfect them or because, more usually, other discoveries and innovations essential to the working of the main ideas had not yet been made.

The fact that an idea is impracticable at the time of its pronouncement does not necessarily detract from its historic interest. Indeed, for more than half a century the history of television consists almost exclusively of plans and suggestions, all of them incapable of immediate practical embodiment and all of them waiting for other inventions and parallel developments which would enable them to be incorporated in a realisable system. But considering these plans, visionary though they may have been initially, it becomes apparent how the broad front of television science progressed, a step or two forward here and there until, in the lapse of time, the whole state of the art was seen to have made considerable progression.

And so, now some of the schemes previously only briefly mentioned are related in detail.

Senlecq's 1879 account of his telectroscope was published in the *English Mechanic*[4]. He wrote:

'The apparatus will consist of an ordinary camera obscura containing at the focus an unpolished glass and any system of autographic telegraphic transmission; the tracing point of the transmitter intended to traverse the surface of the unpolished glass will be formed of a small piece of selenium held by two springs acting as pincers, insulated and connected, one with a pile, the other with the line. The point of selenium will form the circuit. In gliding over the surface, more or less lightened up, of the unpolished glass, this point will communicate, in different degrees and with great sensitiveness, the vibrations of the light.'

Senlecq's scheme for the receiver was similarly simplistic.

'[The receiver] consists of a tracing point of blacklead or pencil for drawing very finely, connected with a very thin plate of soft iron, held almost as in the Bell telephone, and vibrating before an electromagnet, governed by the irregular current transmitted in the line. This pencil, supported on a sheet of paper arranged so as to receive the impression of the image produced in the camera obscura, will translate the vibrations of the metallic plate by a more or less pronounced pressure on that sheet of paper. Should the selenium tracing point run over a light surface the current will increase in intensity, the electromagnet of the receiver will attract to it with greater force the vibrating plate, and the pencil will exert less pressure on the paper. The line thus formed will be scarcely, if at all, visible; the contrary will be the case if the surface be obscure, for, the resistance of the current increasing, the attraction of the magnet will diminish, and the pencil, pressing more on the paper will leave upon it a darker line.'

Senlecq thought these proposals could be simplified by dispensing with the electro-magnet and 'collecting directly on the paper, by means of a particular composition, the different gradations of tints proportional to the intensity of the electric current'. His ideas were naive, but by 1881 he seems to have appreciated some of the practical difficulties that had to be overcome for he published details of a further method in which the transmitter screen was to comprise a large number of very small selenium cells.

Senlecq appears to have been impressed by his own efforts of 1877/78 for he wrote later: 'Since then the apparatus has everywhere occupied the attention of prominent electricians, who have striven to improve on it.' He went on to record that as a result of the widespread publicity in all the 'Continental and American Scientific journals' Ayrton, Perry, Sawyer (of New York), Sargent (of Philadelphia), Brown (of London), Carey (of Boston), Tighe (of Pittsburg) and Graham Bell had proposed designs based on his own ideas.

The reference to Bell[5] concerned a rumour that the inventor of the telephone had deposited in a learned institution a sealed package containing 'the first results of a new and very remarkable instrument first conceived by him during his sojourn in England in 1878'. Actually, Bell never engaged in the above field of research. It appears that a hoax story about a diaphote for seeing by telegraphy was published in the *Boston Transcript* in February 1880 and subsequently was repeated in both American and British newspapers including the august journal *Nature*[16]. Shortly afterwards this story was linked to Bell's work at that time and inspired at least two eminent scientists to give some thought to seeing by electricity. The explanation of the rumour is that Bell and his coworker Tainter deposited, at the Smithsonian Institution, early in March, a sealed tin box with a model of their photophone[17], on which they were working at that time, and an account of its development, so as to establish priority of invention without divulging their line of research prematurely[27]. To Bell's amusement, news of his mysterious box appearing alongside the diaphote hoax touched off indignant claims by several inventors that Bell had stolen the idea of seeing by telegraph from them.

The supposed deposition of Bell's invention brought a swift response from Ayrton and Perry who wrote a letter to *Nature*[5], dated 22 April 1880, with the intention of diminishing any claim which might be put forward on behalf of Bell, rather than with the aim of publicising a definite invention of their own. They were keen to prevent a monopoly in an invention which was 'really the joint property of Willoughby Smith, Sabine and other scientific men, rather than of a particular man who has had sufficient money and leisure to carry out the idea'.

Ayrton and Perry took the precaution of stating their views on how distant vision could be achieved. Their transmitter would be based on a mosaic of small, separate squares of selenium, each piece being connected to a corresponding receiver element. These elements would be shutters, of the magnetic needle type, which would be controlled by the transmitted currents so that by their movements they would open or close apertures through which light would pass to illuminate the back of small squares of frosted glass.

A more promising arrangement, said Ayrton and Perry, was suggested by Professor Korn's experiments. Each receiver square would be made of silvered soft iron and would form the end of a core around which would be placed a coil. The surface formed from a multiplicity of these squares would be illuminated by a beam of light, polarised by reflection from glass, and the reflected beam would be viewed after having passed through an analyser. Consequently, the light flux received by the eye from each square would depend upon the rotation of the plane of polarisation of the light beam, produced by the iron core-coil unit.

Ayrton and Perry concluded their letter by stating: 'It is probable that Professor Graham Bell's description may relate to some plan of a much simpler kind than either of ours; but in any case it is well to show that the discovery of the light effect of selenium carries with it the principle of a plan for seeing by electricity.'

Their suggestions clearly represented an attempt to characterise a distant vision system, unlike Senlecq's which, in modern terms, was an example of a facsimile system. Furthermore, they depicted the image of a distant object as being transmitted by 'a mosaic of electricity', an expression which, suitably interpreted, was to become of increasing importance with the introduction of electronic cameras.

Ayrton and Perry's proposals to utilise Dr Kerr's discovery[18] of the rotation of the plane of polarised light reflected from the pole of a magnet was received rather sceptically and scornfully by a Mr J E H Gordon, who mentioned some experiments which he had carried out in this field in a letter[19] to *Nature*, published on 29 April 1880. An extract from this letter is given below.

'I used an electromagnet consisting of an iron bar two feet four inches long and $2\frac{1}{4}$ inches diameter, surrounded by 70 lbs of wire and excited by ten Grove cells.

'The total double rotation produced, not by slightly altering the resistance, but by reversing the current, was never more than 26′ (twenty-six minutes of arc).

'To see this at all with a very delicate Jellett analyser, it was necessary for the observer to increase the sensitiveness of his eye by sitting in total darkness for some ten minutes before each observation.

'Your readers can judge what chance of obtaining visible changes of illumination there would be with "little" magnets and mere variations in a current not powerful enough to fuse a selenium resistance.'

Obviously, the problem of distant vision was not going to be solved easily. But this was an age of great discoveries and inventions, and inventors were not too dismayed by such practical difficulties. In particular, Ayrton and Perry were not rebuffed by Gordon's views and reiterated their belief in the feasibility of their second plan in a further letter[20] to *Nature*: 'We still have no doubt that with a certain proper arrangement of the apparatus not only the effects observed by Dr Kerr but others of the Faraday polarisation of light effects might be practically made use of.' They then compared the advances made in a related field to lend weight to their arguments. 'For it must be remembered that the actual electric currents now used to transmit articulate speech are only one forty-millionth per cent as strong as those necessary to work even a delicate telegraph relay, whereas it required several Grove's cells to show in a decided way the old experiment of the sound emitted by an iron bar on being magnetised.'

Aryton and Perry's April 1880 letter to *Nature* may have stimulated an obscure inventor, Denis Redmond of Dublin, to draw attention to his plan in a letter to *The Times* (13 May 1880). He referred to a 'relay of peculiar construction' but gave no details and pointed out that he had not patented his apparatus. Redmond's ideas were first stated in a letter published in the *English Mechanic*[12]. His suggested transmitter was similar to that advanced two months later by Ayrton and Perry and also that put forward four months later by Carey. The provenance of Redmond's plan seems clear from his description of the experiments on which he was then engaged. 'By using a number of circuits, each containing selenium and platinum arranged at each end, just as the rods and cones in the retina, the selenium end being exposed in a camera. I have succeeded in transmitting built-up images of very simple luminous objects . . . '

This announcement elicited three letters shortly afterwards, including one asking for the results of the experiments, but nothing was ever published on these.

Redmond recognised the practical difficulties involved in implementing a non-scanning system for he also considered a single circuit scheme—but without success: ' . . . an attempt to reproduce images with a single circuit failed through the selenium requiring some time to recover its resistance.' He was nevertheless aware of the importance of persistence of vision and wrote: 'The principle adopted was that of the copying telegraph, namely, giving both the platinum and selenium a rapid synchronous movement of a complicated nature, so that every portion of the image of the lens should act on the circuit ten times in a second, in which case the image would be formed just as a rapidly-whirled stick forms a circle of fire. Though unsuccessful in the later experiment, I do not despair of yet accomplishing my object, as I am, at present, on the track of a more suitable substance than selenium.' Apart from his later letter to *The Times*, nothing more was heard from the Dublin inventor.

Willoughby Smith's discovery of the photoconductive property of selenium in 1873 provided the basis for nearly all the distant vision schemes until the disclosures made by Elster and Geitel, in the period 1889 to 1913, led to the evolution of a practical and sensitive form of photocell.

A notable exception was Middleton's mosaic of thermocouple elements. In May 1880 Middleton, a tutor of St John's College, Cambridge, wrote to *The Times*[13] and drew attention to the fact that in April he had read a paper before the Cambridge Philosophical Society, describing an instrument for transmitting pictures.

Middleton, like Redmond, pointed out the analogy between his instrument and that of the human eye and, also equated his conducting system with the optic nerve. He gave the following short account of his notions:

> 'A lens is used to throw on a plate or suitably curved receiving plate (inclosed in a camera) the image of any object. The receiving plate of the

camera is composed of thermopile elements, ground to a smooth surface, and having their posterior faces put in electrical communication by a system of wires, with a somewhat similarly constructed plate. The heating, etc., effect of the image on the first plate generates currents of electricity, which flow through the wire system, and on reaching the second thermopile plate are reconverted into heat, etc., according to the law discovered by Peltier, the amount of heat, etc., being directly proportional to the amount of electricity.'

A feature of Middleton's propositions was the possibility of obtaining either positive or negative images. Furthermore, 'these images can be either viewed directly or by reflected light (after the fashion of the Japanese mirrors and projected on a screen), or by suitable apparatus they can be obtained as a photograph, a thermograph, or chemiograph . . . '.

Middleton's plan was criticised in *Design and Work*[21]: 'We fear that the paper is a dream of the future, rather than a statement of the realities of the present', the critic observed. But the inventor was not abashed: 'The instrument described was the outgrowth of some discoveries in electricity and heat, etc., which I made in trying to carry on the work of Seebeck, Peltier, Thompson and others. And an account of these latter would ere this have been published had not ill health and the want of money to make certain experiments connected with the research delayed my paper on the subject.' Nothing further was written on his proposals.

Senlecq continued to pursue the problem and in February 1881 published details[22] of a new scheme in which the transmitter screen consisted of a large number of very small selenium cells. He was now aware of some of the difficulties which had to be overcome to achieve a satisfactory result and opined: 'Some experimenters have used many wires, bound together cablewise, others one wire only. The result has been on the one hand confusion of conductors beyond a certain distance, with the absolute impossibility of obtaining perfect insulation, and, on the other hand, an utter want of synchronism. The unequal and slow sensitiveness of selenium obstructed the proper working of the apparatus.'

The sluggishness of the selenium cell was to tax the inventiveness of scientists and inventors for about 50 years[23].

In his new arrangement, Figure 3.6, Senlecq connected each cell of the transmitter screen to a type of linear distributor through which contact could be made with the single-line wire by means of a falling slider, the disposition being such that each cell was scanned once only. The receiver comprised a multicellular mosaic of fine platinum wires, each joined to contacts on a distributor plate which was associated with a clockwise mechanism intended to ensure synchronism with the falling slider at the transmitter. Thus, each platinum wire was to be rendered momentarily luminous in proportion to the light flux at the corresponding selenium cell.

Senlecq seems to have had equivocal views on the feasibility of his design, for at the beginning of his paper he noted, 'these contacts ought to work the apparatus, and to insure the perfect isochronism of the transmitter and receiver'; whereas

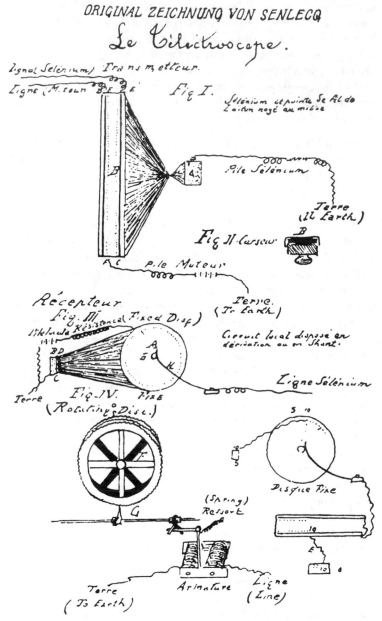

Figure 3.6   Original drawings of Senlecq, 1881. His ideas for 'Le telectroscope' were based on the use of a mosaic of selenium cells scanned by a linear commutator

Source: *English Mechanic and World of Science*, (829), 11th February 1881, pp. 534–535

towards the end of it he confidently claimed, 'as may be seen, the synchronism of the apparatus could not be obtained in a more simple and complete mode . . . '.

Whether actual results were attained is not known, but Senlecq stated: ' . . . we can obtain a picture, of a fugitive kind it is true, but yet so vivid that the impression on the retina does not fade during the relatively very brief space of time the slide occupies in travelling over all the contacts.'

Willoughby Smith's discovery certainly had one effect; it fired man's imagination. Coming as it did in the same decade which saw the invention of the telephone and the phonograph, by Bell and Edison, respectively, the prospect of seeing by electricity as well as hearing by electricity was a powerful incentive for inventors, of whom there was no shortage in the latter quarter of the nineteenth century.

Not all these inventors were competent, however; there was the usual pronouncement of success quite unsubstantiated by factual detail. During an address at Christchurch in 1883 the Reverend Mr Gilbert, while speaking of the telephone, asked his audience if they would be astonished if he were to tell them that it was now proved to be possible to convey by means of electricity vibrations of light, 'to not only speak with your distant friend, but actually to see him'. He went on to relate how a Dr Gridrah, of Victoria, had invented an electroscope to achieve this purpose and how this wonderful instrument had been demonstrated in Melbourne on 31 October 1882 in the presence of some forty scientific and public men, and was a great success. 'Sitting in a dark room, they saw projected on a large disk of white burnished metal the race course at Flemington with its myriad hosts of active beings. Each minute detail stood out with perfect fidelity to the original, and as they looked at the wonderful picture through binocular glasses it was difficult to imagine that they were not actually on the course itself and moving among those whose actions they could so completely scan.'

The 1880 issue[24] of *Design and Work* contains an article describing two methods for seeing by electricity advanced by G Carey (of the Surveyor's Office, City Hall, Boston). They are of special interest because of the drawings and description given. Both methods envisaged the reception of the image on chemically prepared paper, and so, like Senlecq's suggestions, they were suited for facsimile reproduction rather than for distant vision. There was also an idea for a system which would have enabled an observer to have seen a moving image if the ideas had been realisable.

Figures 3.7 and 3.8 illustrated his paper. The disc, P, is drilled through, perpendicularly to its face, and the numerous small holes are filled with selenium. Wires from the selenium elements pass through a similar disc at the receiver. A chemically prepared paper is placed between discs C and D so that the image of an object projected upon disc P is printed upon the paper. Figure 3.7*f* from his paper shows another receiving instrument having platinum or carbon points to reproduce a luminous image.

The transmitter of Carey's first 1880 scheme was very similar to that described by Aryton and Perry and, as with the latter's proposal, utilised a receiver each point on the screen of which was connected to the corresponding point on the transmitting screen. None of these authors calculated the number of elements or wires which would be necessary in their nonscanned systems: this matter was to be given some thought a few years later by Shelford Bidwell.

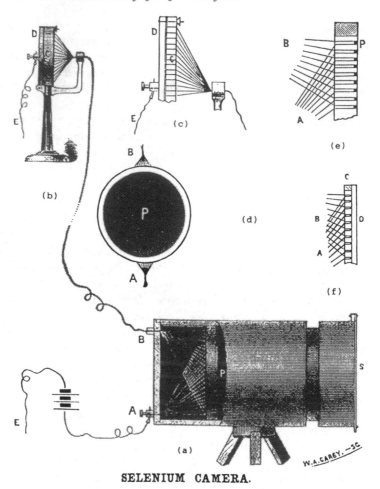

SELENIUM CAMERA.

Figure 3.7   *Carey's selenium camera, 1880. It was proposed to project an image on to a multi-cellular mosaic of selenium cells, each of which was to be connected by a conductor to corresponding points at the receiver where the image was to be reconstituted on a chemically treated sheet of paper by electrochemical decomposition*

Source: *Design and Work*, **8**, 26th June 1880, pp. 569–570

Carey's second design is of greater interest because it incorporated the elements of a practical mechanism for scanning the image of the object, see Figure 3.8*a,b,c*. Referring to Figure 3.8*a*, a clockwork mechanism rotates the shaft K causing the arm L and wheel M to describe a circle of revolution. The screw N being fastened firmly to the wheel, turns as it rotates on its axis and so draws the sliding piece P and selenium point, disc or ring B, towards the wheel. These two motions cause the point, disc or ring to describe a spiral line upon the glass, TT.

Thus, Carey clearly expressed a way of spirally scanning a projected image and transmitting the photocurrents to the receiver via a pair of conductors, rather than

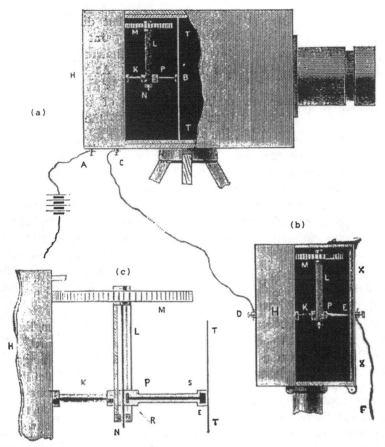

Figure 3.8   *Carey's second design, 1880. This proposal incorporated an idea for exploring the projected image by means of a selenium cell, S. The cell would traverse a spiral path, on the plate, TT, using a clockwork mechanism*

Source: *Design and Work*, **8**, 26th June 1880, pp. 569–570

a multiconductor cable: but he seems to have overlooked the need for a system to synchronise the receiver to the transmitter. The spiral scanning principle, however, was a most important one and was later used until the mid-1930s.

A similar scanning motion was mentioned by W E Sawyer in a letter dated June 1880 to the editor of *Scientific American*[7]. Sawyer was well aware of the formidable nature of the problem to be solved and was not sanguine about the inventions of others, particularly those involving a mosaic of selenium elements. 'There is no likelihood of any plan of this kind ever being reduced to practice, for some of the difficulties in the way of all the plans are insuperable . . . ' Sawyer perceived four reasons for his opinions:

1   the action of light upon selenium in changing its electric conductivity was slow;

2   the most delicate apparatus was unlikely to indicate a change in resistance by
     the projection of light upon merely a selenium point;
3   the reproduction of an image—'one even so small as to be projected upon a
     square inch of surface'—with any accuracy 'would necessitate that this surface
     should be composed of at least 10 000 insulated selenium points, connected
     with as many insulated wires leading to the receiving instrument';
4   isochronism was unattainable.

Sawyer's transmitter was to consist of a coil of fine selenium wire housed in a
darkened case having a diameter of about three inches. Light was to be admitted
into the case to illuminate the selenium wire by means of a fine tube, which starting
at the periphery of the circle would draw concentric imaginary spiral lines until it
reached the centre. This scanning path was identical to that given in Carey's article
(published a week before Sawyer's paper). But whereas Carey gave no indication of
the speed of the scanning process, Sawyer noted that the time for one scan, from the
periphery to the centre of the circle, would have to be such that the impression made
upon the retina would not have ceased until the centre was reached. He was the first
person to appreciate the importance of persistence of vision in the generation of an
image by a scanning process.

Sawyer mentioned that the principles 'and even the apparatus' of his invention
were known in 1877 although he did not give any practical details (unlike Carey) of
how the motion was to be achieved. It remained for Nipkow to propose in 1884 a
simple way of scanning an image based on a rotating disc having a set of apertures
lying along a spiral path.

Sawyer's receiver included a pair of fine platinum points, spaced closely together,
connected with the secondary wires of a 'peculiar induction coil', the primary side
of which constituted a part of the main transmitter-receiver circuit. In operation, the
inventor said: 'It is obvious that as the first spark between the receiving platinum
points would not have ceased to affect the retina until the last spark . . . an exact
image of the object before the transmitter would be reproduced before the eye of the
observer placed at the darkened chamber of the receiver.'

Of the early workers in the field of distant vision Sawyer seems to have been one
of the most pragmatic. Not for him the 'vivid' pictures of Senlecq: 'The trouble is to
make the selenium sufficiently active, and to get the isochronous motion. Perhaps
some of your readers may like to try their hands at rapid synchronism,' he rather
forlornly wrote at the end of his letter.

Maurice LeBlanc, on the other hand, contented himself with a consideration of
the possible solutions for achieving 'la transmission électrique des impressions
lumineuses'. Much of his paper[8] in *La Lumiere Electrique* (1880) consists essentially of
a catalogue of the effects and devices which he thought might prove useful. The photoconductive property of selenium, the photovoltaic cell of Becquerel, the thermo-
electric effect and mechanisms using the pressure of radiation (sic) were mentioned,
Figure 3.9. Of these the most original were the latter. One receiver was to comprise
an egg-shaped enclosure, blackened on the inside, containing a very small, flexible,
blackened piece of steel mounted near a coil. 'Elle sera plus ou moins repousse par

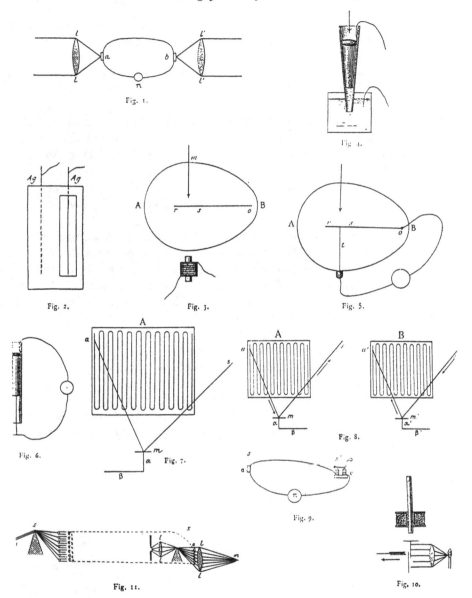

Figure 3.9   Diagrams illustrating the ideas of Le Blanc, 1880. He suggested: a procedure for
systematically analysing and synthesising images (7 & 8); several transducers for
converting variable light fluxes into variable electrical quantities, based on: the
photoconductive property of selenium (1 & 9); the photovoltaic effect (2); and the
pressure of radiation (3, 4, 5 & 6); a receiver light valve modulator (9 & 10) to
convert a changing electric current into a changing light flux; and a system of colour
television (11). The motion of the scanner was dependent on two orthogonal
mechanical actions

Source: *La Lumière Electrique*, t.11, 1st December 1880, pp. 477–481

le rayon tombe en in suivant son intensité et sa refrangibilité,' wrote Le Blanc. Consequently, an electromotive force will be generated, he said.

No practical details or calculations were given.

The invention of the telephone by Alexander Graham Bell on 2 June 1875 and the prospect of 'hearing by electricity' may have inspired a number of the early distant vision workers to consider the parallel object of seeing by electricity. Just as Bell's telephone provided a transducer for converting sound energy into electrical energy, so it was hoped that Willoughby Smith's discovery of the photoconductive effect would provide the means of enabling distant vision to be achieved. An added impetus to this work was probably provided by Bell and Tainter's work on photo-phones[17]. This was a device which allowed sounds—both musical and vocal—to be transmitted to a distance by the agency of a beam of light of varying intensity and a selenium cell.

In the articulating photophone[25], Figure 3.10, a mirror, M, reflected a beam of light through a lens, L, and, if desired for the purpose of experimentally cutting off the heat rays, through a cell, A, containing alum-water, and cast it upon the trans-mitter, B. This comprised a small disc of thin glass, silvered on the front, about the size of the diaphragm of a telephone and mounted in a frame with a flexible india-rubber tube approximately sixteen inches long leading to a mouthpiece. A second lens, R, interposed in the beam of light after reflection from the transmitter rendered the rays parallel. The receiver consisted of a parabolic mirror, C, which served to concentrate the beam and to reflect it down upon a selenium cell, S, which was placed in the circuit of a battery, P, and a pair of telephones, T.

In operation the lenses were so adjusted that when the mirror, B, was flat, (i.e. not vibrating), the projected beam was focused onto the receiving instrument. However, when a speaker spoke into the transmitting apparatus the mirror disc, B, was set into

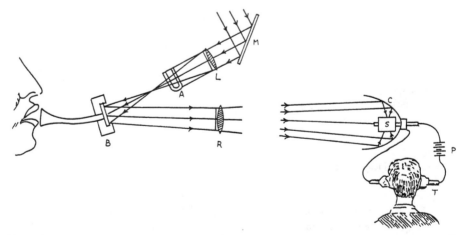

Figure 3.10   Bell's photophone enabled speech to be transmitted by the agency of a light beam. It was demonstrated on 19 February 1880. Bell considered the photophone to be his greatest invention

Source: *Scientific American*, **44**, (1), 1 January 1881, pp. 1–2

vibration thereby making it alternately slightly convex and concave and caused the focus of the transmitted beam to vary. With careful adjustment of the apparatus it was hoped that the received electric current waveshape would be a replica of the sound variations at the transmitter.

Bell and Tainter achieved their desired objective on 19 February 1880 in Bell's laboratory at no. 1325 L Street, Washington[26]. The inventor of the telephone seems to have been greatly elated by his success for he wrote[27], (to his father): 'Can imagination picture what the future of this invention is to be . . . We may talk by light to any visible distance without any conducting wire . . . In warfare the electric communications of an army could neither be cut nor tapped. On the ocean communication may be carried on . . . between vessels . . . and lighthouses may be identified by the sound of their lights. In general science, discoveries will be made by the photophone that are undreamed of just now . . . The twinkling stars may yet be recognised by the characteristic sounds, and storms and sun spots detected in the sun.'

By 26 March 1880 Bell and Tainter had transmitted audible effects over a distance of 82 m and on 1 April Tainter sent a message 213 m from the top of the Franklin School to a window of the L Street laboratory. Another sealed box was deposited in the Smithsonian Institution[28].

In the same year Bell and Tainter devised not only a variety of selenium cells and a parabolic reflector but also more than fifty methods of varying a light beam. These included magnetic fields which affected polarised light, lenses of variable focus, variable apertures and adaptations of Koenig's manometric capsule. Some of these ideas were used in future distant vision equipment, for example, Weiller's 'telephone à gaz' was based on Koenig's capsule. Bell and Tainter also devised a simple process for transforming selenium into the required crystalline state in a few minutes instead of the elaborately accepted procedure which took forty to sixty hours.

As early as May 1880 Bell had considered the photophone to be sufficiently well developed for him to offer it to the National Bell Telephone Company. William H Forbes, the president, accepted it for the company but noted rather reservedly: 'Whether this discovery ever approaches the telephone itself in practical importance or not, it is no less remarkable and a thing which we should be glad to possess.'

Forbes's caution was justified by subsequent events. The photophone could only be employed in clear air conditions and its range was small. By 1893 the limit of transmission was still no more than about 200 m, although in 1897 an American Telegraph and Telephone Company engineer succeeded in increasing the range to several miles[29] by using an arc, which responded to slight variations in current, instead of Bell's mirror diaphragm. He also improved the efficiency of the selenium cell receiver.

But in 1897 Marconi succeeded in sending radio signals several miles. The photophone was thus doomed. Yet, Marconi notwithstanding, Bell in 1898 held the photophone to be his greatest invention, and in 1921, less than a year before his death and in an age of intercontinental radiocommunications, he told an interviewer: 'In the importance of the principles involved, I regard the photophone as the greatest invention I have ever made; greater than the telephone.'

Although Bell was the first person to construct a photophone in 1880, the concept dates from June 1878. In that month a letter[30] appeared in the journal *Nature*, signed by a J F W, stating that the writer had looked in vain for any account 'of experiments with the telephone or phonoscope, inserted in the circuit of a selenium (galvanic) element. One is inclined to think that by exposing the selenium to light, the intensity of which is subject to rapid changes, sound may be produced in the phonoscope. Probably by making use of selenium, instead of the tube-transmitter with charcoal, etc., of Professor Hughes, and by exposing it to light as above, the same result may be obtained'. This letter may have stimulated Bell, for in a lecture before the Royal Institution, in 1878, he announced the possibility of hearing a shadow fall upon a piece of selenium included in a telephone circuit.

The importance of the photophone in the context of the history of television lies in its use of the selenium cell. Here was a practical application of Willoughby Smith's discovery which seemed, at that time, to have a great future. The photophone was a relatively simple device which could be developed and constructed by inventors and possibly lead to advances in the design of the all important optico-electric transducer. A number of patents for selenium cells were sealed following Bell's disclosure of the photophone, and this may also have led to the interest in distant vision shown by a number of persons in 1880.

In one respect Bell and Tainter's work probably raised some false hopes. Their experiments showed that selenium cells could respond to light variations having a frequency in the audio range. This was possible notwithstanding the known sluggishness of the cell because the light intensities used were sufficiently high for a cell output to be produced even though only a small region of the cell's characteristic was being utilised. In distant vision applications much lower light intensities were incident on the device and hence it was necessary to make use of the entire change of cell resistance rather than a part of it as with the photophone. Unfortunately, these changes required a time period which was incompatible with the scanning period.

The history of the photophone is quite short and of no appreciable significance in the development of electrical engineering, but at the time at which Bell and Tainter conducted their experiments the photophone was the first successful application of the selenium cell. The value of the photophone in the history of distant vision stems from this fact and the likely encouragement it gave to the early workers struggling to obtain an image of a distant object by electrical means. It is pertinent to note that Senlecq, Aryton and Perry, Middleton, Redmond, Le Blanc and others made their suggestions in the same year, 1880.

Because many of these suggestions were hopelessly ill conceived and naive there was a need for some experimental evidence which would show the feasibility or otherwise of the ideas being advanced. Shelford Bidwell and Aryton and Perry in February and March 1881, respectively, gave demonstrations of the photoconductive property of selenium as it might be applied in a distant vision system. Very sensibly, these three experimentalists reduced their apparatuses to the simplest forms possible and succeeded in showing the desired effects. At the same time, the enormity of the problems to be overcome were realistically highlighted with the probable consequence that future systems tended to be much more pragmatic than the conjectures of 1880.

Aryton and Perry's demonstration[31] was made using single elements of their multielement, multiconductor scheme, and Shelford Bidwell restricted his practical tests[32] to showing that a selenium cell could be employed in a form of picture telegraphy apparatus[33]. This was exhibited at a meeting of the Physical Society on 26 February 1881 and was described in *Nature* on 10 February 1881.

In a development of the latter equipment a picture not more than two inches square was projected by a lens upon the side of a small rectangular box containing a selenium cell. The box was completely closed except for a small pinhole and was capable of moving up and down, through a distance of two inches, and at the same time laterally through a distance of 1/64 inch so that the pinhole passed successively over every point of the focused image. The receiver was slightly modified from Bakewell's form and enabled an image to be produced on a piece of paper which had been soaked in a solution of potassium iodide, Figure 3.11.

With this apparatus Shelford Bidwell transmitted 'simple designs in black and white, painted upon glass, and projected by a magic lantern. The image of a butterfly with well defined marks upon its wings, and a rude drawing, in broad lines, of a human face' were among the objects which he 'most successfully' reproduced.

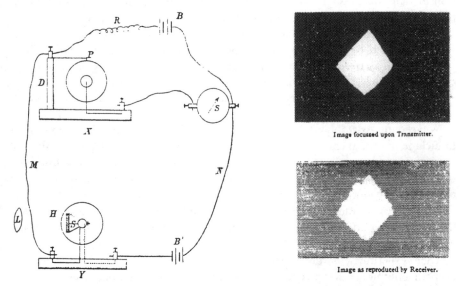

Image focussed upon Transmitter.

Image as reproduced by Receiver.

Figure 3.11   *Shelford Bidwell's transmitter, Y, and receiver, X, of 1881. The scanning cylinders were mechanically coupled to ensure synchronous operation. At the transmitter simple geometrical designs cut out of tinfoil were projected by a magic lantern onto the cylinder. A small aperture in it enabled some light flux to be incident on a selenium cell, S. The receiver's cylinder was covered with a sheet of paper soaked in a solution of potassium iodide. Simple designs could be reconstituted in a series of closely spaced lines*

Source: *Nature*, **23**, 10 February 1881, pp. 344–346

This was certainly a start in seeing by electricity utilising selenium cells, although it was not television. Nevertheless, the equipment manifestly illustrated the limitations of such cells for this purpose; limitations which were not overcome until the mid-1920s. 'Slow rotation is essential,' wrote Bidwell, 'in order both that the decomposition may be properly effected and that the selenium may have time to change its resistance. The photophone shows that some alteration takes place almost instantaneously with a variation of the light, but for the greater part of the change a very appreciable period of time is required.' This rather ominous statement was to be confirmed again and again in the years ahead. For television systems, a minimum scanning rate of ten frames per second was necessary whereas for picture telegraphy single frame scanning—which might take 20 minutes for a single scan—was appropriate. The extent of the difficulty facing inventors was obviously revealed by Bidwell's work.

Aryton and Perry gave a demonstration to the Physical Society of their first idea for distant vision probably to vindicate the proposal which they had made. They desired to show the successful reproduction on a receiving screen of every change of illumination on one square of the sending screen. Their shutter was an elliptical, blackened aluminium disc (suspended in a blackened tube) of a special type of galvanometer and made an angle of 45° with the tube's axis. In this state no light transmission was possible but when the disc was deflected through 45° all the light passed through and formed an image of the square on a screen. Attached to the shutter was a small magnet, making an angle of 67.5° with it, and the two items were suspended by a silk fibre about one-twentieth of an inch in length. These particular angles were selected so that, first, all variations in intensity of illumination could be produced with a small motion of the shutter and, secondly, so that the magnet should always be in its most sensitive position in the coil through which passed the picture currents.

No accounts exist showing whether Aryton and Perry extended their experiment to include multielement arrays of squares. However, they did suggest a method of putting, say, thirty or forty selenium cells on a revolving arm, and hence disposing of a large number of cells, in order to transmit a complete picture which would also obviate the difficulty arising from abnormal variations of selenium.

They were very much aware of the need to reduce the number of wires between the transmitting and receiving stations and mentioned that in practice a telegraph engineer would avail himself of the principles of multiplex telegraphy.

Their experiment hardly advanced the new art: it was carried out with one element only under static conditions and did not illustrate what would happen, particularly at the receiver, under rapidly varying light states. Moreover, it failed to show the action which would result when a large number, say 10 000, of magnetic elements were in close proximity to each other at the receiver.

Of the two scientists, Perry was certainly aware of the advantages of distant vision. Later he wrote (concerning 'the people of one hundred years hence'): 'They will probably speak to one another at a distance without any artificial connection between them . . . They will probably be able to see one another's actions at great distances, just as if they were close together.' These remarks, made before the pioneer work of Hertz on electromagnetic waves[34] was widely known, were quite prophetic.

Aryton and Perry's and Bidwell's demonstrations probably had some effect in terminating the absurd claims and rumours which followed the prospect of seeing by electricity. One of the most blatant of these had been reported in the New York papers and concerned an instrument invented by a Dr H E Lick of Bethlehem, Pennsylvania[14].

Lick's apparatus, which he called a diaphote, was stated in the press[35] to have 'the power of showing in a mirror at one end the image of any object placed in front of a corresponding mirror at the other end'. The mirrors were composed of selenium and chromium at the transmitting end and selenium and silver iodide at the other end. Each mirror was built up from a number of small plates, as in Carey's instrument, and corresponding mirrors at the two ends were linked together. With the transmitting mirror placed in the focal plane of a camera it was expected that an image would be reproduced by the other mirror. It seems that Dr Licks was not disappointed, for: 'A public exhibition of this ingenious instrument took place very recently at Reading in the United States. The receiving mirror was taken down to a room below the hall in which the spectators were assembled and various objects such as an apple, a penknife, a dollar, a watch, part of the printed handbill, etc., were successively placed in front of it and immediately became visible to the audience; and when at length the head of a live kitten was thus seen by telegraph, the enthusiasm of all present was wrought to a frenzy.'

English newspapers took a more cautious view of Dr Lick's apparatus: 'This reads well and in the interests of science we hope it is all true. We remember though that a year or so ago . . .' observed one paper. 'Meanwhile we must accept it with a grain of salt,' stated another. 'For with all its undoubted prodigies of invention America is the land of sham discoveries and bogus engines; and it is our own mournful experience that the New York papers are—well, not quite immaculate in their verity, It is much easier to fabricate a fraudulent imitation of an invention, or concoct an account of it, with a big sensation heading than to do the real thing.' (*The Globe*, 30 April 1880).

The first and only, so far as is known, systematic practical examination of the subject undertaken in the 19th century was that carried out in 1882 by Ll B Atkinson.

In 1882 he was a student at King's College, University of London, and performed his experiments under the direction of Professor Grylls Adams who during the period 1875 to 1877, had been engaged on a study of the properties of selenium and was acquainted with the technique of fabricating cells. Atkinson was thus well placed to conduct an enquiry into the possible devices and ideas which had been advanced by others.

The accounts that exist of Atkinson's contribution are those which he related in a letter[36] to the *Electrical Review* in 1889 and during the discussion in 1924 of a paper by Campbell Swinton on 'The possibilities of television'. Atkinson's apparatus was exhibited[37] before the Television Society (UK) on 5 March 1929 and is now preserved in the Science Museum, London.

Atkinson found from his experiments and the experiments of others that selenium was unsuitable as an opticoelectrical transducer 'on account of its great variation in sensitiveness with varying battery powers, and even in course of time by molecular changes, while its high resistance made the currents available very small'.

In a review article published in 1927 on 'The selenium cell: its properties and applications' the author, G P Barnard, stated that the intensity of illumination, I, on the selenium cell was related to the cell's resistance, R, and was equal to $A(1/R - 1/R_0)^n$ where R_0 is the resistance of the cell when $I = 0$, and where A and n are constants[1]. The exponent n was found to be equal to 1 by Stebbins, 2 by Fournier d'Albe, Rosse, Adams, Berndt and Minchin, 3 by Hopkins, and 4 by Rankine. Athanasiadis obtained a completely different formula, as did Hesehus.

Despite the surprising range of formulae to represent the same property of the cell, these formulae do show that the sensitivity of the cell (defined as dR/dI) increases as the resistance of the cell increases. Bell and Tainter (1880) used cells having a dark resistance of about 1200 Ω, but Fournier d'Albe's cells[38] (1914) had a dark resistance of approximately 23 000 Ω. High resistance cells were desirable from the point of view of sensitivity, but as Atkinson observed the resulting currents were small. Because electronic amplifiers had not yet been invented, these currents were required to actuate directly, or via an electromechanical relay, the electro-optical transducer at the receiver. Relays could be used in picture telegraphy apparatus but their speed of working was too slow for television purposes.

Atkinson also noted: 'Coming to the receiver, I am aware that more sensitive or louder speaking magnetic telephones are made now [1889] than when I tried my experiments, but I was unable to produce, by means of the magnetic telephone any variation in a gas flame, with the movement of the plate, produceable by variation in current, such as a selenium transmitter, or even a microphone (acoustic) transmitter will give.' Another method which he tried but 'which was quite hopeless, because it needed currents and forces which were inapplicable', was the rotation of the plane of polarisation of light reflected from the face of a magnet—the method suggested by Aryton and Perry.

Atkinson did not limit his work to the use of selenium cells but also tried a transmitter comprising a mixture of carbon and sulphur. 'In this case the luminous variation was first transformed into a heat variation, and owing to the sensitiveness of this mixture to heat, as affecting its resistance, into an electrical variation. Again, however, the variation was neither powerful enough nor rapid enough. Similar experiments with thin metal films, after the manner of Professor Langley's bolometer, were unsuccessful. Coming to the receiver, and abandoning the magnetic telephone, I tried the motograph or chalk-cylinder telephone of Edison, and in this case the gas flame movement could be obtained with the variations that could be produced in a microphone (acoustic) transmitter . . . But here the defects of the gas flame receiver are apparent. It is not sufficiently dead-beat, the position of the gas flame depends not on the position of the telephone membrane, but on the rate of change of its position . . .

'Of other magnetic forms I tried an electromagnet with a light, stiff armature, carrying a thin plate, pressing against a slightly convex plate, so as to give a series of

Newton's rings. Viewed by monochromatic light (sodium flame), the centre spot is black or light by a variation in position of half a wavelength of yellow light. The chief defect of this form is that the illumination is scarcely intense enough to sufficiently impress the retina for the purpose in view. I also tried forms in which a small mirror was strung in a tightly stretched wire, so as to have a very small time of oscillation, the movement of the magnetic armature being highly magnified by several different devices. They did not prove, however to be sufficiently dead-beat.'

Atkinson's highly interesting letter ended with two suggestions which he thought would be worthy of investigation. First, the utilisation of electrochemical transmitters and, secondly, the use of a Geissler tube, 'more especially a stratified tube, in a receiver'.

The employment of electrical discharges in gases for distant vision receivers had not previously been made, and as discharge tubes featured in the investigations of the American Telegraph and Telephone Company and those of Baird, *inter alia*, Atkinson's suggestion was quite augural.

His first recommendation related to the employment of small cells in which silver bromide was formed by electrolysis. The action of light on this salt produced another salt, thereby altering the e.m.f. of the cell. This idea was presumably based on the recent discovery in 1887 by Arrhenius[39] that some silver halides conduct electricity more readily when illuminated than when they are not so exposed. Arrhenius's observation was in the same subject area as Becquerel's discovery of 1839, namely, liquid photoelectric cells, but although a reference to such cells for television was made in the 1920s they were never adopted: Elster and Geitel's investigations were to lead to a more practical and satisfactory type of cell.

Atkinson, subsequent to 1882, never returned to the subject of his first researches except to mention, in 1920 in his presidential address[40] to the Institution of Electrical Engineers, the need for further research on, first, some action of light which would act instantaneously and enable a current variation to be set up and, secondly, some massless method of illuminating or darkening a surface, which could be varied by an electric current.

References

1   BARNARD, G.P.: 'The selenium cell: its properties and applications' (Constable, London, 1930)
2   SMITH, W.: A letter to Latimer Clark, *J. Society of Telegraph Engineers*, 1873, **2**, pp. 31–33
3   SALE, Lt.: 'The action of light on the electrical resistance of selenium', *Proc. R. Soc.*, **21**, pp. 283–285
4   SENLECQ, C.: 'The telectroscope', *Les Mondes*, 16 January 1879, **48**, pp. 90–91. Also, *English Mechanic and World of Science*, 1879, (723), p. 509
5   AYRTON, W.E., and PERRY, J.: 'Seeing by electricity', *Nature*, 22 April 1880, p. 589
6   CAREY, G.R.: 'Seeing by electricity', *Sci. Am.*, 5 June 1880, **42**, p. 355. See also reports on 'The telectroscope', *Sci. Am.*, 17 May 1879, **40**; and 'Carey's diaphote', *English Mechanic and World of Science*, 18 June 1880
7   SAWYER, W.E.: 'Seeing by electricity', *Sci. Am.*, 12 June 1880, **42**, p. 373
8   LE BLANC, M.: 'Etude sur la transmission électrique des impressions lumineuses', *La Lumière Electrique*, 1 December 1880, **11**, pp. 477–481
9   W.L.: 'The telectroscope, or seeing by electricity', *English Mechanic and World of Science*, 21

April 1882, **35**, (891), pp. 151–152. Also, letter to *Nature*, 27 June 1936, **137**, p. 1076

10 PAIVA, A. de.: 'La téléscopie électrique, basée sur l'emploi du selenium' (A.J. da Silva, Porto, 1878)

11 PEROSINO, C.M.: 'Su d'un telegrafo ad un solo filo', *Atti Acad. Sci. di Torino I, cl. Sci. Fis. Math. Nat.*, March 1879, **14**, p. 4a

12 REDMOND, D.D.: 'Seeing by electricity', letter (15374), *English Mechanic and World of Science*, 7 February 1879, p. 540. See also BOLTON, H.E.: 'Seeing by electricity', letter (No. 15429), 14 May 1880, p. 235; GLEW, F.H.: 'Electric telescope', letter (No. 15429), 21 February 1879, p. 586; MORSHEAD, W.: 'Electric telescope', letter (No. 15430), 21 February 1879, p. 586

13 MIDDLETON, H.: 'Seeing by telegraph', letter to the editor, *The Times*, 24 April 1880. See also 'Seeing by electricity', *English Mechanic and World of Science*, 30 April 1880, **31**, pp. 177–178

14 Reports in Design and Work, 1880, p. 283 and p. 437; and *The Times*, 24 April, 24a12f

15 Report in *La Lumiere Electrique*, 1880, **2**, p. 140

16 Reports in *Nature*, 15 April 1880, **21**, p. 576 and *English Mechanic and World of Science*, 30 April 1880, p. 177

17 Report on 'Bell's photophone', *Nature*, 4 November 1880, **23**, pp. 15–19

18 KERR, J.: 'A new relation between electricity and light: dielectrified media birefringent', *Phil. Mag. J. Sci.*, November 1875, S.4, **50**, (332), pp. 337–348 and pp. 446–458

19 GORDON, J.E.H.: 'Seeing by electricity', a letter, *Nature*, 29 April 1880, p. 610

20 Aryton, W.E., and PERRY, J.: 'Seeing by electricity', letter, *Nature*, 13 May 1880, p. 31

21 Report on 'The diaphote', *Design and Work*, 15 May 1880, pp. 437–438

22 Senlecq, M.: 'The telectroscope', *English Mechanic and World of Science*, 11 February 1881, (829), pp. 534–535

23 Ref. 1, pp. 96–109

24 Report on 'Seeing by electricity', *Design and Work*, 26 June 1880, **8**, pp. 569–570

25 ANON: 'Bell's photophone', *Sci. Am.*, 1 January 1881, **44**, (1), pp. 1–2

26 BRUCE, R.V.: 'Bell' (Gollancz, London, 1973), p. 336

27 Ref. 26, p. 337

28 Ref. 26, p. 338

29 Ref. 26, p. 342

30 J.F.W.: Letter, *Nature*, 13 June 1878, **18**, pp. 169–170

31 Report on 'Seeing by electricity', *Nature*, 3 March 1881, **23**, pp. 423–424

32 BIDWELL, S.: 'Tele-photography', *Nature*, 10 February 1881, **23**, pp. 344–346

33 BIDWELL, S.: 'On telegraphic photography'. Report of the British Association for 1881, transactions of section G, pp. 777–778

34 HERTZ, H.: 'Ueber sehr schnell elektrische Schwingungen', *Ann. Phy.*, 1887, **31**, pp. 421–448

35 Report in *The Times*, 24 April 1880

36 ATKINSON, L.B.: 'Seeing to a distance by electricity', *The Telegraphic Journal and Electrical Review*, 13 December 1889, p. 683

37 Report on 'Television. A brief account of recent events', *Electr. Rev.*, 15 March 1929, p. 494

38 ALBE, E.E.F.d'.: 'On the efficiency of selenium as a detector of light', *Proc. R. Soc. Lond. A, Math. Phys. Sci.*, 1914, **89**, pp. 75–90

39 ARRHENIUS, S.: 'Ueber das leitungsvermogen beleuchteter Luft', *Ann. Phys.*, 1888, **33**, pp. 683–693

40 ATKINSON, L.A.: 'Inaugural address', *J. Inst. Electr. Eng.*, 1920, **59**, pp. 15–16

Chapter 4

Persistence of vision and moving images (1825–c. 1900)

Persistence of vision is a *sine qua non* for any television system in which the scene being televised is analysed and the associated reconstituted image is synthesised by scanning beams of light or electrons. It is necessary that the rate at which the images are presented to an observer be sufficient to ensure that they do not appear as a series of still images. The rate must be such that the eye-brain combination fuses the individual images into a flickerless picture.

When the earliest 'seeing by electricity' schemes were proposed in the late 1870s much empirical work had been undertaken on the construction of models to demonstrate the creation of moving images from the rapid showing of discrete drawings and photographs (see Table 4.1). Also, Fox Talbot in 1834 had propounded a law which characterised the flicker effect[1]. Later television research, carried out by the Radio Corporation of America in the 1930s, further quantified[2,3] the conditions necessary for the comfortable viewing of moving images (see Chapter 18).

The phenomenon of persistence of vision[4,5] had first been observed many centuries before its importance to motion film projection and television had been appreciated. Lucretius mentioned the fact of persistence of vision—but only in connection with the images seen in a dream. Ptolemy, in his treatise on optics, described a disc, one part of which was coloured, and declared that if it was rapidly revolved the whole disc would be of that colour. Alhazen also referred to the phenomenon and Leonardo da Vinci gave an account of it. But it was not until the 19th century that the principle of persistence of vision was really applied—first, to toys and popular scientific exhibits and later to cinematography.

The London Stereoscopic Company early in the century was one of the first to produce a toy, known as la Toubie éblouissante when made in France, which illustrated the effect. It consisted of a piece of wire bent into the outline of one side of a symmetrical object, for example a vase, which was set in a hollow, vertical, metal spindle. When this was revolved the figure appeared complete. Some years later, in 1825, Dr John Ayrton Paris, a practitioner of medicine in Penzance, invented a simple device, to which he gave the scientifically sounding name of thaumatrope,

Table 4.1   The era of optical toys and models

1826	thaumatrope (Paris)
1832	phenakistiscope/stroboscope (Plateau/Stampfer)
1834	zoetrope or wheel of life (Horner—popularly introduced in 1860 by Desvignes)
1850	heliocinegraphe (Uchatius—a modified phenakistiscope
1850	lantern wheel of life (Ross) } animated lantern slides
1850	choreutoscope
1859	omniscope (du Mont)—a phenakistiscope with stereoscope
1860	stereoscopic zoetrope (Desvignes)
1860	modified phenakistiscope (Shaw)
1861	stereotrope (Shaw)
1861	kinematoscope (Sellers—a stereoscopic viewer)
1861	suggestions for making sequence photographs (du Mont)
1863	stereoscopic phenakistiscope (Bonelli and Cook)
1864	suggestions for cameras with multiple lenses (du Hauron)
c. 1870	stereoviewer for moving pictures (Wheatstone)
1870	phasmatrope (Heyl)

which seemed to show that it was possible for two drawings or pictures of different objects or scenes to be exhibited in the same place at the same time. Dr Paris described the working of the toy in a publication called 'philosophy in sport and science in earnest; being an attempt to illustrate the first principles of natural philosophy by the aid of popular toys and sports'. The plaything itself was made by William Phillips of George Yard, Lombard Street, London, and was available for purchase from April 1825. The little contrivance consisted of a paper disc, each face of which presented a different drawing or picture, attached to strings so it could be rapidly rotated about an axis coplanar with the disc. When this was carried out the two scenes merged together to form a single image: for example, a bare tree acquired leaves, a bald man put on a wig, a rider mounted his horse, and so on.

There seems little doubt that the thaumatrope created considerable interest, for the august Royal Institution was persuaded to sell the toy for seven shillings and six-pence from its Albemarle Street residence. The idea of a moving picture was one of much novelty and attracted the attention of several investigators including Sir John Herschel.

Also in 1825 Dr Roget, the Secretary of the Royal Society, discovered that the spokes of a rotating wheel appeared stationary when seen through a series of verti-cal slots, and shortly afterwards Faraday demonstrated, by means of an apparatus called Faraday's wheel, that when two cogged wheels, each having the same number of teeth, rotated about a common axis at equal speeds but in opposite directions, the eye perceived a stationary image of one wheel only. This observation can be explained using the notion of persistence of vision, for the brain receives a strong impression of the teeth of the two wheels when they coincide but only a very weak stimulus when the cogs of one wheel pass the spaces of the other.

In 1828 a Belgian philosopher and scientist, Joseph Antoine Ferdinand Plateau, (later a professor of physics at Ghent University), made the same discovery. Two years after this he patented an invention which used rotating wheels, spokes and cogs and combined the principles of the devices of Paris, Roget and Faraday with those of his own experiments. He called the invention a phenakistiscope: it was the progenitor of all the subsequent and more complicated forms of motion picture apparatus.

In its English form, made by a T T Bury and variously named the fantascope, Figure 4.1, phantasmascope, magic disc, kaleidorama and McLean's optical illusion or magic panorama, the toy was constructed as follows: 'The edge of a large pasteboard disc [was] deeply notched to form cogs between each of which [were] painted figures in a number of positions. By means of a nut and screw the disc [was] attached to a spindle and rapidly revolved in front of a mirror. The image in the looking glass performed a series of convincing movements, graceful or grotesque according to the drawing. A dancer pirouetted, a frog leapt over a rat, a devil turned somersaults, while a horseman galloped round and round as hard as he could go, a swallow darted about a rose bush . . . '

Great interest was shown in the phenakistiscope and it was so much in demand that 'it became urgently necessary to accommodate a number of viewers instead of one'. Plateau applied himself to the problem in 1849 (six years after he had become totally blind), and devised an apparatus which was the first to demonstrate the

Figure 4.1   A fantascope (also called a phenakistoscope) by T T Bury (published by Ackermann and Co, London, in 1833)

Reproduced by permission of the Science Museum

projection of a continually moving image. He placed 16 images (in a progressive series) around the periphery of a glass disc, which was then rotated, and in front of this he rotated an opaque disc, having four slots, at a speed four times greater than the glass disc and in a reverse direction to it. All the parts of the disc except that showing the erect image were screened off. Plateau's design depicted a devil blowing at a fire and it produced so striking a sensation that the inventor was encouraged to use photographs instead of drawn pictures. In 1852 he secured a series of photographs of a workman using a pestle and mortar, though 'the results, when these photographs were combined by means of the phenakistiscope, showed how impossible it was to obtain a naturalistic effect of movement from a number of posed attitudes'.

A modified form of this instrument was introduced by an Austrian Lieutenant named Franz Uchatius in 1851 and called by him the heliocinegraphe. Two years later he demonstrated his lantern wheel of light at the Vienna Academy of Sciences. In this the images were painted near the periphery of a transparent disc which remained stationary: however, by means of an ingenious contrivance of lenses and a revolving limelight, the succession of images could be projected onto a screen.

A B Brown, in 1869, also invented a projecting phenakistiscope. His instrument is important for it incorporated two devices essential to the development of cinematography—a Maltese cross picture transport mechanism[6] and a shutter which prevented the pictures from being projected when they were being moved. The Maltese cross arrangement enabled the images to be moved intermittently—as in modern motion film practice—so that with the single-bladed shutter the projecting phenakistiscope allowed 'smoothly moving painted figures of blacksmiths working at an anvil, negro boys diving into a lake, horses jumping over a hurdle, urchins sliding on a pond or pole vaulting [to appear] in full movement and almost life size upon the screen'.

Many inventions based on the principle of persistence of vision were put forward during the Victorian era—the Daedelum or wheel of life, the zoetrope, the praxinoscope, Figures 4.2 and 4.3, the viviscope, the tachyscope, the Rudge projector and the zoogyroscope among others—so that when inventors throughout the world turned their attentions to the problem of combining photography with motion to achieve cinematography, in the 1885 to 1895 decade, the basic requirements for the solution of the problem were well known.

Both television and cinematography depend on the rapid display of still images to achieve the illusion of continuous movement. The same principles of persistence of vision apply to the two media forms and they were recognised to apply by the earliest workers in the distant vision field. Le Blanc[7], for example, in 1880, described his system of scanning using a mirror mounted on two vibrating arms A and B and wrote: 'Nous pouvons supposer, de plus, le mouvement vibratoire des lames A and B assez rapide pour que toute la surface de A nous paraisse uniformement éclairée, grâce à la persistence des impressions lumineuses.'

All scanners used in television must be capable of scanning their fields of view several times per second. In the earliest practical low definition systems of the 1920s the rate was about ten frames per second but when high definition systems were

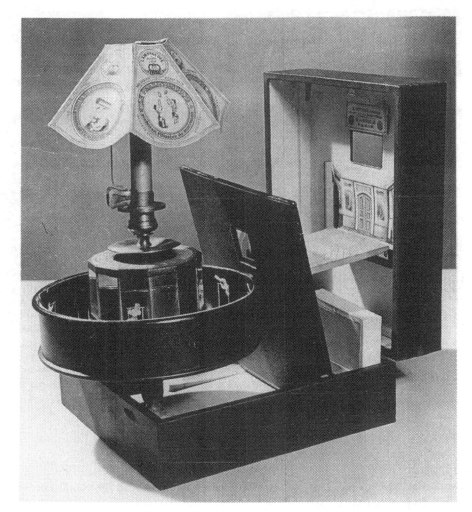

Figure 4.2   *The praxinoscope of c. 1877*

Reproduced by permission of BFI Stills, Posters and Designs

introduced the complete picture rate (for reasons given in Chapter 18) was either 25 Hz or 30 Hz. And, as in cinematography, special means had to be introduced to reduce the flicker effect.

Interestingly, the first proposals for seeing by electricity and cinematography were made at nearly the same time: Figuier had described his telectroscope in June 1877 and Wordsworth Donisthorpe, an English barrister, had filed a patent[8,9] for a moving picture camera on 9 November 1876. But while motion pictures took only a few years to develop—they were first shown in 1893—the gestation period of television was to be almost half a century. Hence, when public television broadcasting services commenced in the late 1920s, much experience and expertise on the

Figure 4.3   Reynaud, in 1877, patented a method, based on the praxinoscope principle, of projecting a series of pictures onto a large screen. He replaced the limited number of pictures which could be accomodated in the standard praxinoscope by a long roll of paper and was able to present extended entertainments to large audiences. He called his system the 'Theatre Optique'

Reproduced by permission of BFI Stills, Posters and Designs

production of films had been gained which could be applied to the production of television programmes.

Donisthorpe's invention had for its object[9]:

> 'to facilitate the taking of a succession of photographic pictures at equal intervals of time, in order to record the changes taking place in or the movement of the object being photographed, and also by means of a succession of pictures so taken of any moving object to give to the eye a presentation of the object in continuous movement as it appeared when being photographed.'

At that time magazine dry-plate cameras were being introduced and Donisthorpe described a version in which plates were to be changed rapidly, the exposure taking place while they were stationary. Positives were to be printed on a long roll of paper and viewed in rapid succession. No clear idea as to how the necessary intermittent viewing was to happen was given in the patent.

Subsequently in a letter to *Nature* dated 24 January 1878, following Edison's 1877 invention of the phonograph, Donisthorpe wrote[8]:

'By combining the phonograph with the kinesigraph [Donisthorpe's invention] . . . the life-size photograph shall itself move and gesticulate . . . the words and gestures corresponding as in real life.

'Each picture as it passes the eye is instantaneously lighted up by an electric spark. Thus, the picture is made to appear stationary while the people or things in it appear to move as in nature.

'I think it will be admitted that by this means a drama acted by daylight or magnesium light may be recorded and reacted on the screen or sheet of a magic lantern, and with the assistance of the phonograph the dialogues may be repeated in the very voices of the actors . . . '

Donisthorpe did not achieve success at this early date but in the late 1880s he collaborated with W C Crofts in designing another motion picture camera and projector which were patented on 15 August 1889. Only a few frames, taken at eight to ten frames per second, in Trafalgar Square, London, in 1890, survive but they indicate that Donisthorpe and Croft's approach seemed to work well. Of particular interest is Donisthorpe's attempt to obtain financial backing, a matter which was to be experienced by some of the television pioneers. He approached Sir George Newnes who[10]:

'submitted the matter to two 'experts' selected, by Sir George Newnes, to pronounce on its merits. One I afterwards learnt was an artist, a painter who was as ignorant of the physical sciences as Noah's grandmother, and the other was, I believe, a magic-lantern maker.

'I need hardly [say] that both these 'experts' reported adversely. They agreed that the idea was wild, visionary and ridiculous, and that the only result of attempting to photograph motion would be an indescribable blur.

'What could Sir George Newnes do in the face of such 'expert' testimony?'

Meanwhile, the English photographer E J Muybridge[11,12] had acquired some considerable fame in analysing motion by sequence photography. At a demonstration on 15 June 1878 the press and other visitors saw 12 photographs, made in about 0.5s, of a horse in movement. Since Muybridge used the wet-plate process, the result were little more than silhouettes, but nevertheless the successive movements of the horse were recorded with considerable accuracy. (It is perhaps pertinent to note that the earliest televised images, by J L Baird, C F Jenkins and D von Mihaly, were silhouettes.)

The following year the number of cameras was increased to 24 and by the end of the year Muybridge had obtained hundreds of photographs of animals, birds and athletes in motion. His financial sponsor, L Stamford, a former Governor of

California and the President of the Central Pacific Railroad, had now spent, it is believed, approximately $42 000 since he commissioned Muybridge in 1872 to photograph a record-breaking trotter of Stanford's called Occident.

The success of Muybridge's sequence photographs spread throughout America and Europe and the images were reproduced in many of the leading magazines. Articles in the press in 1879 suggested that the photographs should be mounted in a zoetrope so that the motion of the horse could be synthesised. One London magazine implemented such a system and exhibited it in its office window. Realistic reproductions of the original movements were created and soon afterwards Muybridge designed a projector, based on a similar principle, which he first called the zoogyroscope, but later, in 1881, renamed the zoopraxiscope[13].

Muybridge's zoogyroscope was finished in the autumn of 1879 and was first described in the *Alta California* newspaper on 5 May 1880:

> 'Mr Muybridge has laid the foundation of a new method of entertaining the people, and we predict that the instantaneous photographic magic lantern zoetrope will make the rounds of the civilised world.'

During 1881 and 1882 he toured Europe, lectured to learned societies and, in the spring of 1882, demonstrated his zoopraxiscope to the Royal Society, the Royal Academy, the Society of Arts and the Royal Institution of London. Everywhere he met with marked interest and enthusiasm.

The *Photographic News*[14] reported on the Royal Institution lecture as follows:

> 'Mr Muybridge exhibited a large number of photographs of horses galloping, leaping etc . . . By the aid of an astonishing apparatus called a zoopraxiscope, which may briefly be called a magic lantern run mad (with method in the madness), the animals walked, cantered, ambled, galloped and leaped over hurdles in a perfectly natural and lifelike manner.
>
> 'I am afraid that, had Muybridge exhibited his zoopraxiscope three hundred years ago, he would have been burned as a wizard . . . '

The publication of Muybridge's work stimulated others to investigate sequence photography (see Table 4.2). E J Marey[15], of France, in the 1880s pioneered the science of chronophotography (the analysis of movement) using a single sequence camera, and O Anschutz, a Prussian photographer, in 1887 devised the electro-tachyscope, a type of viewing apparatus.

Marey's 1882 camera was in the shape of a gun, with an eccentric cam rather than a Maltese cross to achieve the intermittent motion, and took 12 exposures, the duration of each being approximately 1/720 of a second, on a circular glass plate. The glass plate was inherently unsuited for recording long sequences of action, but in 1885 a development occurred which transformed the taking and reproduction of serial images. In that year George Eastman, an American manufacturer of dry plates, with W H Walker, designed a roll holder to replace the conventional camera

Table 4.2   The era of animated photographs

1872–1873	work on sequence photography (Muybridge)
1873	book on animal location (Marey)
1874	sequence photography (Janssen)
1876	suggestions for taking sequence photographs (Donisthorpe)
1877–1904	work on sequence photography (Muybridge)
1878	12 sequence photographs of a horse (Muybridge)
1879	24 sequence photographs of a horse (Muybridge)
1879	zoogyroscope—a sequence photograph projector (Muybridge) renamed zoopraxiscope in 1881
1882	sequence photograph gun (Marey)
1882–1891	multilensed cameras for sequence photography (Londe)
1883–1887	work on sequence photography (Anschutz)
1887	electrotachyscope (Anschutz)

plate holder. The new device enabled a paper roll coated with a gelatine emulsion to be used. Four years later Eastman introduced a thin, tough, flexible transparent celluloid film to supersede the paper film in the roll holders. This was the essential invention which made cinematography possible[16].

In 1888 Marey began experimenting with moving film and evolved what is deemed to be the first successful cine camera. At first rolls of paper were employed but when celluloid film became available Marey worked with the new material. The roll of film passed from one spool to another and was stopped in its motion, for the exposure, by an electromagnet mechanism. The images were about 9 cm by 9 cm in size and as the rolls of film were four metres in length only approximately 40 exposures could be recorded. Each exposure lasted 1/1000 of a second and the camera was capable of taking up to 50 pictures per second.

Marey was primarily a scientist and was uninterested in pursuing his investigations for general entertainment. He had endeavoured to devise means to study animal motion and in this he had been eminently successful.

Anschutz's sequence negatives were printed[17] as 90 mm by 120 mm transparencies and fixed around the circumference of a large steel disc, Figure 4.4. As the disc rotated a Geissler discharge tube produced a brief intense flash of 1/1000 of a second duration every time a transparency passed in front of it and behind the viewing aperture. An electric switch mounted on the disc ensured that the flashes were synchronised with the appearance of the transparencies. These followed each other at about 1/30 of a second and, because of the persistence of vision, created the illusion of a continuously moving picture.

100 of the machines were manufactured by the Siemens and Halske Company and were widely distributed throughout Europe and America. In December 1892 the electrotachyscope arrived in London under the name 'an electrical wonder', and was operated by a penny-in-the-slot mechanism. According to the *Amateur Photographer*[18], the impression gained from viewing the transparencies was so lifelike that 'we think this new wonder will become a very good thing'. But not everyone

Figure 4.4   Anschutz's electrical tachyscope of c. 1892

Source: *Scientific American*, 16th November 1889, p. 303

was convinced. In London's Strand one lady was overheard to say to another: 'It's a show of moving figures. It's awfully stupid.'

Among others who in the 1880s participated in the advancement of the display of moving images mention can be made of William Friese Green, an English professional photographer, and Louis-Aime-Augustin Le Prince, a Frenchman who worked in New York as the manager of a chain of panoramas. They both secured patents[19,20] for their inventions, in 1889 and 1888, respectively.

However, it was the 'Wizard of Menlo Park', Thomas Alvar Edison[21], who provided the idea and the means which made cinematography practical. Once again it was Muybridge who supplied the stimulation. He had lectured in New Jersey on 25 February 1888 and on the 27[22] visited Edison to discuss with him the possibility of combining the zoopraxiscope with the phonograph. Edison was intrigued with the notion and set his assistant W K L Dickson, a Scotsman who had emigrated to the US in 1879, to work on the project.

In advancing their ideas Edison and Dickson were influenced by the work of Anschutz on his electrical tachyscope, and in a patent dated 20 May 1889 Edison and Dickson used the same general arrangement of continuous movement and momentary light flashes in their viewing device, the kinetoscope.

Little success seems to have been achieved by the two inventors until after Edison's meeting, in August 1889, with Marey at the Paris Exposition. Marey showed[23] him his new roll film camera and the results which had been obtained with it. On his return Edison applied for a caveat which covered the use of a roll of film and mentioned the crucial innovation of perforations along the edge of the film. By their means, and a toothed sprocket wheel driven by an intermittent escapement, a positive drive could be obtained which would ensure the correct registration of successive frames.

An experimental kinetoscope viewer was shown to the the National Federation of Women's Clubs on a visit to Edison's laboratory on 20 May 1891. Limited production began in the summer of 1892 and full manufacture in 1893. By the beginning of April 1894[24] a batch of ten kinetoscopes had been sent to a firm, Holland Brothers at 1155 Broadway, New York, where the first kinetoscope parlour opened on 14 April 1894. The viewer came to Europe in the same year and was on exhibition in August in the Boulevard Poissoniere in Paris, and in October at 112 Oxford Street, London. By the

Figure 4.5   *Edison's kinetoscope of 1894*

Reproduced by permission of the Science Museum

Figure 4.6   A televisor of the late 1920s (compare with Figure 4.5)

Reproduced by permission of Radio Rentals Ltd

end of the year kinetoscope parlours had been opened all over Europe and North America. More than 1000 machines were sold before their popularity declined.

The coin-in-the-slot kinetoscope, Figures 4.5 and 4.6, was a viewing cabinet which permitted one person at a time to see a film of around 50 feet in length. Each frame of the continuously moved film was seen momentarily by light transmitted by a slit in a revolving shutter placed between the film and an electric lamp. The picture rate of 46 frames per second restricted the viewing time to about 15 seconds. The subjects shown included wrestling, Highland dancing, a trapeze act and the strong man Eugene Sandow.

Strangely, Edison, who acquired 1093 patents during his life, failed to see the potential of the kinetoscope[25,26], even though it was initially successful in Europe and America, and regarded it as a novelty which in time would pass out of fashion. He failed to patent it in Great Britain, to the advantage of Robert W Paul[27], a London scientific instrument maker, who soon began making copies of the machine. When Edison retaliated by attempting to deny Paul the use of films for his kinetoscopes he gave Paul an incentive to construct his own moving film camera and associated projector. Paul abandoned Edison's use of continuously moving film and devised a mechanism based on the utilisation of two seven-star Maltese crosses to transport the film intermittently past the projector lens.

The projector was in use in 1895 and received its first public demonstration on 28 February 1896 before the Royal Institution. In the following month Paul began

showing films at the Alhambra, Leicester Square, London. The nightly presentations were popular and continued for the succeeding four years.

Almost simultaneously with these events the Lumiere brothers[28], Auguste and
Louis, who had patented in France, on 13 February 1895, a combined camera,
printer and projector, had given on 20 February 1896 the first showing of their
cinematographe in London at the Marlborough Hall, Regent Street[29]. The Empire
Theatre of Varieties in Leicester Square soon contracted to include the cinematographe in its shows, and it featured in their performances from 9 March 1896.
In the US the machine was exhibited at Koster and Bial's Music Hall on 34th Street,
New York, and subsequently at halls throughout the country. The Lumiere agents
took the cinematographe all over the world during 1896.

And in Germany Max and Emil Skladanowsky projected films with their own
projector, touring Europe in 1896.

Great interest was shown by the general public in these first cinema entertainments
and this led to a rapid growth in the motion picture industry. By the end of the 19th

Table 4.3   The era of motion film photography

1885	sensitised paper rolls—the gelatin coating could be stripped for making transparencies (Eastman)
1888	Kodak camera using 70 mm wide paper film introduced (Eastman)
1888	patent granted for a multilens camera using paper negative film (Le Prince)
1888	single lens camera using a roll of paper film (Le Prince)
1888–1890	camera for taking 20 pictures per second using paper film—primarily for sequence photography (Marey)
1888–1889	patents on optical equivalent of phonograph (Edison)
1889	celluloid film (Eastman)—a key development for practical 'movies'
1889	patent on film perforation (Edison)—a key development for practical 'movies'
1889–1890	kinesigraph camera and projector (Donisthorpe)
1889–1891	camera and viewer using perforated film (Edison)
1889–1893	camera for taking about four or five pictures per second (Friese Green)
1890	patent for a twin-lens camera for taking sequence stereoscopic photographs granted (Varley)—two or three pictures per second
1892	commercial introduction of kinetograph camera and kinetoscope viewer (Edison)
1892–1893	patents for cinematographe, an instantaneous photographic apparatus (Bouly)
1895	patent for a cinematographe (Lumiere Brothers)
1895	intermittent motion projector (Paul)
1895	intermittent motion camera and projector (Acres)
1896	twin-film projector (M and E Skladanowsky)
1896–to date	by the end of 1896 the motion film industry was a thriving business

century many different cameras and projectors, utilising a variety of film sizes, were available and cinemas had opened in many large towns and cities[30] (see Table 4.3).

When this brief outline of the beginnings of cinematography is compared with the history of television it will be readily apparent that certain factors were common to the evolution of the two entertainment medias. The comparison shows that the *modus operandi* of the early television pioneers—particularly Baird (in the UK), Jenkins (in the US) and Mihaly (in Germany)—was consonant with those of the early moving film pioneers. As a consequence, some of the adverse criticisms which they experienced, in endeavouring to develop a new media form, were perhaps unjustified.

The early history of moving photographic images illustrates the apparent need for:

1   patent protection (cf. Edison and Paul);
2   financial sponsorship (cf. Donisthorpe and Muybridge);
3   demonstrations, even when a particular apparatus/system is in a rudimentary state of development;
4   good publicity, by widespread exhibitions;
5   approval by learned scientific societies (e.g. the Royal Institution);
6   early commercialisation of apparatus/system (e.g. the 100 electrical tachyscopes of Anschutz and the 1000 kinetoscopes of Edison);
7   new materials (e.g. glass plates → paper film → celluloid film);
8   stimulation of others (cf. influence of Muybridge on Marey, Anschutz and Edison);
9   new ideas (e.g. film perforation to give uniform film registration, use of intermittent film motion);
10   convergence of standards (e.g. use of 35 mm film);
11   sustained public interest.

It will be seen later that each of these factors, without exception, either influenced the development of television or was thought to be important in its advancement. Indeed, the parallel between the beginning of the two industries may be further extended. Both histories can be shown to comprise three distinct periods from the initiation of the earliest notions to the prolonged commercial success of the medias. For cinematography the periods are the eras of:

1   optical toys and models which illustrated persistence of vision (c. 1824 to c. 1877);
2   animated photographs (c. 1872 to c. 1888);
3   35 mm cine film cameras and projectors (c. 1896–).

For television the periods are the eras of

1   speculation (1877 to c. 1922);
2   low definition television broadcasts (1926 to c. 1935);
3   high definition television broadcasts (1936–).

In each case, once the basic scientific principles had been practically established, the growth times from restricted/rudimentary moving images to high definition images were relatively short, approximately eight and ten years respectively. And in both cases the transitions were made possible by the availability/realisation of new materials, *viz.* 35 mm perforated cine film and photoelectric mosaics capable of being electronically scanned. When the manufacturing processes for these materials had been validated, progress towards viable public entertainment systems was quick.

References

1 WALSH, J.W.T.: 'Photometry' (Constable, London, 1926) pp. 67–72
2 ENGSTROM, E.W.: 'Determination of frame frequency for television in terms of flicker characteristics', *Proc. IRE*, April 1935, **24**,(4), pp. 295–310.
3 KELL, R.D., BEDFORD, A.V., and TRAINER, M.A.: 'Scanning sequences and repetition rate of television images', *Proc. IRE*, 1936, **24**, (4), pp. 559–575
4 THOMAS, D.B.: 'The origins of the motion picture' (HMSO, London, 1964)
5 CORK, O.: 'Movement in two dimensions' (Hutchinson, London, 1963) Chapter 8, pp. 121–136
6 Article on motion pictures, *Encyclopaedia Britannica* 1963, **15**
7 BLANC, M. Le.: 'Étude sur la transmission électrique des impressions lumineuses', *La Lumiere Électrique*, 1 December 1880, **11**, pp. 477–481
8 DONISTHORPE, W.: 'Talking photographs', a letter, *Nature*, 24 January 1878, **18**, p. 242
9 DONISTHORPE, W.: 'Apparatus for taking and exhibiting photographs'. British patent 4344, 9 November 1876
10 COE, B.: 'Muybridge and the chronophotographers' (Museum of the Moving Image, London, 1992) p. 45
11 HAAS, R.B.: 'Muybridge—man in motion' (University of California Press, Berkley, 1976)
12 MUYBRIDGE, E.: 'Method and apparatus for photographing objects in motion'. US patent 212 864, 4 March 1879
13 MUYBRIDGE, E.: 'Animal locomotion; an electro-photographic investigation of consecutive phases of animal movement' (Lippincott, Philadelphia, 1887)
14 Ref.10, p. 19
15 MAREY, E-J.: 'La photographie du movement: les méthodes chronophotographiques sur plaques fixes et pellicules mobiles' (Carre, Paris, 1892)
16 MUSSER, C.: 'History of the American cinema. The emergence of cinema' (Charles Scribner, New York, 1990)
17 ANSCHUTZ, O.: 'Projektions Apparat für stroboskopisch bewegte Bilder'. German patent 85 791, 5 November 1894
18 Ref.10, p. 36
19 GREEN, W.F.: 'Photographic printing apparatus'. British patent 4956, 29 March 1889
20 PRINCE, L-A-A Le.: 'Animated photographic pictures'. US patent 376 247, 2 November 1886
21 CLARK, R.W.: 'Edison, the man who made the future' (Macdonald and Jane's, London, 1977) Chapter 9, pp. 170–179
22 Ref.21, p. 172
23 Ref.21, p. 175
24 Ref.21, p. 176
25 EDISON, T.A: 'Apparatus for exhibiting photographs of moving objects'. US patent 493 426, 24 August 1891
26 EDISON, T.A.: 'Kinetographic camera'. US patent 589 168, 24 August 1891
27 Ref.4, p. 30
28 LUMIERE, A., and LUMIERE, L.: 'Appareil servant à l'obtention et à la vision des épreuves chronophotographiques'. French patent 245 031, 13 February 1895
29 Ref.10, pp. 52–53
30 SKLAR, R.: 'Film. An international history of the medium' (Thames and Hudson, London, 1993)

Chapter 5
Distant vision (1880–1920)

As noted previously, the portrayal and the putative portrayal of illusions and images had attracted the attention of magicians, charlatans and pseudoscientists. There seemed to be a popular demand for visual displays and exhibitions of the unexpected as part of the social fabric of living, a demand which can now be fulfilled by the various media forms. Even great writers were not immune from referring to magical illusory effects. An interesting example occurs in Sir Walter Scott's 'My Aunt Margaret's mirror', which was published as one of the Waverly novels in 1825[1]. In this story the heroine, in an endeavour to locate an unfaithful husband, consults a physician who has a reputation as a conjurer. She is led into a room containing a very tall and broad mirror, and then:

> 'Suddenly the surface assumed a new and singular appearance. It no longer simply reflected the objects placed before it, but, as if it had self-contained scenery of its own, objects began to appear within it, at first in a disorderly, indistinct and miscellaneous manner, like form arranging itself out of chaos; at length, in distinct and defined shape and symmetry. It was thus that, after some shifting of light and darkness over the face of the wonderful glass, a long perspective of arches and columns began to arrange itself on its sides, and a vaulted roof on the upper part of it; till, after many oscillations, the whole vision gained a fixed and stationary appearance, representing the interior of a foreign church . . . '

An analogous scene occurs in the third act of George Bernard Shaw's play[2] 'Back to Methuselah' which is set in the year 2170 AD (see Chapter 9).

The possibility of achieving, at a future date, the means by which images of people or events could be transmitted from one place to another was enhanced by the ideas, crude and simplistic as they were, which had been propounded in the 1870s. These notions stimulated writers and cartoonists to evoke fantasies showing, perhaps, the eventual outcome of 'seeing by electricity'. Sir Walter Scott's allusion to television was similar to the cartoon published in the magazine Punch in 1879 (Figure 3.3, see also Figure 5.1).

And in 1881 the editor of the Electrician mused on the possibility of distant vision in a report on the 1881 Electrical Exposition, Paris[3]:

> 'The telephotograph of Mr Shelford Bidwell even gives us the hope of being able, sooner or later, to see by telegraph, and behold our distant friends through the wire darkly, in spite of the earth's curvature and the impenetrability of matter. With a telephone in one hand and a telephote in the other an absent lover will be able to whisper sweet nothings in the ear of his betrothed, and watch the bewitching expression on her face the while, though leagues of land and sea divide their sympathetic persons.'

The most remarkable examples of prophecy in this field were given by A Robida in an engrossing, and prescient, book entitled 'The XXth century, the conquest of the regions of the air' published in 1884[4]. In this Robida foretells (for the year 1945) how the 'telephonoscope' will impact on people's lives:

> 'Among the sublime inventions of the XXth century and the thousand marvels of an age rich in magnificent discoveries, the telephonoscope may be considered as one of the most surprising, one of those inventions which bring the fame of scientists nearer to the stars.
>
> The old electric telegraph, that childish application of electricity was dethroned by the telephone and then by the telephonoscope, which is the supreme and final development of the telephone. The oldtime

Figure 5.1   A 19th century impression of two-way television in the year 2000 AD

Source: National Archives, Washington

telegraph allowed us to understand a correspondent at a distance, but the telephonoscope allows us both to see and hear him at the same time.

The invention of the telephonoscope was received with the greatest delight. The apparatus was attached to the instruments of all telephone subscribers who desired it, on payment of a supplementary charge. Dramatic art found in the telephonoscope an opportunity for immense prosperity. The theatrical performances transmitted by telephone became all the rage when it was also possible to see the performers as well as hear them.

Theatres had thus, besides the ordinary number of spectators in the building, a number of listeners-in and spectators in their own homes connected to the theatre with the wire of the telephonoscope. Here was a fresh source of gain and box-office receipts. No limit to profits now, 'no house full' limits to a theatre! When a show enjoys a big success, in addition to the three or four thousand spectators in the theatre, fifty thousand spectators are in their homes, and fifty thousand at least in other countries of the world.

The Universal Company of Theatrical Telephonoscopy, founded in 1945, has now 600 000 subscribers scattered in all countries of the world. This company centralises the wires and pays the dues to the theatres for the reception of their performances and programmes.

The apparatus consists in a simple plate of crystal let into a wall of the apartment or placed like a mirror over any piece of furniture. The subscriber without disturbing himself, sits down before the mirror or plate, chooses his theatre; switches on the communication and enjoys the show.

The telephonoscope, as the word indicates, allows us both to see and hear. Dialogue and music are transmitted as with ordinary telephone lines, but at the same time the stage, with all illumination, its decor, its actors appear on the glass screen as clear as anything seen in direct vision. The performance can be witnessed with both ears and eyes. The illusion is complete and absolute. It is like being in a box at the opera or theatre.'

Of the cartoons included in Robida's 1884 book all those shown in Figure 5.2 have their modern counterparts. Distance learning (cf. the Open University), instantaneous images of war, reports from foreign correspondents, televised opera and videoconferencing are now everyday matters of fact. However, more than 40 years were to elapse before Robida's inspirations could be implemented. Many developments in pure science and technology relating to photoelectricity, thermionic and secondary electron emission, the physics of phosphors, electromagnetic wave propagation, electronics and radio generally were necessary before the first flickering, crude televised images could be reproduced. But in one branch of television, that of analysing and synthesising an image by mechanical scanners, progress was made in the 1880s. Indeed, contemporaneously with the publication of Robida's book, a

Figure 5.2a, b, c & d In his book 'The 20th century, the conquest of the regions of the air', published in Paris in 1884, A Robida correctly forecast some applications of 'seeing by electricity' (television). Figure 5.2a shows a family watching a televised report from a battlefield. Figure 5.2b illustrates a televised opera production being presented in a gentleman's drawing room. Figures 5.2 c & d display aspects of two-way television

German inventor had patented a scanner which, for a few years in the late 1920s and early 1930s, was widely used in low-definition television systems.

The 1880–1900 era was characterised by a new realism among the proponents of distant vision schemes. No longer, in the main, were hopelessly ill-conceived and naive systems advanced. Multiconductor, nonscanning suggestions were abandoned, only to be raised for a brief moment by Shelford Bidwell in 1908. The inventors and scientists of this period had a clear conception of the principles of scanning. During the formative years, in the 1920s, of low-definition television, using mechanical scanning, the various scanners used were generally those advanced by Nipkow (1884), Weiller (1889), Brillouin (1891) and Szczepanik (1897). And the first mention of line and frame scanning dates from 1880.

Le Blanc's article[5] in *La Lumiere Electrique* (1880) contains drawings which unambiguously show two vibrating arms, each carrying a small mirror, for the execution, by line and frame scanning, of picture analysis and synthesis; although these would not have led to linear frame scanning as he stated but rather to Lissajous figure scanning.

As usual with most of the early schemes no details were given by Le Blanc but the idea was adopted in 1926 by Rtcheouloff[6], who confirmed the mode of scanning.

Figure 5.2 continued

In a letter[7] to the *English Mechanic and World of Science* (1882), signed W L, the author (whose identity, William Lucas, was revealed in 1936) proposed another method of scanning using 'a pair of ordinary achromatic prisms, one placed vertically and the other horizontally, . . . each capable of turning about its axis'. Again, no details were given—' for the simple reason that I have not, as yet, fully worked them out. The motions required are somewhat involved and the fact that they must be synchronous and have a high speed still further complicates the mechanism by which they must be produced'. Several workers found various types of rotating prisms attractive for scanning and subsequently patents were taken out for their use by Jenkins, Zworykin, Vorobieff, Westinghouse and de Wet.

WL's contribution was not confined to scanning, for he advanced a novel idea to employ a pair of Nicol prisms to modulate the light from a fixed source according to the variations of luminosity on the sending end photocell. This method depended on the fact that, if a beam of light is projected through a pair of Nicol prisms onto a

(ABOVE) *Telephonoscopique* kiosks are to be found on every street-corner in the better neighbourhoods throughout France in the mid-twentieth century. Not only can you speak with your distant friends and loved ones in all the corners of the Empire, you can also see them at the same time.

Figure 5.2 continued

screen, the intensity of the luminous spot so produced varies from a maximum to a minimum as one prism turned through an angle of 90°. WL thought that the angular displacement of one of the prisms could be varied by a mechanism actuated by an electromagnet which was itself to be operated by the modulated electric currents from the photocell. No record exists that such a device was ever worked in a successful demonstration of television (its time constant would have been much too long except for a very low definition system), but the idea is interesting as it gives an indication of the ways by which inventors were trying to solve the problem of converting a variable electric current to a variable light flux. Also, WL's proposal may have inspired the suggestion, first made by Nipkow two years later, to interpose a Kerr cell between two crossed Nicol prisms in order to achieve the same effect.

Of all the early scanning systems, that conceived by Paul Nipkow, who at the time was a student of sciences in Berlin, was to achieve the greatest fame. In his German patent[8] dated 6 January 1884, which one historian has designated as the 'master television patent', the inventor delineated his television system. This was based on

(ABOVE) Frustrated and enraged at obtaining so many crossed visual lines. a man hurls a chair at *le télé*. An observation by Robida that while tremendous strides may be made in technology. human behaviour changes not in the slightest.

Figure 5.2 continued

scanners which were to be employed by individual inventors and large manufacturing organisations alike, in many countries, until the 1930s. The success of Nipkow's method was undoubtedly due to its simplicity and, at a later stage of the development of television, to its adoption by Baird for many of his experiments.

Figure 5.3 depicts Nipkow's scheme. The picture analysing and synthesising scanning discs were pierced by 24 apertures arranged at equal angular displacements along a spiral line. Each of the apertures had the shape of a picture element and allowed a light flux corresponding to the brightness of a picture element to be incident onto the selenium cell, S, via the condenser lens. With this mode of scanning a given aperture scanned a given line, and after one rotation of the disc the whole projected image had been scanned. The rapid flyback of the scanning beam after each line and each frame was carried out automatically.

At the receiving end the image signal currents were fed to the light modulator PGA via a single line wire. The principle of operation of this modulator was founded on the magneto-optical Faraday effect[9]. The magnetic field established by the current flowing in the coil, G (situated between crossed Nicol prisms P and A in the glass tube), rotated the plane of polarisation of the light emitted from the source.

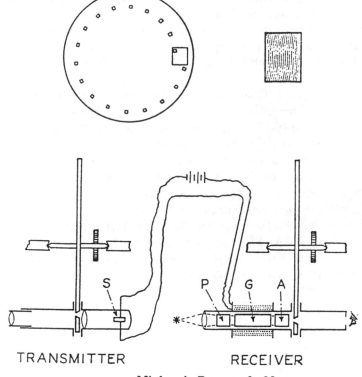

TRANSMITTER   RECEIVER

Nipkow's Patent of 1884.
S.   Selenium cell.
P.   Polarising prism.
G.   Flint glass.
A.   Analysing prism.

Figure 5.3   *Nipkow's 1884 proposal. His apertured scanning disc was used in television systems until c. 1938*

Source: German patent no. 30105, 6 January 1884

Consequently, the brightness of the image element seen behind the analyser, A, varied as the current varied. The complete image was synthesised by a combination of the receiver scanning disc and the phenomenon of persistence of vision.

Initially, Nipkow wished to achieve synchronism of the two scanning discs by means of clockwork mechanisms, but later he recommended the use of two phonic wheels, of the type described by La Cour 1878, for this purpose. Their speeds were controlled by tuning fork oscillators.

In common with most of the early distant vision workers, Nipkow made no attempt to put his ideas into practice. At the time, 1884, their realisation was bound to fail, if only for the fact that the light modulator would have needed about 10 W of control power.

An appreciation of this truth may have induced Nipkow to propose his alternative receiver light modulating device. In October 1885, in an article[10] on 'Der Telephotograph und das elektrische Telescop', he expounded his concept of a light controlling element based on a telephone receiver, see Figure 5.4. Possibly he was influenced by the success of Bell and Tainter's photophone. In essence the telephone, T, of Figure 5.4 acts in the inverse mode compared to the transmitter of the photophone (cf. Figure 3.10.) Surprisingly, most historians seem to have overlooked this article: certainly the ideas contained in it were retrogressive.

Nipkow's first suggestion for a light modulator stemmed from Faraday's 1845 discovery that the rotation of the plane of polarised light transmitted through a slab of silica borate of lead glass was influenced by a magnetic field. His diary comment states[9]: ' . . . and thus magnetic force and light were proved to have relation to each other. This fact will most likely prove exceedingly fertile . . . '

Later in 1875, Dr John Kerr, mathematics lecturer of the Free Church Training College, Glasgow, outlined the experiments which he had carried out on electrostatic birefringence of optical media[11], something which had been sought by Faraday.

Using plate glass as the medium Kerr noted the optical effect took 'a certain time (apparently 30 s)' to reach its full intensity. But with liquids such as carbon

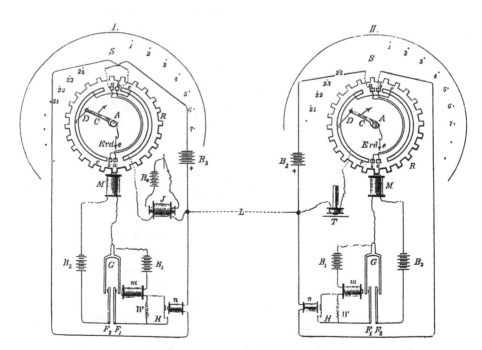

Figure 5.4   Nipkow's suggested method of synchronising two rotating discs based on the utilisation of phonic motors and tuning forks

Source: *Elekrotechnische Zeitschrift*, October 1885, pp. 419–425

bisulphide, the 'electrostatic force and the birefringent power increase together; they also vanish simultaneously'.

Henry Sutton[12] was the first person to apply the Kerr effect to the problem of transmitting optical images by the aid of telegraph wires. He also invented a new name for this subject and advocated calling it 'telephany', and the electro-optic instrument the 'telephane'[13], from the analogy between the transmission of luminous information by electricity and the transmission of acoustic information by the telephone. He clearly understood the nature of the problem, for he wrote: 'We have to take an optical image, seen as a surface, translate it into a line of consecutive varying electrical currents, and by means of these produce an effect as a surface, having the characteristics of the original image.'

Figure 5.5 illustrates his concept for accomplishing this specification. L is a photographic objective of the rapid type, producing an intensely illuminated aerial image at AA. DD is a Nipkow disc revolving at a 'fixed rate of not less than 650 r.p.m. under the control of La Cour's phonic wheel and fork apparatus as in the Delaney multiplex system'. C is a small piece of lamp black, selenium or other substance, the resistance of which may be varied by heat or light. 'Lamp black compressed is probably the most suitable.' So far, Sutton's apparatus is similar to that of Nipkow. But in the design of the receiver the two inventors had different views.

In Sutton's system, a beam of light from a source, S, passes through a pair of Nicol prisms, P, A (between which is placed a small glass cell holding a drop of carbon bisulphide and containing electrodes K, K), and is received and magnified by the eyepiece M, M before being viewed by an observer. Significantly, Sutton stated: 'The presence of a translucent screen at X, X, would be fatal, owing to the delicate nature of the desired effect.'

He appears to have appreciated the main disadvantage of utilising the Kerr cell; a disadvantage which was to lead eventually to an abandonment of its application, for he wrote: 'With regard to the receiver . . . the actual quantity of light required to reach the eye may be very small when received optically: in fact, so small as to have no power of illumination on a translucent screen.'

This point was confirmed by later workers. In particular E H Traub, Foreign Secretary of the Television Society, reporting[14] on the exhibit of Tekade at the 1936 Berlin Radio Exhibition wrote:

'Tekade . . . are today the only exponents of mechanical optical television receivers in Germany. Two receivers were exhibited, but only one was shown working. This was a 180-line receiver giving a picture of about 2 ft × 2 ft projected onto a glass screen from a double mirror screw used in conjunction with a rotating condenser (lens drum). An arc lamp and sealed-off Kerr cell were used and the details and geometry of these pictures were remarkably good, and it was quite impressive to see the line structure while standing a few feet away, and the brightness was reasonable.'

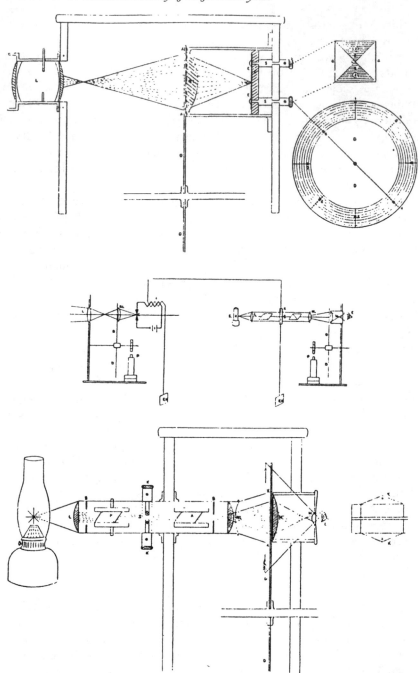

Figure 5.5   *Sutton's 'seeing by electricity' scheme. The drawings illustrate the transmitter and receiver (centre) and the receiver (bottom)*

Source: The Telegraphic Journal and Electrical Review, **27**, 7 November 1890, pp. 549–551

However, during the discussion following Traub's paper, he mentioned:

> 'With regard to Tekade, perhaps I did not point out that this was not a home set. It was really not a big success at all, as the picture was not bright enough. This was unfortunate. The power output on the video set was definitely very high—two 100@W valves in parallel. This must of necessity be so when you are using the Kerr cell for this number of lines.'

Sutton's opinion was thus vindicated by later progress, but notwithstanding this drawback, Sutton thought his proposal offered a fair approximation to the solution of a very difficult problem. Actually, much use was made of the Kerr effect by Karolus in the 1920s; and Baird, too, experimented with the cell.

Another system which was given much attention by later innovators was one advanced by Lazare Weiller[15,16] in 1889. His invention (Figure 5.6), according to his paper, was stimulated by the experiments of Lissajous, who had investigated the vibratory motion of bodies by optical methods. In several of these experiments Lissajous had attached small mirrors to the tines of tuning forks and had observed his now famous figures delineated on an opaque screen by reflecting a beam of light from the two mirrors in cascade.

Weiller generalised Lissajous's arrangement and proposed the use of a drum fitted with a number of tangential mirrors, each successive mirror being orientated through a small angle so that, as the drum rotated, the area of the image was scanned in a series of lines and projected onto a selenium cell. At the receiver an identical mirror drum, synchronised to the transmitter drum, reflected light from a telephone à gaz, to which the signal currents were applied, and created a raster on an opaque screen. The telephone à gaz was an ingenious solution to the problem of establishing a variable light source. It combined the telephone and the capsule of Konig, and was utilised with success by Giltay and Ruhmer in their work on photophones, although it was never incorporated into a practical distant vision receiver.

The mirror drum was to be an essential element in mechanically scanned television systems for many years. It featured in television schemes in the US, in the UK and in Germany. By 1932 television equipment based on the Weiller drum scanner was commercially available for the home constructor, and by 1934 Fernseh AG of Germany was manufacturing mirror drums and mirror screws to a precision of ten seconds of arc in the positioning of the mirrors.

The drum was used by Rosing, Baird, EMI, Karolus and others. It was an alternative to the Nipkow disc for a considerable period even though it was not until c. 1930 that the relative efficiency of the two scanners was theoretically evaluated—the early papers on television are notably lacking in quantitative details and analysis. Nevertheless, it can be shown that when the number of lines per image is greater than 36 the aperture disc of Nipkow is more efficient than Weiller's mirror drum.

For indirect scanning (see Chapter 8) it can be deduced that the aperture disc and the mirror drum have practically the same efficiency when the number of lines is less than 85: for a larger number the mirror drum has an efficiency which decreases rapidly as the number increases.

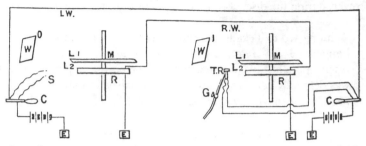

L.W., line wire ; R.W, regulator wire ; L₁, L₂, lenses ; M, mirror apparatus ;
R, regulators ; O, object ; I, image ; S, selenium transmitter ; C, commutators ; G, gas ; T.R, telephone receiver ; E, earth

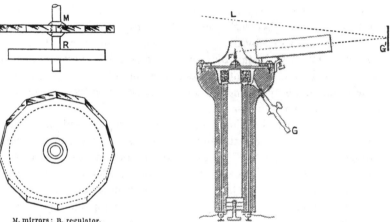

M, mirrors ; R, regulator.
30 revolutions per second ; 300 glasses on the sides instead of 12.

L, lunette ; F, flame : G, gas ; G₁, glass.

Figure 5.6   *Weiller's mirror drum system, 1889. The mirror drum scanner was used by Baird,
Alexanderson, Karolus and others in the 1920s and 1930s. When the number of
lines per picture is less than 36, the mirror drum is more efficient than the Nipkow
disc. The schematic diagram of the system, the mirror scanner, and telephone type
receiver are illustrated*

Source: *Le Genie Civil*, t.xv, 12th October 1889, pp. 570–573

Weiller made it clear in his paper that he had not fabricated his phoroscope, even
through Ll B Atkinson had used mirror drums for scanning in 1882. Atkinson's work
was not published at that time and so Weiller is generally regarded as the inventor
of this type of scanner.

In the year, 1889, of Weiller's invention, Jules Verne wrote an article[17], which
appeared in the *New York Forum*, in which he imagined a scene supposed to happen
in the year 2889 AD. A machine which he called the phonotelephotograph featured
in the imagery.

'. . . Francis Benett awoke that morning in a bad humour. His wife had
been in France for eight days now, and he felt rather lonely. It may seem

incredible, but in all the ten years of their married life, this was the first time that Mrs Edith Benett, professional beauty, had been absent so long from her husband.

'As soon as he was aroused and fully awake, Francis Benett started to operate his phonotelephoto machine, the wires of which ended in his own house in the Champs-Elysees quarter.

'The telephone completed with the telephotographer another conquest of our civilisation! If the transmission of words by means of electric currents is a very old idea, it is only recently that we have been able to transmit images. A precious discovery for which Francis Benett is not the only one to bless the inventor when he sees a picture of Mrs Benett reproduced in the telephotographic mirror, despite the enormous distance between them.

'A delightful sight! A little tired after the dance and the theatre, Mrs Benett is still in bed. Although it is nearly mid-day there in France, she is asleep with her pretty head sunk in the lace of the pillow'.

Jules Verne's pessimism ('it is only recently . . . ', c. 2889 AD) was obviously extreme although the prospect of seeing by electricity in 1889 was not bright.

A pragmatic view of the need for visual images was given in the same year by E H Hall, Jr, the innovative vice president of American Bell's long distance subsidiary, AT&T. He opined[18] that both sounds and sights were important in dialogues:

'We all appreciate the advantage of speaking face to face above all other methods of communicating thought because ideas are then conveyed in three ways—by the words, by the tone and by the expression and the manner of the speaker. Until Bell invented the telephone only the first method was available—first by letter and later by the rapid letter—the telegram. It is perhaps fair to say that we obtain our impressions from all three of those things in equal proportions. I would not set bounds to the possibilities of inventive genius. Some day we may see as well as hear our distant friends when we communicate with them by the telephone.'

Hall's prospect of the future was implemented by his company in 1930.

The last decade of the nineteenth century produced only a few proposals for a distant vision system. Inventors were possibly disenchanted with the subject following the complete lack of progress since the discovery of the photoconductive effect in 1873. Many ideas had been put forward but although the competence of some of the originators was considerable the realisation of a practical scheme seemed remote.

For a M Brillouin the possibility of obtaining a successful solution seemed hopeless. Brillouin's paper[19] in the *Revue General des Sciences* (1891) was one of the first to contain an elementary analysis of the relation between picture definition and corresponding speed of signalling. A similar simple calculation was, surprisingly, to give rise to a considerable amount of discussion in the semitechnical press in the late 1920s.

In carrying out his analysis Brillouin assumed that the transmitted image should be viewed at a distance of 30 to 40 cm and that the width of each scanning line should be not more than one twentieth of a millimetre. He then correctly deduced that an image 4 × 4 cm needed 640 000 dots to delineate it; all of which had to be illuminated in one tenth of a second. Consequently, the transmission apparatus had to respond to changes in less than 1/5 000 000 of a second, in round numbers, Brillouin noted. This was much too small for Brillouin, who was the Maître de Conferences de Physique à l'École Normale Superieure, and so he modified his objective. He subsequently dealt with the problem of sending photographic images to a distant place by means of electric currents. This of course, was the domain of picture telegraphy rather than television, but as his apparatus and ideas formed the basis of some later developments using orthogonal discs, lensed discs and galvanometer modulators, it is important to discuss his proposals, Moreover, his paper is valuable because it gives a contemporary view of the limitations of several previous notions.

Thus on Weiller's, Nipkow's and Sutton's schemes he made the following points:

1   The capsule and flame telephone (Weiller) . . . cannot light up brightly enough, moreover it cannot guarantee that the illumination of the object and the flame are in proportion; it is insensitive and is difficult to regulate as a result of the small amplitude of the vibrating membrane.

2   The polished membrane telephone (Nipkow) . . . but it seems highly questionable that the membrane remains sufficiently plane in equilibrium not to be constantly misaligned; moreover, the changes in the curvature produced by the passage of current only result in variations of illumination between a maximum value and a minimum of more than zero; the image will always be saturated in light.

3   The rotation of the plane of polarisation of light (Sutton) . . . has in its favour an extremely fast reaction time; but it needs quite a strong electric current for any noticable rotation, and, as with the cross Nichols, the brightness is a minimum or zero, a rotation of a given angle proportional to the intensity of the current results only in an increase in brightness proportional to the square of the intensity of the current, and as a result also proportional to the square of the illumination of the object being reproduced, which changes its character entirely.

In addition to these limitations, Brillouin felt that Nipkow's disc did not permit fineness or brilliance of reproduction and Weiller's 360 mirror cylinder was impractical to construct with precision. He did not state whether he had arrived at his conclusions by practical tests but, as he was a scientist of some standing, it is likely that his comments were soundly based[20]. Indeed, they confirmed the negative result which had been obtained a few years earlier by Atkinson.

In the face of these difficulties Brillouin advanced an entirely new method of scanning and an apparatus for converting the received currents so as to give a luminous spot a brightness which was proportional to the electric current, His scanner consisted of two circular lensed discs arranged one behind the other so that the two circumferences of the centres of the lenses of each disc were orthogonal. Each disc held ten lenses and their rotational speeds were in the ratio of one thousand to one. The result was that the image of the object was scanned in a series of very close, parallel arcs.

Brillouin gave no indication as to how he arrived at his solution to the scanner problem, but he may have been led to his proposal by an analogy. In the last decade of the nineteenth century photography was well established and cameras readily available. Camera obscuras had been known for many centuries and, as mentioned previously, had utilised initially a small hole in the side of a box or closed room to give the desired effect. Later, G Cardano in 1550 had introduced a lens in place of the hole to increase the brilliance of the projected image. Bearing in mind that Nipkow's scanner 'ne permet ni finesse ni éclat', according to Brillouin, it may be that he introduced lenses in place of Nipkow's apertures after the practice of Cardano. But, because of the different sizes of the holes and lenses, a simple substitution would have entailed a reduction in the number of lenses which could be employed, and this would have led to a lack of finesse in the scanning process. Brillouin presumably adopted the two disc arrangement as a consequence. Both Jenkins and Baird experimented with lensed discs in the 1920s but did not achieve much success with them.

By 1892 the basis for many of the mechanical scanning methods used in the twentieth century and particularly the 1920s had been laid. Surprisingly, despite the brilliant theoretical researchers of some of the nineteenth century scientists, and despite the continental education system which stressed the importance of theoretical and mathematical studies, no detailed theoretical study of any aspect of distant vision or picture transmission was made in that century. And yet the physical basis for an investigation of, for example, the scanning problem was certainly known: this simply depended on certain laws and theorems in the fields of geometrical optics and photometry (see Appendix 3).

Probably the most important aspect of Brilloiun's conception concerned the means for producing a luminous spot, the brightness of which was proportional to the received picture current.

Previous suggestions made by Nipkow, Weiller, Sutton and others (but with the exception of that put forward, and demonstrated, by Aryton and Perry) were not sufficiently sensitive to give the anticipated effect. This was due to the high resistance of the selenium cell circuit limiting the received current. A similar problem had been faced by the advocates of the transatlantic cable scheme, in the 1860s. Then, professor W Thomson devised his mirror galvanometer to detect the extremely weak currents and for a time this instrument remained the only one capable of allowing a message to be received. Brillouin, as a physicist (as were Aryton and Perry), possibly appreciated from his training and scientific knowledge that only a mirror galvanometer would be suitable for indicating the minute currents which he hoped could be used to synthesise a picture.

Thomson developed his form of mirror galvanometer in 1858[21]. It consisted of a tiny circular glass mirror, to the back of which were cemented two or three pieces of steel watch spring, flattened and hardened, and permanently magnetised. The mirror was hung by a silk fibre in a horizontal tube, a little larger than the mirror itself, and the fibre, which was quite short, passed through a hole in the upper part of the tube. It was secured by a drop of wax or cement. This assembly fitted into the centre of a coil through which the current to be measured passed. Outside the coil an adjustable magnet or pair of magnets gave a suitable controlling field: in addition a beam of light from a lamp was reflected from the mirror onto a scale. Thomson's objective in designing the instrument was to achieve as sensitive a galvanometer as possible, consistent with a very small inertia of the suspended parts, and an absence of friction. His instrument was used for several years. Later, wishing to record the received electrical impulses instead of merely showing their effect to the watchful eye of a skilled observer, Thompson devised a substitute in which he inverted the function of the magnet and the coil—the coil being the moveable part and the magnet the fixed piece. The coil, therefore, was made very light, and the magnet, which was now stationary, could be extremely powerful. This design formed the basis of Thomson's siphon recorder of 1867 and was the earliest example of the moving coil type of galvanometer, sometimes called the d'Arsonval type by those unfamiliar with the history of the instrument.

The first form of the mirror galvanometer clearly inspired Ayrton and Perry. A comparison of the descriptions of the two instruments shows that Aryton and Perry replaced the mirror with an aluminium disc, the orientation of which in a cylindrical tube controlled the luminous flux from a lamp.

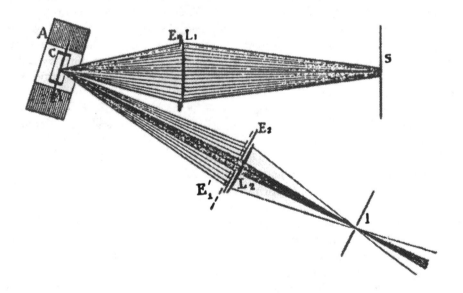

Figure 5.7   Brillouin's system of 1891. His suggestion for a receiver included a novel application of the galvanometer to control the brightness of the scanning light spot

Source: Revue Générale des Sciences, (2), 30th January 1891, pp. 33–38

Brillouin's suggestion for a receiver included a novel application of the galvanometer to control the brightness of the scanning spot so that a linear relationship existed between the received current and image brightness.

His galvanometer comprised a permanent or electromagnet and a coil assembly and was employed as shown in Figure 5.7. Light from a hole, S, of one or two millimetres in diameter was focused to form a real image on the suspension of the galvanometer by lens L_1. The rays were then reflected from the concave mirror, M, to form a smaller fixed image, I, of the hole, S, using the lens L_2: hence, by means of the fixed masks E_1 and E_2 placed in the light path, Brilloiun was able, in principle, to control the intensity of illumination of the image, I. An image of the mask E_1 was formed, by the concave mirror, M, slightly in front of the mask E_2 so that when the galvanometer indication was zero the image of the apex of the triangular opening in E_1 coincided with an edge of the opening in E_2, as illustrated in Figure 5.8. Then, as the received current increased the image of E_1 moved across E_2 and allowed an increased luminous flux to fall on the image spot, I. Using this 'galvanométrique Deprez-d'Arsonval, ou plutôt dans la partie galvanométrique d' un siphon recorder de Sir W Thomson' (Brillouin), a linear relationship between angle of deflection and current could be obtained, and so with his shaped masks 'l'éclat de l' image I était donc proportionel a la déviation et par suite a l'intensité du courant'.

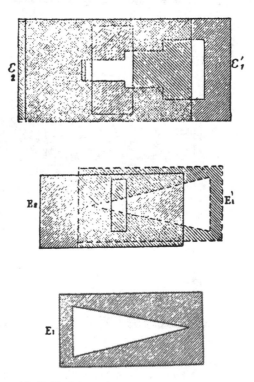

Figure 5.8   *The masks used in Brillouin's receiver*

Source: Revue Generale des Sciences, (2), 30th January 1891, pp. 33–38

Brillouin did not construct any of this apparatus, although he thought it was capable of realisation by an experienced instrument maker.

One of the difficulties which faced the early inventors of television was the slow response of selenium and hence its insensitivity to rapid changes of light intensity. It

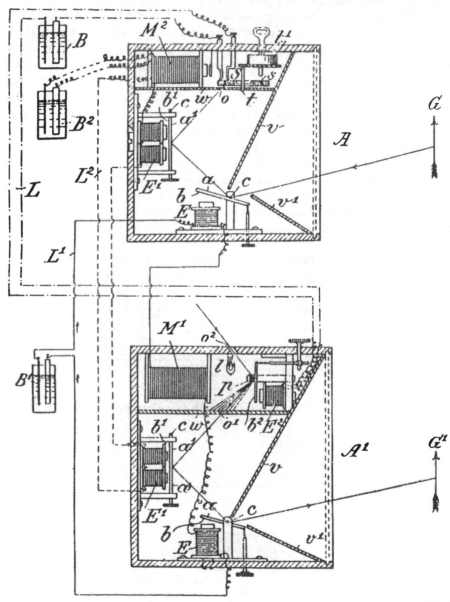

Figure 5.9   Schematic diagram of Szczepanik's apparatus of 1897. It included oscillating mirror scanners

Source: British patent no. 5031, 24 February 1897

was perhaps this characteristic of the photoconductive cell which tended to discourage interest in schemes for distant electric vision during the last ten years of the nineteenth century.

An attempt to overcome the sluggishness of selenium was advocated in 1897 by Szczepanik[22]. He recommended the construction of a disc type selenium cell in which a thin annular layer of selenium was sandwiched between two concentric brass rings. This was to be slowly rotated about its axis by a clockwork motor so that it continually exposed a fresh surface to the small aperture of the cell. In this way it was hoped that the cell would be capable of responding to higher frequencies than had previously been possible.

Szczepanik suggested scanning the image of the object by means of a pair of oscillating mirrors as shown in Figure 5.9. His notion seems to have been based on the scanning ideas of Le Blanc, but whereas Le Blanc used one mirror, mounted on a structure consisting of two arms, which could vibrate in a plane parallel to the plane of the mirror, Szczepanik proposed to separate the functions of line scanning and frame scanning. Possibly he felt that the two motions could be more easily controlled than with Le Blanc's design. The transmitting and receiving oscillating mirrors were to be synchronised by connecting the controlling electromagnets at the transmitter directly to the corresponding electromagnets at the receiver.

In writing a history of the development of distant electric vision in the nineteenth century the historian is hindered by the paucity of available manuscript material. Many of the seeing by electricity schemes were only sketchily reported. Moreover, most of these gave rise to only one paper or article rather than to a series of papers. As a consequence a writer on the subject is limited, in the main, to published, uncritical, papers and patents and must endeavour to interpret inventors' claims in the light of subsequent events. A particularly interesting example of extravagant and ill-founded claims for a distant vision apparatus concerns Szczepanik's equipment.

In 1899 a Mr Cleveland Moffett stayed in Vienna for a fortnight and during this period 'saw much of Szczepanik and learnt things about him and about his inventions', the distance-seer being one of them, which he later related in an article[23] published in 1899 in *Pearsons Magazine*.

According to Moffett's account, Szczepanik's initial experiments were carried out using Nipkow's discs, 'perforated with a great number of small holes (not less than forty thousand)', a disc of selenium as the transmitting element and a receiving light beam modulator which consisted of 'two steel plates brought very close together, so that only a thin band of light could pass between them. The series of electric impulses acting upon the two steel plates caused them to open and close so that the band of light between them was constantly changing in intensity'.

Szczepanik's apparatus was put to the test in the early autumn of 1896, according to Moffett. Accompanied by a Herr Schmidt, an electrical engineer, and a Herr Kleinberg (whose name is included with Szczepanik's in a 1897 patent), the inventor set up 'one of the dark boxes' in the Karlplatz before the Karl's church, while the other 'was placed in Herr Kleinberg's house in Bartenstein Street, in a room where there was a telephone. The connection between the two boxes was made by the city telephone wires . . . The distance by wire between the boxes was about two miles'.

In his demonstration Szczepanik employed a photographic plate to record the received image. As described by Moffett: 'It showed a photograph of the Karl's church, not a very good one it is true, but unmistakable for anyone who knew the building and absolutely convincing as a demonstration that a method of transmitting vision had at last been discovered or at least the foundation of such a method.'

If Moffett had terminated his report with this account of the test there might have been some future debate as to whether or not Szczepanik achieved his objective, but the author's remaining paragraphs leave no room for uncertainty in the audacity of Szczepanik's claims. These concerned his vibrating mirror scheme. With this apparatus the inventor stated he had transmitted vision signals over a distance of sixty kilometres. He told Moffett: 'Whatever comes over the wires will be projected plainly for everyone to look at—colours, movements and all, just as in life. We can show a charge of cavalry, the finish of a race, the launch of a battleship, the movements of a street and similar scenes, without end, and each as it is actually transpiring at some distant point.'

In his 1897 patent the Polish inventor stated: 'This invention relates to a method and an apparatus for transmitting a picture or representation of an object . . . and rendering it visible in its natural colours at a distant place by electrical means.' However, Szczepanik's system could not have worked as he described it in the patent and so his statement to Moffett that he had transmitted vision over a length of sixty kilometres cannot be true. His selenium disc could have responded to variations in the light flux incident onto the cell, but Szczepanik wrote: 'The rays are converted into electrical currents of various strengths according to the colour of the light emitted from the several picture points.' His apparatus, Figure 5.9 included no means, at the transmitter, for analysing the incident light into its spectral components, but rather oddly the receiver used a moving prism for this purpose. The transmitted currents from the selenium cell energised an electromagnet at the receiver which produced 'a movement of the prism in such a manner that only a ray of similar kind of the light falling from a source of light onto the prism and becoming decomposed thereby can reach the synchronously oscillating mirrors at the receiving station. The reflecting action of these mirrors is such as to cause the picture point transmitted to the eye of the observer to be visible to him at the place and in the colour corresponding to the point at the transmitting station'. Szczepanik evidently did not appreciate the need to send signals corresponding to both the intensity and the colour of the light reflected and scattered from an object. His receiver was founded on incorrect principles and could not have functioned as he described.

Moffett stayed in Vienna for two weeks and saw much of Szczepanik but nothing, it seems, of the distant-seer. The inventor told him it could not be shown to the public until the Paris Exhibition in 1900 as Kleinberg and he would forfeit 1 000 000 francs by the terms of a bond to a French syndicate who had contracted for the exhibition rights. 'They had bound themselves under a 2 000 000 franc forefeit, which sum they had deposited in the hands of a third party, to make all necessary arrangements for exhibiting the telectroscope at the exhibition including the construction of a building capable of seating from eight to ten thousand people.'

Szscepanik saw no difficulty in projecting the telectroscope pictures onto a screen—'it will be as simple as working an ordinary lantern'.

Moreover, the inventor saw no hindrance in making vast profits from his enterprise with Kleinberg. The French syndicate was to have 40 % of the profits and Szczepanik and his financier friend 60 %. 'You can estimate what the profits are likely to be in six months with several representations a day at three francs a head,' he told Moffett. 'Six million francs is putting it low.'

The journal *Electrical Engineer*[24] in July 1898 briefly mentioned the financial aspects of Szczepanik's invention:

> 'According to the testimony of eye witnesses the transmission of pictures is perfect, and the central committee of the Paris Exposition of 1900, where the invention will be shown for the first time to the public, have guaranteed the young inventor 3 000 000 francs. At least, "that's what they say". We are not aware of the existence of a central committee of the Paris Exposition.'

The journal had previously published several reports[25] on the invention of 'the gentleman with the jaw-breaking name' and had taken to task the writer of an article in the *Daily Telegraph* who had mentioned that the Polish schoolmaster had 'perfected [the selenium cell's] use in an apparatus by which we are to see by electricity'.[26]

> 'We like dogmatism—we are dogmatic ourselves at intervals, but often that is when discretion is absent, and only valour is present. The *Telegraph's* correspondent has a little discretion, though his or her valour is abundant. There may be a schoolboy—we can hardly realise that there exists a student, much more a schoolmaster or a professor, who could dogmatise as follows: "There is no real difficulty about converting rays of light into electrical current impulses. That can be done in a variety of ways, and one has to choose". We are at once reminded of a somewhat cynical poet named Pope, who suggested that "Where ignorance is bliss, 'tis folly to be wise", as having provided a sentiment applicable to these utterances of the *Telegraph*.'

References

1 SCOTT, W.: 'My Aunt Margaret's Mirror' (Black, Edinburgh, 1871) pp. 378–379
2 SHAW, G.B.: 'Back to Methuselah. A metabiological pentateuch' (Constable and Company, London, 1924)
3 Editorial: 'The electrical exhibition at Paris', *Electrician*, 3 December 1881, pp. 40–41
4 ROBIDA, A.: 'The XXth century, the conquest of the regions of the air' (Paris, 1884)
5 Le BLANC, M.: 'Etude sur la transmission électrique des impressions lumineuses', *La Lumiere Electrique*, 1 December 1880, **11**, pp. 477–481
6 RTCHEOULOFF, B.: 'Improvements in and relating to television and telephotography'. British patent 287 643, 24 December 1926
7 W.L.: 'The telectroscope, or seeing by electricity', a letter (no. 19941), *English Mechanic and World of Science*, 21 April 1882, (891)

8 NIPKOW, P.: 'Elektrische teleskop'. German patent 30 105, 6 January 1884
9 FARADAY, M.: 'On the magnetisation of light', *Phil. Trans.*, 1846, p. 1
10 NIPKOW, P.: 'Der telephotograph und das elektrische teleskop', *Elektrotechnische Zeitschrift*, October 1885, pp. 419–425
11 KERR, J.: 'A new relation between electricity and light: dielectrified media birefringent', *Phil. Mag.*, November 1875, S4, **50**, (332), pp. 337–348 and 446–458
12 SUTTON, H.: 'Telephotography', *The Telegraphic Journal and Electrical Review*, 7 November 1890, **27**, pp. 549–551. Also reproduced in *Scientific American Supplement*, 24 January 1891, (786), p. 12645
13 E.R.: 'Le problème de la téléphanie, d'après M. Henri Sutton', *La Lumière Electrique*, 1890, **38**, pp. 538–541
14 TRAUB, E.H.: 'Television at the Berlin Radio Exhibition, 1936', *J. Television Society*, 1936, p. 186
15 WEILLER, L.: 'Sur la vision à distance par l'électricité', *Le Genie Civil*, 12 October 1889, **XV**, pp. 570–573
16 H.W.: 'Sur la vision à distance par l'électricité par L. Weiller', *La Lumière Electrique*, 16 November 1889, **34**, pp. 334–336
17 See also: VERNE, J.: 'Le château des Carpathes' (Hamilton, London, 1963, Evans, I.O. (Ed.))
18 HALL, E.H.: Letter, *Electrical Review*, 21 September 1889, p.9
19 BRILLOUIN, M.: 'La photographie de objets a très grande distance par l'intermediaire du courant électrique', *Revue Generale des Sciences*, 30 January 1891, (2), pp. 33–38
20 BLONDIN, J.: 'Le téléphote', *La Lumière Electrique*, 1893, **43**, pp. 259–266
21 EWING, J.A.: 'The work of Lord Kelvin in telegraphy and navigation', *J. IEE Eng.* 1910, **44**, pp. 538–571
22 Szczepanik, J., and KLEINBERG, L.: 'Method and apparatus for reproducing pictures and the like at a distance by means of electricity'. British patent 5031, 24 February 1897
23 MOFFETT, C.: 'Seeing by electricty', *Pearson's Magazine*, 1899, pp. 490–496
24 Report on 'The Szczepanik telectroscope', *The Electrical Engineer*, 7 July 1898, **26**, (531), pp. 14–15
25 Reports in The Electrical Engineer: 4 March 1898, p. 257; 11 March 1898, pp. 304–305; 5 April 1898, p. 449; 22 April 1898, p. 483; 3 June 1898, p. 675; 1 July 1898, p. 3; 29 July 1898, p. 129
26 Editorial: 'Next please', *The Electrical Engineer*, 11 March 1898, pp. 304–305

Chapter 6

A possible way forward
(1900–1920)

At the turn of the century the principles which delineated the capture and reproduction of both static and moving images were well known. The science and the
practice of photography were soundly based and photographs were a feature of
everyday life. Cinematography had been established in the 1890s and was progressing rapidly. The Lumiere Brothers had become the first persons to give a public
exhibition of moving pictures, at the Grand Cafe in Paris on 28 December 1895[1], at
which an admission fee had been charged, and soon afterwards, on 20 February
1896, their films had been shown at the Royal Polytechnic Institution in London.
Also in 1896, the first film production unit in the world had been founded by George
Melies[2]. His films—he produced about 4000—were to be considered classics of
cinematography. In one of these, the famous 'Journey to the moon', trick photography was used for the first time. By the end of the century many different cameras
and projectors, using a variety of filmgauges, were on the market and cinemas had
opened in numerous large towns in several countries.

On the electrical transmission of static pictures and photographs much had been
undertaken following the original ideas of Bain and Bakewell and the early efforts at
commercialisation by Caselli, Meyer and d'Arlincourt. Willoughby Smith's 1873
disclosure on the photoconductive property of selenium had given further impetus
to the advancement of facsimile reproduction and among the post 1873 developers
mention must be made of Cowper (1879), Senlecq (1879), Edison (1881), Gray
(1893) and Amstutz (1893)[3].

After 1900, Ritchie (1901), Korn (1902), Carbonelle (1907), Berjonneau (1907),
Belin (1907), Thorn-Baker (1907) and Semat (1909) were among those who endeavoured to further picture telegraphy[4]. Their ideas and those put forward in the last
quarter of the 19th century led to the electrical transmission and reproduction of
half-tone photographs, as a newspaper service, in the first decade of the 20th
century.

Much work was carried out by Professor A Korn[5] of Munich on phototelegraphy
during this period and this led to the successful reception in October 1907 of a

Table 6.1   *Summary of the principal proposals made during the first decade of the 20th century for 'seeing by electricity'*

Date	Name	Transmitter scanner	Receiver scanner	Transmission link
1902	O von Bronk[8] Germany	mirror drum	commutator	signal wires
1902	M J H Coblyn[9] France	rotating cylinder having helicoidal slots	same as transmitter scanner	signal wires
1903	A Nisco[10] Belgium	a form of commutator— a steel blade passes over a static ebonite cylinder through the surface of which copper pins project slightly	slotted cylinder	signal wires
1904	W von Jaworski and A Frankenstein[11] Germany	mirror wheel with rotating sequential colour disc	same as transmitter scanner	signal wires
1906	G P E Rignoux[12] France	two vibrating mirrors, scanning orthogonally, mounted on two tuning forks (500 Hz and 10 Hz)	same as transmitter scanner	two wires for the signal and four wires for synchronisation
1906	F Lux[13] Germany	nonsimultaneous system	nonsimultaneous system	multiwire
1907	N Rosing[14] Russia	two mirror drums scanning in orthogonal directions	cold cathode-ray tube, electron beam magnetically deflected	two signal wires and synchronising wires
1908	J Adamian[15] Germany	apertured disc	apertured disc	single wire
1908	S Bidwell[16] United Kingdom	nonsimultaneous system	nonsimultaneous system	multiconductor, 90 000 wires— diameter of cable 8" to 10"

Synchronisation means	Opticoelectrical transducer	Electrooptical transducer	Remarks
not stated	three selenium cells and associated red, green and blue filters	an array of Geissler tubes and associated red, green and blue filters	an early proposal for colour television
few details given	selenium cell	lamp and light valve of variable shutter type based on Blondel oscillograph	not patented
not stated	selenium mosaic, the wires from which connect with the transscanner	spark gap	an impractical scheme
not stated	selenium cell	spark light source; spark gap modulated by vibrating wire connected to telephone diaphragm	use of rotating discs, having sequential red, blue and green filters, at transmitter and receiver for sequential colour television
receiver scanners connected directly to transmitter scanner	selenium cell	modulated arc	
not necessary— simultaneous transmission	array of selenium cells each controlling the amplitude of an oscillator	array of vibrating resonant reed light shutters and light source	the n transmitter signals at f_1, f_2, ... are received and applied to the array of light shutters each tuned to one of the frequencies f_1, f_2, ...
deflection coils of receiver cathode-ray tube fed by currents generated at the transmitter	selenium cell, or photoelectric cell, or photovoltaic cell	electron beam, of c.r.t., modulated by video signals determines spot brightness	
not stated	not stated	two Geissler tubes adapted to emit differently coloured lights	objective: to produce a picture having, to a certain extent, some natural colouring
not necessary— simultaneous transmission	array of 90 000 selenium cells, c. 8' square	array of 90 000 light valves	estimated cost: £1 250 000 (in 1908); size of receiver c. 4000 ft³

Table 6.1 continued

Date	Name	Transmitter scanner	Receiver scanner	Transmission link
1908	A C and L S Andersen[17] Denmark	apertured band or disc	apertured band or disc	single wire
1906	M Dieckmann and G Glage[18] Germany	a rotating disc having 20 brushes arranged in a single spiral	electron beam in a cathode-ray tube	signal wires and synchronising wires
1909	E Ruhmer[19] Germany	none—simultaneous system	none—simultaneous system	multiconductor
1909	G E P Rignoux and A Fournier[20]	none—simultaneous system	none—simultaneous system	multiconductor
1910	G E P Rignoux and A Fournier[21]	(i) array of n relays & commutator (ii) mirror drum	mirror drum mirror drum	two wires two wires
1910	A Ekstrom[22] Sweden	spiral scanning mirror	spiral scanning mirror	signal wires
1910	M Schmierer[23] Germany	two comutators acting on row and column elements of array so that only one element at a given instant is connectec in circuit	same as transmitter scanner	signal wires
1910	G H Hoglund[24] US	a pair of apertured discs: the discs rotate in opposite directions	same as transmitter scanner	signal wires
1910	A Sinding–Larsen[25] Norway	two vibrating mirrors mounted on two tuning forks (10 kHz and 100 Hz)	same as transmitter scanner	narrow tube with strongly reflective inner surface

Synchronisation means	Opticoelectrical transducer	Electrooptical transducer	Remarks
electrically actuated clockwork	selenium cell with prism and rotating colour filters	lamp with moveable graded density filter and rotating colour filters	objective: use of rotating sequential filters at transmitter and receiver to enable colours of object to be reproduced
varying voltages from linear potentiometers	none; the 20 wire brushes of the transmitter contacted the object	fluorescent screen of receiver c.r.t	not a television system—essentially a picture telegraphy system
not necessary— simultaneous transmission	5 × 5 array of selenium cells each connected to a sensitive relay; 25 signals transmitted; no half tones	5 × 5 array of light modulators (sensitive) galvanometers) plus light source	system demonstrated on 26 June 1909; Ruhmer planned to construct a 10 000 element array (at a cost of Fr 6M) for the Brussels International Exposition
not necessary— simultaneous transmission	array of selenium cells (n)	n small galvanometers each with miniscule shutter	system demonstrated in 1911; no half tones reproduced
synchronous motors at transmitter and receiver	array of selenium cells (n)	lamp & light modulator based on Faraday effect	simple black and white figures reproduced; no half tones
not stated	single selenium cell	light source and light valve of Brillouin type	use of indirect scanning
	array of light sensitive elements	array of light producing elements	an n-line television system would require n^2 elements for a 1:1 aspect ratio
synchronous motors at transmitter and receiver	selenium cell	'speaking' arc	one disc has arcuate slots; the other has a 'plurality of slots extending in a stepped line outwardly from the centre of the disc'
transmitter and receiver driving elements coupled in series	not necessary since the light is transmitted directly from the transmitter	not necessary—the light is received directly from the transmitter	Use of optic tube for tube for transmission. (cf. modern optic fibre.)

photograph in the French office of the weekly paper L'Illustration which had been transmitted from Berlin. Then on 7 November 1907 a Paris-London service was inaugurated when a picture of King Edward VII was sent electrically to the London office of the *Daily Mirror*. This service was later extended to Manchester but was not an unqualified success for the two systems of telegraphy then in use, Morse and Baudot, gave marked induction effects on the lines and at times it was possible for experts actually to read messages on some of the photographs which were transmitted[6].

The manager of the *Daily Mirror* phototelegraphic department during the formative period of the above venture was Thorn-Baker. He had studied with Korn in Paris in 1907 and on his return had developed an apparatus which he called a telectrograph. This was first employed by the *Daily Mirror* in July 1909. Subsequently, many hundreds of photographs were sent by the instrument on the Paris-London and London-Manchester routes but, like its predecessors, its period of utilisation was short; the very high costs of hiring the necessary telephone lines led to a termination of the service[7].

With the growth of photography, cinematography and facsimile transmission it was to be expected that seeing by electricity would continue to attract interest. The subject had a wide appeal and during the first decade of the new century inventors in Germany, France, Belgium, Russia, the United Kingdom, Denmark, Sweden, the US and Norway advanced notions for its practical realisation, see Table 6.1[8-25]. Not all the proponents of the divers schemes tried to reduce their ideas to practice and of the persons listed in Table 6.1 only Lux[13], Rosing[14], Dieckmann and Glage[18], Rignoux and Fournier[20,21] and Rhumer[19] engaged in laboratory work.

The word 'television' was coined at the beginning of the century when Constantin Perskyi read a paper[26] titled 'Television' at the International Electricity Congress, on Friday 25 August 1900. Eight years later the British Patent Office introduced a new subject title—television—in its system of patent classification. The first patent taken out, on 24 December 1908, under the new group name was one due to A C Anderson and L S Anderson[17].

Despite the lack of really effective progress and achievement during the thirty-five year period following Willoughby Smith's announcement there were some people who thought television was almost at hand.

According to a telegram from the Paris correspondent of *The Times*, dated 28 April 1908, the problem of distant electric vision was engaging the attention of M Armengaud, who 'firmly believed that within a year as a consequence of the advance already made by his apparatus, we shall be watching one another across hundreds of miles apart'[27].

This was too much for Shelford Bidwell[16] who as noted previously had carried out some elementary experiments in 1881 on seeing by electricity, and in a letter which he sent to the editor of *Nature* he opined: 'It may be doubted whether those who are bold enough to attempt any such feat adequately realise the difficulties which confront them.' In particular, Shelford Bidwell emphasised the impracticability of such proposals because of the formidable nature of synchronising, by mechanical means, the receiver to the transmitter of a high definition system.

He calculated that a received image 50 mm by 50 mm would need 150 000 elements if its definition were to be as satisfactory as that presented to the eye by a good photograph, and that necessarily the number of synchronised operations to be undertaken by the transmitting and receiving equipments would be 1 500 000 per second. For a quality comparable with that of a coarse half-tone newspaper picture, Figure 6.1, the number of elements would be about 16 000 and the video frequency would be 160 000 Hz. Even this would be 'widely impracticable, apart from other hardly less serious obstacles which would be encountered', said Bidwell. He mentioned that the number of operations might be greatly diminished by employing an oscillating or rotating arm carrying a row of sensitive cells, but commented: 'For a coarse grained picture 50 mm square 120 of these would require 120 line wires, and [this] would also introduce a new series of troubles.'

Although the problem was superficially incapable of solution on the lines indicated, Shelford Bidwell felt that there was no reason beyond that of expense why vision should not be electrically extended over long distances. His plan was

Figure 6.1   Coarse screen half-tone photograph

essentially one put forward by Carey in 1880, which was based on the structure of the eye. 'The essential condition is that every unit area of the transmitter screen should be in permanent and independent connection with the corresponding unit of the receiving screen.'

In this way, thought Bidwell, the difficulties due to synchronisation could be resolved without any grave complexity apart from that arising from the multiplication of components. He estimated the cost of such a scheme, assuming the transmitting and receiving stations to be 160 km apart, the received picture to be 50 mm square and the length of a picture element or pixel to be 1/6 mm. 'Of each of the elementary working parts—selenium cells, luminosity controlling devices, projection lenses for the receiver and conducting wires—there would be 90 000. The selenium cells would be fixed on a surface about 2.6 m square, upon which the picture would be projected by an achromatic lens (not necessarily of high quality) of 1.0 m aperture. The receiving apparatus would occupy a space of about 130 m³, and the cable connecting the stations would have a diameter of 20 cm to 25 cm.' Bidwell's estimate for such a design was £1 250 000. By an application of the three colour principle he thought it would be possible to present the picture in natural colours. The cost would then be £3 750 000.

Bidwell was not alone, in the 1900–1910 decade, in considering a nonscanning television scheme. Table 6.1 shows that Lux, Rhumer and Rignoux and Fournier also gave some thought to such a *modus operandi*. Interestingly, they all angaged in experimental work on television during this decade.

Lux's transmitter[13] comprised a 10 × 10 array of selenium cells the resistances of which separately controlled the amplitudes of 100 oscillators operating at frequencies $f_1, f_2 \ldots f_{100}$. At the receiver a similar array of 100 vibrating reed shutters, each tuned to one of the frequencies $f_1, f_2, f_3 \ldots f_{100}$, varied the light flux from a suitable source. It seems that Lux's intention was to multiplex the 100 generated signals so that only a pair of conductors would be required in the transmission path.

Ruhmer's apparatus[19] was demonstrated on 26 June 1909. He used a 5 × 5 array of selenium cells each of which was connected in series with a sensitive relay. The relays functioned when the associated cells were exposed to light and enabled currents to be sent to the 5 × 5 array of light modulators (sensitive galvanometer-type variable shutters) and light sources of the receiver. With this system the transmitted signals were telegraphic in character, that is, ON or OFF, and so half tones could not be reproduced. Ruhmer's work hardly advanced the technology of television, even though simple geometric figures were reproduced, and was little more than an extension of the rudimentary experiments of Bidwell and of Ayrton and Perry which were performed in February and March 1881. A contemporary report mentioned that Ruhmer planned to construct a system with 10 000 cells, at a cost of $1 250 000, for the Brussels International Exposition.

On Rignoux and Fournier's early work on nonscanning television, only one brief account (published in 1910) seems to be extant. Few details were described but it appears that some results were obtained[28]:

The inventors worked with various very simple objects; the best results were obtained with the letters of the alphabet cut out of cardboard or painted on sheets of white iron; the first were glued straight onto the selenium screen; the second projected onto the screen with a powerful lens.

A paper[21] presented in 1914 gave an account of Rignoux's later telephote apparatus, see Figure 6.2. A mosaic of selenium cells was still used but whereas the earlier system utilised a wire for each cell the 1914 scheme employed a collector which reduced the number of transmission wires to two for the vision signal and one for the synchronising signal.

In operation an image of an object was focused onto the mosaic of 64 cells each of which was connected in series with an electromagnetic relay across a common battery supply. The illuminated cells, only, enabled the relays to close and thereby allowed a current to be sent to the receiver via the rotating arm of the collector. At the receiver the sequential currents were applied to a Faraday effect modulator. A Weiller mirror wheel scanner and a light source and modulator reconstituted the image. In this way Rignoux succeeded in reproducing faint images of the letters of the alphabet: H, T, L and U. No half tones could be synthesised because of the on-off characteristic of the relays[29].

Nothing further was heard of Lux's, Ruhmer's and Rignoux and Fournier's efforts in the field of nonscanning television systems. Clearly, television was not going to be developed along the lines put forward by Bidwell and others. On the other hand, it seemed inconceivable that mechanical methods could be found which would allow 160 000 synchronised operations per second to be performed. The solution to this dilemma was first given by A A Campbell Swinton in a letter[30] to *Nature* which was a comment on Bidwell's letter of 4 June 1908.

Campbell Swinton is an important figure in the early history of television and so a few biographical details[31,32] about him are appropriate. Born in 1863, he was the third son of Archibald Campbell Swinton (who had been professor of civil law in the University of Edinburgh) and his wife Georgiana. The lineage of the Swintons can be traced from as far back as one Edulf, first Lord of Bamburgh, who accepted the overlordship of Alfred the Great in 886 AD, and includes many men of distinction.

Alan Campbell Swinton was not an academic and in his 'Autobiographical and other writings' he expressed a disdain for formal education: 'I never passed or tried to pass a single examination of any description . . . examinations are a woeful waste of time. I am myself a great believer in self education.' He attributed the successes and recognition which he achieved to the 'cultivation of scientific friends and to attending lectures at institutions such as the Royal Institution'.

At school Swinton showed a strong interest in subjects—such as photography, electricity, telegraphy and the like—which were amenable to practical experimentation, and much preferred engaging in activities of such a nature rather than games which he loathed. In 1878 he was enrolled at Fettes College, Edinburgh but found the strictly academic curriculum uncongenial. On this period he recollected in later life: 'I still look upon the greater part of the three years I was at Fetters with horror.'

He left school at 17 and spent the next nine months in France, first in Le Havre and then in Clermont Ferrand. A visit to the 1881 Electrical Exhibition in Paris was a memorable occasion. There he saw the latest devices and machinery and the first electric tram.

When he returned home in 1882 Swinton began an apprenticeship with Sir William Armstrong, the armaments manufacturer and shipbuilder, at the Elswick works and yards, Newcastle upon Tyne. Here he found life much to his liking and in 1884 had a book published[33] on 'Elementary principles of electric lighting'—a considerable accomplishment for a young man of 21.

Shortly after he had completed his apprenticeship in 1887 Swinton moved to London and established himself as an electrical contractor and consulting engineer. One of his early professional occupations was the installation of electric lighting in many large town and country houses. Several years later, concurrently with his business work, he carried out experimental investigations, in his own laboratory, on the discharge of electricity through a Crookes tube. He began publishing papers on the results of his experiments with cathode rays from 1896, the year of Roentgen's discovery of X-rays. Such was Swinton's practical skill that a few days after the announcement of the discovery he was able to take an X-ray photograph of the bones in his hand. This photograph was reproduced in *Nature* in January 1896.

Swinton was fascinated by the phenomena associated with the Crookes tube and carried out many original investigations on cathode rays and X-rays. These formed the basis for numerous papers which were published in more than a dozen scientific journals. He was also a prolific writer of letters to newspapers and magazines and commented on issues of the day.

His response to Bidwell's letter was to suggest the use of two cathode-ray tubes in which the beams of electrons would be synchronously deflected by the varying fields of two electromagnets, placed at right angles to each other, and energised by two alternating currents of widely differing frequencies, so that the moving extremities of the two beams would be swept synchronously over the whole of the image surfaces within the 0.1 s necessary to take advantage of visual persistence. He observed[31]:

> Indeed, so far as the receiving apparatus is concerned, the moving cathode beam has only to be arranged to impinge on a sufficiently fluorescent screen, and given suitable variations in its intensity, to obtain the desired result.
>
> The real difficulties lie in devising an efficient transmitter which, under the influence of light and shade, shall sufficiently vary the transmitted electric current so as to produce the necessary alterations in the intensity of the cathode beam of the receiver, and further in making this transmitter sufficiently rapid in its action to respond to the 160 00 variations per second that are necessary as a minimum.
>
> 'Possibly no photoelectric phenomenon at present known will provide what is required in this respect, but should something suitable be discovered, distant electric vision will, I think, come within the region of possibility.'

Three years later, in November 1911, Campbell Swinton, in his Presidential Address[34] to the Röntgen Society, elaborated on this statement and described an idea for an electronic camera; an idea which was to be implemented in the 1930s

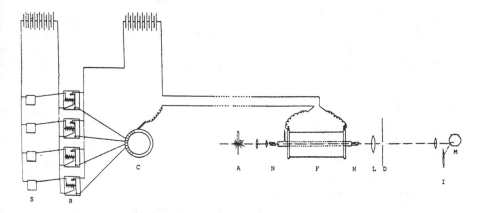

Figure 6.2   *Diagram of Rignoux's telephote. Selenium cells, S, electromechanical relays, R, commutator, C, arc lamp, A, Nicol prisms, N, Faraday polarised light rotator, F, projection lens, L, diaphragm, D, mirror drum, N, image screen, I*

Source: *Comptes Rendus*, **159**, 1914, pp. 301–304

and which was founded on the utilisation of a cathode-ray tube in the transmitter as well as in the receiver.

Modern high-definition television would not be possible without some version of the cathode-ray tube, a device which owed its origin as a practical laboratory instrument to Braun in 1897[35]. Braun's tube was not the outcome of a stroke of genius but rather it represented the consequence of an enquiry into the conduction of electricity in gases which commenced with the work of William Watson[36], an eminent English scientist of the eighteenth century.

Watson's experiments were made possible by a discovery which, according to Professor Tyndall[37], put all former ones in the shade and which Dr Priestly[38] called the most surprising yet made in the whole business of electricity. This was the storage of electric charge in a capacitor, called a Leyden jar after the name of the place where the discovery was originated. It was first reported by von Kleist, the discoverer, and dean of the cathedral in Camin, in a letter[39] dated 4 November 1745, to a Dr Lieberkuhn at Berlin.

Watson was particularly interested in the conduction of electricity through rarefied gases and vacuo and communicated his results in 1752 in a paper to the *Philosophical Transactions*. In his experiments he used a glass tube, one metre in length and 75 mm in diameter, fitted with a fixed brass plate at one end. A movable plate could be inserted into the tube and caused to approach the fixed plate. When the glass cylinder was evacuated and the electrodes electrified, there was 'a most delightful spectacle'. Priestly has written[40] about Watson's ventures 'to see the electric matter in its passage through this vacuum; to observe, not as in the open air, small brushes or pencils of rays, an inch or two in length, but coruscations of the whole length of the tube, and of a bright silver hue'.

Nearly a century later the great Faraday investigated[41] the discharge of electricity through rarefied gases (including air, hydrochloric acid gas, coal gas, hydrogen and nitrogen) with characteristic care and attention to detail. 'It would seem strange,' he wrote, 'if a theory which refers all the phenomena of insulation and conduction, i.e. all electrical phenomena, to the action of contiguous particles, were to omit to notice the assumed possible case of a vacuum.' Fortunately, for the development of electrical science, vacuum pumps were available to Faraday for von Guericke had produced the first air pump in 1654 and this had been followed by similar pumps of Boyle, Hawksbee and Smeaton having increased efficiencies. Von Guericke's apparatus had been made of glass but later oil-air pumps (in which the space below the piston was filled with oil to eliminate the air that would otherwise have been trapped) were made of metal. The first mercury pump was constructed by Swedenborg in 1722 and later versions were manufactured by Baader in 1784, Hindenburg in 1787, Edelkrantz in 1804 and Patten in 1824. Geissler's mercury pump of 1855 was capable of creating a vacuum of 0.05 mm of mercury but this was not available to Faraday, who carried out his experiments at the reduced pressure of 6.5 inches of mercury for some and 4.4 inches for others.

It was during this series of experiments that Faraday discovered the dark space which is named after him. With his great wisdom and insight into physical problems, he regarded the phenomena of the discharge tube as an admirable and important field for research. 'The results,' he observed, 'connected with the different conditions of positive and negative discharge will have a far greater influence on the philosophy of electrical science than we at present imagine, especially if, as I believe, they depend on the peculiarity and degree of polarised condition which the molecules of the dielectrics concerned acquire.' This was a prophetic statement as later events were to show.

Geissler's vacuum pump of 1855 and Ruhmkorff's induction coil of 1851 made it possible for much larger potential differences to be applied to the terminals of the discharge tube and also allowed much diminished pressures to be obtained. The improved vacuum tubes (often known as Geissler tubes) were utilised in the investigations of Plücker[42], which he reported in 1858. He noted visible stratifications in the discharge and found that they could be modified by a magnet placed outside the tube. In addition the diffused light seen near the negative electrode was found to be concentrated by the action of the magnet and Plucker concluded that the glow consisted of 'lines and light which, proceeding from the separate points of the positive electrode, coincide with magnetic curves'.

Several weeks later, Hittorf[43], who was one of Plücker's students, stated that the glow was due to some kind of rays which were given off in straight lines from the negative electrode or cathode. Goldstein, who also observed that the cathode rays, as they were called, were given off in a direction normal to the surface of the cathode, noted that a shadow of an obstacle, placed in the path of the rays, was cast on the wall of the evacuated vessel[44].

Much attention was given to the study of gaseous discharge by various investigators, but it was Sir William Crookes who carried out a systematic experimental investigation to elucidate their properties. His tubes were constructed by a skilled instrument maker named Gimingham and were probably the finest that had been made. With these Crookes performed many experiments and published several papers in 1878 and 1879[44,45]. In his Bakerian Lecture of 1878 he described in detail the different appearances of the discharge as the pressure was reduced from 68 mm to 0.078 mm of mercury. He discovered what is now known as the Crookes dark space and speculated that it was a region in which the cathode rays had a free path before colliding with the gas molecules—the blue glow being caused by these collisions.

A few years prior to the above lecture, Cromwell Varley, in 1871, had advanced[46] the idea that the cathode rays were negatively charged particles but with only partial proof. However, Jean Perrin, in 1875, confirmed[47] Varley's hypothesis. At about the same time Clerk Maxwell published his important 'Treatise on electricity and magnetism' (1873)[48] and introduced the idea of 'one molecule of electricity'. G Johnstone Stoney[49], a year later, went further than Maxwell and made an estimate of the elementary charge. He enunciated his views at a scientific meeting in 1874, but these were not adequately published until 1881. It was Johnstone Stoney who first suggested in 1891 the name electron for the natural unit of electricity. By then Schuster in a series of investigations[50] carried out in Manchester from 1884 to 1890 had determined the ratio of the charge-to-mass of the particles which comprised the cathode rays. He achieved this by deflecting the charged particles, with the aid of a magnetic field, into a circular path and obtained a value of e/m of about 1.1×10^6 C/kg.

By the early 1890s the problem of the conduction of electricity in gases was assuming great importance. Lord Kelvin, normally austere and conservative in speculative matters, noted in 1893[51]: 'If the first step towards understanding the relations between ether and ponderable matter is to be made, it seems to me that the most hopeful foundation for it is knowledge derived from experiments on electricity in high vacuum.' Lord Kelvin's opinion seems to have had an effect for within the next few years remarkable advances were made. Sir J J Thomson, in 1894, began a series of brilliant enquiries on cathode rays which culminated in the determination of the charge, mass and velocity of the particles in the rays[52]. For the accuracy of measurement he had in mind Thomson found it necessary to construct a discharge tube of the form shown in Figure 6.3. A and B were thick metal discs which filled the cross section of the tube and thus divided the tube into two main portions. A discharge could be created in the left-hand end by the application of a large potential difference to the cathode, C, and anode A. The electrode, C, was flat, with its plane perpendicular to the axis of the tube, and consequently the cathode rays travelled down the tube, through the slits in A and B, towards the far end in a direction parallel to the tube's axis. An indication of the presence of the electron beam was given by the fluorescence at P caused by the bombardment of the glass by the electrons.

The electrodes D and E, together with a suitable source of voltage, enabled an electric field to be established which deflected the beam in an appropriate direction. In addition a deflection could be generated by means of a magnetic field applied perpendicularly to the axis of the tube.

Thomson's tube clearly satisfied the two criteria necessary for a use to be made of cathode rays for measurement purposes; namely, the separation of the electron beam from the other phenomena of the discharge, and a means for measuring the deflection of the rays with certainty and accuracy. His investigations, using the above apparatus, made use of Plucker's observation of 1859 that cathode rays are deflected by a magnetic field having a component perpendicular to their direction of travel and Goldstein's demonstration of 1876 that the rays are influenced by a suitable electric field. But it was the combination of these individual discoveries, together with his considerable skill as an experimenter, which permitted Thomson to design and make a cathode-ray tube that would serve as an accurate measuring instrument.

The credit for the invention of the cathode-ray tube as a commercial instrument is usually given to Ferdinand Braun. Professor Braun, then at the University of Strasbourg, published the first account[35] of his tube in 1897. It was much larger than Thomson's, although simpler in design and differed from it in a number of details: the use of only one diaphragm B (see Figure 6.4 and 6.5) having a circular hole, (2 mm diameter), instead of a slit; the introduction of a fluorescent screen, S, consisting of a mica plate coated on one side with a fluorescent substance; and the use of magnetic deflection only.

With this tube the screen showed a bright round spot of light where the electrons bombarded the screen, instead of a line, and thus it was possible to measure the deflection of the spot in any direction across the screen.

Braun employed the apparatus—which was constructed by Franz Muller of Bonn, the successor of Geissler—for investigating phase relations associated with polarised electrolytic cells[53]. Two years later, in 1899, Zennek, in his work on radio circuits and the propagation of radio waves, utilised an improved form of Braun tube, again made by Muller, which incorporated two separate diaphragms, in place of one, and a metal plate, with a hole in it, as the anode rather than the small electrode in a side tube. The effect of these changes was to give a finer electron beam although the resulting decrease in spot brightness led initially to the use of long exposure times (ten minutes) when the screen traces were photographically recorded.

Zennek, who was one of Braun's assistants (and later President of the Deutsche Museum in Munich) has described[54] the impact of Braun's invention on his own work: 'When Braun brought out his tube, I was very enthusiastic about it. It was exactly what I had wanted for a long time, a device with which one sees what is going on in the current circuit. Later, I considered it a sport to discover as many possibilities of application as possible.'

A tube similar to Zennek's and manufactured by A C Cossor Ltd was utilised by MacGregor-Morris, in 1902, for determining the maximum value of an alternating current[53]. Ryan, in 1903, described a modification of Braun's original design because he had found that the tubes on the market were altogether too small to be of practical use. He required the fluorescent screen to be double the size (i.e. 15 cm) of the existing tubes. After many trials Muller-Uri of Braunschweig succeeded in delivering two tubes with screens 125 mm in diameter but the larger tubes proved unsatisfactory at the outset, giving an intermittent electron beam. The defect was due to corona and external leakage of the high voltage supply and was overcome by the use of a thick jacket of solid insulation (ebonite discs sealed on with paraffin wax) about the cathode end of the tube.

Much work was undertaken by experimenters on the development and application of cathode-ray tubes following Braun's publication in 1897, and so it was perhaps inevitable that someone would suggest the employment of such a tube in a television system.

Boris Rosing was the first person, on 1907, to put forward a seeing by electricity scheme incorporating a cathode-ray tube but, prior to this date, M Dieckmann and G Glage in a German patent application[18] dated 12 September 1906 had described a 'Method for the transmission of written material and line drawings by means of cathode-ray tubes'. Figure 6.6, taken from the patent, illustrates the apparatus.

The tracing point, for example a pencil, was attached to a slider, the guide for which carried by two further sliders capable of moving on two supports mounted at right angles to the first guide. Consequently, the movement of the tracing point was mechanically resolved into two mutually perpendicular components. These were then converted into resistance changes—using the wires f and l—which caused corresponding changes in the currents passing through the wires.

At the receiver the two transmitted currents excited two orthogonal sets of deflection coils, associated with the Braun tube, and thereby established two magnetic fields. The resultant field deflected the electron beam, and the associated luminous spot on the fluorescent screen, in accordance with the motion of the tracing point at the transmitter. Dieckmann and Glage found no difficulty in transmitting drawings and written words in a few seconds with their method.

In an adaptation of their cathode-ray tube receiver the inventors designed a small dynamo which provided currents for the deflection coils so as to allow the spot of light to trace a 1.25 inch square raster, of parallel and equidistant lines, in one tenth of a second.

At the transmitter a fine metal brush, moved synchronously with the luminous spot, and passed over a sheet metal pattern, Figure 6.8. When the brush contacted the conductor pattern, a current was sent to the receiver and there used to excite the coils. The magnetic field created deflected the electrons to such an extent that none

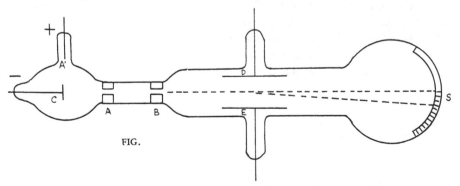

Figure 6.3   *Diagram of the cathode-ray tube which J J Thomson used in the 1890s to determine the ratio e/m for the electron. The cathode, C, anode, A', collimating apertures, A and B, deflecting electrodes, D and E, and the screen, S, are shown*

Source: *Phil. Mag.*, **44**, October 1897, pp. 293–316

passed through the hole in the diaphragm. Consequently, those parts of the raster which corresponded with the metal pattern were obliterated, or as Dieckmann and Glage wrote 'the production of an exact copy of the pattern in black on a luminous field. As the entire copy was produced [in one tenth of a second] it followed every movement impressed on the pattern'.

This was not television, of course, but a form of autographic telegraphy. Nevertheless, as Dr Dieckmann noted: '[The] experiment shows that the cathode ray is well worth the attention of inventors in search of apparatus destitute of inertia.'

Boris L'vovich Rosing was born on 3 April 1869 in St Petersburg. Following a gymnasium education he was admitted, in 1887, to the faculty of physics and mathematics, St Petersburg University, and from there he graduated in 1891. Two years later after further training, he defended his dissertation for the candidate's degree and then joined the teaching staff of the Technological Institute, St Petersburg. Initially Rosing's research activity was in the field of magnetism[55].

According to Dr V K Zworykin[56] (who invented the iconoscope and was a former student under Rosing), Rosing was not aware of the work of Dieckmann and Glage. Zworykin has not mentioned in his reminiscences why Rosing chose to work on television, but as he was an assistant professor (in charge of students' experimental work) at the Institute, it is possible that he was familiar with the ideas of his countrymen, M Vol'fke and A A Polumordvinov, who had had patents sealed on the electrical transmission of images, in 1898 and 1899 respectively.

In addition it may be that Rosing knew about the phototelegraphy system of P I Bakmet'yev[57], which was described, in 1885, in the Russian journal *Elektrichestvo*. This seems to have been the first account of a distant vision scheme in a Russian publication. Bakmet'yev's notion was to use a single selenium cell to scan, spirally, the image plane, and a corresponding single light source, again scanning spirally, to reconstitute the image.

With the considerable interest being shown by experimenters in cathode-ray tubes at the turn of the century it is likely that Rosing's imagination was stirred to consider the problem of seeing by electricity using the new apparatus.

One year after Dieckmann and Glage's patent was published, Rosing applied for a British patent on a 'new and improved method of electrically transmitting to a distance real optical images and apparatus therefor'. In 1911 he submitted two

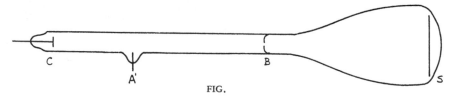

FIG.

Figure 6.4   *F Braun is usually credited with the invention of the cathode-ray tube as a commercial laboratory instrument. The cathode, C, anode, A', collimating aperture, B, and screen, S, are indicated.*

Reproduced by permission of the Science Museum

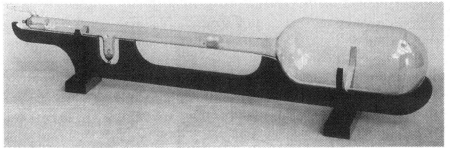

Figure 6.5 Braun's cathode-ray tube, c. 1897

Reproduced by permission of the Science Museum
Source: Annalen der Physik, **60**, 1897, p. 552

further applications, again to the British Patent Office, on the subject of 'electrical telescopy'.

Rosing's 1907 patent[14] indicates that he was knowledgeable about some previous work on this subject because he stated that the raison d'être for his patent was to obviate the defects of receiving equipment which were 'insufficiently mobile and sensitive' for the purpose. Moreover, his patent made reference to two German scientific works (Liesegang's 'Beitrage zum Problem des elektrischen Fernsehens' of 1891, and Winkelmann's 'Handbuch der Physik), and it contained proposed solutions to certain aspects of the problem which had been advanced in three patents published in 1906.

Figure 6.9 illustrates Rosing's apparatus. An image of the object or picture, 3, was cast by the lens, 4, and the two polyhedral mirrors, 1 and 2, upon a photoelectric cell, 5, 'such, for example, as a light-sensitive layer of selenium, a photoelement such for example as the element of Becquerel with chloride of silver, or an actinoelectric element, such as the element of Elster and Geitel with sodium amalgam . . . ' The mirrors rotated about mutually perpendicular axes with differing velocities so that all the points in the plane of the picture or field of view were successively scanned.

At the receiver Rosing reconstituted the image of the object with a Braun tube. Magnetic deflection of the electron beam was employed and the coils for this purpose were excited by currents, controlled by resistors, which varied depending on the movements of rubbing contacts, 7, applied to each face of the polyhedral mirrors. Variations in the electric field established between the plates, 16, by the photoelectric signals caused variations in the number of electrons which passed through the diaphragm, 13, and hence changes in the brightness of the screen.

Rosing's patent of 1907 contained an important new feature, namely, the suggestion of using a photoelectric cell of the emissive type. The patent does not describe the construction of it but, in an article[58] published in 1911, E Ruhmer stated that it consisted essentially of a glass bulb containing rarefied hydrogen or helium and a sodium, potassium, caesium or rubidium amalgam cathode together with a platinum anode. The advantage of such a cell was its effectively instantaneous response to a light stimulus, but against this most worthwhile feature the photoemissive cell generated only very weak currents. It is significant to note that selenium cells were

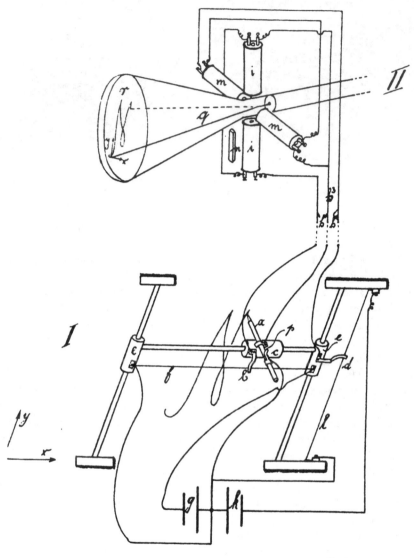

Zu der Patentschrift

№ 190102.

Figure 6.6   *The illustration shows a diagrammatic representation of the apparatus of Dieckmann and Glage (1906)*

Source: German patent no. 190 102, 12th September 1906

still being used as late as the 1920s for television development, even though valve amplifiers had been available from approximately 1912.

In an unpublished interview[56] Zworykin has related how Rosing and he had tried to copy the Elster and Geitel cell using potassium hydrate but without any real

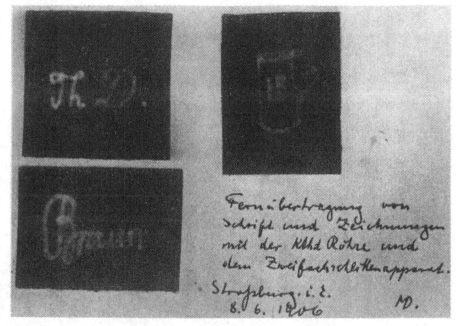

Figure 6.7   *Images photographed from the screen of a Braun tube using the system of Dieckmann and Glage*

Source: Archir fur Post-und Fernmeldewesen, August 1953, p. 272

success, and in 1910 Rosing had reverted to the practice of previous workers and utilised selenium cells. These, too, proved troublesome because of their sluggishness. Nevertheless, Zworykin confirmed that Rosing did obtain some results: '. . . in my time, stationary geometrical figures, very fuzzy, like a triangle and some kind of distorted circle and so on . . .' His success was achieved with the aid of a 'very, very intense light' and no amplification. Moving images could not be depicted because 'the sensitivity was not enough and the selenium cell was very laggy'.

Some of Rosing's laboratory notebooks exist and in number three the following entry occurs: 'On 9 May 1911, a distinct image was seen for the first time, consisting of four luminous bands'[59].

Zworykin was an assistant to his professor for two years from 1910 to 1912. At the end of 1910 the position which had been reached was that Rosing had constructed his polyhedral mirror wheels while during Zworykin's stay the cathode-ray tube with gas focussing had been completed. It is likely that this tube was in working order by the early part of 1911 because Rosing applied for another patent[60] on 4 March 1911 and in reading this it seems he had carried out experiments on the tube which showed the fundamental limitation of gas focussing, namely, the difficulty of controlling the intensity of the beam. As a consequence, Rosing put forward an original solution to the problem, a solution which was to be used by Thun[61], von Ardenne[62], and Puckle and Bedford[63] approximately two decade later. Instead of intensity mod-

Figure 6.8   *Photograph of a model of Dieckmann and Glage's apparatus in the Deutsches Museum, Munich*

Reproduced by permission of the Deutsches Museum

ulating the beam, Rosing introduced the concept of velocity modulation in which the beam travels at a relatively low speed over the lighter parts of the picture and at a relatively high speed over its darker portions. In Rosing's words, 'the time of action of the signal upon the eye of the observer corresponds to the intensity of the light signals at the transmitting station'. Unfortunately, his scheme would not have worked as he described it for the light signal must also modulate the speed of the transmitting scanner if accurate reconstitution of the original scene is to be obtained at the transmitter.

Possibly Rosing tried out his ideas on velocity modulation only to experience difficulties with the synchronisation of the transmitter and receiver scans. It is significant to record his conclusion on the problem of synchronisation, which he gave in the patent of March 1911: 'It has however been found impracticable to obtain the necessary synchronism . . . in view of the enormously high speeds of the mechanism which have to be dealt with.' His solution was to replace the cathode-ray tube receiver by a pair of galvanometer mirror scanners, the movements of which were 'directly and completely subordinated to the corresponding mechanism at the transmitting station . . .' This arrangement of electromechanical oscillographs was used by D von Mihaly[64] for several years, but without success. If anything, therefore, Rosing took a retrograde step with this patent.

Two days after the above patent was sealed Rosing obtained further protection for some of his ideas. In his latest patent he described a device for interrupting or chopping the light beam falling onto the photocell. Essentially, the photoelectric current amplitude modulated a higher frequency signal. Again, the principle was used by later investigators, including Baird during his early work in the 1920s.

After 1911 Rosing did not make any significant contributions to the television problem, although he maintained an interest in the subject. His book on 'The electrical telescope (sight at a distance)' was published in 1923[65]. He certainly appreciated the benefits which television would bring and, in an article in the French journal *Excelsior*, he wrote[66]:

'The range of application of the telephone does not extend beyond human conversation. Electrical telescopy will permit man not only to commune with other human beings, but also with nature itself. With the "electric eye" we will be able to penetrate where no human being penetrated before. We shall see what no human being has seen. The "electric eye" fitted with a powerful lamp and submerged in the depths of the sea, will permit us to read the secrets of the submarine domain. If we recall that water covers three quarters of the earth's surface, we readily realise the infinite extent of man's future conquests in this portion of his domain, till now inaccessible to him. From now on and in all future times we can imagine thousands of electric eyes travelling over the floor of the sea seeking out scientific and material treasures; others will carry out their explorations below the earth's surface, in the depths of craters, in mountain crevices and in mine shafts. The electric eye will be man's friend, his watchful companion, which will suffer from neither heat nor cold, which will have its place on lighthouses and at guard posts, which will beam high above the rigging of ships, close to the sky. The electric eye, a help to man in peace, will accompany the soldier and facilitate communication between all members of human society.'

Rosing's place in the history of television stems from his advocacy of using a cathode-ray tube as the receiving element in a system of distant vision, of interrupting at a high frequency the signals from the photoelectric cell, of resolving the focussing problem by means of velocity modulation and of incorporating photoemissive cells in such systems.

Campbell Swinton, to whom reference has been made, and who was the first person to suggest an all-electronic television system, could not have been influenced by Rosing's early work in 1907 for Rosing's patent (27 570) was not accepted and

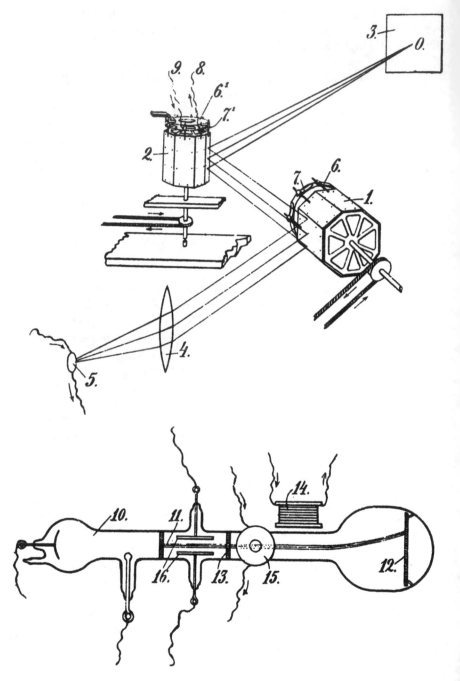

Figure 6.9   *Diagrams of Rosing's apparatus of 1907. Mirror-drum scanning was used at the transmitter and cathode-ray scanning was employed in the receiver*

Source: British patent 27 570, 13 December 1907

published by the Patent Office until 25 June 1908, whereas Campbell Swinton's letter to *Nature* was published on 18 June 1908.

In a paper[67] published in Modern Wireless in 1928 Campbell Swinton gave something of the background which led him to put forward his 1908 ideas:

'It was only a few years after the introduction of the cathode-ray oscillograph, by Braun in 1897, that I first thought of the possibility of producing practical television by means of instruments working on this cathode-ray principle and, in order to study the matter, I obtained from Germany one of Braun's tubes, with which I made many experiments showing the rapidity and precision with which the cathode-ray beam could be deflected both magnetically and electrostatically.'

Subsequently, during the period 1903 to 1904, Swinton, with his assistant J C M Stanton, tried some experiments on a cathode-ray transmitter using the known dependence of the resistance of selenium on visible light.

'A metal plate, one surface of which was covered with selenium, was mounted in a vacuum tube, so that the end of a cathode-ray beam from a suitably placed electrode could, by electromagnetic deflection, be caused to traverse the coated plate, while at the same time, the bright image of an electric arc was thrown on the selenium surface by means of a lens.'

Campbell Swinton hoped that the variation of the resistance of the selenium signal plate would cause the current flowing in the circuit—comprising the signal plate, the electron beam and the power supply—to vary and be capable of giving an indication on a sensitive galvanometer. No reliable results could be achieved notwithstanding some assistance given by Professor G M Minchin, a noted authority on the subject of light-sensitive cells.

Swinton wrote in 1926, in a letter[68] to *Nature*, that he had also tried experiments in receiving with a Braun tube 'but in its then hard form it proved very intractable'. This statement gives some insight into the problems which Rosing must have faced and which led him to consider velocity modulation.

Further work was abandoned but, when Shelford Bidwell's paper[16] on 'Telegraphic photography and electric vision' was published in *Nature* on 4 June 1908, Swinton felt compelled to point out the impracticability of the scheme proposed.

It is interesting to note that the above experiments were repeated in 1937–38 by J Strange and Dr H Miller under the direction of Dr (later Professor) J D McGee, (all of Electric and Musical Industries Ltd)[69]. They obtained pictures which were of good quality, although they were troubled by lag, which is serious in this type of tube. Selenium was one of the light-sensitive materials used and it gave successful results.

Swinton's 1911 presidential address[34] gave a complete description of his ideas together with a diagrammatic illustration of his scheme, see Figure 6.10. In this both the transmitting and receiving cathode-ray tubes were of the Braun type and were fitted with cold cathodes and anodes. Although Wehnelt had suggested in 1905 coating a platinum filament with lime so that the necessary electron emission could be obtained with a relatively low anode-cathode voltage, Campbell Swinton in his original proposal showed a tube with a cold cathode. This needed a potential difference of about 100 000 V between the anode and cathode structures to give an appropriate beam current. In both tubes the electron beams were deflected by magnetic fields, using electromagnets excited from a.c. generators. The line scan was at the rate of 1000 Hz and the frame scan at 10 Hz. Because of the sinusoidal nature of the output of the generators the electron beams would have described Lissajous figures on the screens of the transmitting and receiving cathode-ray tubes—a form of scan first put forward in 1880 by Le Blanc.

Campbell Swinton's choice of frequencies is intriguing as he advocated the adoption of 200 lines per frame. This was certainly ambitious in concept because the first demonstration of rudimentary television in 1926 used 30 lines per frame. Even the first trials of systems with cathode-ray tubes, in the early 1930s, by EMI and by RCA did not employ a line standard of more than 150 lines per picture. The figure of 200 was probably reached from a consideration of the number of dots used in printing photographs with the aid of process blocks rather than from any analysis of the television image and the bandwidth of the system required to transmit it.

In the receiving tube the electrons were to produce a luminous image by striking a fluorescent screen and the intensity of the image was to be controlled by applying a voltage, derived from the transmitting tube, to a pair of deflecting plates. With regard to the receiving apparatus, there was nothing of a particular novel nature. The novel aspect of Swinton's conception lay in the original approach to the generation of the image signal at the transmitter.

He conceived the idea of a mosaic of photoelectric elements onto which the image of the object or scene would be projected by means of a lens and which would be scanned on the side away from the image side by a beam of electrons controlled by line and frame a.c. voltages. In the transmitter cathode-ray tube the gas tight screen, J, was to be formed from a number of small insulated metallic cubes of a metal such as rubidium, which was strongly active photoelectrically, so that a clean metallic surface would be presented to the electron beam on one side and to a suitable gas or vapour, say, sodium vapour, on the other. A metallic gauze screen, L, parallel to J was to be placed in front of the screen, J.

Campbell Swinton's account of the supposed operation of the system follows, because some discussion took place in 1936 on whether the integrating feature of the iconoscope was implicit in his writings (see Appendix A.1):

'As the cathode rays oscillate and search out the surface of J they will impart a negative charge in turn to all the metallic cubes of which J is composed. In the case of cubes on which no light is projected, nothing further will happen, the charge dissipating itself in the tube; but in the case of such of those cubes as are brightly illuminated by the projected image, the negative charge imparted to them by the cathode rays will pass away through the ionised gas along the line of the illuminating beam of light until it reaches the screen L whence the charge will travel by means of the line wire to the plate O of the receiver. This plate will thereby be charged—will slightly repel the cathode rays in the receiver; will enable these rays to pass through the diaphragm P, and impinging on the fluorescent screen H will make a spot of light. This will occur in the case of each metallic cube of the screen J which is illuminated, while each bright spot on the screen H will have relatively exactly the same position as that of the illuminated cube of J. Consequently, as the cathode-ray beam in the transmitter passes over in turn each of the metallic cubes of the screen J, it will indicate by a corresponding bright spot on H whether the cube in J is or is not illuminated, with the result that H, within one-tenth of a second, will be covered with a number of luminous spots exactly corresponding to the luminous image thrown in J by the lens M, to the extent that this image can be reconstructed in a mosaic fashion. By making the beams of cathode rays very thin, by employing a very large number of very small metallic cubes in the screen J, and by employing a very high rate of alternation in the dynamo G, it is obvious that the luminous spots of H of which the image is constituted can be made very small and numerous, with the result that the more these conditions are observed the more distinct and accurate the received image.

'Furthermore, it is obvious that, by employing for the fluorescent material on the screen H something that has some degree of persistency in its fluorescence, it will be possible to reduce the rate at which the synchronised motions and impulses need take place, though this will only be attained at the expense of being able to follow rapid movements in the image that is being transmitted.

'It is further to be noted that, as each of the metallic cubes in the screen J acts as an independent photoelectric cell, and is only called upon to act once in a tenth of a second, the arrangement has the obvious advantages over other arrangements that have been suggested, in which a single photoelectric cell is called upon to produce the many thousands of separate impulses that are required to be transmitted through the line wire per second, a condition which no known form of photoelectric cell will admit of.

'Again, it may be pointed out that sluggishness on the part of the metallic cubes in J or of the vapour in K in acting photoelectrically, in no way interferes with the correct transmission and reproduction of the image, provided all portions of the image are at rest; and it is only to the extent that portions of the image may be in motion that such sluggishness can have any prejudicial effect. In fact, sluggishness will only cause changes in the image to appear gradually instead of simultaneously.'

This, then, was the scheme of Campbell Swinton: a most remarkable one when it is borne in mind that at the time it was enunciated 'radiocommunication was in its

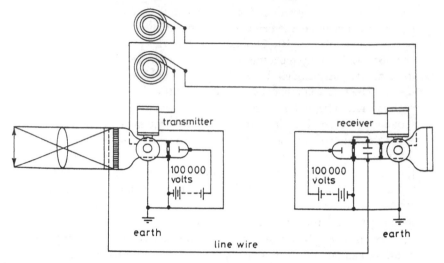

Figure 6.10   In his presidential address to the Roentgen Society in 1911, Campbell Swinton described his scheme for television using nonmechanical scanning means at both the transmitter and receiver

Source: *Journal of the Roentgen Society*, **8**, (30), January 1912, pp. 1–5

infancy, radio valves practically unknown, vacuum technology very primitive, photoelectric cells very inefficient'.

Campbell Swinton never attempted to construct a working model of his 1911 transmitter, and he fully appreciated the difficulties which would have to be surmounted before it could be made to work. With typical honesty he remarked: 'It is an idea only . . . Furthermore, I do not for a moment suppose that it could be got to work without a great deal of experiment and probably much modification.'

Campbell Swinton continued to advance his concept until his death in 1930. He modified some of the details of the scheme by including hot cathode tubes and valve amplifiers but, in the main, his system still incorporated the 1911 principles. He retained an abiding opinion that successful television would only come about by the use of nonmechanical scanning. He was not sanguine about the financial return to an individual who attempted to pursue the solution, but 'if we could only get one of the big research laboratories, like that of GEC or of the Western Electric Company, one of these people who have large skilled staffs and any amount of money to engage on the business, I believe they would solve a thing like this in six months and make a reasonable job of it'. 'There are, at any rate, no theoretical objections to the scheme,' he wrote in his 'Autobiographical and other writings'[31].

Dr J D McGee, the leader of the team at Electric and Musical Industries Ltd which evolved the emitron camera tube, has written: 'Modern television owes much to the researches and achievements of many distinguished workers, but in essence it has been developed upon the fundamental lines first put forward by Campbell Swinton'[70].

Since his proposals were made apparently realistic by work which had been progressing in pure physics, it is now convenient to consider some of the results that had
been obtained, and which were about to come to fruition during the first and second
decades of the new century, in this field.

References

1 THOMAS, D.B.: 'The origin of the motion picture' (HMSO, London, 1964) p.30
2 ECO, V., and ZORZALI, G.B.: 'A pictorial history of inventions' (Weidenfeld and
   Nicolson, London, 1962) p. 272
3 BOYER, J.: 'La transmission telegraphique des images et des photographies' (Paris, 1864)
4 THORNE-BAKER, T.: 'The telegraphic transmission of photographs' (Constable,
   London, 1900)
5 KORN, A., and GLATZEL, B.: 'Handbuch der Photo-telegraphie und Teleautographie' (Leipzig, 1911)
6 Ref.4, p. 54
7 THORNE-BAKER, T.: 'Wireless pictures and television' (Constable, London, 1928)
   Chapter XI, pp. 141–166
8 VON BRONK, O.: 'Verfahren und Vorrichtung zum Fernsichtbarmachen von Bilder
   bzw Gegenständen unter vorübergehander Auflösung der Bilder in parallele
   Punktreihen'. German patent 155 528, 12 June 1902
9 COBLYN, M.J.H.: 'La vision à distance par l'électricité', *L'Eclairage Electrique*, December
   1902, **33**, pp. 433–440
10 NISCO, A.: 'La vision a distance par l'électricité', *Electro*, October 1903, **11**, pp. 153–154
11 VON JAWORSKI, W., and FRANKENSTEIN, A.: 'Verfahren und Vorrichtung zur
    Fernsichtbarmachung von Bildern und Gegenständen mittels Selenzellen,
    Dreifarbenfilter und Zerlegung des Bilden in Punktgruppen durch Spiegel'. German
    patent 172 376, 20 August 1904
12 RIGNOUX, G.P.E.: 'Appareil destiné à transmettre a distance les images des objets'.
    French patent 382 535, 10 December 1906
13 LUX, F.: 'Der elektrische Fernseher', *Bayerisches Industrie und Gewerbeblatt*, January 1906,
    38, (2), pp. 13–19
14 ROSING, B.L.: 'New or improved methods of electrically transmitting to a distance real
    optical images and apparatus therefor'. British patent 27 570, 13 December 1907
15 ADAMIAN, J.: 'Improvements in electrically controlled apparatus for seeing at a dis-
    tance'. British patent 7219, April 1908
16 BIDWELL, S.: 'Telegraphic photographic and electric vision, letter, *Nature*, 4 June 1908,
    pp. 105–106
17 ANDERSEN, A.C., and ANDERSEN, L.S.: 'Improvements in apparatus for electrically
    transmitting images of natural objects to a distance'. British patent 30 188, 24 December
    1908
18 DIECKMANN, M., and GLAGE, G.: 'Verfahren zur Uebertragung von Schriftzeichen
    und Strichzeichnungen unter Benützung der Kathodenstrahlrohre'. DRP 190 102,
    September 1906
19 Ref.3, pp. 80–82
20 KUBICKI, C.D.: 'La télévision', L'Industrie Électrique, 1910, **19**, pp. 80–83
21 RIGNOUX, G.P.E.: 'Dispositif pour la vision à distance', *Comptes Rendus*, 1914, **159**, pp.
    301–304
22 EKSTROM, A.: 'Anordning for of verforande af bilder pa afstand'. Swedish patent
    32 220, 24 January 1910
23 SCHMIERER, M.: German patents 234 583 of 10 April 1910, 229 916 of 30 April 1910
    and 234 601 of 8 July 1910
24 HOGLUND, G.H.: 'Mechanism for electrically transmitting and reproducing images'.
    USA patent 1 030 240, 18 April 1910
25 SINDING-LARSEN, A.: 'Improvements in and relating to the transmission of pictures
    of moving pictures'. British patent 14 503, 20 June 1910

26 PERSKYI, C.: 'Television', *Electrician*, September 1900, **45**, pp. 820–822
27 Paris correspondent: a telegram to *The Times*, 28 April 1908
28 Ref.20, p. 80
29 ARAPU, R.: 'The telegraphic apparatus of Georges Rignoux', *Scientific American Supplement*, May 1915, (2055), **79**, p. 331
30 CAMPBELL SWINTON, A.A.: 'Distant vision', letter, *Nature*, 18 June 1908, p. 151
31 CAMPBELL SWINTON, A.A.: 'Autobiographical and other writings' (Longmans, London, 1930)
32 BRIDGEWATER, T.H.: 'A A Cambell Swinton'. Royal Television Society monograph, (1), 1982
33 CAMPBELL SWINTON, A.A.: 'Elementary principles of electric lighting' (Crosby Lockwood, 1885)
34 CAMPBELL SWINTON, A.A.: Presidential Address, *J. Roentgen Society*, **8**, (30), January 1912, pp. 1–5
35 BRAUN, F.: 'Ueber ein Verfahren zur Demonstration und zum Studium des Zeitlichen Verlaufes Variabler Ströme', *Ann. Phys.*, 1897, **60**, p. 552
36 WATSON, W.: 'An account of the phenomena of electricity in vacuo with some observations thereupon', *Phil. Trans.*, 1752, **147**, p. 362
37 MOTTELAY, J.P.: 'Bibliographical history of electricity and magnetism' (Charles Griffin, London, 1922)
38 PRIESTLY, J.: 'History and present state of electricity' (London, 1767)
39 Ref.37, p. 173
40 Ref.38
41 FARADAY, M.: 'Experimental researches in electricity', **1**, paras. 1529, (p.487) and 1554, (p.494). Also see para. 1523 (p. 485) (London, 1839–55)
42 PLUCKER, J.: 'On the action of the magnet upon the electrical discharge in rarified gases', *Phil. Mag.*, 1858, **16**, pp. 119–135, 408–418
43 HITTORF, J.W.: 'Ueber die Elektricitätsleitung der Gase', *Ann. Phys.*, 1858, **136**, pp. 1–30, 197–234
44 MEYER, H.W.: 'A history of electricity and magnetism' (MIT Press, 1971) p. 227
45 CROOKES, Sir W.: 'On the illumination of lines of molecular pressure and the trajectory of molecules', *Phil. Trans.*, 1879 **170**, pp. 135–164
46 VARLEY, C.: 'Some experiments on the discharge of electricity through rarified media and the atmosphere', *Proc. Roy. Soc.*, 1871, **19**, pp. 236–242
47 PERRIN, J.: *Comte Rendu*, 1895, **121**, pp. 1130–1136
48 MAXWELL, J.C.: 'Treatise on electricity and magnetism' (Clarendon Press, Oxford, 1892)
49 JOHNSTONE STONEY, G.: 'On the physical units of nature', *Phil. Mag.*, **11**, p. 384
50 SCHUSTER, A.: 'Experiments on the discharge of electricity through gases. Sketch of a theory', *Proc. Roy. Soc.*, 1884, **37**, p. 317. Also 1887, **42**, p. 371
51 Quoted *in* TURNER, D.M.: 'Makers of science, electricity and magnetism' (OUP, London, 1927) p. 163
52 THOMSON, J.J.: 'Cathode rays', *Phil. Mag.*, October 1897, **44**, pp. 293–316
53 MACGREGOR-MORRIS, J.T., and Mines, R.: 'Measurements in electrical engineering by means of cathode rays', *J. IEE*, 1925, **63**, p. 1074
54 ZENNECK, J.: 'Eine Methode zur Demonstration und Photographie von Stromcurven', *Ann. Phys.* 1899, **69**, p. 838
55 GOROKHOV, P.K.: 'History of modern television', *Radio Engineering*, June 1961, **16**, pp. 71–80
56 ZWORYKIN, V.K.: Taped interview (unpublished) with GRM Garratt and W. Baker, 3 May 1965, Science Museum, UK
57 BAKHMET'YEV, P.I.: 'The new telephotography', *Elektrichestvo*, 1885, (1), pp. 1–7
58 RUHMER, E.: 'An important step in the problem of television', *Sci. Am.*, December 1911, **105**, p. 574
59 ROSING, B.L.: Notebook 3, 1911, Archives of the A.S.Popov Central Museum of Communication. Quoted in Ref.55
60 ROSING, B.L.: 'Improvments relating to the transmission of light pictures in electrical telescopic and similar apparatus'. British Patent 5486, 4 March 1911
61 THUN, R.: 'Method of and apparatus for transmitting pictures'. British patent 355 319, 18 May 1929

62 VON ARDENNE, M.: 'Television method'. British patent 397 688, 6 October 1931
63 PUCKLE, O.S., and BEDFORD, L.H.: 'A velocity modulation television system', *J. IEE*, 1934, **75**, pp. 63–82
64 VON MIHALY, D.: 'Das Elektrischen Fernsehen und das Telehor' (M. Krayn Verlag, 1923)
65 ROSING, B.L.: 'The electrical telescope (sight at a distance)—the approaching problems and prospects' (Academy, Petrograd, 1923)
66 ROSING, B.L.: Article in the French journal *Excelsior*, c. 1910. (Quoted by A. Korn and B. Glatzel in 'Handbuch der phototelegraphie und Telautographie')
67 CAMPBELL SWINTON, A.A.: 'Television by cathode rays', *Modern Wireless*, June 1928, pp. 595–598
68 CAMPBELL SWINTON, A.A.: Letter, *Nature*, 1926, **118**, p. 590
69 STRANGE, J.W., and MILLER, H.: 'The electrical reproduction of images by the photoconductive effect', *Proc. Phys. Soc.*, 1938, **50**, pp. 374–384
70 MCGEE, J.D.: 'Campbell Swinton and television', *Nature*, 17 October 1936, pp. 674–676

Chapter 7

Developments of importance to television

During the period of the late 19th and early 20th centuries when inventors were endeavouring to seek solutions, using known techniques and devices, to the difficulties posed by the implementation of distant vision systems, other scientists were engaging in fundamental physical researches. Some of these were to have the most far reaching consequences for the progress of television.

The work of Heinrich Hertz[1,2] in 1887 provides a convenient and suitable base from which the influence of these researches may be considered.

It was in this year that Hertz performed his classic experiments on the effect of electrical discharges in one oscillatory circuit upon another similar, but separate, circuit. A Ruhmkorff coil, excited by a battery, caused primary sparks to traverse a gap. A second smaller coil produced secondary sparks, about 1 mm in length, which bridged the gap in a Reiss spark micrometer. In his work Hertz investigated the effect of the parameters of the first circuit on the length of the spark in the micrometer gap. He observed: 'In all of the experiments described, the apparatus was arranged so that the spark of the inductor was visible from the position of the spark at the micrometer gap. If this condition was altered, the same qualitative results were obtained, but the lengths of the secondary sparks appeared to decrease.'

Subsequently, Hertz concluded that ultraviolet light from one spark gap could enhance the passage of sparks across a second gap. He did not pursue this subject further and wrote: 'For the present I limit myself to the presentation of these established facts without attempting to advance a theory of how such observed phenomena could occur.'

During the following year Wiedmann and Ebert confirmed Hertz's results— particularly those which showed that the effect was confined to the negative terminal of the irradiated gap.

In the same year, 1888, Hallwachs demonstrated[3] that under the influence of ultraviolet radiation 'negative electricity' left a negatively charged body and followed the electrostatic lines of force. Also, in 1888, Righi found[4] that a polished metal plate and metal grid configuration was capable of producing a current under

HORRORS OF THE FUTURE: THE "RADIO EYE."

Figure 7.1   Horrors of the future: the radio eye. From Popular Wireless Weekly, 1922, well before any television pictures had been seen

the action of light. He termed the configuration a photoelectric cell. Righi employed a quadrant electrometer in his experiments but in 1890 Stoletow utilised[5] a high resistance galvanometer and an external source of electromotive force and was able to show that a small continuous current flowed from the grid to the plate (in the positive, conventional sense) when the polished plate was irradiated with light.

The next great advance in photoelectricity was made by the two famous co-workers, Elster and Geitel[6]. They had observed that, of all the metals which had been studied for photoelectric sensitivity, aluminium, magnesium and zinc appeared to give the best results. Hence, they considered that, because these metals were all electropositive, it seemed reasonable to expect similar or better results from metals which were more electropositive than those mentioned above. Elster and Geitel therefore proceeded to investigate the alkali metals, particularly sodium and potassium. Unfortunately, these chemically active elements were found to react almost instantly with air and water vapour to form oxides and hydroxides which were relatively insensitive. Nevertheless, they persevered with their study of these two metals and having noticed previously that an amalgam of zinc could be utilised with much greater satisfaction than zinc alone they decided to investigate the photoactivity of amalgams of the alkali elements. Success followed. After much preliminary work they discovered that a fresh dilute amalgam of either sodium or potassium was many times more sensitive than zinc amalgam.

By 1890 Elster and Geitel had published[7] a detailed account of the manufacture of a sodium amalgam photoelectric cell in which the metal electrodes were enclosed in an evacuated, glass vessel. They wrote: 'As may be judged from the description, no provision has been made to admit light of short wavelength. A window of quartz or similar material is not necessary, as light transmitted by the glass proves to be sufficient.'

At about this time, (1890) the nature of the negative electricity which left the negative electrode of the cell was not known. Necessarily, much work was undertaken by physicists on the elucidation of its characteristics. Elster and Geitel, Lenard and Merritt and Stewart, *et al.* showed beyond doubt, by 1900, that the photoelectric current was due to the emission of electrons from the negative electrode of the cell.

Other investigations enabled the fundamental laws of photoelectricity to be derived and work on the photoelectric sensitivity of metal surfaces allowed much improved cells to be manufactured. The early cells tended to deteriorate with time unless their cathode surfaces were occasionally renewed, but in 1904 Hallwachs constructed a cell for photometric purposes which was found to be constant in operation over a period of several months. This consisted of an evacuated vessel having a copper plate coated with black oxide as a cathode, but it was not sensitive to visible radiations.

Elster and Geitel continued to make strides and observed that the hydride crystals of sodium and potassium were more responsive than the metals themselves. They also evolved a sensitising procedure which marked a new era in the development of photoelectric cells. If a glow discharge was excited in an alkali cell filled with hydrogen, then the cathode surface was transformed into a colloidal state and became as much as one hundred times more sensitive than the pure element.

And so, when Rosing, and Campbell Swinton, embarked upon their considerations of the seeing by electricity problem, it was logical that they should endeavour to use or incorporate the results which had been obtained in the above field. Campbell Swinton's 1911 notion of a mosaic screen comprising a very large number of minute photocells, rather than a single cell, was outstanding, for it overcame the objection of all previous schemes which used a single cell. Whether Campbell Swinton fully appreciated the integrating property of the mosaic is discussed in Appendix A.1.

Apart from the unavailability of the techniques needed to implement his ideas, there was still one important component of any television system which had not advanced, by 1911, to the stage of general usefulness, namely, the triode amplifier. Nevertheless, progress was being made in the field of thermionics and a later event, the Great War of 1914–1918, was to act as a spur to the enhancement and universal application of de Forest's audion invention.

The science of thermionic emission developed side by side what that of photoelectric emission and the science of the conduction of electricity through gases. Great discoveries were made in these fields, during the last quarter of the nineteenth century and the first decade of the twentieth century, which had important consequences for the furtherance of television broadcasting.

The first systematic investigations in thermionics were carried out by Elster and Geitel during the years 1882 and 1889[8]. They studied in detail the charge acquired by an insulated metal plate, mounted close to a metallic filament within a glass bulb, under different conditions of filament temperature and gas pressure. In the same period, at the Philadelphia Exhibition of 1884, Edison exhibited a discovery[9] which he had made while investigating incandescent lamps, and which is now known as the Edison effect. Neither Edison nor Sir William Preece[10], who subsequently performed some experiments on this effect, gave any explanation of the phenomenon, nor was any practical application made of it.

Later, in 1890, Professor J A Fleming showed[11] that, when the negative leg of a heated carbon loop filament was surrounded by a cylinder of either a metal or an insulating material, the Edison effect almost disappeared. Other experiments of a similar nature demonstrated that the action was due to the passage of negative electricity from the incandescent filament to the cold electrode and corroborated the findings of Elster and Geitel.

Ensuing investigations by Thomson in 1899[12] indicated, in the case of a carbon filament glowing in hydrogen at a very low pressure, that the negative electricity was given off by the filament in the form of free electrons. This conclusion applied too to the electric current emanating from a lime-covered platinum cathode (Wehnelt, 1905)[13].

The first application of the phenomena associated with thermionic emission was Fleming's 1904 device for rectifying alternating voltages[14]. It consisted of a carbon filament incandescent lamp provided with a separate insulated electrode in the shape of a flat or cylindrical metal plate, or another carbon filament sealed into a bulb. Later, in 1908, Fleming noted that much improved results were obtained when the valve was constructed with a tungsten filament and an insulated coaxial, cylindrical, copper anode[15].

The next step in the evolution of the thermionic valve was contributed by Dr Lee de Forest[16] in 1906 when he introduced a third electrode into the rectifying valve, sited between its filament and anode structures. The principle of grid control had been used previously by the German physicist Lenard[17] for studying the motion and nature of the electrons liberated from a zinc cathode by ultraviolet light, but Lenard had not conceived of its use for the detection or amplification of wireless signals.

Surprisingly, de Forest[18], despite his doctoral training, did not fully understand the principles of the triode[19] and, although he applied it to the detection of radio signals in 1907[20], he undertook few scientific experiments with the audion in the five-year period after its invention. Initially he believed the presence of some residual gas was essential for the correct functioning of the triode. However, in 1914, E H Armstrong[21] showed that the presence of gas in a valve was not fundamental for its performance, and I Langmuir developed methods for producing high vacuum or hard valves.

A consequence of de Forest's lack of insight was that his soft audions were nonuniform in performance and proved less satisfactory than some other detecting devices such as the electrolytic, magnetic and crystal types. Commercial users of wireless receivers found the audion superior to these detectors, but de Forest's triodes needed such constant attention and such frequent adjustment of the anode potential and filament current that they preferred the earlier forms of detector[19].

In 1912 de Forest, while working for the Federal Telegraph Company, undertook some new experiments with the audion, based on its amplifying property. He obtained a patent[22] on 'A method of and apparatus for amplifying and reproducing sounds', in January 1914. His amplifying system had the 'distinguishing feature that a circuit arrangement [was] employed which [permitted] the use of two or more amplifiers connected up in cascade and the employment of a single lighting battery'. He utilised the cascaded stages as telephone repeaters and exhibited his scheme to the AT&T Company in October 1912. The following spring the company purchased the telephone repeater rights for $50 000 and later paid $90 000 for the radio rights to the triode[19].

The position in England regarding de Forest's audion patent was rather different to that which prevailed in the US because, although the inventor had taken out a British patent in 1908, he allowed it to lapse on account of nonpayment of the renewal fees due in January 1911. Hence, the audion was freely available in England and the Fleming patent, of 16 November 1904, became a master patent[21].

The Fleming patent was held by the Marconi Company and so further advancements in valves took place largely in that company's laboratories. The two principal workers were H J Round and C S Franklin[23]. They had access to the German Lieben-Reisz tube and concentrated their efforts on the manufacture and application of hard valves, with the aid of Langmuir's vacuum techniques. Because of their company's interests, Round and Franklin directed their attentions towards the solution of practical problems: hence the use of valves developed faster in England than in the US.

The First World War gave an enormous impetus to the utilisation of valves in signalling systems and so stimulated advances in technique that by 1918 triodes could be manufactured to cover a wide power range and were suitable for both receiving and transmitting purposes; their theory and operation were both thoroughly understood.

Accordingly, when interest in television was revived a few years later by scientists and inventors in the UK, the US, France, Germany and elsewhere, the basic components of a distant vision system were well known. The principles of scanning an object or image by means of apertured discs, lensed discs and mirror wheels had been expounded by Nipkow, Brillouin and Weiller, *et al.*; methods and apparatus for synchronising two nonmechanically coupled scanners had been suggested by many workers and demonstrated in facsimile transmission systems, much development work had taken place on photoelectric cells and now, at the end of the 1910–1920 decade, the means for amplifying the weak currents obtained from these cells seemed to be available.

There was another property of triode valve circuits which undoubtedly hastened the time when seeing by electricity became a reality. In 1912 de Forest[24] discovered that the triode valve could be employed in an oscillator to generate electromagnetic waves in addition to acting as a detector and as an amplifier. This was to be a finding that was to prove of immense significance in the history of sound and television broadcasting. The Marconi spark apparatus, the Poulsen arc and the Alexanderson alternator, were all expensive and cumbersome, but with de Forest's valve generator and the stimulus provided by the Great War, the progress of continuous wave radiocommunications rapidly moved forward to the stage where commercial sound broadcasting could be seriously contemplated shortly after the cessation of hostilities in 1918. And if sound signals could be propagated by radio, then surely vision signals too could be transmitted.

The development of large power valve transmitters was first shown to be within the realms of practical accomplishment when AT&T in August 1915 used such a transmitter to send speech signals from Arlington, Virginia, to Darien in the Panama Canal zone, 2100 miles distant[25]. About one year later the same company communicated with Paris and employed a bank of 500 valves, each having a capacity of 15 W, in its Arlington transmitter.

During the 1914–18 war, valve transmitters were used principally for communications between the ground and aircraft in flight over France and, in this work, the Marconi Company played a most valuable role. Later, the valve transmitter quickly outsted the spark, the arc and the high frequency alternator types of transmitters and by 1921 the Marconi Company had a valve transmitter with a rated output of 100 kw installed at Caernarvon, North Wales.

The first broadcasters were the radio amateurs or 'hams', whose hobby it was to communicate with other radio hams[26]. Their activities were curtailed in the United Kingdom during the 1914–18 period so that no interference with essential communications would occur, but in 1919 the restrictions were lifted. In November of that year, the Postmaster General stated that he was willing to consider applications from British subjects having the necessary knowledge and skill for licences to own and operate radio transmitters provided that they had definite objectives of scientific value or general public utility.

The activities of the amateur in the US were not curbed by the commencement of hostilities and hence the climate there favoured the establishment of regular broadcasting. In Pittsburg, the Westinghouse Electric and Manufacturing Company owned station KDKA which was used for experimental purposes[26]. The station's broadcasts were so popular that a local store sold receiving sets to people who wished to listen. This popularity induced Westinghouse to establish a regular broadcasting service in 1920, as it provided an outlet for the activities of the company's radio section, which had been developed during war, and enabled the firm to make use of its manufacturing capacity in radio apparatus. Additionally, a source of revenue for the service came from local advertisers who were prepared to pay for transmission time.

Westinghouse's initiative touched off an immediate boom in radio broadcasting and in the formation of transmitting stations. Initially there were no restrictions, no licences and the outcome was chaos. The experience gained during this formative period in sound broadcasting clearly showed the need for a framework of regulations.

In the United Kingdom progress was more cautious and orderly. The early radiotelephony accomplishments of America gave it a lead in the world's markets for equipment and, in order to counter this, the Marconi Company secured a licence for, and constructed, a 6 kW station at Chelmsford. The station was intended to broadcast on a worldwide basis and to show potential customers that in technical developments the British company was not lagging behind its American counterparts and that only Government restrictions prevented a more widespread usage of broadcasting in Britain. The station was followed by another of 15 kW on the same site and, from 23 February to 6 March 1920, two daily half-hour programmes of news and music were broadcast, using the wavelength allotted to Marconi's Poldhu

Table 7.1   *Summary of the principal proposals made during the 1910 to 1920 decade for television*

Date	Name	Opticoelectrical transducer	Transmission scanner	Transmission link
2.03.1911 (30.11.1911) British patent	B Rosing[28] Russia	photoelectric receiver —not specified	two rotating mirror drums scanning in orthogonal directions	two signal wires plus four synchronising wires
7.11.1911 (lecture) January 1911 (paper)	A A Campbell Swinton[29] UK	mosaic of photosensitive cubes (e.g. of rubidium) in contact with sodium vapour	electron beam, of cold cathode-ray tube, magnetically deflected	signal wire and two synchronising wire
25.06.1914 (25.06.1915) British patent	S L Hart[30] UK	any form of photosensitive cell; a selenium bridge is mentioned	a rotating multilens drum, the axis of which oscillates through a small angle; the axes of the lenses are tilted with respect to each other	one wire for both the vision and the synchronising signal
13.07.1914 (presentation) 27.07.1914 (paper)	G Rignoux[31] France	a mosaic of 64 selenium cells (each of which has a relay in series with it)	a rotating commutator samples the relay field (the relays operate when the selenium cells are illuminated)	two signal wires plus four synchronising wires
01.04.1915 (08.12.1915) French patent	A Voulgre[32] France	a glass ampoule containing an amalgam of sodium and rubidium	the scanning system uses three moving belts—each of which has transverse slots—and a rotating slotted disc (the slots being radial)	two signal wires
07.12.1917 (16.10.1923) US	A M Nicolson[33] UK	a photocell—not specified	a small oscillating mirror is supported by two orthogonal wires and scans spirally by means of two electro-magnets	a radio link is described
18.02.1919 French patent	D von Mihaly[34] Hungary	a mosaic of selenium cells (each of which has a coil in series with it) and a common battery	a rotating coil has a voltage induced into it when it scans the bank of coils associated with the selenium cells	two signal wires
10.09.1919 (25.07.1922) US patent	H K Sandell[35] US	a linear array of n selenium cells	A single mirror drum scans the image over the linear array of cells (the width of the scanner is the same as the width of the linear) array of cells	n transmitters and n receivers; transmitter-receiver combination is tuned to a different carrier frequency
18.08.1920 Russian patent	C H Kakourine[36] Russia	a photoelectric cell	a Nipkow disc	a radio link
24.08.1920	H C Egerton[37] US	a photoelectric cell	a single oscillating mirror driven by two orthogonal vibrating motors	a wire or radio link for the combined (vision and synchro-nising) signal

Electrooptical transducer	Receiver scanner	Synchronisation means	Remarks
not specified	two mirror oscillographs, the axes of which correspond to the axes of the mirror drums	the mirror drums generate synchronising signals which drive the receiver's oscillographs	the patent is wholly concerned with synchronising means and states: 'It has been found impracticable to obtain necessary synchronism . . . in view of the enormously high speeds of the mechanisms . . .'
the brightness of the luminous spot on the fluorescent screen of a c.r.t. is modulated by the electron beam	the electron beam of a cold c.r.t.; the beam is magnetically deflected	a common source of line and frame scanning signals for both the transnit and receive	see chapter
a discharge tube and an external electromagnet, excited by the vision signal, which deflects the line of the discharge	the same as the transmitter scanner	line and frame pulses are generated and combined with the vision signal	a form of interlacing is described; this seems to be the first patent to state a method for combining the vision, line and frame signals; no evidence of experimentation.
An arc lamp and a Faraday effect modulator which uses carbon tetrachloride	a single mirror wheel	not described	images of the letters H, T, L and U were reproduced on a screen; no half-tines could be reproduced and the images were faint; The commutator ran at 450 r.p.m.
a mercury vapour lamp fed from the output of a transmer the input to which is the vision signal	the same as the transmitter scanner	not described; the motors driving the scanners at the transmitter and receiver must have constant speeds	a rather impractical scheme; an attempt to ease the problem of synchronisation; no evidence of experimentation
the brightness of the luminous spot on the fluorescent screen of a c.r.t. is modulated by the electron beam	the electron beam of a c.r.t. plus two pairs of deflecting plates	the synchronisation and vision signals modulate a carrier wave; they are separated at the receiver	see chapter; no evidence of experimentation
a bank of lamps, each of which is connected to a coil	a rotating coil connected to the signal wires causes the relays of the bank of lamps to operate	a pendulum escapement driven through a worm gear	see chapter; much experimental work was undertaken by Mihaly
a linear light source and a linear array of *n* small mirrors each of which is deflected by a small electromagnet	a single mirror wheel causes the light reflected from the mirrors to be reflected onto the screen	not described	see chapter; the width of the scanners is the same as the width of the linear arrays; no evidence of experimentation
an electromagnetically operated shutter	a Nipkow disc	not clear from patent	an impractical, naive scheme as described in the patent; no evidence of experimentation
a lamp, plus a screen of varying transparency, and a movable mirror controlled by the vision signals	same as transmitter scanner	synchronising signals generated at the transmitter and sent over the link to the receiver	the picture area could be scanned in groups of scannings, each group consisting of two scannings in succession, the different scanning of a group traversing different paths; no evidence of experimentation

station[21]. Then the broadcasts were stopped and the company's licence withdrawn on the grounds that they might interfere with essential services.

However, the time was now ripe for sound broadcasting—the techniques were available and public demand was growing—and if conditions in the UK did not favour a rapid development programme the same state of affairs did not exist elsewhere in several European countries. By 1920 sound broadcasts were being transmitted from The Hague, Paris and other western European stations: these could be received in the United Kingdom. Radio amateurs, in particular , were vociferous in their demands for a regular service so that they could proceed with their experimental work. Manufacturers of valves and telephony equipment were also keen for broadcasts to occur: many of them had established costly development and production facilities for such devices and apparatuses during the war and wished to have a continuing market for their goods. It is significant that, of the six guaranteeing firms of the future British Broadcasting Company, three were valve makers (GEC, BTH and Metropolitan Vickers).

Negotiations between the British Post Office and the Wireless Society were held in 1921, the Post Office stating its willingness to licence the Society, but not the Marconi Company, for transmissions. Eventually, in January 1922 the Marconi Company was authorised to transmit 30 minutes of telephony each week from its Writtle station[27]. Much discussion took place in 1922 between the Post Office and the major manufacturers of radio apparatus on the formation of a broadcasting company. By 18 October 1922 a scheme was agreed by the manufacturers and on 1 November 1922 the issue of broadcast receiving licences was started. Broadcasting commenced in London on 14 November, in Manchester on 15 and in Birmingham the next day.

Essentially, the birth of the new form of entertainment took place in 1920 because the conditions necessary for its success were opportune at that time. The discovery of the amplifying and oscillating characteristics of valve circuits in 1912 and their subsequent rapid improvement for military purposes provided the foundation for the design of the receivers and transmitters which were needed to create a broadcasting service. Economic considerations acted as a catalyst but, most important, there was a strong demand from a section of the public for sound broadcasting transmissions. The early service proved satisfactory in reception, sets could be bought for a few or many pounds, the signals were capable of being received by a majority of the population and at a cost it could afford; and so broadcasting went from strength to strength.

Undoubtedly the growth of commercial radiotelephony and domestic broadcasting influenced the progress of television. Whereas only a few new schemes for seeing by electricity were put forward in the 1911–1920 period, see Table 7.1, during the next decade television was to become a reality.

References

1 HERTZ, H.: 'Ueber sehr schnell elektrische Schwingungen', *Ann. Phy.*, 1887, **31**, pp. 421–448
2 HERTZ, H.: 'Ueber einen Einglass des ultravioletten Lichtes auf die elektrische Entladung', *Ann. Phys*, 1887, **31**, pp. 983–1000

3 HALLWACHS, W.: 'Ueber den Einfluss des Lichtes auf electrostatisch geladene Körper', *Ann. Phys.*, 1888, **33**, pp. 301–312

4 RIGHI, A.: 'On some electrical phenomena provoked by radiation', *Phil. Mag.*, 1888, **25**, pp. 314–316

5 STOLETOW, M.A.: 'Sur les courants actino-électrique dans L'air raréfié', *Journal de Physique*, 1890, 2nd series, **9**, pp. 468–473

6 ELSTER, J., and GEITEL, H.: 'Ueber die Entladung negative elektrische Körper durch das Sonnen und Tageslicht', *Annalen der Physik und Chemie*, 1889, **38**, (12), pp. 497–514

7 ELSTER, J., and GEITEL, H.: 'Ueber die Verwendung des Natrium amalgames zu licht-elektrischen versuchen', *Annalen der Physik und Chemie*, 1890, **41**,(10), pp. 161–176

8 SMITH-ROSE, R.L.: 'The evolution of the thermionic valve', *J. Inst. Electr. Eng.* 1918, **56**, pp. 253–265

9 DYER, F.L., and MARTIN, T.C.: 'Edison; his life and inventions' (Harper Bros, London, 1910)

10 Ref.8, p. 253

11 FLEMING, J.A.: 'On electrical discharge between electrodes at different temperatures in air and in high vacua', *Proc. Roy. Inst.*, 1890, **47**, pp. 118–126

12 THOMSON, J.J.: 'On the masses of the ions in gases at low pressures', *Phil. Mag.*, 1899, **48**, p. 547

13 WEHNELT, A.: 'On the discharge of negative ions by glowing metallic oxides and allied phenomena', *Phil. Mag.*, 1905, **10**, pp. 80–90

14 FLEMING, J.A.: 'Improvements in instruments for detecting and measuring alternating electric currents'. British patent 24 850, 16 November 1904

15 FLEMING, J.A.: 'Improvements to instruments for detecting electric oscillations'. British patent 13 518, 25 June and 10 December 1908

16 DE FOREST, L.: US patent 879 532, application date 29 January 1907

17 LENARD, P.: 'Erzeugung von Kathodenstrahlen durch ultravioletteslicht', *Ann. Phys.*, 1900, **2**, pp. 359–375

18 DE FOREST, L.: 'The reflection of short Hertzian waves from the ends of parallel wires'. PhD thesis, Yale University, 1899

19 MACLAURIN, W.R.: 'Invention and innovation in the radio industry' (Macmillan, London, 1949) pp. 70–87

20 DE FOREST, L.: 'Improvements in space telegraphy'. British patent 1427, 21 January 1908

21 STURMEY, S.G.: 'The economic development of radio' (Duckworth, London, 1958) Chapter 2, p. 33

22 DE FOREST, L.: 'Method of and apparatus for amplifying and reproducing sounds'. British patent 2059, 24 June 1913

23 Ref. 21, p.34

24 DE FOREST, L.: 'Radio signalling system'. US patent no. 1 507 016, 23 September 1915; and 'Wireless telegraph and telephone system'. US patent 1 507 017, 20 March 1914

25 GUY, R.F.: 'AM and f.m. broadcasting', *Proc. IRE*, May 1962, p.812

26 Ref. 21, p. 137

27 Ref.21, p. 143

28 ROSING, B.L.: 'Improvements in electrical telescopy and the like'. British patent 5259, application date 2 March 1911

29 CAMPBELL SWINTON, A.A.: 'Presidential Address', *J. Roentgen Society*, January 1912, **8**, (30), pp. 1–5

30 HART, S.L.: 'Improvements in apparatus for transmitting pictures of moving objects and the like to a distance electrically'. British patent 15 270, application date 26 September 1914

31 RIGNOUX, G.: 'Dispositif pour la vision a distance', *Comptes Rendus*, 1914, **159**, pp. 301–304

32 VOULGRE, A.D.J.A.: 'Dispositif d'appareils permettant la television et al telephotographie'. French patent 478 361, application date 16 September 1915

33 NICOLSON, A.Mc.L.: 'Television'. US patent 1 470 696, application date 7 December 1917

34 VON MIHALY, D.: 'Synchronisation d'installations fonctionnant des sensibles à la lumière pour la transmission électrique à distance des images'. French patent 546 714, application date in France 25 January 1922, application date in Hungary 18 February 1919

35 SANDELL, H.K.: US patent 1 423 737, application date 10 September 1919; also covered by a British patent on 'Improved process of and apparatus for telegraphically reproducing pictures and the like' issued to the Mills Novelty Company, patent 200 643, application date 16 May 1922

36 KAKOURINE, C.H.: Russian patent 144, application date 18 August 1920

37 EGERTON, H.C.: 'Television system'. US patent 1 605 930, application date 24 August 1920

Part II
The era of low-definition
television 1926 to 1934

Chapter 8
The breakthrough,
J L Baird and television (the 1920s)

Shortly after Guilieglmo Marconi's death in 1937 Professor Sir Ambrose Fleming, who had been a consultant for Marconi's Wireless Telegraph Company, wrote an appreciation of him and stated:

> 'In the first place, he was eminently utilitarian. His predominant inter-est was not in purely scientific knowledge *per se*, but in its practical application for useful purposes. He had a very keen appreciation of the subjects on which it was worthwhile to expend labour in the above respect.
> 'He had enormous perseverance and power of work. He was not discouraged by initial failures or adverse criticisms of his work. He had great powers of influencing others to assist him in the ends he had in view. He had remarkable gifts of invention and ready insight into the causes of failure and means of remedy. He was also of equable temperament and never seemed to give way to impatience or anger. He also owed a good deal to the loyal and efficient work of those who assisted him.'

Fleming's eulogy could also have been applied almost word for word to Baird. The two inventors had much in common. Baird's plans for television were ambitious and extensive as were those of Marconi for marine wireless communications[1]. Both inventors commenced their experiments in private houses and, initially, neither Baird nor Marconi had any substantially novel ideas to put forward for the solution of their problems, Figure 8.1. Moreover, after their early successes, both inventors, with their associates, endeavoured to establish a vigorous policy of commercial expansion. Companies were formed in the United Kingdom and overseas and the two inventors encounted some criticism about their business dealings. Furthermore, both Marconi and Baird displayed, during the formative periods of their organisa-tions, blind spots to progress.

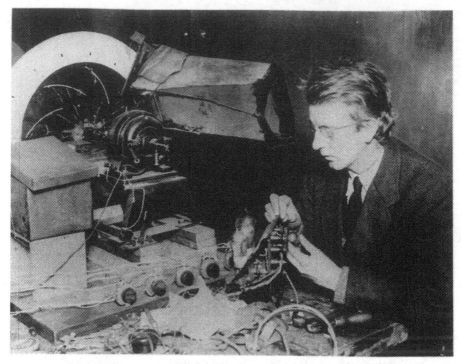

Figure 8.1   Baird working in his lodgings on his early television apparatus. According to Moseley (Baird's staunchest supporter from 1928), the first televisor Baird devised 'had the ingenuity of Heath Robinson and a touch of Robinson Crusoe'. Baird described it as having the saving grace of simplicity

Reproduced by permission of the Glasgow Herald

John Logie Baird was born in 1888, 14 years after the birth of Marconi, in the town of Helensburgh, a small seaside resort situated approximately 22 miles north west of Glasgow[2].

Like the Italian pioneer, Baird was brought up in a comfortable middle-class, professional household. His father was a minister of the local church and an intellectual of some merit[3]. He was 47 years older than John Logie, in an era when any man over 50 had become an elderly, pompous figure. Here, again, a similarity exists between the two inventors, for Guilieglmo's father was 48 years old when the radio experimenter was born.

Baird's acquired interest in science, while a schoolboy, seems to have been self generated for no form of science was taught at the school, Larchfield, in Helensburgh which he attended in his youth. Nevertheless, he engaged in various experiments and projects in his spare time. He installed electric lighting in his home, The Lodge, at a time when such an event could make news in the local press; he constructed a small telephone system so that he could easily contact his friends; he tried to make a flying machine; and he fabricated selenium cells.

This curiosity in science appears to have greatly influenced Baird's post-Larchfield education, for he rejected his father's request to enter the ministry[4] and enrolled at the Royal Technical College, Glasgow, in 1906 to follow a course in electrical engineering. Eight years later Baird was awarded an associateship of the College. An examination of the course curriculum shows that the subject timetables for the second and third years of the mechanical engineering and electrical engineering programmes (following a common first year) were almost identical. The only difference between the two curricular occurred in the third year when students had to choose between taking a laboratory class in either electrical engineering or mechanics[5].

This fact had an important bearing on Baird's work on television, for he had a penchant for designing and inventing devices which had a mechanical basis rather than an electrical foundation. Baird displayed considerable ingenuity and innovativeness in the fields of optics and mechanics and produced many patents on aperture disc, lens disc and mirror drum scanning mechanisms, but only a few on electronic devices or systems. Electronics was not Baird's forte. *Prima facie*, it would seem that neither electric telegraphy nor wireless telegraphy formed part of the diploma programme.

Baird was 26 when he left the Royal Technical College[6]. He tried to enlist in August 1914 and when he was declared unfit for service entered Glasgow University as a BSc student. Possession of the associateship award of the RTC entitled a holder to take the appropriate final year degree examination of the university after a period of six months' attendance. Baird spent an enjoyable session at the university but did not sit the examinations.

Subsequently, he obtained work as an assistant mains engineer, at 30 shillings per week, with the Clyde Valley Electrical Power Company. This job entailed the supervision of the repair of any electrical failure in the Rutherglen area of Glasgow, whatever the weather, day or night.

Throughout his life Baird was subjected to colds, chills and influenza which necessitated lengthy periods of convalescence. His studies at the Royal Technical College 'were continually interrupted by long illnesses' and his numerous absences from his employment as an assistant mains engineer because of illness militated against any promotion in the company. Because of this he disliked the job and eventually resigned.

Actually, Baird's departure from the CVEP Company was hastened by his entreprenurial exploits during the period 1917–19. In 1917 boot polish was difficult to obtain. Baird seized the opportunity to enliven his existence by registering a company and employing girls to fill cardboard boxes with his own boot polish[3]. This venture possibly escaped the notice of his employers, but the next did not.

Baird had always suffered, and always did suffer, from cold feet, and on the principle of capitalising on one's deficiencies, he devised an undersock—consisting of an ordinary sock sprinkled with borax. He arranged its commercial exploitation with such a degree of business acumen and skill that, when he sold the enterprise twelve months later, he had made roughly £1600—a sum of money which would have taken him 12 years to earn as an engineer with the CVEPC. He employed the first

FOOTBALL BY WIRELÈSS.

A Hastings resident is experimenting with a wireless invention to enable us to see sporting events at our own firesides. What a boon this might be to football referees!

Figure 8.2   Cartoon published in the Hastings and St Leonards Observer, 26th January 1924.
It refers to Baird's early television experiments which he carried out in Hastings from
1923

women bearers of sandwich boards seen in Glasgow and also constructed a large
model of a tank—plastered with posters about the efficacy of the Baird undersock in
providing comfort for soldiers' feet—which was trundled about the streets of the city[7].

Solid scent was added as a sideline to Baird's main business. However, all of this
was too much for the managers of the CVEPC, and their erring engineer was given
an ultimatum; either he had to give up his business interests or he had to leave the
company. Baird chose the latter alternative[3].

By 1919 the future television pioneer appeared to be on the threshold of a lucrative commercial life. Unfortunately, continuous good health was not a blessing
which had been bestowed on Baird and, during the winter of 1919–20, he suffered
a cold which entailed an absence of six weeks from his venture. He decided to sell
out and, following glowing accounts from a friend of the possibilities which seemed
to exist in the Caribbean, travelled to the West Indies. His stay there was short and
unprofitable and he returned to London in September 1920[3].

Figure 8.3   *Baird with the television apparatus which he presented to the Science Museum. Note the double spiral lens disc. The inventor is holding 'Stooky Bill', a ventriloquist's model, which he used in his early experiements*

Reproduced by permission of the Glasgow Herald

Again he set about establishing a trading business and bought two tons of Australian honey, at a giveaway price, and did a brisk trade selling it in 28 lb tins. Baird added the sale of fertilisers and coir-fibre dust to his interests and his enterprise prospered. Another illness caused him to remain in bed for several weeks, 'the business meanwhile going to bits', and when his cold did not improve he sold his undertaking[3].

Later in 1922, on returning to good health, Baird purchased two tons of resin soap and so once more another business gamble was started. Baird's Speedy Cleaner was soon being sold to hotels, boarding houses, ship's chandlers and street barrow boys. The concern flourished so much that Baird imported large quantities of soap from France and Belgium and formed a limited liability company, with £2000 authorised capital, with two associates. But again illness compelled him to sell out and convalesce. He went to Hastings where a friend from childhood lived[3].

Although there is some slight evidence that Baird had an interest in television in 1913, his life's work effectively commenced during the winter of 1922–23 which he spent in Hastings. Television development was not initially in Baird's mind when he settled there, for in his autobiographical notes he related how, when his health improved, he attempted to invent a pair of boots having pneumatic soles, and also to produce a glass safety razor. It is likely that his attraction to television was stirred by his reading of an article[8], 'A development in the problems of television', by N Langer in the *Wireless World and Radio Review* issue of 11 November 1922. Langer's paper was optimistic in tone and endeavoured to indicate the lines along which a solution to the television problem could be found.

Whatever the source of his inspiration, Baird realised the difficult nature of the task: 'The only ominous cloud on the horizon,' he wrote[9], 'was that, in spite of the apparent simplicity of the task, no one had produced television.'

It is interesting to recall that Marconi was led to pursue his life's work when he read, while holidaying at Biellese in the Italian Alps, an obituary describing Hertz's experiments. He subsequently said that, as a result of the article, the idea of wireless telegraphy using Hertzian waves suddenly came to him: 'The idea obsessed me more and more and, in those mountains of Biellese, I worked it out in imagination.'

Like Baird, Marconi considered the solution to his posed problem to be essentially simple and seems to have been surprised that it had not been solved by others. 'My chief trouble,' he noted, 'was that the idea was so elementary, so simple in logic, that it seemed difficult for me to believe that no one else had thought of putting it into practice. Surely, I argued, there must be much more mature scientists than myself who had followed the same line of thought and arrived at an almost similar conclusion.'[1]

Both Marconi and Baird started their experiments in private residences. Marconi had two large rooms at the top of the Villa Grifone set aside for him by his mother, and Baird made use of various rooms which he rented when staying in Hastings and elsewhere.

Neither Baird nor Marconi had any particularly original suggestions to put forward at the outset of their investigations and both experimenters modelled their schemes on the ideas of others. Marconi's earliest transmitter was still based on a coil and spark gap (as used by Hertz, although the design of the spark gap had been slightly changed to incorporate an improvement due to Righi). Baird's earliest vision apparatuses were based on proposals which had been advanced by Nipkow and others in the late 19th and early 20th centuries.

Both Baird and Marconi commenced their investigations at opportune times for, in addition to the ideas which had been put forward, the technology existed for narrowband television in the one case and narrowband wireless communication in the other.

Consequently, when Baird decided to apply himself to the problem of a practical television scheme, the solution seemed to him to be comparatively simple. Two optical exploring devices rotating in synchronism, a light sensitive cell and a controlled varying light source capable of rapid variations in light flux were all that were required, and these appeared to be already, to use a Patent Office term, known to the art.

Baird's principal contemporaries in this challenge were C F Jenkins of the US and D von Mihaly of Hungary. Other inventors were patenting their notions on television at this time (1923), but only Jenkins, Mihaly, Baird and a few others were pursuing a practical study of the problem based on the utilisation of mechanical scanners.

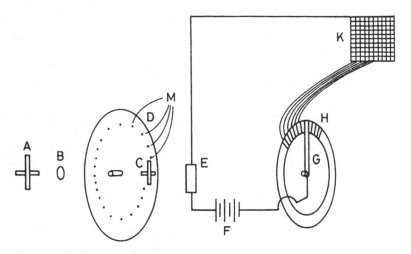

Figure 8.4   *Baird's first patent was filed on 26 July 1923. It describes a method of displaying a televised image using a screen of small lamps. The concept formed the basis for Baird's large-screen reproduction of images of the Derby, on 28 July 1930, at the Coliseum Theatre, London*

Source: British patent 222604, 26 July 1923

The approaches of the three inventors to their tasks were individualistic. Jenkins was a well known inventor and a person of considerable means. He had produced important inventions in the field of cinematography and was able to design and manufacture apparatus of some complexity. His early rotary scanners consisted either of specially ground prismatic discs or costly lensed discs. Mihaly, an experienced patent expert and engineer, used an oscillating mirror scanner, together with tuning forks and phonic motors for synchronising purposes. Baird's approach necessarily had to be entirely different to those of his rivals. He had little money, no laboratory facilities for the construction and repair of equipment, no access to specialist expertise and no experience of research and development work in electrical engineering. He had to carry out his experiments in the unsuitable conditions of private lodgings. Still, undaunted by the formidable difficulties which faced him, he started his investigation by collecting bits and pieces of scrap material and assembling them into a system. The constraints imposed by his financial state severely limited the type of work which he could carry out, but nevertheless he pursued his objective with dogged determination, ingenuity and resourcefulness. A Nipkow disc could be made from a cardboard hat box, the apertures could be formed using a knitting needle, electric motors could be bought cheaply from scrap metal merchants and bull's-eye lenses could be obtained at low cost from a cycle shop.

The exact sequence of experiments and investigations which Baird undertook in 1923 is not known, but an associate has written that during his contact with Baird the inventor always used a Nipkow disc scanner in one of its many different forms, either at the transmitting end or at the receiving end of the television link. Baird utilised disc scanners for many years, and even when the London station was established in 1936 one of the essential items of equipment provided by Baird Television Ltd was a spotlight scanner of the Nipkow type.

Nipkow's invention had an inherent limitation which precluded its use for a purpose which Baird held to be of some importance, namely, cinema television[10]. Figure 5.3 illustrates the principle of operation of a Nipkow disc scanner. In its simplest form the disc consists of a flat, thin, circular piece of metal, or other suitable material, pierced by a number of small apertures, equal in number to the number of scanning lines required, and arranged to lie along a single turn spiral. Given a disc of diameter D and a number n of scanning lines, the circumferential distance between any two apertures is D/nk where k is the aspect ratio (assuming horizontal line scanning). Thus, for a disc 1.0 m in diameter the size of the aperture scanned is only approximately 8.1 cm by 6.4 cm, for an aspect ratio of 5:4 and a sequential scan of 30 lines. Hence, although such a disc could be employed to analyse the image of an object or scene to be televised it was not possible to contemplate its use as an image synthesiser in a large screen system, bearing in mind the controllable light sources then available.

Baird's solution, which he advanced in his first patent[11], was based on an idea suggested by Ayrton and Perry, by Redmond, by Middleton and by Carey in 1880 and by Selecq in 1881—the idea of a mosaic of reproducing elements. The patent, titled 'A system of transmitting views, portraits and scenes by telegraphy or wireless telegraphy', describes the use of an analysing Nipkow disc and a receiver mosaic of incandescent lamps. Baird envisaged the disc to be provided with a series of 18 small holes, each 3 mm in diameter and each circumferentially separated from the next by 25 mm. At the receiver a rotating brush commutator was to be employed to switch the received signal to the appropriate lamps of the mosaic, whence the varying brightness of the lamps would reproduce the image, and persistence of vision would cause the whole reproduced image to appear simultaneously on the screen of lamps.

Baird actually tried to implement this scheme in 1923, for the first known published report of his work, which was given in *Chambers Journal*[12] (November 1923), describes some of the characteristics of the apparatus. The aperture disc scanner was 50 cm in diameter and the image frame measured 5 cm square. A rotation speed of 20 revolutions per second was adopted for the frame scanning rate and at the receiver the signals were taken to the fulcrum of an arm, with a copper brush at the end, which rotated around a ring of tiny contacts. These contacts were connected in sequence to a number of lamps of only 3 mm in diameter mounted in a picture frame. No indication was given in the report of the number of lamps, or holes employed, nor was any reference made to the performance of the system. The reporter of the article seemed to have an optimist's outlook for he/she wrote: '. . . we may shortly be able to sit at home in comfort and watch a thrilling run at an international football match, or the finish of the Derby.'

Large-screen television was later, in July 1930, publicly demonstrated by Baird[13], and in 1927 the American Telephone and Telegraph Company[14] displayed the visual equivalent of a public address system. The essential difference between the two equipments lay in the design of the receiver light source: AT&T utilised a special

Figure 8.5   *Baird demonstrating his apparatus, in his Hastings workroom, to W Le Queux (left) and C Frowd (right). The date is early 1924*

Reproduced by permission of R M Herbert

type of multielectrode glow discharge lamp, whereas Baird used easily obtainable electric lamps.

By using a screen of lamps, the individual lamps may have a considerable light decay time constant, whereas with a single source of illumination the source must vary instantaneously with changes of applied voltage. The 1923 scheme would have needed n times n lamps for a square picture scanned by n lines, but by mounting the lamps spirally on a rotating disc, Baird was able to reduce the number of lamps needed for viewing to n. Each lamp was positioned on the receiver disc in the same place as the corresponding aperture in the transmitter disc and the lamps connected to a commutator at the centre of the disc. This arrangement was described by Baird in a paper published in May 1924[15], but as with most of Baird's early writings there was an absence of practical details in the account. Nevertheless, the equipment or a similar version to it was seen by a resident of Hastings who described the crude images it reproduced.

Baird was at this time in urgent need of money. He therefore gave a demonstration of his apparatus to the press and managed to get a mention in the *Daily News* (15 January 1924). A friend of Baird's father saw the *Daily News* paragraph, and mentioned it to him, with the consequence that Baird received a much required present of £50[2].

Another person who probably saw the few lines in this paper was Mr Odhams of Odhams Press. He and Baird met on several occasions and, although no financial support was obtained from Odhams, Baird did receive help in another way. Both W Surrey Dane, subsequently a joint manager at Odhams, and John Dunker, editorial chief, were interested in Baird and his work and gave him much needed publicity and encouragement. A few months later the editor of *Wireless World and Radio Review* was able to mention: 'A good deal of popular interest has been aroused by the experiments in television recently conducted by Mr Baird.'[2]

The concept of television had a popular appeal and in February 1924 the *Radio Times* carried an article[16] headed: 'Seeing the world from an armchair. When television is an accomplished fact'. After expounding certain aspects of wireless broadcasting, such as international broadcasting, the transmission of wireless waves, fading and so forth, the writer posed the question 'What will be the next stage?' He went on: 'The answer seems to be television. We have encircled the earth with our music and speech, will the next year enable us to see around the earth with our eyes? Eminent scientists have progressed far along the road at the end of which will be discovered the secret of television, or simply, seeing by wireless.'

The writer, whose name is not known, depicted Fournier D'Albe's vision of the future:

> 'It is highly probable, he [d'Albe] is reported to have said, that we shall be able to sit in, say, the Albert Hall and actually watch the Derby or the 'Varsity Boat Race, or a Naval Review, or a prize fight in America or, for that matter, a battle. I mean, watch a moving picture of any of these things on a screen, at the moment they are happening . . . As we know now that wireless waves can be relayed also indefinitely, I see no reason why in ten years time we should not be able to see what is happening on the other side of the globe. It is only a matter of effort in research, and if the public interest is there the effort will be there.'

Baird's future colleague Moseley and his future business partner Hutchinson were to excite this interest in large measure. The writer ended his article by stating that J L Baird had succeeded in transmitting the outlines of objects and that C F Jenkins had reproduced his moving hand on the screen. 'These experiments indicate the miraculous linking up of the whole earth by wireless in the not too distant future.'

One well known person who lived in Hastings at this time was the novelist William Le Queux. He was very interested in radiocommunication and had carried out some radio experiments in Switzerland in 1924 with Dr Petit Pierre and Mr Max Amstutz[17]. Le Queux's fame and interests resulted in him being elected the first president of the Hastings Radio Society in 1924[18]. The inaugural lecture to the newly formed society was given by Baird on 28 April 1924 when he talked on 'Television'. The report[19] of the lecture in the local press indicated that Baird was still using selenium cells.

Le Queux attended various demonstrations and he was eager to help Baird, but all his money was tied up in investments in Switzerland. He did, however, write an article[20] for the *Radio Times*, in April 1924, with the title 'Television—a fact'. The article is important as it gives some inkling of the progress which Baird was making at this stage of his work. After noting the successful transmission of outline images by Jenkins and Baird, Le Queux wrote:

'In both cases, however, the receiving and transmitting machines were mechanically coupled. Mr Baird has now succeeded in overcoming the great synchronising difficulty and has successfully transmitted images between two totally disconnected machines, synchronism being accomplished with perfect accuracy by comparatively simple and inexpensive apparatus.'

Baird was still experimenting with small lamps in his receiver, but there is no doubt that these could produce crude images:

'My fingers moved up and down in front of the transmitting lens were clearly seen moving up and down on the receiving disc, and so forth. It remains now to transmit detailed images and a machine[15] to do this has already been designed.'

In his work Baird, like some other experimenters, used a selenium cell. As noted previously, the time lag of this type of cell had proved a source of constant concern to distant vision workers. Various attempts had been made to compensate for the lag and Baird, in his early experiments, gave the subject much thought. His patent[21] of 12 March 1924 specifies a method which he was to employ for about a year in order to overcome the difficulty.

Baird outlined the application of this invention in his first paper as follows:

'In the transmitter an image of the object to be transmitted is focused on a disc rotating at a speed of approximately 200 r.p.m. The disc is perforated by a series of holes staggered around the circumference. In the experimental apparatus described four sets of five holes were thus arranged: in proximity to this disc revolved a serrated disc at some 2000 r.p.m., and on the other side of this and in line with the focused image of the object to be transmitted there was a single selenium cell connected to a valve amplifier.'

Baird's rotating serrated disc was a light chopper, Figure 8.6, a device which is still used in certain applications. It allows the picture information to be sent by means of a modulation of a carrier signal and its utilisation eased the problem of low frequency amplification.

Baird wrote about his method as follows:

'The use of the rapidly revolving serrated disc overcomes [the time lag of the selenium cell], as the actual resistance of the cell at any instant is not of consequence, it is the pulsations which are transmitted.'

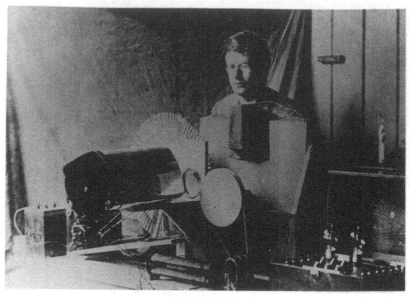

Figure 8.6   J L Baird with his apparatus. Note the serrated disc which was utilised as a light chopper

Reproduced by permission of R M Herbert

Although Baird did not patent this important idea until 12 March 1924, a photograph published in the 19 January 1924 issue of the *Hastings and St Leonards Observer* clearly shows the serrated disc incorporated in his apparatus.

Baird gave much thought in 1924 and 1925 to the problem of improving the unsatisfactory performance of selenium cells. He was at this time faced with a dilemma. The photoconductive selenium cells which he was utilising were moderately sensitive, easy to manufacture but suffered from the serious drawback mentioned above and thereby gave rise to blurred images. On the other hand, photoemissive cells had a much superior transient response but their sensitivity was extremely low, and most probably below the capacity of Baird's earliest amplifiers. Baird chose to persevere with the use of selenium cells and noticed that when a cell was exposed to a change of light flux the current would change rapidly at first and then more slowly. By employing a rate of change of current signal in addition to the cell's current he tried to minimise the effect of the lag.

This solution to the problem of reducing the inertia effect of selenium cells was put forward as a claim in Baird's patent[22] of 21 October 1925.

Prior to this date many investigations had been undertaken by research workers on the dependence of the inertia effect of such cells on the duration and strength of the incident light flux, on the heat treatment of the selenium, on the thickness of the photoconductive layer, on the previous illumination, on the colour of the incident light, on the purity of the selenium and the nature of the electrodes, on the voltage applied to the cell and on the cell's temperature.

Of particular concern to television workers was the behaviour of a selenium cell exposed to continually changing radiation. This point had been studied by Nisco, Glatzel, Romanese, Bellati and Majoranda[23].

How much of this work was known to Baird in 1923–1924 is a matter for conjecture, but as he was living in the nonuniversity town of Hastings at that time it seems unlikely that he had easy access to the learned journals in which the findings of the above workers were published.

Numerous attempts had been made by experimenters to minimise the undesirable property of the selenium cell during the development of picture telegraph systems and of these those Szczepanik[24] (1895), Korn[25] (1906), Zavada[26] (1911) and Cox[27] (1921) were of some importance.

Baird's solution to the problem had a characteristic simplicity and consisted in adding to the output current of the selenium cell a current proportional to the first derivative of the output current. Figure 8.7 taken from the patent illustrates the result of adding these two currents together. In this patent the inventor described several circuit arrangements, for accomplishing the desired effect, using passive circuit elements.

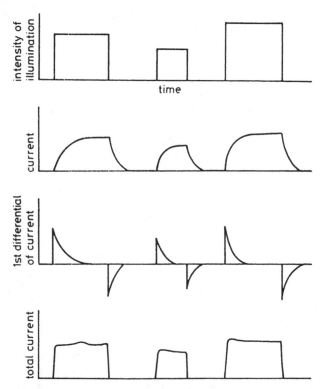

Figure 8.7   Baird's patent 270 222 describes a method for improving the resolution of television systems based on the use of selenium cells. If the first derivative of the cell's current is added to the cell's current an improvement of image quality is realised

Source: British patent 270 222, 21 October 1925

A similar solution was employed by Dr F Gray, of the American Telephone and Telegraph Company, and was explained by him in a memorandum[28], dated 27 February 1926, on 'Correcting for the time lag in photosensitive cells'. Gray wrote:

'We may consider the light intensity as the integral of a series of sine and cosine components. For good transmission, the cell must transform these components into alternating electric current components of the same relative amplitude and phase relation.

'Cells that exhibit time lag do not meet this requirement. For certain cells, however, the error in phase relation of the transformed components is negligible over the frequency region concerned in picture transmission. Such cells behave as if they transformed low frequency components into current much more effectively than they transform the high frequency components. It is therefore possible to approximately correct for the time lag of such cells by introducing in the circuit an element that will equalise the components in the proper manner. This can be done either by an element that will attenuate low frequencies more than high frequencies, or by an element that will amplify high frequencies more than low frequencies.'

Gray used a simple CR circuit in his amplifier to enhance the gain at high frequencies.

One of the major problems which had to be solved in any system of television by wireless using Nipkow discs, Weiller drums or other mechanical scanners at the transmitting and receiving stations was the need to maintain the two scanners in synchronism. Baird worked on this problem in 1923–1924 and on 17 March 1924 applied for a patent[29] with the title 'A system of transmitting views, portraits and scenes by telegraphy or wireless telegraphy'. Baird devised a suitable scheme in an idiosyncratic way; his solution was the simplest, most robust and probably the cheapest which could be used but—it worked.

He wrote:

'To obtain isochronism an alternating current generator is coupled to the shaft of the transmitter and current from it controls the speed of a synchronous motor driving the receiving machine. To obtain synchronism the driving mechanism is rotated about the spindle of the receiver until the image comes correctly into view.'

Although this method enabled the transmitting and receiving discs to be mechanically decoupled, the solution, of necessity, involved the transmission of a low frequency alternating current in addition to the signal currents which were of a higher frequency. Baird was not at this time working on the problem of transmitting both the signal and synchronising currents as a modulation on a carrier wave. This had to wait several years for a solution. There is no doubt that Baird's method worked, as Le Queux was able to confirm in his *Radio Times* article.

The above patent further illustrates Baird's gift for reducing a problem to its simplest statement and then finding an equally simple solution. Not for Baird were the complicated, fragile and expensive mirror scanning systems of Szczepanik and Mihaly or the more robust but still costly mirror drums of Weiller. Baird was short of finance and had to find answers to his difficulties using almost literally the inventors' beloved sealing wax, string and glue. He could not have engaged in research work on cathode-ray oscilloscope systems but had to make do with the meagre and crude facilities at his disposal. It is to his credit that he achieved any results at all when faced with such daunting hurdles.

From Hastings Baird moved in 1924 to a small attic at 22 Frith Street, London, found for him by a Mr W E L Day who had a financial stake in Baird's television activities. Here he continued his researches, which he narrated in a paper[30] published in a January issue of the *Wireless World and Radio Review* for 1925.

Baird was at this time attempting to transmit an image of an actual object, not by having a source of light behind it, but by reflecting light from it. This was a problem of quite a different order of difficulty to the transmission of shadowgraphs, as Baird explained:

> 'With shadowgraphs the light-sensitive cell is only called upon to distinguish between total darkness on the one hand and the full power of the source of light, possibly several thousand candle power, on the other.
>
> 'In the transmission, however, of actual objects even where only black and white are concerned, the cell has to distinguish between darkness and the very small light, usually indeed only a small fraction of a candle power, reflective from the white part of the object. The apparatus has therefore to be capable of detecting changes of light, probably at least a thousand times less in intensity than when shadowgraphs are being transmitted.'

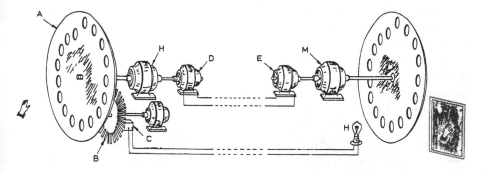

Figure 8.8   *Baird's television system, as described by him in his paper published in the Wireless World and Radio Review (January 1925). At this time Baird was using single spiral lens discs for scanning*

Figure 8.8 shows the disposition of Baird's apparatus. Regrettably, Baird gave no clue in his paper as to the nature of either his receiver lamp or the light-sensitive cell except to mention that the latter was neither a photoemissive nor a selenium, but 'a colloidal [fluid] cell of my own invention . . .' He did mention the method of synchronisation, the same as quoted in his patent of 17 March 1924.

Baird was very modest in reporting the success he had achieved:

'The letter H, for example, can be clearly transmitted, but the hand, moved in front of the transmitter, is reproduced only as a blurred outline. A face is exceptionally difficult to send with the experimental apparatus but, with careful focussing, a white oval, with dark patches for the eyes and mouth, appears at the receiving end and the mouth can be clearly seen opening and closing.'

The sequence and detail of the continuous experiments that Baird pursued during the 1923–1926 period are not on record, for Baird kept no regular notes[31]. He did not keep a laboratory notebook and the few articles which he published were usually of an elementary character. J D Percy, who later worked with Baird, has said:

'. . . it is true, I think, to say, that his mind worked too quickly for the satisfactory recording of his myriads of ideas. Explanations to those of us who worked with him was amplified only by rough sketches on the backs of old envelopes, by scrawling diagrams on walls (the Long Acre Laboratories were literally covered in these) or even occasionally on the tablecloths of restaurants and inns.'

Apart from the casual press notices in the newspapers the general public first became aware of television in April 1925. Selfridge's had been on the look out for an attraction for their birthday week and George Selfridge had visited Baird. The consequence was that Baird was offered £20 a week to give each day three demonstrations of seeing by electricity to the public in Selfridge's store. Baird, of course, was not so concerned with the publicity value of this enterprise; rather it was the case that he could not refuse the weekly cheque, without which it would have been difficult for him to have carried on[32].

A portrayal of the apparatus which Baird used was given in *Nature* (4 April 1925)[33]. Baird was now utilising a neon lamp in his receiver rather than the arrangement outlined in his 1924 article. Otherwise, his television apparatus was similar to his earlier set up which he mentioned in the *Wireless World and Radio Review*.

The crudeness of the images hinted at in the handout which the public were given in Selfridge's was confirmed in the report printed in *Nature*: 'Mr Baird has overcome many practical difficulties, but we are afraid that there are many more to be surmounted before ideal television is accomplished.'

Baird had few competitors in 1924–1925 in the United Kingdom, and hence, whenever television was mentioned in the press, Baird's name tended to be coupled with the report. He was the only experimenter who was able to give demonstrations of equipment and to show the transmission of crude outlines. However, during the month that Baird gave his Selfridge shows, Dr Fournier d'Albe, who had made a name for himself by his invention of the optophone—an instrument which allowed blind persons to read—was giving a private demonstration of his television apparatus at his laboratory at Kingston-upon-Thames[34]. He patented his ideas in January 1924 in a patent[35] titled: 'Telegraphic transmission of picture and images' and a few months later (October) adapted the equipment to record and reproduce sounds[36]. D'Albe's method was doomed to failure. He experimented with a non-

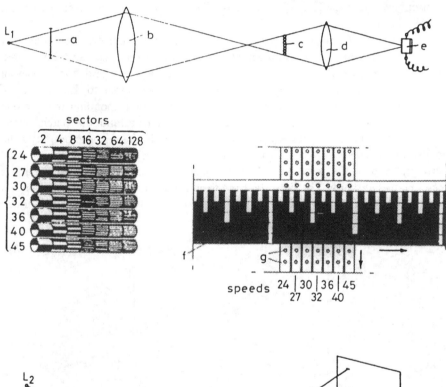

Figure 8.9   Diagrammatic representation of Dr Fournier d'Albe's television scheme, patented in January 1924

Source: British patent 233746, 15 January 1924

scanning system for transmitting information about the luminosities of the elements of a picture. His apparatus belongs to the same group as those of Carey and Lux.

The general arrangement of d'Albe's original method is shown in Figure 8.9. An essential feature was the inclusion of a means for breaking up the light. This consisted of seven transparent cylinders each divided into seven sections which were split up by opaque parts as indicated by black lines. Each section had twice as many subsections as the preceeding section and the cylinders rotated at different speeds. The effect of this was to produce alternate 'brightenings and darkenings in the light transmitted across it'. The light was then collected by a lens and projected onto a selenium cell.

'The medley of electrical impulses produced in the selenium' was subsequently 'converted into sound by any of the methods usual for this purpose' and transmitted to the receiver. Here the sound waves excited a loudspeaker and 'the medley of sounds' was analysed by a set of resonators equal in number to the number of patches in the original picture or image, each resonator responding to only one pitch. The resonators were so constructed as to produce a luminous patch on the screen when a note of their own pitch was contained in the acoustic output of the loudspeaker. This was achieved by having reeds, to which were attached small mirrors, fixed to the resonators so that when the resonators sounded the reeds vibrated in sympathy. The light reflected from the reeds was then directed onto a screen in such a way as to correspond to the original patches of the picture transmitted, Figure 8.10.

Figure 8.10   Dr Fournier d'Albe and his television receiver. The amplified sound transmitted acts on a group of acoustic resonators, each of which produces a patch of reflected light on a ground glass screen when its note is produced by the transmitter

Reproduced by permission of the BBC

The apparatus which d'Albe exhibited on 18 April 1925 was a modification of this scheme. The seven cylinders were now replaced by a revolving siren disc provided with thirty concentric holes. An image of the disc, onto which an image of the object was formed, was projected onto a transmitting screen studded with thirty small selenium tablets, arranged so that each tablet was exposed to a different audio frequency. The tablets were connected in parallel to the input of a two-valve amplifier, and the sound produced in a loudspeaker was allowed to act upon thirty compound resonators.

It was claimed that as the response occurred within a twentieth of a second the apparatus transmitted some six hundred signals per second. D'Albe's demonstration was reported in *Nature* but no indication was given in the account of the quality of the reproduction. The writer, though, did state that 'the complete transmission of an object such as a changing face requires at least ten thousand signals per second' and that therefore 'there is still a considerable gap to be filled'.

D'Albe hoped to do this by increasing the number of resonators and their selectivity or, in the last resort, by transmitting over more than one wire or on more than one wavelength.

Shortly after the demonstrations at Selfridge's store the first of the companies associated with Baird was formed. This was called Television Ltd, and was registered on 11 June 1925.

The object[37] of the company was to 'purchase or otherwise acquire from any person or persons lawfully entitled to dispose thereof the whole or any part of the right, title and interest in, or appertaining to the inventions relating to "A system of transmitting views, portraits and scenes by telegraphy or wireless telegraphy" comprised in and covered by the registered patents 222 604 and 230 576; also the provisional specifications 4800, 6363 and 6774 of 1924 and 48 and 911 of 1925 . . .'

Baird and Day were the subscribers of 20 founders' shares and the nominal capital of the company was £3000. The capital was made up of 2900 ordinary shares of £1 each and 2000 founders shares of one shilling each. The registered office of Television Ltd was initially at 22 Frith Street, W1, and the first directors were Baird and Day. Later, the company moved to Motorgraph House, Upper St Martins Lane, WC (15 February 1926) and then to 133 Long Acre, WC (9 February 1928).

Following the establishment of Television Ltd, Baird and Day entered into an agreement with the firm (on 12 June 1925), the pertinent points of which were:

1   'The Vendors shall sell and the Company shall purchase the whole of the Vendors' right, title and interest in or benefit of the Patents and Inventions . . .
2   'The consideration for the said sale shall be the sum of £100 which shall be paid and satisfied by the allotment to the Vendors or their respective nominees of 2000 Founders shares of 1s. each in the capital of the company—(1000 shares to J L Baird and 1000 shares to W E L Day—and £500 to be paid by the company after £15000 in shares had been sold and paid for at which time the directors, J L Baird and W E L Day, may draw upon the company for all or any part of this sum (£500) as they mutually agree.'

Figure 8.11   Captain O G Hutchinson, Mr J L Baird and Mr S A Moseley in Berlin when the German company, Fernseh AG, was formed

Reproduced by permission of Radio Rentals Ltd

W E L Day, a successful person in the wireless and cinema business, had bought in 1923 (following an advertisement placed in *The Times* by Baird) a one third interest in Baird's invention for £200.

The precise sequence of experiments which Baird carried out during the months after the Selfridge demonstrations is not known, but on 2 October 1925 he noticed that, when he viewed the head of the dummy, which he used as an object, in his receiver, there was light and shade; he had achieved crude television by reflected light with tone graduation.

This success posed a dilemma for Baird. He urgently needed publicity and funds, but feared that there was a danger of his system of television being exploited and developed by powerful firms. Three months were to elapse before Baird had the courage to demonstrate publicly his results.

Fortunately, during this period of indecision, Baird met, by chance, Captain O G Hutchinson whom he had known during his college days and also when he was selling soap.

Hutchinson became Baird's business partner towards the end of 1925 and together they embarked upon a programme of expansion which was to cause Baird and his work to become alienated from certain sections of the scientific community—principally the British Broadcasting Corporation—because of the publicity methods adopted by Hutchinson.

Hutchinson's first task was to obtain some much needed financial assistance for Baird's efforts. This he did by reorganising the nominal capital of Television Ltd,

Figure 8.12   The Baird television apparatus, which is in the Science Museum, London

Reproduced by permission of the Science Museum

and persuading various persons and bodies to take up shares in the company. Day's interest in Television Ltd was purchased by Hutchinson, and Day resigned his directorship on 16 December 1925[37].

Altogether, 43 persons had a financial stake in Television Ltd less than one year after Hutchinson became Baird's business partner.

Larger premises were acquired and in February 1926 Baird moved his apparatus from the small room he rented at 22 Frith Street to Motograph House in Upper St Martins Lane.

Prior to this transfer Baird and his friends had arrived at a solution to the dilemma which confronted them. A compromise was adopted—only selected individuals would be given a demonstration of television. Invitations were sent out to members of the Royal Institution, and to *The Times* to represent the press, '. . . these would give dignity and importance to the occasion'[2].

The chosen date for the first public demonstration of television was 26 January 1926. *The Times* report for 28 January was the only press statement[38] obtained first hand of this historic event.

'Members of the Royal Institution and other visitors to a laboratory in an upper room in Frith Street, Soho, on Tuesday saw a demonstration of apparatus invented by Mr J L Baird . . .

'For the purpose of the demonstration the head of a ventriloquist's doll was manipulated as the image to be transmitted, though the human face was also reproduced. First on a receiver in the same room as the transmitter, and then on a portable receiver in another room, the visitors were shown recognisable reception of the movements of the dummy head and of a person speaking. The image as transmitted was faint and blurred, but substantiated a claim that through the televisor, as Mr Baird has named his apparatus, it is possible to transmit and reproduce instantly the details of movement, and such things as the play of expression on the face.'

Baird's achievement may be weighed by considering the work of the British Admiralty Research Laboratory (ARL) on television[39]. The objective of the Admiralty in conducting experiments in this field was 'for spotting at sea with the use of aeroplanes'. A university-trained research scientist and others commenced the investigation in 1923. In January 1925, Dr C V Drysdale, the superintendent of the laboratories, described the problem as difficult but felt that it could be solved with 'money and staff'. Approximately 17 months later he had modified his opinion and referred to the extreme difficulty of finding a practical solution. An inspection of ARL's television equipment, which included a photoelectric cell made by the National Physical Laboratory (NPL), was undertaken by two representatives of the Air Ministry on 27 May 1926, and during this an image of an object consisting of a grid of three bars of cardboard, each about 6.35 mm in width and 6.35 mm apart, was transmitted. Although transmitted light was employed the object 'could just be recognised at the receiving end, but the reproduction was very crude'.

News of Baird's success travelled fast, for two days after the demonstration to members of the Royal Institution, the Press Photonachtrichten Dienst of Berlin wrote to the British General Post Office and requested details of the Baird system[40].

The prospect which faced Baird in 1926 was much more daunting than that which faced Marconi in 1895–1896. Marconi had the advantage that his invention had an immediate application in military and naval operations and, when his demonstrations before service officers proved successful, his future seemed assured. Additionally, Marconi obtained valuable support for a short but critical period from W Preece, the engineer in chief of the British Post Office, who had himself been experimenting on signalling through space without wires.

On the other hand, Baird's invention had no immediate application to warfare or safety, and he received no patronage from the one body that could assist him, the BBC.

And so, in order that funds could be obtained from the general public, an indication had to be given that investment in the new form of broadcasting was worthwhile.

Hutchinson was keen to start broadcasting television, notwithstanding the faint and blurred image, and applied on 4 January 1926 to the Postmaster General (PMG) on behalf of Television Ltd for a licence to transmit from London, Glasgow, Manchester and Belfast[41].

After much anxious delay and correspondence the PMG gave his permission for the company to install wireless experimental transmitting stations at Motograph

House, and at the Green Gables, subject to the power and wavelength being fixed at 250 W maximum and 200 m respectively[42]. The licences for 2TV and 2TW were dated 5 August 1926. They were granted on condition that transmission would not take place during broadcasting hours of the London station.

Meanwhile, through the long period of procrastination, Baird had continued his experiments and demonstrations, including one which was reported[43] in *Nature* by Dr Alexander Russell, FRS, the principal of Faraday House. He was agreeably surprised by the considerable progress made in solving the television problem and stated:

> 'We saw the transmission by television of living human faces, the proper gradation of light and shade, and all movements of the head, of the lips and mouth and of a cigarette and its smoke faithfully portrayed on a screen in the theatre, the transmitter being in a room in the top of the building. Naturally the results are far from perfect. The image cannot be compared with that produced by a good kinematograph film. The likeness, however, was unmistakable and all the motions are reproduced with absolute fidelity . . . This is the first time we have seen real television and, so far as we know, Mr Baird is the first to have accomplished this marvellous feat.'

Acknowledgement of Baird's achievement came from a number of sources. The *New York Times*[44], 6 March 1926, gave a whole page to the subject and said: 'No one but this Scottish Minister's son has ever transmitted and received a recognisable image with its gradation of light and shade . . . Baird was the first to achieve television'. Later, in September 1926, the *Radio News* (of the US) sent a reporter to investigate Baird's claims: 'Mr Baird has definitely and indisputably given a demonstration of real television . . . It is the first time in history that this has been done in any part of the world.'[45]

In his early experiments Baird had great difficulty in reducing the intensity of the light used to illuminate his subjects without impairing the results achieved by his equipment. The photocells that were available in 1926 were small gas-filled cells which had a high ambient noise level. This, together with the parasitics introduced by the then dull emitter valves which were used in the amplifiers, caused the received picture to have a very poor signal/noise ratio[31]. Baird accordingly set out to try and select cells with a colour response which matched the luminosity-wavelength characteristic of his floodlights and this led him to experiment with various coloured lamps and filters. It was during these trials that the lamps were masked, as an experiment, with wafer thin ebonite sheets so that all visible light was cut off and only infrared radiation played on the subject. Much to Baird's surprise, the picture was not only visible at the receiver, but the signal/noise ratio, since he was using red sensitive cells, was surprisingly good. Baird called his discovery noctovision, and thought it had great potential, particularly to penetrate fog[46], Figure 8.13.

For normal television the use of infrared radiation had certain disadvantages. First, it was generally inconvenient for the subject to sit in total darkness and,

Figure 8.13   J L Baird with his noctovisor, a device for seeing in the dark. It used infrared radiation

Reproduced by permission of Radio Rentals Ltd.

secondly, with the early type of photoelectric cells the correct colour tones were difficult to achieve; red appeared as white and blue did not appear at all. The effect, therefore, was to give a rather ghostly aspect to the image of the person being televised.

Hence, although when noctovision was demonstrated at the British Association meeting in Leeds in 1927 it proved such a popular scientific exhibit that the police had to be called in to regulate the queues, it really represented no advance for television.

An alternative approach to the use of infrared radiation to shield a subject from the intense light of the floodlights was the use of the spotlight method which Baird patented[47] in January 1926. Figures 8.14*a* and *b* show the floodlight and spotlight systems, respectively.

In any method, such as that shown in Figure 8.14*a*, which depends upon scanning an image, as formed by a lens, of the object, the efficiency of the system is ultimately limited, for any given size of image that can be scanned, by the ratio of the aperture to the focal length of the lens.

Experiments showed that with the best lens then available to form a 25 mm by 25 mm image, it would be necessary to illuminate a subject with a 16 000 candle

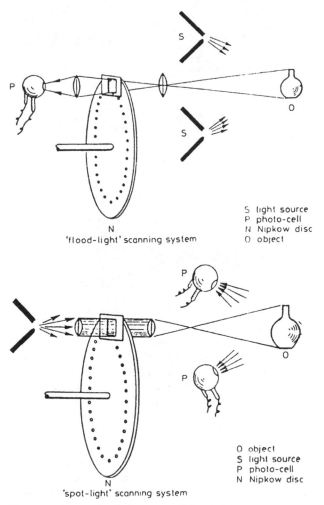

S light source
P photo-cell
N Nipkow disc
O object

N
'flood-light' scanning system

O object
S light source
P photo-cell
N Nipkow disc

N
'spot-light' scanning system

Figure 8.14   Diagrams showing floodlight and spotlight television

power arc at a distance of about 130 cm in order to secure an image bright enough for the photoelectric cell to give an output current above the noise level of the amplifier[48].

Baird reversed this process as follows[47]: 'At the transmitter the scene or object to be transmitted is traversed by a spot of light, a light-sensitive cell being so placed that light reflected back from the spot of light traversing the object falls on the cell.'

This method of scanning permitted two very large gains to be made in the amount of light flux available for producing the photoelectric current. First, the transient nature of the light allowed a very intense illumination to be used without inconvenience to the subject and, secondly, the optical efficiency of the system was not limited by the apertures of the lenses to hand but could be increased by utilising large photocells and more than one cell, all connected in parallel.

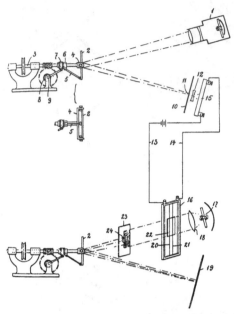

Figure 8.15   Ekstrom's spot light scanning system of 1910. Light from an arc, 1, was focused to a spot, 11, onto a slide/transparency, 10. Scanning was achieved by the oscillating mirror, 2, and its associated drive (3–9). At the receiver the image signal from the selenium cell, 15, controlled the light valve, 16, which, together with the graded density filter, 23, gave a variable light flux from the arc, 17, in accordance with the image signal. The transmitter and receiver scanners were identical

Source: Swedish patent 32220, 24 January 1910

Baird applied for provisional patent protection for this important principle on 20 January 1926 and submitted the complete specification on 18 November 1926. His patent was accepted on 20 April 1927. These dates are of some relevance, for on 6 April 1927, Electrical Research Products Inc (assignees of F Gray), applied for patent security, for the same principle, in the United States, and made an application for a similar patent in the United Kingdom on 18 January 1928. Baird's application was thus made before that of the American company, but notwithstanding the resemblance of the first two claims of Baird's patent with the first claim of the ERP patent, the latter was not invalidated by the Comptroller General of the Patent Office. The relevant claims are as follows.

Baird:

1   In a television or like system using a high intensity of illumination of the object whereof an image is to be transmitted, the method of illuminating the object with an intensity so high that *if the illumination were continuous on any one part the object would be damaged (i.e. burned)*, which method consists in traversing over the object a spot of light of the desired intensity.

2 In a television or like system the combination with means for illuminating the objects set forth in claim 1, of a light-sensitive cell so placed that light reflected back from the spot of light traversing the object falls on the cell.

ERP:

1 A television or like system in which the object, or subject is scanned by means of a beam of intense light which is moved rapidly so as to cause a cyclic point-by-point illumination of the object—the light reflected from the object being received directly on a photoelectric cell or cells of wide aperture, there being no obstruction between the cell and the object or subject.

The italicised section of Baird's claim introduced a weakness in its validity which presumably permitted ERP's claim to stand. Shoenberg of EMI was to use this point in discussions with the Television Committee in 1934[49].

Apart from the inherent value of the spotlight principle, the idea is important as it probably stemmed from practice in the field of picture telegraphy and therefore represents another application of the techniques of this branch of electric telegraphy to television. Several proposals had previously been made by inventors for scanning a transparency by a spot of light so that the transmitted light influenced a photo-sensitive cell. Korn had advanced this method for phototelegraphy purposes in 1904[50] and in 1923 the Westinghouse Electric and Manufacturing Company had included in one[51] of its patents an arrangement whereby the spot of light produced on the screen of a cathode-ray tube by fluorescence was used to scan a moving film and thus cause a varying signal to be generated by a photocell placed behind the film.

Actually, although both Baird and ERP were probably inspired to put forward their spotlight patents by events which had taken place in picture telegraphy, the original claim for the principle was made by Ekstrom in a Swedish patent dated 24 January 1910. Neither Baird nor ERP made reference to this little known (in 1926–27) patent, and it is unlikely that they knew of its existence.

The spotlight system was to be used by Baird's company for the next ten years and was an essential feature of the company's installation at Alexandra Palace in 1936.

Baird desperately needed money and facilities for his work. 'Luckily for him,' wrote one commentator[2] (S A Mosely), 'television was born at a time when company promoting was running rather wild. Speculators, greedy for fat profits and quick returns, were ready to gamble on very slim chances.'

Baird Television Development Company was established in April 1927[52]. The object of the company was essentially 'to develop commercially the Baird television and other inventions', and to achieve this the company acquired from Television Ltd the sole right to exploit the inventions in the United Kingdom and a 66.66% interest in the net proceeds arising from and payable to Television Ltd in respect of the sale and exploitation of the foreign and colonial rights. The purchase price was £20 000 and this was paid and satisfied by the allotment to Television Ltd of 400 000 fully paid up deferred ordinary shares of 1 shilling each.

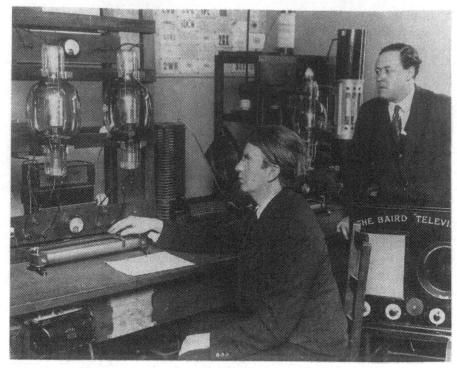

Figure 8.16   Baird transmitted televised images across the Atlantic Ocean in 1928. He is seen here with B Clapp, one of the engineers of the Baird Television Development Company

Reproduced by permission of the Glasgow Herald

A nominal capitalisation of £125 000 was set for the size of the company divided into:

100 000 preferred participating ordinary shares of £1 each   £100 000
500 000 deferred ordinary shares of 1 shilling each   £25 000

Each of the two classes of shares carried the same voting rights and as 400 000 of the deferred ordinary shares had to be allotted to Television Ltd the latter company had an important stake in the new enterprise. When the shares of the new company became available for purchase there was no shortage of buyers.

New premises were procured in Long Acre, in the west central district of London, to extend the existing laboratory facilities for the additional staff who were recruited.

Baird's requirements for substantial capital to develop television may be appreciated by noting that the AT&T Company approved the expenditure of $308 100 on low definition television developments from 1925 to 1930 (inclusive), and a further $592 400 on other aspects of television from 1931 to 1935 (inclusive).

Three adventitious factors aided Baird's work in the period following his October 1925 demonstration and prior to the introduction of the 30-line experimental television transmissions in 1929. First, the appointment of O G Hutchinson as Baird's business partner in late 1925; secondly, the state of the stock market in 1927; and thirdly, the association of S A Mosely with Baird, and the companies in which he had an interest, from 1928 to 1933.

The engagement of Hutchinson enabled Baird to concentrate full time on laboratory work. As a consequence he was able to devote all his energy and inventive skills to the achievement of 'firsts' without the need to attend to business matters: he was able to demonstrate a number of applications of his basic scheme before any other person or industrial organisation succeeded in doing so. These demonstrations attracted favourable comments from several notable scientists (including Professor Ambrose Fleming, FRS, Dr Alexander Russell, FRS, principal of Faraday House and Professor E Taylor-Jones who held the chair of natural philosophy at Glasgow University), many newspapers (including the *New York Times*) and two Postmaster Generals (Sir William Mitchell-Thompson and Sir Herbert Samuel).

As an illustration of these comments the *New York Times*, on 11 February 1928, noted[54] (apropos Baird's transatlantic television experiment): 'His success deserves to rank with Marconi's sending of the letter 'S' across the Atlantic, the first intelligible signal ever transmitted from shore to shore in the development of transoceanic radio telegraphy. As a communication, Marconi's 'S' was negligible; as a milestone in the onward sweep of radio, of epochal importance. And so it is with Baird's first successful effort in transatlantic television . . . All the more remarkable is Baird's achievement because he matches his inventive wits against the pooled ability and vast resources of the great corporation physicists and engineers, thus far with dramatic success. . . .

Unfortunately, Hutchinson's business and publicity methods were justifiably regarded with some suspicion and concern in certain quarters, notably the BBC, and these caused Baird to suffer some criticism.

There is little doubt that Hutchinson engaged in gross exaggeration to advance Baird television. For example, after a demonstration of television to a reporter of the *Daily Express* on 7 January 1926, Hutchinson wrote[55] to the Secretary of the British Post Office and stated that the Television Company was 'having made 500 receiving sets . . .'. Although this statement cannot easily be denied, the first receivers for sale to the public appeared in February/March 1930. Nevertheless, it has to be said that Hutchinson brought the accomplishments of Baird to the attention of the populace and this led to the flourishing formation of two new companies.

Moseley's association with the Baird companies was of appreciable importance to the furtherance of the low-definition television system in the UK. With his knowledge of finance, journalism and publicity and his contacts in Fleet Street, Parliament and elsewhere, Moseley brought much valuable expertise to bear on the side of the television pioneer. He played a central and crucial role in the development of television in Britain prior to his departure from Baird Television Ltd in 1933[56].

Marconi was one of the first to foresee the commercial possibilities arising from the work of himself and the early radio pioneers. He worked energetically throughout his life to progress and extend wireless communications. His primary interest, however, was in the technical development of the subject and his knowledge of business methods was such that, when he managed both the commercial and innovative aspects of his company, considerable difficulties arose. But when Godfrey Isaacs, a born businessman, was appointed managing director, thereby allowing Marconi more time to further his research and development work, the company prospered[1]. Marconi, like Baird, had a well balanced personality and worked easily with others: Isaacs and he formed an effective team, each respecting the other's skills, to the advantage of their common interest.

A similar parallel can be inferred from the relationship between Baird and Moseley: Baird was quite content to allow Moseley to look after his business and financial matters and a bond of friendship was established between them, shortly after Moseley's introduction to Baird in 1928 which lasted until the latter's death in 1946. The part executed by Moseley in promoting Baird's work was stressed in the BBC's letter[57] of regret to him (on hearing of his resignation in 1933): 'Although there has not always been agreement either in policy or in method, it should be recognised that your consistently active advocacy has been an important, perhaps the decisive, factor in the progress that Baird Television has made to date.'

On the other hand, Marconi's principal rivals in the US, de Forest and Fessenden, were not so well adjusted personally as either Baird or Marconi and did not team up with such skilful operators as Moseley or Isaacs. Marconi, de Forest and Fessenden were associated with the three most important early American communication firms, namely Marconi Wireless Telegraph Company of America, de Forest Wireless, and National Electric Signalling[58]. Only the Marconi companies which were well managed survived, although each of the three communication concerns was organised around one outstanding inventor.

The year 1928 was one of much activity for the Baird organisation. Not only was Baird effecting a number of firsts, the company was growing. Commander W W Jacomb was appointed chief engineer of the Baird Television Development Company and under his direction the first home receivers were designed and built and the transmitting equipment rationalised for the first time. Among the technical assistants were B Clapp, T H Bridgewater, J Denton, D R Campbell, J D Percy, A F Birch, J Collier and J C Wilson.

Life in the laboratories was hectic; no sooner was one of Baird's ideas past the experimental stage than he was on to the next, leaving Jacomb and the staff to perfect earlier ideas as best they could. Hours and holidays meant nothing to Baird and time was always too short for the implementation of the many thoughts which occurred to him. Baird was the kingpin of the organisation and after the formation of the public company the directors were much worried about this: if anything happened to Baird, who did not enjoy good health, the company would collapse.

The solution was to insure Baird for the sum of £150 000 for one year at a premium of £2000[2].

Baird and the directors of the Baird companies made a number of errors of judgement in pursuing their objectives[59]. First, they failed initially to appoint a sufficient number of high-calibre research scientists and engineers. The importance of this point is well illustrated by the successes which the highly qualified research teams of RCA and EMI, led by Drs Zworykin and Engstrom and by Mr Schoenberg, respectively, achieved in the 1930s. Although the staff recruited by the two Baird organisations were keen and eager to progress the development of television, they lacked the formal university research training which many of Schoenberg's and Engstrom's staffs possessed. In this respect, Baird's approach to television development differed from the plan of action of Marconi to the advancement of radiocommunication. Marconi surrounded himself from an early stage with a group of very able engineers and technicians, including Dr W H Eccles, Dr Erskine Murray, W W Bradfield, A Gray, C S Franklin, H J Round *et al.* By 1900, there were 17 professional engineers in the Marconi company in the UK. Electric and Musical Industries had the brilliant A D Blumlein as one of its first research engineers and engaged many university trained research workers: the research department included Dr L Klatzow, Dr B M Miller, Dr B M Crowther, Dr J D McGee, Dr H G Lubszynski, Dr W Stewart Brown, Dr L F Broadway, Dr E L C White, C O Browne, F Blythen, G E Condliffe *et al.* and many junior technical staff.

Secondly, insufficient importance was attached to the value of consultants. Marconi fully appreciated their worth and employed them for many years. He sought out some of the most promising university scientists who were interested in wireless and appointed them as consultants; Professor Ambrose Fleming and Professor M Pupin were two of the notable of these. Marconi was not hesitant about engaging staff with a greater intellectual and technical competence than his own. One of his most famous demonstrations, the transatlantic transmission, owed much to Fleming's work. It was Fleming who designed the transmitters, each of which incorporated an ingenious circuit having two spark gaps operating in cascade at different frequencies. Scientists of the proven ability of, say Watson-Watt or Appleton, could have had a most beneficial effect on Baird's work, particularly in the use of the short wave bands for television.

With Jacomb in charge of the development of the basic Baird system of television, Baird was able to devote time to the achievement of more firsts. After the demonstration of daylight television in June 1928 he went on to show colour television, for the first time anywhere in the world, on 3 July 1928[60]. For this Baird used transmitting and receiving Nipkow discs with three spirals and associated filters, one for each of the primary colours, a single transmission channel and effectively three receiver light sources (giving red, blue and green outputs). These sources were switched sequentially by a commutator so that only one lamp was excited at a time, Figure 8.17.

A report[61] in *Nature* described the operation of the apparatus:

> 'The process consisted of first, exploring the object the image of which is to be transmitted, with a spot of red light, next with a spot of green light and finally with a spot of blue light. At the receiving station a similar process is employed, red, blue and green images being presented in rapid succession to the eye . . . [The transmitter] disc revolves at ten revolutions per second, and so thirty complete images are transmitted every second—ten blue, ten red and ten green.
>
> 'At the receiving station a similar disc revolves synchronously with the transmitting disc, and behind this disc, in line with the eye of the

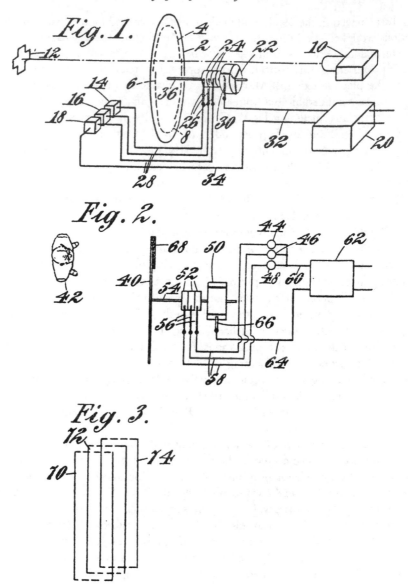

Figure 8.17   Schematic diagrams of Baird's colour television system, taken from his British patent 321 389, application date 5 June 1928. The system used sequential three-colour scanning. Fig. 1 of the patent shows the transmitter and Fig. 2 the receiver. Three-spiral Nipkow discs were utilised—each of the spirals being associated with a primary colour filter

observer, are two glow discharge lamps. One of these lamps is a neon tube and the other a tube containing mercury vapour and helium. By means of a commutator the mercury vapour and helium tube is placed in circuit for two thirds of a revolution and the neon tube for the remaining third. The red light from the neon tube is accentuated by placing red filters over the view holes for the red image. Similarly, the view holes corresponding to the blue and green images are covered by suitable filters. The blue and green lights both come from the mercury helium tube, which emits rays rich in both colours. The coloured images we saw which were obtained in this way were quite vivid. Delphiniums and carnations appeared in their natural colours and a basket of strawberries showed the red fruit very clearly.'

Marconi and Baird had the same policy with regard to their technical contributions. They applied for patents on every method and apparatus which they devised. In this respect they were inventors and innovators rather than scientists. The traditions of pure science did not allow university scientists, for example, to seek commercial gain for their work.

Baird had to devote a great deal of time and labour to the acquisition of a patent holding which would place his companies in a favourable position commercially. Until about 1930–31, he engaged in this task almost single handedly. From the start of his work in 1923 to the end of 1930 Baird applied for 88 patents; the number of patents originating from other members of the Baird companies in the same period totalled four.

Not surprisingly, Baird had little time for writing scientific papers and occupying himself in field trials of the type demonstrated by the AT&T Company in April 1927. He tried to anticipate every likely development and application of the new art. Daylight television, noctovision, colour television, news by television, stereoscopic television, long-distance television, phonovision, two-way television, zone television and large-screen television were all demonstrated by Baird during a hectic period of activity from 1927 to 1931.

A former Baird engineer has said[62]: 'Too much time was spent on adventurous sidelines and in exploiting the 30-line system—largely in pursuit of publicity which, rightly or wrongly, was considered necessary for the attraction of public interest and capital; too little time on essential technical improvements.' However, when an inventor wishes to further the commercial prospects of his ideas, and there are competitors about, it is essential that he should protect his basic methods and apparatuses by patents as quickly as possible: once these have been safeguarded, technical enhancement can take place at a more leisurely pace.

Certainly, there were many industrial organisations interested in television by the late 1920s. The Western Electric Company, Westinghouse Electric and Manufacturing Company, The Radio Corporation of America, General Electric, the American Telephone and Telegraph Company, the Jenkins Television Corporation and others in the US; Telefunken, Fernseh A G and Telehor A G in Germany; and in the UK several companies were considering pursuing investigations in this field, including both the HMV and the MWT companies.

Baird's strategy in the 1920s was clearly to secure as many firsts as possible, not only for publicity purposes but to give his companies commercial bargaining power.

He was not exploring adventurous sidelines wholly for good public relations, but as a means to an end. It is not always possible to foresee the likely outcome of an invention or the manifold applications of a basic idea.

By June 1928, the Baird Television Development Company had been operating for about 14 months and during that period the private company, Television Ltd, which was solely responsible for the exploitation of the foreign and colonial rights in the Baird inventions, had (as stated by Hutchinson) 'been more or less inundated with enquiries from practically every foreign and colonial country', principally owing to the publicity achieved by the long distance television and other demonstrations. It became obvious to the directors of both Television Ltd and the Baird Television Development Company that if Baird's system of television was to be properly developed and exploited throughout the world, it would be necessary to form an international company to take over the exploitation of the foreign and colonial rights. Accordingly, on 25 June 1928, a British public company was registered under the name of Baird International Television Ltd with the object of acquiring from Television Ltd, the vendors, the rights and interest of that undertaking in the Baird inventions and patents. The intention was to form or collaborate in the formation of overseas manufacturing and marketing companies and it was stated that negotiations 'with interests of first importance were in hand for the commercial exploitation of the inventions relating to facsimile telegraphy and phonovision'[63].

The initial capital of Baird International Television Ltd was £700 000, divided into 1 400 000 A shares of 5/- (five shillings) each, and 1 400 000 B shares of 5/- each. The company acquired from Television Ltd 300 000 deferred ordinary shares of 1/- in the capital of Baird Television Development Company Ltd and, in addition, the remaining one third interest of Television Ltd in the net proceeds arising from the sale or exploitation of the foreign or colonial rights in the Baird inventions. The purchase price paid by the international company was £350 000, which was satisfied by the allotment to Television Ltd of the 1 400 000 B shares of 5/- each, credited as fully paid up.

The Directors of the new company were:

> The Rt Hon Lord Ampthill (chairman)
> Sir Edward Manville
> Lt Col George Bluett Winch
> Mr John Logie Baird
> Captain Oliver George Hutchinson

The international company made a public issue of 1 000 000 of the A shares of 5/- each, at the issue price of 6/- per share. The issue was heavily over subscribed and the subscription lists closed on 26 July[64]. With regard to the 1 400 000 B shares alloted to Television Ltd, that company (so long as it held not less than 750 000 of those shares) had the right to nominate and appoint one half of the Board of Directors of the international company, including the chairman, with a second or casting vote but this right was never exercised by Television Ltd.

Of course, with the formation of Baird Television Development Company Ltd and Baird International Television Ltd, the directors of both companies had a responsibility to their shareholders. The latter no doubt anticipated that their shares would not only appreciate in value but would also earn dividends. Hence, there was a fairly urgent need for low-definition television broadcasting to commence in the UK and various overseas countries.

In many respects Baird's early commercial activities and difficulties mirrored those which had been experienced by Marconi. Both inventors were keen to capitalise on their initial successes, and as their work could only be furthered by obtaining funds from the public, companies were created. The Wireless Telegraph and Signal Company was established in 1897, the year following Marconi's visit to England. Television Ltd was registered in 1925, the year of Baird's first demonstration of rudimentary television. When further companies were floated these incorporated the name of the inventor in each case. None of the public companies was initially rewarding for their shareholders: Marconi's Wireless Telegraph Co Ltd did not pay dividends from 1897 to 1910[65]. Both inventors felt that the creation of overseas interests was necessary for the advancement of their plans. In 1899 the Marconi Wireless Telegraph Company of America was set up and was followed by the Marconi International Marine Communication Co Ltd in 1900.

The original capital of the Wireless Telegraph and Signal Company was £100 000[66], that of the Baird Television Development Company was £125 000, and both were largely subscribed by wealthy individuals. Presumably, in each case, the subscribers wanted a speculative investment in the new communication systems. Marconi was handsomely rewarded for his enterprise and received £15 000 in cash and 60% of the original stock in exchange for nearly all his patent rights[66]. He was just 23 years of age at that time. Similarly, Baird Television Development Company acquired the sole rights to exploit the inventions of Baird in the UK by paying a purchase price of £20 000 to Television Ltd[37] which had Baird as one of its founder members.

Moreover, both Marconi and Baird encountered difficulties in advancing their commercial interests. Aggressive tactics were adopted by the two companies to further their objectives, and much goodwill towards Marconi and Baird was dissipated later by this approach. Sir William Preece, who initially championed Marconi, said in 1907[67]: 'I have formed the opinion that the Marconi Company is the worst managed company I have ever had anything to do with . . . Its organisation is chiefly indicated by the fact that they quarrel with everybody.'

Notwithstanding Baird's successes and the good publicity which these achieved, there was a cloud on the horizon; the British Broadcasting Corporation was not impressed by Baird's work[68].

The lack of enthusiasm shown by the BBC towards Baird's low definition system was a source of much concern and frustration to Baird and his supporters, and resulted in delays in the execution of their plans.

The monopolistic position of the corporation during the early years of the Baird companies was, of course, a considerable obstacle. Essentially, the BBC was not interested in participating in the advancement of television on the basis of a system which could not reproduce images of, say, a test match at Lords or tennis at Wimbledon; the BBC considered that low-definition television was inappropriate to its service. As a consequence, the BBC's policy towards Baird's work was necessarily negative in outlook and did not conduce to the rapid advancement of Baird's aspirations.

Patronage and encouragement are important factors in the early development of an invention. Marconi initially was fortunate in this respect. In America and elsewhere facilities for television broadcasting were given by several broadcasting stations from 1927, but in Britain Captain P P Eckersley, the chief engineer of the BBC, opposed the use of the BBC's stations for this purpose[68]. He was in a powerful position to influence the adoption or rejection of television broadcasting and his view was that ' . . . a radical discovery is necessary before television will be practicable, just as the valve made broadcasting possible'. He was not opposed to television, *per se*, but opposed to the adoption of a system that might not fulfil the hopes which had been made for it. 'Now if television were perfected,' he commented, 'that would be a different proposition. There would be, I believe, a very popular demand for the BBC to take it up. But in its present form it would be useless for us to do anything. We might just as well have inaugurated a broadcasting system twenty years ago with the Poulsen arc as the nucleus of our transmitting equipment.' For Eckersley the Baird system was not capable of development and, as chief engineer, he did not, presumably, wish to be responsible for sending out a poor transmission. He thought that it was unfair to say that wireless broadcasting was as undeveloped as television when it started: ' . . . the spoken word was perfectly intelligible, music was rough and this has been improved, but on the basis of the spoken word alone, the Writtle transmissions [of 1922], for instance, were entirely successful[69].

The nature of a television service was clear in his mind: it was not to consist only of 'heads and shoulders' but of two men standing talking together, of a lot of men playing football, of a liner arriving at Plymouth, of topical events and so on, Figure 8.18:

> 'If Baird can show us the interior of a room with the people in it fairly
> clearly that would have a different service aspect, we could do plays, but
> if his plays are going to be silhouettes exchanging places with one
> another as they speak, I doubt if that has service value.'

Eckersley had his supporters, among whom Campbell Swinton was to be particularly vociferous, both privately and publicly. In a letter[70] to *The Times* dated 20 July 1928 he felt some 'comments should be made on the many, and in some cases absurd, prognostications that have appeared during the last few weeks in the daily press on this important subject'. His arguments were based on the very large band width required for successful public television compared with that adopted by Baird and other low-definition television workers. As an illustration he estimated that a ten inch by 16 inch photograph of the Eton and Harrow match in an issue of *The Times* contained a quarter of a million dots and that if this picture were transmitted by television at a rate of transmission of 16 per second, following cinematograph film practice, a rate of transmission of 4 000 000 dots per second would need to be realised. 'Such achievements,' he wrote, 'are obviously beyond the possible capacity of any

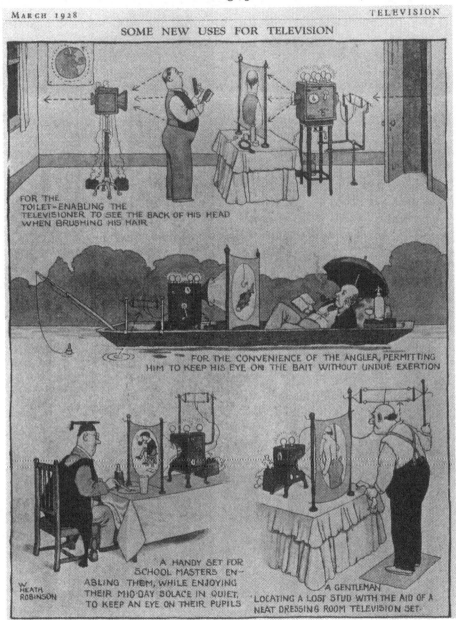

Figure 8.18   Some new uses for television as seen by W Heath Robinson (March 1928)

Reproduced by permission of the Science Museum

mechanism with material moving parts, and this view, which I have personally been inculcating in scientific circles for many years, has recently been endorsed by no less an authority than Sir Oliver Lodge, himself a notable pioneer inventor in wireless telegraphy.'

The first demonstration[71] of Baird's system to the GPO took place on 18 September 1928 at Long Acre and at the Engineers' Club about 600 yards away. Two tests were conducted, one in which a line transmission circuit was used for a short transmitter-receiver distance in the company's offices at Long Acre, and another which involved a radio link, on a wavelength of 200 m, between the same building and the Engineers' Club in Coventry Street, London. The demonstrations showed the facial images of several individuals while they carried on a conversation or sang, and these were reproduced by the receiver to give an image size about 3½ inches by 2 inches. This was viewed by a lens (approximately five inches in diameter) to enable a magnification of nearly two to be achieved.

Of the two tests, that which utilised the line circuit gave the better results. Angwin, of the GPO, observed: ' . . . the faces were shown with features outlined as clearly as they would have been seen from reflection in a metal reflector.' Twelve images per second were televised and so some flicker was produced but this did not seem to be objectionable. The synchronisation method and control were 'definitely superior in efficiency and simplicity to that used by other systems' and only required two external handles for speed and phase adjustment.

The wireless television transmission experiment was less satisfactory because of some apparent interference in the radio channel from local sources. The result was

Figure 8.19   Stereoscopic television apparatus which was used to give demonstrations at the Glasgow British Association meeting in 1928. Mr Collier, an engineer, is seated on the right

Reproduced by permission of Radio Rentals Ltd

a tendency to swinging of the image in the plane of observation and the flicker effect was more pronounced. Angwin concluded his report by stating:

'The system merits consideration from the simplicity of the receiving apparatus and the possibility of improvements if further developed . . . An experimental trial from one of the BBC's stations with observations on a set of the model it is proposed to sell to the public would, I consider, be desirable to test out the quality of the reproduction that might be expected under normal broadcasting conditions.'

The BBC was not prepared to accede to this course of action and insisted on a demonstration to its senior staff, who had not attended the 18 September tests. The date chosen for the next demonstration of Baird's apparatus was 9 October 1928.

On 8 October Eckersley wrote[72] a long memorandum on 'Suggested attitudes towards television' which was seen by the members of the visiting party prior to the demonstration. By this action he had rather unfairly endeavoured to prempt an adverse decision against Baird's system. He mentioned that the BBC's Control Board would see a demonstration of the head and shoulders of a man and that it would be extremely interesting and quite likely better than what they had been led to expect but, he added brusquely:

'if Control Board feel that this would justify a service then let us go ahead, but I warn everyone that, in my opinion, it is the end of their development not the beginning and that we shall be forever sending heads shoulders. Are heads and shoulders a service? Has it any artistic value? Is it not, in fact, simply a stunt?'

Eckersley had clearly made up his mind before the trial. As the BBC's chief engineer he was not only concerned about the quality of the transmission but also had to consider the implications of the service if the Control Board thought that the system had merit, and here there were difficulties. He wrote:

'It must be remembered that in effect an extra wavelength must be sacrificed for television. It has been pointed out that an extra wavelength is almost impossible as we have so few as it is, and I think to take up so much ether in the broadcasting band to give the small picture of the head and shoulders of a man to people who can afford sets is rather ridiculous.'

Eckersley's memorandum had the hoped for results. The BBC's senior staff were unimpressed by the display of televised images and so on 17 October 1928 the board of the corporation recommended that an experimental transmission through a BBC station should not be undertaken—at present. A notice was issued to the press[73].

The immediate effect for Baird's companies was a slump in the value of their shares which fell back from a recent high of 11s 3d, for the five shilling shares, to 6s

3d on the day of the statement[74]. On the stock exchange television shares had had a very good run and the share capital of the Baird companies had gone to a large premium[75].

J L Baird took the BBC's verdict with a certain degree of stoicism[76]. 'I regard the decision of the BBC to grant no facilities to television as a challenge which I mean to take up.' The battle of words was about to begin.

Baird and his associates were understandably apprehensive at this time as reports were being published in the press of the successes of other individuals and firms. The AT&T Company had broken, in 1927[14], Baird's monopoly to demonstrate true television and it was known that both the Radio Corporation of America and the General Electric Company were conducting experiments in television transmission. In Germany Mihaly had announced[77] in July 1928 that he was ready to manufacture sets which would bring 'into the drawing room a horse race or a boxing match so perfectly that the jockeys' caps will be distinctly visible and you will be able to recognise the faces of the horses'. He told a reporter of the *Daily News* that he intended to start a television company in London in a few weeks 'with a system which would be different from Bairds and far more effectual'. The sets would have one valve and would sell for £20 each.

The opposition of the BBC to the use of its stations by Baird has its parallel in the history of the Marconi organisation[1]. Although Preece, the chief engineer of the British General Post Office, had given Marconi much needed support in the early stages of his work, Austen Chamberlain, as Postmaster General, had taken quite a different attitude. He saw the Marconi company as a potential competitor of the government-controlled telegraph industry and, at first, stubbornly refused to allow the Marconi overseas service to utilise the Post Office's telegraph lines. Later, an agreement between the Marconi company and the GPO was signed on 11 August 1904 and facilities for wireless telegraphic traffic were granted to it.

Following the issue of the BBC's press release on 17 October 1928 much lobbying and discussion of Baird's case took place, both publicly in the national newspapers and privately by correspondence. The details have been given in the author's book 'British television, the formative years' and will not be repeated here, suffice to mention that a further demonstration of the Baird system was given to a group of Members of Parliament on 5 March 1929.

Soon afterwards (on 28 March 1925) the Postmaster General published in *The Times* the letter[78] which he had sent, apropos the demonstration, to the secretary of the Baird Television Development Company. He confirmed his earlier opinion that the Baird system was capable of producing, with sufficient clearness to be recognised, the features and movements of persons posed for the purpose at the transmitting point, although he added that it was not yet practicable to transmit a scene or programme which required a space of more than a few feet in front of the transmitting apparatus. In the PMG's view the system represented 'a noteworthy scientific achievement' but had not reached a stage of development which would merit the inclusion of television programmes within broadcasting hours. He was anxious that facilities for further development should be granted and mentioned that he would assent to a station of the BBC being used for this purpose outside broadcasting hours. The PGM thought that it was probably essential that television should be accompanied by speech and that two transmitters would be needed, but pointed out that a second transmitter would not be available until the new station at Brookmans Park was completed, possibly in July. In the meantime the company was to open negotiations with the BBC on the financial and other arrangements which might be necessary.

These transactions dragged on for almost six months until 30 September 1929, when the corporation transmitted its first experimental television broadcast. Particulars of the transactions are given in copious detail in the author's book, *op. cit.* They will not be reiterated here except to note the points which were agreed between Hutchinson and officials of the BBC, on 11 September 1929[79].

1   the transmission should commence on Monday, 30 September from the Oxford Street transmitter, and later from the Brookmans Park transmitter;
2   the transmission times would normally be from 11.00 a.m. on Monday to Friday inclusive, but after 31 October, when it was anticipated that additional periods would become available, the morning transmissions might be replaced occasionally by transmission after midnight, or at other times outside programme hours as mutually agreed;
3   the extent of any interference would not be greater than that given by a music transmission as normally radiated from programmes;
4   the transmission would take place from television studios on Long Acre and the Baird Company would be responsible for renting all lines;
5   a BBC engineer would be allowed reasonable access to the television transmitter;
6   the company would install and maintain one televisor at Savoy Hill, one at the BBC receiving station and one at the GPO;
7   the BBC would answer technical queries from home constructors;
8   the BBC reserved the right to curtail or discontinue any particular transmission should it conflict with the BBC's own programme, on a particular day;
9   transmissions would not be curtailed, discontinued or at any time altered by the BBC except with three months' notice given in writing.

For this concession the Baird Company was to pay £5 per half hour and reimburse the BBC for any capital cost involved.

Thus, although the September 1928 demonstration of the Baird television system was considered to be sufficiently satisfactory for the Post Office to agree, as far as it was concerned, to the use of a BBC station for further experiments, the start of the experimental service had been delayed by a full year.

In his published letter of 28 March 1929[78] the PMG advised the Baird companies to press on with experiments on a much lower band of wavelengths. His counsel was to prove highly significant in the later history of television. Unfortunately for the companies, J L Baird did not immediately initiate an urgent programme of research and development work in this region of the electromagnetic spectrum, but concentrated instead on achieving some further successes with the 30-line low-definition system. In retrospect, the position of Baird vis-à-vis EMI Ltd in the 1930s might have been much improved had he appreciated earlier than he did the limitation of the 30-line system and the need for a television service operating on a higher definition standard. If Baird had produced a 150 or 180-line picture system (say), working on short waves, by 1931, the future of his three companies probably would have been more secure. Later events were to show that Baird made the change to the higher standard too late for his companies to surpass the impressive EMI and M-EMI television systems, and yet the writing was on the wall for the 30-line system, even in 1929. Eckersley and others had given their views on head and shoulders television broadcasts and the medium waveband did not allow, owing to bandwidth limitations, any extension to a better definition system, and hence an improved and more varied picture. In addition, the Post Office had indicated to the Baird Television Company in 1929 that they had no objection to them experiementing for demonstration purposes on 75 m or 50 m. The 75 m band was reserved, by the Washington Convention for Broadcasting, for experimental work, although the 50 m band was allowed for broadcasting (Baird had used a wavelength of 45 m for his transatlantic experiments but only for a 30-line image).

There were certain unknowns in working in the above wavebands and these may have influenced Baird's determination to continue with his imperfect system. In 1929 there was insufficient information available to show how far either of the 50 m or 75 m wavebands would be effective inside the London area, although it was known that the presence of steel-framed buildings had a deleterious influence on the reception of short waves.

Following the formation of Baird International Television, the company made some progress with regard to the exploitation of the American, Canadian and Mexican rights in the Baird inventions; it had prepared heads of agreement for the formation in the Argentine of a company to further the inventions in that country and other South American states and was pursuing every opportunity to secure the establishment of the Baird system in various European countries.

Naturally, the directors of Baird International Television Ltd and of Baird Television Development Company Ltd were keen for their companies to earn profits but without the needed facilities for television broadcasting in 1927–28 the directors had not been able to complete their licence arrangements, with members of the radio manufacturing industry, to make and sell Baird televisors.

After the establishment of BIT Ltd in June 1928 it had become increasingly obvious to the directors of the two public companies that the functions of these companies overlapped and that it was desirable, for economic reasons, that the companies should be amalgamated[80]. Sir Mark Webster Jenkinson, an expert on such matters, was consulted and he submitted a scheme of amalgamation to the joint boards of directors which, after discussion, was adopted and approved by them. Meetings of the companies were held in April 1930 and the scheme for amalgamation was approved by the shareholders and became effective. The world interests in Baird television were thus vested in the new company which was called Baird Television Ltd. The capital of that company was £825 000 divided into 2 100 000 preferred ordinary shares of 5/- each, entitled to a ten per cent noncumulative preferential dividend and 40 % of the surplus profits, with preferential rights in the event of winding up, and 1 200 000 deferred ordinary shares of 5/- each entitled to 60 % of the profits, after payment of the noncumulative ten per cent preference dividend on the preferred ordinary shares. The 2 100 000 preferred ordinary shares and 200 000 of the deferred ordinary shares were alloted to the shareholders (other than Television Ltd) in the public companies. Television Ltd received from Baird Television Ltd 1 000 000 deferred ordinary shares of 5/- each in exchange for its previous holding of 1 400 000 B shares in Baird International Television Ltd. It might appear that Television Ltd, in assenting to the scheme of amalgamation and in accepting the exchange of shares, had given up shares of the nominal value of £100 000 without adequate compensation. However, it was thought that the right of participation in the profits of the 1 000 000 deferred ordinary shares was likely to be much greater than the previous right of participation attaching to the 1 400 000 B shares in the international company.

The right of appointment of one half of the Board of Directors, including the chairman, was also given up and no rights in regard to the appointment of the directors were attached to the 1 000 000 deferred ordinary shares.

According to Hutchinson, the cash assests of Baird Television Ltd, at the time of the merger, were 'considerable, and at the current rate of expenditure, without allowing in any way for income, [were] sufficient for the needs of the company for some considerable time to come'.

To further its objectives of exploiting the Baird system, negotiations were carried out in many countries of the world, including some British Dominion and Crown colonies. Television transmitting and receiving apparatuses were sent to Australia, South Africa and Canada, erected and demonstrated. Many requests (said Hutchinson) were received from other countries, including the Scandanavian countries and 'in almost every case facilities [were] offered for experimental broadcasting of television under the Baird system'.

In Belgium, consultations had progressed, by November 1929, to such an extent that detailed proposals had been put forward which would have resulted in the installation of Baird equipment in the principal broadcasting station, Radio Belgique: and an experimental station on the outskirts of Brussels, under the control of the Société Belgique, would have been made available for music and speech transmissions simultaneously with the television transmissions.

In France many discussions with leading radio manufacturers had taken place by the end of 1929. If an agreement had been obtained with these manufacturers television broadcasting facilities would have been provided through Radio Paris and several other important French broadcasting stations. Hutchinson has recorded that the directors of Baird Television Ltd gave these matters their careful consideration, but ultimately came to the conclusion that having formed a strong alliance in Europe through the creation of the German company Fernseh AG (see Chapter 11), in which Baird Television Ltd had a 25 % stake, their immediate policy should be to exploit the rights in the British Dominions and the Crown colonies. For these reasons the negotiations in Belgium and France were prorogued.

All of this activity demanded that Baird Television should have a strong patent holding in its system of television. Large sums were therefore necessarily expended

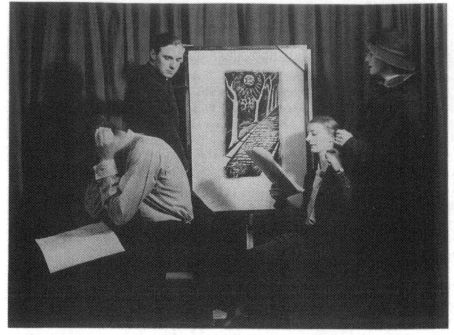

Figure 8.20   The first television play to be produced in the UK was 'The man with a flower in his mouth' by Pirandello. L Sieveking and S A Moseley were the producers. The photograph shows (left to right) Sieveking, C Denis Freeman, Mary Eversley and Lionel Millard (July 1930)

Reproduced by permission of the BBC

in protecting the Baird inventions in the principal countries of the world and by 1930 the company felt that its system was adequately covered by patents.

It is apparent from an examination of the early histories of the Baird and Marconi endeavours that Baird wished to emulate the success which Marconi, following his visit to the UK in 1896, had achieved. There is a parallel between the formations of the wireless telegraph companies and the television companies associated with the two inventors. More particularly, there is a close similarity between Baird's international aspirations and those of Marconi[1].

On 22 November 1899, an American company was to have paid the Wireless Telegraph and Signal Company $7 000 000 for its patent rights. In the event this transaction never materialised; instead the original American company was later merged with the parent company with the title of Marconi Wireless Telegraph Company of America. This was to become part of the Radio Corporation of America.

In Europe the Italian Ministry of Marine had already confirmed, in May 1898, that the Italian navy would adopt the Marconi system. Two years later, on 25 April 1900, the Marconi International Marine Communications Company Ltd was formed. A Belgian associated company came into being on 26 October 1900 when the Cie de Télégraphie sans Fils (Belge) was created to develop and operate the Marconi system on the Continent. Eighteen months later, on 24 April 1903, a further continental foothold was established with the registration of the Cie Francais Maritime et Coloniale de Télégraphie sans Fils of Paris.

Another valuable contract was obtained on 24 July 1903 when an agreement was signed between the Marconi Company and the British Admiralty for the general use of the Marconi system in the Royal Navy.

Marconi's eagerness to install his equipment in the United Kingdom and overseas was brought about by the knowledge that rival systems of wireless telegraphy had made their appearance, notably, those of Popov-Ducretet, Slaby-Arco and the Siemens-Halske-Braun combine, but also others. Even by 1900 the French naval authorities had veered towards the Popov-Ducretet system and the German navy had adopted that of Slaby-Arco. Interestingly, all these rival systems suffered from the same defects, and all were seeking the master invention which would give the individual system a decisive lead over its competitors.

Baird's position during 1929–30 was similar to that of Marconi in 1899–1900. Rudimentary apparatuses had been devised for narrowband television and narrowband wireless telegraphy and the potential for technical and commercial development seemed vast. Marconi had undoubtedly realised many of his hopes by 1929–30 (aided, of course, by a strong patent holding), and so with this in mind it is likely that Baird and his directors felt that the pursuance of a similar strategy would be lucrative.

Thus, the Baird Television Company's objectives were the build up of its patent stock and the establishment of a strong international and economic position. An example[80] of the value of such a position will illustrate this point. In the 1920s the Mackay Company, of the US, made an application to the Federal Radio Commission for an allocation of a series of wavelengths to enable it to operate a transoceanic communication system. At the hearing of the application the RCA and its affiliated companies argued that it would be wrong to grant any competitor rights in the transoceanic field because the RCA, by virtue of its worldwide system, was much better able to carry on those communications. Nevertheless, the FRC granted wavelengths to the Mackay Company, and subsequently the company approached 30 countries in an attempt to establish such communications. It found only one country, Spain, which had not tied itself up with the RCA: hence a competing company could not enter into this business. Hutchinson has stated that the policy of Bairds was to secure a similar international and economic arrangement and towards the end of the 1920s this was being steadily implemented in conjunction with the patent position.

The power and influence of the Radio Corporation of America and its affiliated companies in the US were well recognised by the Baird interests. Hutchinson visited[80,81] the States in September/October 1930 and had talks with Mr O Schairer, the patents manager of the RCA. The result of this visit impressed the directors of Baird Television Ltd that if some proper arrangement could be made with the RCA and its affiliated companies for the exploitation of the Baird patents, trade marks and improvements in the US, with reciprocal rights in regard to new inventions and improvements, such an arrangement would be to the ultimate, and possibly immediate, benefit of the RCA and its affiliate companies and to Baird Television Ltd.

In the furtherance of its international policy, Baird Television Ltd had exhibited its system in New York City on 2 September 1929[82]. Vision and sound signals had been sent by wire from the Paramount Building to a laboratory at 44th Street. Later on 20 December 1929 a demonstration had featured Mayor J Walker of New York[83].

Hutchinson's wish to visit RCA seems rather curious. Although it is apparent how RCA's worldwide interests could aid the British company the converse is not so obvious. On the one hand Baird Television Ltd was committed in 1930 to low-definition television based on mechanical scanning at the transmitter and receiver; on the other hand Zworykin's intentions were to seek an all-electronic solution to the television problem. He had read a paper on his kinescope[84] (a cathode-ray tube used in a television receiver) on 18 November 1929, and station KDKA had been broadcasting daily 60-line motion picture images, using the Conrad television system, from 25 August 1929[85]. Whether Hutchinson was shown RCA's television developments during his visit or whether he knew of their work from other sources is not known.

Schairer suggested that Hutchinson, on his return to England, should approach the Gramophone Company, in which RCA Victor had a stake. Hutchinson accepted this advice and on 8 October 1930 sent a letter[86] to the secretary of the British company. Two days later Hutchinson, A Clark (the chairman of the Gramophone Company) and B Mittel (the director of production) had a general talk on the scope of the Baird companies. Clark was invited to visit Baird Television Ltd and on 14 October he and two colleagues (B Mittel and W B Brown) discussed with Hutchinson at BTL's laboratories the position of television in America. The Gramophone Company's senior staff did not attend the meeting in a state of ignorance because, the previous day, they had been to their production department and had seen a demonstration of television reception based on the company's recent work[86].

A further visit to BTL was made by G E Condliffe, of the Gramophone Company (HMV), on 16 October 1930[87]. He was given demonstrations of Baird's noctovision and film scanning systems, and inspections of their transmitter and studio equipment during a broadcast performance. He was not impressed. 'Their scheme as it exists at present is crude, and except for head and shoulders pictures [sic], has no possibility of success.' Unknown to Baird Television, the Gramophone Company, at this time, was working on 150-line television using films. Inevitably their 150-line images were far superior to those achieved using the 30-line system. 'A demonstration of a film showing a boxing match was very poor, and that of a horse race picture so bad that it was totally devoid of interest. The apparatus showed no novelty, and the scheme is incapable of improvement on the present lines.'

From the Gramophone Company's viewpoint the only possible advantage which could be derived from a link with BTL would be the use of any master patents which Bairds held. Such was Hutchinson's eagerness to effect some contract with HMV that he had previously sent the company copies of all Baird's published and pending patent specifications in the UK and the US together with a history of the Baird companies. HMV examined carefully each patent for claims which might subsequently be awkward in operating its own television system. All doubtful cases were referred to its patent agents for an opinion as to the scope or validity of the particular claims, but the conclusion reached was that their transmitting and receiving devices were patent free[87]. 'The Baird company own many patents on variations in disc design, and in regard to the number of discs etc, employed, but they are of no value and are not used by the Baird company themselves. In this connection the Baird stereoscopic and colour scanning devices should be noted. These appear to cover obvious fundamental points. It is unlikely, however, that they will be used for many years, and are of small potential value. A few patents might be worth getting hold of if they were going cheap.'

Further talks between Hutchinson and Mittel and Clark were held on 4 and 17 November. By 14 November Clark had formed an opinion on the Baird position and had written to D Sarnoff, who had become a director of the Gramophone Company in March 1929 and who was president of the Radio Corporation of America, to the effect that he had been unable to find any solid assets. Hutchinson was told nothing further could be discussed until RCA responded to Clark's letter.

Regrettably, EMI's archives, which include the archives of HMV, do not contain Sarnoff's reply. However, there is a memorandum regarding an interview[88], with a Mr W Barrie Abbott, a 'very intimate friend of Baird', held on 26 January 1931. He argued, in a private capacity, for a joining of hands with Baird in the development of television. The next day a Mr A G Clark had a discussion with Mittel about the possibility of an amalgamation of Baird Television Ltd and HMV[86].

In the meantime, Baird Television had brought an action against the Gramophone Company and had alleged 'that it had manufactured exhibited and used at the Imperial College of Science and Technology, Kensington, a certain apparatus in January 1931'—which presumably infringed a patent of the plaintiff company. Hence, when A G Clark had a telephone conversation with Mr Alfred Clark, of HMV, he was told that the company was not prepared to carry on negotiating while the action was still unsettled. With this mild rebuff A G Clark said 'he would see what he could do'. Baird Television did not proceed with the case and costs were awarded against it[89].

A further letter requesting an informal chat, with the prospect of finding some solution, was sent to A Clark by A G Clark on the 9 February 1931. Baird Television was certainly forcing the pace and seemed very anxious for an agreement. Although the available archival material gives no real indication why this should be so, it is possible that BTL could foresee the financial difficulties (which are dealt with at length in Burns, *op,cit.*) which the company would experience later in 1931. This view is supported by the opinion of HMV after another visit from A G Clark on 13 February[86]: 'Not time yet to discuss pooling of patents with company that was going to abandon research. We might also obtain patents from America.' Nothing transpired between the two firms until August 1931.

Any pioneer company clearly must keep ahead of its competitors if it is to remain operationally viable. Baird realised this point in so far as it related to low-definition television, for he expended much thought and effort in modifying his basic system so that it could be utilised for colour television and noctovision *inter alia*.

He failed, however, to appreciate at a sufficiently early stage that 30-line television could never give rise to an all-embracing television broadcasting service. Although Baird's low-definition system was capable of improvement, as evidenced by the successive demonstrations given by the company and later the BBC, the system was effectively restricted to head and shoulders' type shots. Baird persisted in this view until he was almost overtaken by EMI and RCA. He believed that television broadcasts should be transmitted in the medium waveband to ensure a wide coverage of the population, even though the bandwidth available severely circumscribed the definition of the images which could be sent out.

This perspective was shared by the directors of the Baird companies until approximately 1932, and it seems that they overrated the commercial possibilities of 30-line television, failing to recognise that television broadcasting would only become widespread in popular appeal when the Derby, sporting and athletic events, motor racing and so on could be televised to give adequate image detail, as Eckersley had pointed out.

It is possible that, if the company had retained a number of distinguished and eminent radioengineering consultants, it may have heeded the Postmaster General's advice at an earlier stage. It is significant to note that it was Appleton, the radio scientist, who urged Baird to investigate the use of the HF and VHF bands.

Rather interestingly, Marconi too had a blind spot[90]. His dominant urge was to extend his system of wireless communications, but he failed initially to perceive the importance of continuous wave operations in transatlantic working, and thereby did not visualise the benefits of radio telephony. Several of Marconi's contemporary inventors took a different outlook, but Marconi did not share their optimism. He considered the Morse code to be quite suitable for ship communications and for transoceanic signalling, and saw no real need for a wireless telephone. Like Baird, his approach to his work was pragmatic. He was not interested in following scientific investigations in fields which had a doubtful commercial viability and, in this respect, Baird and Marconi lacked some foresight, although both inventors possessed an entrepreneurial spirit. This blind spot was unfortunate at first for the Marconi Company and the furtherance of communications. Luckily for Marconi, some of the important early work on radiotelephony was undertaken by two of his rivals, de Forest and Fessenden, and neither inventor had access to financial or engineering resources or skills comparable to those of the Marconi companies.

During the protracted discussions which led to the setting up of the BBC's experimental television service and during the period of company formation and expansion, J L Baird had been obtaining further publicity with demonstrations of adaptations of his basic 30-line scheme. Stereoscopic television was shown in the Long Acre laboratories for the first time on 10 August 1928: it was also demonstrated at a meeting of the British Association in Glasgow in September 1928[91]. For this application the scanning transmitting and receiving Nipkow discs each had two sets of spiral apertures. Two light sources—one for each spiral—and a photocell bank enabled two time-sequential signals per single rotation of the transmitter disc to be generated. At the receiver a neon lamp and a stereoscopic viewing device, comprising two prisms and two eyepieces, allowed the alternating left and right reconstituted images to be combined by the observer's brain into a three-dimensional picture.

The low-definition experimental service progressed appreciably when the BBC granted Baird Television an additional wavelength so that sound and vision could be synchronised. This demanded the use of both of the Brookmans Park transmitters, one operating on 261 m for vision and the other on 356 m for sound: the first combined transmission was sent out on 31 March 1930[92].

For the wireless correspondent of the *Evening Standard* the reception of the transmission was remarkable: 'It was, so to speak,' he wrote[93], 'a "talkie" by wireless, but a "talkie" that consisted of close-ups.'

The use of the two wavelengths permitted the televising of plays. The first UK television play was Pirandello's 'The man with a flower in his mouth' and was chosen because it had three characters only, namely, Gladys Young, Earle Gray and Lionel Millard. 'They came to Long Acre and were made up in yellow, with navy blue shading around the eyes and nose,' wrote Margaret Baird[4], 'these colours on the face improving the picture.' Again head and shoulders were all that could be observed. Only one face at a time was shown and between each sequence a checkered curtain was drawn across the screen, incidental music filling in the pauses.

'Allowing for such things as the televisor going out of synchronism every now and then and the poor quality of the sound transmission, it must be recorded that the broadcast as a scientific achievement was a success,' wrote *The Times*[94]. 'As a play—well, it left a certain amount to the imagination, but it was good entertainment and certainly an advance on the mere reception by sound,' it continued.

The *Daily Mail*[95] found that the pauses tended to minimise the dramatic interest of the performance 'and constituted one of the problems which it [was] hoped to solve when bigger sets [were] cheaper. But it was certainly startling, as well as helpful to the dialogue, to be able to see their every expression even to the lifting of the eyebrows. We even saw the gestures of their hands—although we had to sacrifice their faces for the time being.'

A select audience in July 1930 in a canvas theatre on the roof at Long Acre also saw the play on Baird's big screen. Senatore Marconi, who had been invited by a director of Baird Television, was a member of the audience but unfortunately his views on the play were not recorded. Baird was much impressed by Marconi's aloof politeness and almost regal manner and described him in words which could equally have been applied to himself[4]:

'Although the invention of no single device of fundamental importance can be attributed to Marconi, it was he who ventured forth like Christopher Columbus and forced upon the attention of the world the existence of a new means of communication.'

In retrospect, this avidity may have lulled the Baird directors into a false sense of security in respect of the standing of low-definition television in general and Baird's system in particular. But curiosity in television was growing rapidly. Mihaly, Fernseh AG and Telefunken (in association with Dr A Karolus) in Germany, the American Telephone and Telegraph Company, the Radio Corporation of America, the General Electric Company, the work of C F Jenkins and of P Farnsworth in the US, and other companies in France and elsewhere, were stimulating people with their displays of television and the opportunities which could follow. Baird Television was finding it increasingly difficult to be first in the field to show something new. Although Baird's large-screen system was based on a patent dated 26 July 1923[11], nevertheless the first demonstration of television to a large audience had been given by AT&T in 1927[14]. The first play broadcast by television was produced in the US by the National Broadcasting Company (NBC) in April 1930, three months before the production of 'The man with a flower in his mouth'. Also, television companies abroad either were not being hampered by a hesitant broadcasting monopoly or were being encouraged in their activities.

References

1   JOLLY, W.P.: 'Marconi' (Constable, London, 1922). Also: BAKER, W.J.: 'A history of the Marconi Company' (Methuen, London, 1970)
2   MOSELEY, S.A.: 'John Baird' (Odhams, London, 1952)
3   BAIRD, J.L.: 'Sermons, soap and television' (Royal Television Society, London, 1988)
4   BAIRD, M.: 'Television Baird' (HAUM, South Africa, 1974)
5   BURNS, R.W.: 'British television, the formative years' (Peter Peregrinus Ltd, London, 1986) Chapter 1, pp.3–4
6   Calender for the year 1914–1915 of the Royal Technical College, Glasgow
7   KEMPSELL, A.: 'The man with many dreams', *Helensburgh Times*, 3 September 1975, p.15
8   LANGAR, N.: 'A development in the problems of television', *The Wireless World and Radio Review*, 11 November 1922, pp. 197–210
9   BAIRD, J.L.: 'Television', *J. Scientific Instruments*, 1927, **4**, pp. 138–143
10 Ref.5, p. 10
11 BAIRD, J.L., and DAY, W.E.L.: 'A system of transmitting views, portraits and scenes by telegraphy or wireless telegraphy'. British patent 222 604, 26 July 1923
12 ANON.: 'Seeing by electricity', *Chambers J.*, November 1923, pp. 766–767
13 MOSELEY, S.A., and BARTON CHAPPLE, H.J.: 'Television today and tomorrow' (Pitman, London, 1934) Chapter 12
14 IVES, H.E.: 'Television', *Bell Syst. Tech. J.*, October 1927, pp. 551–559
15 BAIRD, J.L.: 'An account of some experiments in television', *The Wireless World and Radio Review*, 7 May 1924, pp. 153–155
16 ANON.: 'Seeing the world from an armchair. When television is an accomplished fact', *Radio Times*, 15 February 1924, p. 301
17 ANON.: Report in the *Hastings and St Leonards Observer*, 26 April 1924
18 ANON.: Report in the *Hastings and St. Leonards Observer*, 12 April 1924
19 ANON.: Report in the *Hastings and St Leonards Observer*, 3 May 1924
20 LE QUEUX, W.: 'Television—a fact', *Radio Times*, 25 April 1924, p. 194
21 BAIRD, J.L., and DAY, W.E.L.: 'A system of overcoming the time lag in a selenium or other light sensitive cell used in a television or like system'. British patent 235 619, 12 March 1924
22 BAIRD, J.L.: 'Improvements in or relating to television systems and apparatus'. British patent 270 222, 21 October 1925
23 BARNARD, G.P.: 'The selenium cell: its properties and applications' (Constable, London, 1930)

24 SZCZEPANIK, J., and KLEINBERG, L.: 'Method and apparatus for reproducing pictures and the like at a distance by means of electricity'. British patent 5031, 24 February 1897
25 KORN, A.: 'Photographic transmission', *The Electrician*, 1 March 1907, pp. 765–766
26 ZAVADA, B.: 'Anordnung zur Beseitung der storenden Wirkungen der Tragheit von Selenzellen für telephotographische Zwecke', *E.T.Z.*, 1911, **32**, pp. 1111–1112
27 Report on 'Cox's selenium amplifier for submarine telegraphy', *Electrician*, 1921, **86**, pp. 131–133
28 GRAY, F.: internal memorandum, 27 February 1926, AT&T Company, Warren Record Centre, USA
29 BAIRD, J.L., and DAY, W.E.L.: 'A system of transmitting views, portraits and scenes by telegraphy or wireless telegraphy'. British patent 236 978, 17 March 1924
30 BAIRD, J.L.: 'Television. A description of the Baird system by its inventor', *The Wireless World and Radio Review*, 21 January 1925, pp. 533–535
31 PERCY, J.D.: 'The founding of British television', *J. of the Television Society*, 1950, pp. 3–16
32 TILTMAN, R.F.: 'Baird of television' (Seeley Service, London, 1933)
33 ANON.: Report, *Nature*, 4 April 1925, pp. 505–506
34 ANON.: Report, *Nature*, 25 April 1925, p. 613
35 D'ALBE, E.E.F.: 'Telegraphic transmission of pictures and images'. British patent 233 746, 15 January 1924
36 D'ALBE, E.E.F.: 'Improvements in apparatus for recording and reproducing sound'. British patent 247 629, 23 October 1924
37 Company file, Television Ltd, Public Record Office, Kew Gardens, UK
38 ANON.: Report, *The Times*, 26 January 1926
39 BURNS, R.W.: 'Early Admiralty interest in the detection of aircraft' *in* 'History of electrical engineering', IEE conference publication 1985, pp. 10/1–10/24
40 Press Photonachtrichtendienst, letter to the General Post Office, UK, 28 January 1926, minute 51/1929, file 1, Post Office Record Office, UK
41 HUTCHINSON, O.G.: Letter to the Postmaster General, 4 January 1926, minute 4004/33, file 1, Post Office Record Office
42 PHILLIPS, F.W.: Letter to Television Ltd, 15 July 1926, minute 4004/33, file 2, Post Office Record Office
43 A.R.: 'Television', *Nature*, 3 July 1926, **118**, pp. 18–19
44 ANON.: Report, *New York Times*, 6 March 1926
45 ANON.: Report, *Radio News*, September 1926, p. 283
46 RUSSELL, A.: 'Television' a letter, *Nature*, 5 February 1927, **119**, pp. 198–199
47 BAIRD, J.L., and Television Ltd.: 'Apparatus for the transmission of views, scenes or images to a distance'. British patent 269 658, 20 January 1926
48 GRAY, F.: 'The use of a moving beam of light to scan a scene for television', *JOSA*, March 1928, **16**, pp. 177–190
49 SHOENBERG, I.: evidence to the Television Committee, 17 June 1934, minute 33/4682, Post Office Record Office
50 KORN, A.: 'Ueber Gebe und Empfangsapparate zur elektrischen Fernübertragung von Photographien', *Physikalische Zeitschrift*, 1904, **5**, (4), pp. 113–118
51 GARDNER, J.E., and HINELINE, H.D.: 'Improvements in and relating to television systems'. British patent 225 553 application date (UK) 25 November 1924
52 ANON.: 'Baird Television Development Company, Statutory Meeting', *The Times*, 18 July 1927
53 BURNS, R.W.: 'The contributions of the Bell Telephone Laboratories to the early development of television' (Mansell, London, 1991) History of technology, **13**
54 ANON.: Report, *New York Times*, 11 February 1928
55 HUTCHINSON, O.G.: Letter to the Secretary of the GPO, 11 January 1926, minute 4004/33, file 1, Post Office Records Office
56 Ref.5, Chapters 4–10
57 BBC.: a letter to S.A.Moseley, 21 June 1933, private collection
58 MACLAURIN, W.R.: 'Invention and innovation in the radio industry' (MacMillan, London, 1949) Chapters 2 and 4
59 BURNS, R.W.: 'J L Baird: success and failure', *Proc.IEE*, (9), **126**, September 1979, pp. 921–928

60 ANON.: Report, *Morning Post*, 7 July 1928
61 ANON.: Report, *Nature*, 18 August 1928, **122**, pp. 233–234
62 BRIDGEWATER, T.H.: 'Baird and television', *J. Roy. Television Soc.*, 1967, pp. 60–68
63 ANON.: Report, *Evening Standard*, 25 June 1928
64 ANON.: 'Baird television', *The Financial Times*, 29 June 1928
65 Report of a meeting of Marconi's Wireless Telegraph Co. Ltd., *The Times*, 29 June 1910, p.20
66 BAKER, W.J.: 'History of the Marconi company' (Methuen, London, 1970) p. 35
67 Report of the Select committee on Radio Telegraphic Convention (HMSO, London, 1907) pp. 232–234
68 Ref.5, Chapter 4
69 Ref.5, Chapter 5
70 CAMPBELL SWINTON, A.A.: 'Television methods of reproducing pictures', letter to *The Times*, 20 July 1928
71 ANGWIN, A.S.: 'Baird television demonstration', a memorandum, 19 September 1928, minute 4004/33, Post Office Records Office
72 Chief engineer (BBC).: Memorandum, 8 October 1928, BBC file T16/42
73 BBC statement, 17 October 1928, minute 4004/33, Post Office Records Office
74 ANON.: 'BBC and television. Baird shares slump', *Glasgow News*, 18 October 1928
75 ANON.: 'Share movement', *Daily News*, 18 October 1928
76 ANON.: 'Mr Baird and the BBC. "Inexplicable" attitude to television and the reply', *Evening Standard*, 18 October 1928
77 ANON.: 'Television. Big claim from Germany. 1-valve sets. Demonstration in London soon', *Daily News*, 29 July 1928
78 Postmaster General: letter to Baird Television Development Company, *The Times*, 6 March 1929
79 CARPENDALE, C.: letter to Baird Television Ltd, 11 September 1929, BBC file T16/42
80 HUTCHINSON, O.G.: 'A short history of Baird Television and of the companies concerned in the development and exploitation of that system', EMI Archives, c.1930, pp. 1–17
81 DALE HARRIS, E.P.: a letter to H.G.Grover, RCA Patent Department, 21 October, EMI Archives
82 ANON.: 'Voice and image go together over wire', *New York Times*, 3 September 1929
83 ANON.: 'Walker televised at demonstration', *New York Times*, 21 December 1929
84 ZWORYKIN, V.K.: 'Television with cathode ray tube for receiver', *Radio Engineering*, 1929, **9**, pp. 38–41
85 See Table 12.1 and Chapter 12
86 History sheet on Baird Television, EMI Archives
87 CONDLIFFE, G.: 'Report on Baird Television Ltd'. EMI Archives, 13 November 1930, report G.1
88 Interview with Mr Barrie Abbot on 26 January 1931, EMI Archives
89 ANON.: 'Baird television v. Gramophone', *Financial Times*, 16 March 1931
90 Ref.58, p. 52
91 ANON.: 'Colour television', *Glasgow News*, 5 September 1928
92 MURRAY, G.: Letter to S.A.Moseley, 12 February 1930, BBC file T16/42
93 ANON.: 'First sound television broadcast. Singers seen 10 miles away. Dual transmission by BBC twins', *Evening Standard*, 31 March 1930
94 ANON.: 'First play by television', *The Times*, 15 July 1930
95 ANON.: Report, *Daily Mail*, 15 July 1930

Chapter 9

The approaches of a lone inventor and a chief engineer (the 1920s)

Television development in the US commenced, in a practical way, in 1923—the year in which J L Baird began his television activities—with the work of Charles Francis Jenkins. Jenkins was not Baird's contemporary in age for he was born 21 years before Baird. Unlike the Scottish pioneer, the American inventor had been associated with the reproduction of moving images, by either picture telegraphy or cine film, from an early age.

Jenkins was born on a farm near Dayton in Oregon on 22 August 1867[1]. His parents seem to have been comfortably well off for his education was undertaken at Spiceland Academy and at Earlham College. In 1887, when much experimentation was being carried out by inventors and scientists on the projection of moving images, Jenkins started making experiments with apparatus for recording and reproducing motion pictures. He achieved some success and a few years later, in June 1894, demonstrated his first motion picture machine, which he called a phantascope. During the following year, at the Colton States Exposition in Atlanta, he gave a number of public exhibitions of motion pictures. The phantascope was the protoype of the modern motion picture projector and was manufactured in large numbers: it earned for Jenkins in 1908 the Elliott Cresson gold medal of the Franklin Institute. Later, in 1913, when the Institute awarded him the John Scott medal it was 'in recognition of the value of the invention . . . Eighteen years ago the applicant exhibited a commercial motion picture projecting machine which he termed the phantascope. This was recognised by the Institute and subsequently proved to be the first successful form of projecting machine for the production of life-size motion pictures from a narrow strip of film containing successive phases of motion'.

It is interesting to note that the machine was devised when Jenkins was the private secretary, from 1890 to 1895, to S I Kimball of the life saving service.

Jenkins had a natural creative talent for invention and, during the course of his life, produced more than 400 patents in such diverse fields as aeroplane construction, automobile engineering, facsimile transmission, television and cinematography. His interest in television dates from the early 1890s and, in July 1894, he had

Figure 9.1   Charles Francis Jenkins (1867–1934)

a short paper[2] published in *The Electrical Engineer* on 'Transmitting pictures by electricity', Figure 9.2.

This paper described a multiconductor, nonscanning television system using a mosaic of selenium cells at the transmitter and a corresponding mosaic of filament lamps at the receiver. It would appear that Jenkins was not familiar with the ideas of Carey for the system he described was almost identical to that which Carey had put forward in 1877[3]. Jenkins appreciated the disadvantage of the method nevertheless and wrote: 'The scheme, if practicable when necessary modifications are made, is objectionable in that it contemplates a multiplicity of conductors, but as a basis for study the method has its merits.' The apparatus was never constructed—'I should be glad to learn of the success of such an experiment by some one, as I cannot at present test it myself'—and Jenkins did not return to the subject until 1913.

In the meantime, in 1898, he engineered the first car with the engine in front of, instead of under, the seat; in 1901 in Washington he built the first motor sight-seeing bus; in 1908 he invented the all-paper spiral wound paraffined box for carrying liquids, and applied for a patent on a teleautograph; and in 1911 he designed one of

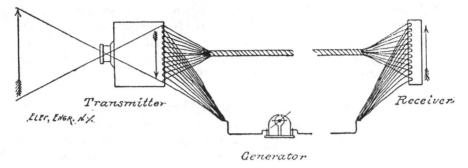

Figure 9.2   Diagram of Jenkins' phantoscope of 1894

Source: *The Electrical Engineer*, 25 July 1894

the first self starters for cars. He made further improvements in automobile engineering in 1912 and later in the field of aircraft construction. He founded the Society of Motion Picture Engineers in July 1916. Its objective was 'advancement in the theory and practice of motion picture engineering and the allied arts and science, [and] the standardisation of mechanisms and practices therein'.

In 1913 Jenkins had a paper[4] published in the magazine *Moving Picture News* on 'Transmission of motion pictures by wire'. Success in this venture eluded him for several years and it was not until 13 June 1925 that he was able to demonstrate[5] his invention for the wireless transmission of radiovision and radio movies.

The broadcasts, on a wavelength of 546 m, were sent out from the Naval Radio Station, NOF, Anacostia and were received in Jenkins' Washington laboratory before a prestigious gathering of government officials. These included navy secretary C D Wilbur, acting secretary Judge Davis of the Commerce Department, Dr G M Burgess, director of the Bureau of Standards, W D Terrell, chief radio expert in the Department of Commerce, Admiral D W Taylor, Captain P Foley of the Naval Research Laboratory, and several other notable persons.

The *Sunday Star*[5] heralded the occasion as the first time in history 'that man has literally seen far away objects in motion through the agency of wireless'.

Although the image broadcast itself was devoid of dramatic interest, being merely a silhouette image of a small model of a windmill with the blades in motion, the event stimulated an awareness for possible future feats: 'I suppose we'll be sitting at our desks during the next war and watching the battle in progress,' commented Wilbur during the test. Thirty years later his forecast was a commonplace occurrence.

The demonstration was of a strictly private nature and, in the words of the inventor, did not pretend to be a show. 'It is merely a scientific test that proves we have attained our goal [sic],' Jenkins told his visitors. 'By making numerous improvements in our sending and receiving machines, we expect to be able shortly to stage a radiovision show with the talent performing at the broadcasting station and the audience watching the performance at the receiving studio miles distant.'

Rash statements, forecasts and aspirations seem to have no boundaries and the simple demonstration at 1519 Connecticut Avenue North West triggered off a reaction which had been paralleled in Frith Street, London, when Baird showed his apparatus to members of the Royal Institution in January 1926. Still, Jenkins had given a test transmission of radiovision using silhouettes and, although the image was 'not clear cut, [it] was easily distinguishable'.

A lens-disc scanner was used in the transmitting and receiving equipments for the demonstration[6]. Jenkins had considered employing apertured Nipkow disc scanners but was aware of their disadvantage: 'Because this scanning disc limits the illumination to the light which can pass through a single one of these tiny holes, a powerful source of light is required for adequate lighting, just as is required in a pin-hole camera, with which it is comparable. As such a powerful light was not available in my laboratory, I put a lens over each aperture in the disc, making the aperture as large as the working area of the lens, and a comparatively small light source, e.g. an automobile headlight lamp, was then quite adequate.'

The discs each included 48, 8.5 dioptre lenses arranged in a spiral, with an offset at the ends of the spiral of about 2.2 cm[7]. In order to accommodate the 42 mm diameter lenses on the 45.7 cm diameter disc, the inventor had had to shape the lenses to an approximately rectangular form. The discs rotated at an angular speed of 960 r.p.m. to give a scanning rate of sixteen frames per second.

A neon corona glow lamp, developed and patented, in 1917, by Moore, of GE, was used for reconstituting the transmitted image, but the type of light sensitive cell employed is not known. Synchronous motors enabled the transmitter and receiver discs to rotate at the same speed, although the received image had to be framed manually.

The television systems designed by Jenkins and Baird in 1925 were similar in concept and probably represented the only practical methods available at that time. The period, from 1920 to 1925, spent by Jenkins in arriving at a partial solution to the television problem, gives some indication of the difficulties which faced the early television workers, although Baird found a limited solution after approximately two years.

Jenkins was a very different type of experimenter to Baird. Whereas the Scottish inventor's first televisor 'had the ingenuity of Heath Robinson and a touch of Robinson Crusoe', the American's models and equipment possessed fine workmanship and engineering appreciation. Not for Jenkins the hat box, darning needle, tea chest approach of Baird to the construction of television apparatuses, the inventor from Oregon had his parts carefully machined and assembled: he had a long background of instrument making based on his motion film projector activities which Baird lacked. Possibly it was this care for quality which delayed Jenkins' demonstration until 1925, whereas Baird, who was never inhibited in the first few years of his television activities by the need to show first-class equipment, successfully reached the same stage of system development after a much shorter period of experimentation.

Jenkins' lens disc scanner was not the first such product of his endeavours, rather it represented the outcome of a fairly long evolutionary development which commenced in 1915. In that year, two years after he had put forward his ideas for the transmission of motion pictures by wire, he invented the prismatic ring, a new device in optics. His objective was to scan ciné films continuously, without the use of a shutter, and for this purpose he needed a scanner which would cause a light ray to travel across the film in a succession of parallel adjacent lines so that the motion of the film coupled with the motion of the scanner would enable all points of every frame to be sampled by the light beam.

In a short paper[8] published in 1920 Jenkins claimed the following advantages for continuously moved, instead of intermittently moved, ciné film:

1   longer life for the film, consequently larger revenue from each print;
2   100 % increase in lighting efficiency, i.e., the same screen lighting with 50 % of the present light source intensity, (because of the absence of the rotating shutter used with conventional cameras and projectors);
3   a noiseless and much less complicated projecting machine;
4   almost unlimited speed which in a camera would open up possibilities for scientific research not available in any other way.

To satisfy these objectives Jenkins developed, in collaboration[9] with the United States Bureau of Standards, a glass disc scanner in which the section of its circumferential ring was ground into a prismatic shape as shown in Figure 9.3a.

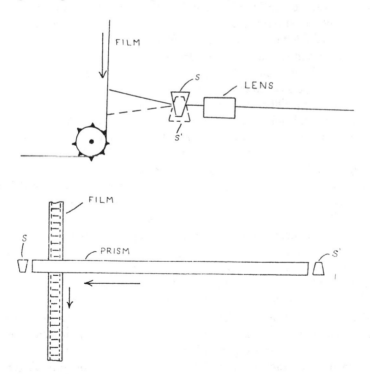

Figure 9.3   *Jenkins' prismatic ring scanner for use with continuously moved, instead of intermittently moved, cine film*

Source: *Trans. Soc. Motion Picture Engs.*, **10**, May 1920 p. 9

Consequently, if this strip of glass was passed across the path of the film, in the light cone and in synchronism with the travel of the film in a projector, the running film would appear stationary on the screen, Figure 9.3b. As Jenkins said: 'The light ray acts as though it were hinged at the location of the prism.'

Later Jenkins constructed band-type prismatic ring scanners[10]. These were formed from a glass ring in such a way that the section of the ring acted as a prism, Figure 9.4. From the point, R, of discontinuity on its circumference to a point, R', diametrically opposite, the ring's section was formed with the base of the prism inward, and from R' to R, for the remainder of the circumference, the section was ground with the base of the prism outward. Consequently, a beam of light incident normally on the interior surface of the ring was undeflected at point R' but was deflected either to the left or right of its original path as the ring rotated.

Jenkins also utilised disc type prismatic scanners in pairs for scanning stationary two-dimensional surfaces, Figure 9.5, as in facsimile transmission[11].

Presumably, he was led to adopt pairs of discs as a scanner in the hope that they would improve the optical efficiency of his system compared with that for an aper-tured disc scanner. When this type of scanner is utilised in the reproduction of motion pictures, the optical system must produce a band of light having a width equal to the width of the film picture and a height equal to the spacing between lines, but with Jenkins's invention the beam of light could be concentrated so that its section corresponded in area to the area of a single picture element.

Using this invention, in a still picture transmission system Jenkins sent his first portraits by radio in January 1922 and on 4 October 1922, after the machine had been improved, he gave an official demonstration[12], with the cooperation of the US Navy and the Post Office Department. Portraits of President Harding, Secretary Denby and several other persons were sent by wire from Jenkins's laboratory in Washington to the Naval Radio Station, NOF, seven miles away, and thence were broadcast by radio to a receiver situated in the Post Office building on 16th Street, Washington. In March 1923 photographs and photomessages were sent by Jenkins's method from NOF to the *Evening Bulletin*, Philadelphia, a distance of 135 miles, and were reproduced in an edition of the paper[13]. They were the first US radio pho-tographs published by any newspaper. The experiments led the US Navy to employ the Jenkins' equipment to broadcast weather maps to ships at sea.

These modest successes encouraged Jenkins to explain publicly that if sixteen pictures per second (ciné film projection speed) could be broadcast then television would be possible. Such an achievement, said a commentator in the November 1922 issue of *Scientific American*[9], would mean that there would be 'no reason why we should not, with the new service, broadcast an entire theatrical operatic perfor-mance so that, instead of going to a movie house for an evening's entertainment, we can turn a switch and see the latest play and hear it spoken at the same time or, by tuning out the play and tuning in the concert watch the operatic singer as well as hear her'.

For the transmission of moving pictures Jenkins discarded his prismatic discs in favour of the scanner shown in Figure 9.6[10]. This consisted of a circular metal disc around the periphery of which were placed a number of small lenses. Rotation of

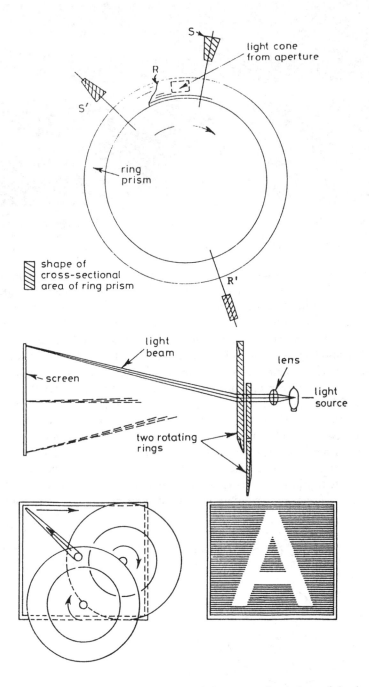

Figure 9.4   Diagrams showing (a) the variation of the cross-sectional shape of the ring prism, and (b) the use of two discs to scan a raster

Source: *Trans. Soc. Motion Picture Eng.*, **10**, May 1920, p.10

Figure 9.5   Photograph of Jenkins' specially ground prismatic disc scanners

Reproduced by permission of the Science Museum

the disc enabled line scanning to be achieved but for frame scanning, instead of arranging the lenses in a spiral formation according to the layout of the apertures in a Nipkow disc, the American inventor utilised the principle of his prismatic disc scanner. Behind each lens Jenkins mounted a small prism and these were so made that the angle of the prisms altered slightly and progressively around the circle of lenses (see also Figure 9.7). In this way, an effect similar to that given by a prismatic disc scanner was obtained. It is not clear why Jenkins adopted this type of scanner for television use rather than his earlier prismatic disc type, but it may that the strength of the latter was not sufficiently great to withstand the larger internal forces generated as a result of the very considerably increased rotation speed necessary for television usage. Another reason, possibly, was the need to improve the optical efficiency of the facsimile apparatus for the above purpose by the employment of lenses, and this could not be done conveniently with prismatic rings and discs.

With his lens-prism scanners Jenkins, in 1923, succeeded in transmitting a picture of 'a shadowy wave of the hand or movement of the fingers'[14]. The transmitting and

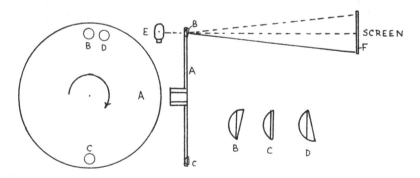

Figure 9.6   Diagrammatic representation of Jenkins' lens-prism scanner

Source: *The Wireless World and Radio Review* 5 May 1926, p. 642

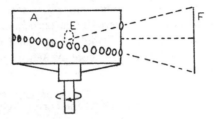

Figure 9.7   *An alternative arrangement to that shown in Figure 9.6*

Source: *The Wireless World and Radio Review,* 5 May 1926, p. 642

receiving apparatuses were driven from a common electric motor, the scanning rate was 16 frames per second and the number of lines per frame was 48. A corona neon glow lamp was used in the receiver.

Later, Jenkins arranged his lenses in a spiral path and so was able to dispense with his use of prisms. It was with lensed disc scanners of this type that Jenkins made his 13 June 1925 demonstration. 'But the disc scanner,' wrote the inventor[6], 'whether aperture disc or lens disc, has physical limitations in practical applications which seem, as at present employed, not to permit very much development.' He based this view on the size of the picture which would be reproduced by a Nipkow disc where the offset of the ends of the spiral determines one dimension of the picture. 'A 36″ diameter disc is required, therefore, for a 2″ square picture,' he noted. 'A 4″ picture would require a 6′ diameter disc—a rather impractical proposition in apparatus for home entertainment, even if it were possible to get power enough out of the house wiring to turn the disc up to speed.'

This latter statement was not an exaggeration. In 1937 Bell Telephone Laboratories designed and constructed a video generator, to test their wide-band-width coaxial cable transmission line, which was based on a 6′ diameter analysing disc. The disc had to be driven by a ten horse power motor[15].

There was another difficulty. The neon glow lamp had to be constructed with a plane cathode having an area slightly larger than the picture area, say, 2.5 by 2.5 inches for a 2.0 by 2.0 inch picture. 'To light this 2.5″ cathode plate requires from 70 to 90 mA of current, necessitating special amplification of currents obtainable from the plate of the last amplifier tube of usual radio sets. And so the proposition as a whole does not look very enticing to the amateur experimenter. And it was for these reasons that I never employed the elementary area apertured scanning disc. The lens disc was the nearest I ever came to it.'

To overcome these difficulties Jenkins invented[16] yet another type of scanner using a hollow cylinder 7.0″ in diameter, 3.0″ in length and 1/16″ in wall thickness. This had 48 scanning apertures, punched or drilled into it, the apertures being arranged in four helical turns. Each turn was 0.5″ apart from its neighbour and the apertures were spaced 2.0″ apart circumferentially round the cylinder. Means were provided for rotating the cylinder about its axis.

With this drum scanner the inventor utilised a four-target cathode glow neon lamp (Figure 9.8) mounted inside the cylinder and coaxially with it so that the cylinder could rotate about the stationary lamp. Between the lamp and the inside surface of the drum scanner Jenkins fixed 48 quartz rods, each rod ending under its particular drum aperture. The four cathode targets were located under each of the four rows of quartz rods and the targets were excited, in turn, via a four-segment commutator, from the radio receiver. Because the movements of the inner ends of the rods were so small, Jenkins was able to reduce the size of the cathodes to about 1/8" by 3/16" and hence the magnitude of the current required to approximately 3 mA to 5 mA. The quartz rods enabled the light loss due to the inverse square law to be avoided; as Jenkins said, ' . . . to discover how effective they are one has but to remove the rods, for no picture can be seen without them.'

Some success was obtained with the drum scanner: 'Magnified, the picture appears about 6" square; and in daily use it has been found that five or six people, the whole family, can very conveniently enjoy the story told in the moving picture.' However, notwithstanding the apparent advantage of the new scanner, the Jenkins Television Corporation (JTC) later reverted to the manufacture of 48-hole Nipkow discs[17].

Jenkin's technical work on television effectively highlighted the value of the Nipkow disc as a scanner, for the American's work led successively by stages to its eventual adoption by the JTC, as shown below.

1   prismatic discs and rings, 1920;
2   prism-lensed discs, 1923;
3   lensed discs, 1925;
4   apertured discs and drums, 1928.

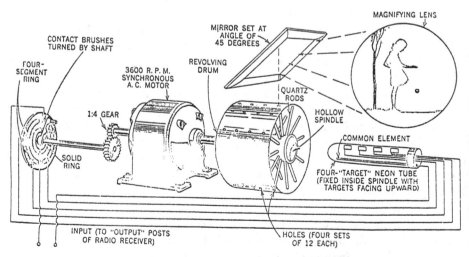

Figure 9.8   Shematic diagram showing the configuration of Jenkins' drum scanner and four-target neon lamp. The 4 × 12 array of quartz rods enabled the light fluxes from the lamp to be transmitted to the 4 × 12 apertures of the drum scanner without appreciable loss. A 48-line picture was reproduced. The quartz rods acted as optical waveguides

Source: Radio News, August 1928, p. 116

Sometimes Jenkins used a combination of different scanners for analysing and synthesising images. On 5 May 1928 he gave a demonstration[18] to members of the Federal Radio Commission of the projection of motion pictures by television, using a 48-hole, 15″ diameter, 900 r.p.m. scanning disc at the transmitter and a drum scanner at the receiver. Silhouette images from 35 mm film were transmitted.

Television broadcasting, on a regular basis, in the US began on 2 July 1928[19] when the Federal Radio Commission issued the first television licence to C F Jenkins, authorising him to broadcast television signals at frequencies of 1605 kHz and 6420 kHz from station W3XK, situated in the suburbs of Washington, DC. The station operated on a radiated power of 250 W initially, but by 1931 the FRC had approved an increase to 5 kW. Jenkins' station W2CXR in New Jersey was also in operation in 1930.

Jenkins' prime objective at this time (1928) was to assist the American radio amateur in acquiring the basic techniques of television reception. Silhouttes only were initially broadcast from W3XK so that the video signals could be generated in the permissible 10 kHz bandwidth. Later, the FRC allocated a 100 kHz band (4.90 MHz to 5.00 MHz) to the W3XK station and this allowed half-tone movies to be shown.

Regarding the production of these programmes, Jenkins has been quoted as saying: 'Our organisation is complete and self sufficient. We write our own scenarios, we build and operate our own movie studio, the only one of its kind in the world; we designed, built and operate our own film developing and printing equipment, and we do our own editing and cutting.'

On 16 January 1930 a public demonstration[20] of the first commercial model of the Jenkins television receiver was held at the Lauter Piano Company, Newark, New Jersey. The receiver, which retailed at $395.00, comprised merely a synchronous motor driven disc and neon tube, mounted in a cabinet having a viewing aperature fitted with a lens. Two additional receivers, not included in the price, were necessary for receiving the sound and picture signals (at 130 m and 187 m). The actual size of the image produced by the neon tube was approximately 2 × 1.5 inches, but this was enlarged by the lens to nearly 4 × 3 inches. One reporter said the synchronism was good; another (in a different geographical location) noted[21] that the receiving disc could not be held closely in synchronism with the transmitter because the 60 Hz supply in New York and New Jersey differed slightly.

During one of the ten to 15 minute demonstrations the feature picture was a prize fight in silhouette. 'The figures of the principals and referee and the outlines of the ring, ropes, towels, etc could be made out quite easily, although the edges were foggy and there was no fine detail sufficient to recognise the faces . . . The picture shown seemed to the writer to possess little or no positive entertainment value except for the novelty factor'[20]. Another observer[21] wrote: 'The half-tone picture reception was very unsatisfactory. Could not identify any of the subjects except the company's trade mark. The silhouette picture was a slight improvement over previous receptions but it could be improved greatly as to definition and general sharpness of outline.' These unsatisfactory broadcasts were due to the very limited bandwidth, 6 kHz, of the picture signals.

Elsewhere in the US and also in the UK, much superior demonstrations of television had been given by 1930. In particular, in April 1927, the Bell Telephone Laboratories had field tested their superbly engineered systems of television using both radio and wire links and small and large-scale viewing screens. Their tests were reported widely and were given great praise by Jenkins in a paper[22] presented in June 1927.

A few months earlier Dr E F W Alexanderson, of the General Electric Company, had described the course of action which he was following to implement seeing by electricity. Baird in England, Mihaly in Germany, Belin in France, Jenkins in the US and several others had all been working more or less single handedly to perfect the art. Now the large industrial organisations were beginning to take an interest in the subject, and were bringing to bear to its problems their expertise and vast resources for research and development. The first to enter the field were the research departments of the General Electric Company and of the American Telephone and Telegraph Company. Later, in the US, Westinghouse and the Radio Corporation of America joined in the struggle to progress a field which appeared to have a bright future. After 1925 the science and technology of distant vision was no longer the preserve of the lone inventor working in some dusty attic or dark basement: he was up against the world's scientific and engineering elite.

The interest during the 1920s of the large electrical companies, in television, stemmed from their involvement in engineering facsimile systems for sending images of documents, photographs and the like from one place to another. Apart from Jenkins, the General Electric Company, the American Telephone and Telegraph Company, the Radio Corporation of America, Marconi's Wireless Telegraph Company of England, Siemens-Karolus of Germany and Belin of France had all devised, prior to their involvement in television, methods for transmitting, at a very low information rate, images to a distant point. Essentially, telephotography is an extremely slow speed version of television, from an electrical engineering viewpoint. In the former process the scanning of the object may take several minutes, but in the latter case the process must be accomplished in less than one tenth of a second. Therein lay the problem of television. Nevertheless, the above mentioned firms had devised various facsimile system components which possibly could be modified or their principle of operation adapted to television. The Kerr cell, for example, was one of these. It was used by Auguste Karolus, in Germany, in both facsimile and television systems. It was also used by Baird and G E. Again, work which had led to the production of photoelectric cells for use in phototelegraphy could be extended to the television field.

An early hint to the general public that there were well endowed companies at work in the US on television was contained in a report[23,24] headed 'Radio Mirror', in the *Daily News* for 30 December 1926. This mentioned an 'intimation from Dr E F W Alexanderson (consulting engineer of the General Electric Company and the Radio Corporation of US) that soon London [might] be able to view New York by television'. The reporter added: 'When perfection has been attained, an operator will be able to stand in New York and transmit to a London cinema, or to a private house where a receiving set is installed, the jostling crowds of Broadway, the millions of electric coloured lamps at night and the sky signs, which will be reproduced on a screen. By the same apparatus it will be possible for the Lord Mayor of London, while telephoning to Mayor Walker of New York, to see (on a silver screen) Mayor Walker in his parlour, and *vice versa.*'

These and other visions of the future stimulated considerable interest among lay persons; for others, even greater feats were likely. 'There may come a time,' wrote Professor A M Low[23], 'when we shall have "smellyvision" and "tastyvision". When we are able to broadcast so that all the senses are catered for, we shall live in a world which no one has yet dreamt about.'

Newspaper reporters were not alone in forecasting what the state of the art would be in decades to come. George Bernard Shaw[25], in his play 'Back to Methuselah', had already described a situation, for the year 2170, which was now entering the realm of practicability. In one scene in the play the head of the British Government holds conferences with his cabinet ministers several hundred miles away. At his desk is a switchboard, and in the background is a screen upon which, when he selects the appropriate key at the switchboard, a life-size picture appears of the man to whom he is talking, a picture which portrays faithfully all the movements of the distant minister.

Alexanderson referred to Shaw's prophecy in a paper which he delivered before the AIEE at St Louis on 15 December 1926. He stated[26]: 'A passage of this sort by a great writer is significant. The new things that civilisation brings into our lives are not created or invented by anybody in particular; it seems to be predestined by a combination of circumstances that certain things are going to happen at certain times. It is often the great writers and the great statesmen who have the first presentiment of what is coming next. Then the inventors and engineers take hold of the ideas and dress them up in practical form.'

A significant factor regarding the participation of the large electrical firms in television was that their developments were centred, in most cases, on outstanding electrical engineers. Dr E F W Alexanderson of the General Electrical Company, Dr H E Ives of the American Telephone and Telegraph Company, Dr V Zworykin and Dr Engstrom of the Radio Corporation of America, Mr I Shoenberg and Mr A D Blumlein of Electric and Musical Industries, Mr L H Bedford of A C Cossor Ltd, *inter alia*, were by any standards most notable contributors to the advancement of their chosen fields.

Ernst Fredrik Werner Alexanderson, Figure 9.9, was born in Uppsala, Sweden, on 25 January 1878 and was educated by private tutors before being enrolled at the state secondary school in Lund, Sweden[27]. Following a year at the University of Sweden in Lund, he moved to the Royal Technical University, Stockholm, and graduated in engineering in 1900. A further year was spent on postgraduate studies in electrical engineering at the Charlottenberg Technical College in Germany.

The crossroads in Alexanderson's early life came in 1901 after he had read C P Steinmetz's book 'Theory and calculations of alternating current phenomena' (1897). Steinmetz was the presiding genius in the General Electrical Company and Alexanderson determined to travel to the US and seek employment with him.

Alexanderson arrived in the United States in the autumn of 1901 and obtained work with the C&C Electrical Company, Garwood, NJ as a draughtsman. A few

Figure 9.9   *Dr E F W Alexanderson with his multiple-light beam mirror drum scanner. The television projector focuses seven spots of light on the screen*

Reproduced by permission of the Schaffer Library, New York

months later, in February 1902, Steinmetz engaged him as a draughtsman, but arranged in early 1903 to have Alexanderson transferred to the testing department. It was here that his talent for innovation and invention became evident. Assigned temporarily in 1904 to the engineering department to write the patent specification of his first invention, he conceived a second invention before the first had been protected. Thereafter he became the holder of 350 patents, the last of which was issued to him in 1973 when he was 95 years old.

Alexanderson, from 1904, soon established himself as the leading authority on the design of high power, high frequency alternators for wireless communications. His original high frequency alternator enabled Fessenden to transmit the world's first long range broadcast of both voice and music in 1906. By 1911 Alexanderson's improved alternator permitted the first transatlantic radio transmission to be successful and during the First World War he perfected a 200 kW alternator which was used by President Woodrow Wilson to send messages to his commanders in Europe. On 20 October 1918 Wilson employed the alternator to deliver the ultimatum to Germany which led to the Armistice.

The success of the alternator led directly to the formation of the Radio Corporation of America by General Electric in 1919 (see Chapter 17). Alexanderson was appointed chief engineer of the new corporation and stayed with RCA until 1925 when he returned to GE.

His association with Fessenden and later from 1912 with J H Hammond and I Langmuir gave Alexanderson an interest in radioengineering. An important outcome of the latter collaboration was the shift of research emphasis at GE from the improvement of electric light bulbs to the design and production of radio tubes and amplifiers. Steinmetz, in 1910, had organised a consulting engineering group at GE and Alexanderson became a member of the group. In 1918 he was named head of the new radioengineering department.

Alexanderson became interested in television during the summer of 1924. In connection with his early experimentation in this field he had engineered a facsimile transmission system which he demonstrated in 1924. By October 1925 GE had studied[28] various ways of implementing a practical scheme of television and had undertaken preliminary tests of two methods, *viz.* the use of a scanning disc with the controllable light source devised by D MacFarlane Moore; and, secondly, the use of a rotating prism for scanning and an oscillograph for light control.

Of the different scanners considered (such as rotating prisms, rotating mirrors, rotating lenses, oscillating mirrors and cathode-ray oscilloscope) the rotating prism was thought to be appropriate for facsimile and general television, and oscillating mirrors for amateur use. Alexanderson's conclusions from this work were that television was 'within reach but several of the necessary elements [had to be] more highly developed'. These were the photocell, the light control and the method of synchronisation.

Alexanderson's multiple beam method was suggested[29] in October 1925 and was described in his 1926 paper[26]. Figure 9.10 shows his apparatus. He was well aware of the need to transmit 'something like 300 000 pictures units per second' to achieve a good image (corresponding to rather more than 100 lines per picture) and indicated the difficulties as follows: 'Besides having the theoretical possibility of employing waves capable of a high speed of signalling, we must have a light of such brilliance that it will illuminate the screen effectively, although it stays in one spot [for] only 1/3 000 000 of a second. This was one of the serious difficulties because, even if we take the most brilliant arc light, no matter how we design the optical system, we cannot figure out sufficient brilliancy to illuminate a large screen with a single spot of light.'

It seems that Alexanderson wanted to follow Bernard Shaw's text for the television picture to be shown life size on a large screen and to attain this purpose had a model television projector built to study the problem.

'Briefly,' he said, 'the result of this study is that, if we employ seven spots of light instead of one, we will get 49 times as much illumination . . . The drum has 24

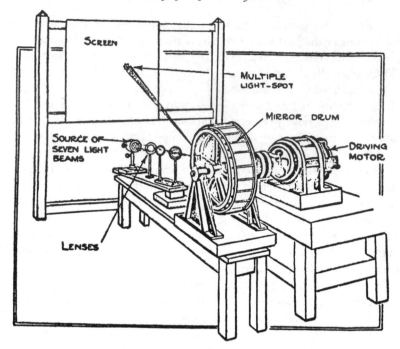

Figure 9.10   Arrangement of Dr Alexanderson's experimental television projector

Source: *Wireless World*, 20 April 1927, p. 479

mirrors and, in making one revolution, causes one light spot to pass over the screen 24 times; and when we use seven sources of light and seven light spots we have a total of 168 light spot passages over the screen during one revolution of the drum. The gain in using seven beams of light in multiple is two fold. First, we obtain a direct increase of seven to one in illumination, and then secure the further advantage that the speed at which each light beam must travel on the screen is reduced at a rate of 7:1 because each light spot has only 24 tracks to cover instead of 168.'

This use of multiple light sources represented the novelty of Alexanderson's method. It was tested, in a simulated way, on 18 September 1926[30].

His drum was approximately 2.5′ in diameter, with a flange about 10″ wide, and had 24 mirrors each measuring eight inches by four inches mounted around the periphery[31]. The drum was directly coupled to a high speed electric motor and the screen used was about 4 × 4 feet. In his experimental projector, Alexanderson arranged his seven light sources close together in a star formation so that the multiple beam, when seen stationary upon the screen, showed seven points of light arrayed like the end section of a piece of seven-stranded wire.

The use of multiple beams had a further advantage. 'Each light beam has to give only 43 000 instead of 300 000 independent impressions per second,' wrote Alexanderson. 'It is easy enough to design a television system that produces about 40 000 picture units per second, but the images are so crude that they would have very little practical value.'

Thus, Dr Alexanderson's plan consisted of reproducing on the screen seven crude pictures, containing only approximately 40 000 picture units each, and so interlacing them optically that the combined effect was that of a single good picture comprising about 300 000 units.

The complete television system would, of course, have required an independent control for each of the seven light sources and, for this purpose, seven photoelectric cells would have had to be arranged in a cluster at the transmitting end and their outputs utilised to modulate a multiplex radio system having seven channels. No details of Alexandersons's transmitter were given in his address and he made no claim to have completely solved the problem of good definition television. 'How long it will take to make television possible we do not know, but our work has already proved that the expectation of attainment is not unreasonable and that it may be accomplished with means that are in our possession at the present day.'

He based this conclusion on an experiment in which, using one light beam, he transmitted a motion picture film at the rate of seven seconds for each picture and argued that the same rate of speed with seven light beams would give one picture per second. Hence, to achieve television would require an increase in the speed of the process of sixteen times. By extrapolating his results, Alexanderson considered that, for the system he had in mind, a bandwidth of 700 000 kHz would be needed and that this could be included between wavelengths of 20 and 21 m.

His work at this time (1926) was far from complete. He had not even hinted at the method of light control in his lecture, but in an internal memorandum[30] dated 18 September 1926 he described his version of Karolus's Kerr cell. Alexanderson had recommended in 1924 the purchase, at a price of $200 000, of the rights to use this cell.

On 19 October 1926 Alexanderson applied for his first television patent[32]. It was based on his early ideas of utilising multiple light sources. The patent showed the transmitting and receiving arrangements for a four-channel scheme and the scanning of the object by a group of four light beams. In addition, the patent illustrated the method of light control, namely, the use of four galvanometers rather than four Kerr cells.

One of the disadvantages of this scheme lay in the need to have a different carrier frequency assigned to each photoelectric cell-transmitter combination. Consequently, Alexanderson's method represented an extension of the conventional systems which were dependent on a single modulated carrier wave for the image signals, but was a contraction of the scheme first put forward by the Mills Novelty Company in 1922[33], whereby the information content of a whole line of a picture was to be transmitted simultaneously. The latter strategy was a reduction of the idea originally advanced by Carey in 1877 in which the light values of all the elements which formed the scene were to be propagated at the same time and not sequentially.

In the Mills Company's proposal, Figure 9.11, the projected picture, 11, was to be scanned by a mirror drum, 12, and the reflected rays allowed to fall on a linear array of selenium cells, 18. A slit in the screen, 16, was to prevent light rays from other than a particular line of the picture being incident on the cells at any given instant of time. At the receiver a similar system was to be used with the exception that the cells were to be replaced by a linear array of light valves comprising mirrors, 37, attached to the armatures of electromagnets, 34, the individual stators of which were to be excited by currents proportional to the corresponding individual photo-conductive cells in the transmitter. Each of the cells was to be connected 'with a separate aerial of different characteristics so that the Hertzian waves transmitted were each of different frequency'.

It is interesting to note from Alexanderson's December 1926 paper[26] that his employers were anxious to develop television although he did not say why. 'If we knew of any way of sweeping a ray of light back and forth without the use of mechanical motion, the solution of the problem would be simplified. Perhaps some such way will be discovered, but we are not willing to wait for a discovery that may never come.'

It could be that GE was aware of the preparations which were being made by AT&T and did not wish to be left behind in the pursuit of a means of television that could be coupled to the vast American telephone network. Certainly, Alexanderson's paper indicates that this possibility was in his mind—a possibility which was to be implemented by AT&T a few months later. Also Alexanderson well

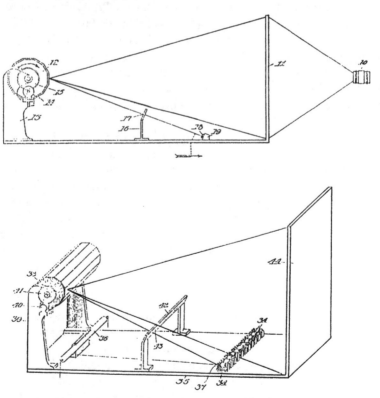

Figure 9.11   *The transmitter and receiver arrangements proposed in the 1922 patent of the ÏMills Novelty Company*

Source: British Patent 200643, 16 May 1922

knew the importance of a strong patent holding. In any new technical venture the acquisition of master patents can give the possessor considerable commercial bargaining power.

A demonstration[34], on 24 December 1926, of his multibeam projector gave very poor results. 'We found that the optical impressions are not entirely uniform when the lines are traced consecutively on the screen, the effect being that of a wave of light passing over the screen. To overcome this we adjusted the mirror so as to break up this wave impression. This is an improvement but there is yet visible a certain tremble of the light.' Planned changes using interlacing were never successfully tested and the multispot scheme was soon abandoned.

In April 1927 the AT&T Company demonstrated[35] its small and large screen systems of television using wire and radio transmission links. The demonstrations highlighted the very appreciable disparity between the achievements of GE and of AT&T and their approaches to the solution of the television problem. On the one hand AT&T employed many scores of scientists, engineers and technicians on their television work and tests; on the other hand, GE's work was conducted on a much smaller scale. Its results reflected this disparity. Alexanderson witnessed a test[36] of the GE television system, on 21 May 1927, arranged by P A Kober and R D Kell, in which a 48-line image was projected onto a small screen and the sending and receiving machines were kept in synchronism by hand. 'We could clearly distinguish the image of a hand, the opening and closing of the fingers.' The source of light for reproduction was a high frequency mercury light which Kober had brought from the Harrison Tube Plant. Two of the arcs were to have been used but as only one was to hand the actual image was produced by tracing every other line. Although the image quality was inferior to that shown by AT&T and others, Alexanderson felt that 'the test made was a very encouraging demonstration of the general operativeness of [their] system of television'. Kell[37], who was on the staff of the testing department, had been assigned to work in Alexanderson's television laboratory sometime in April 1927. He was soon to be supervising the project.

Kell does not seem to have favoured any of Alexanderson's multiple light source schemes. In a demonstration[38] given on 13 January 1928 from GE's research laboratories' broadcasting station in Schenectady, the type of equipment employed was that which was being used elsewhere (Figure 9.12). Very many multifarious schemes of seeing by electricity had been put forward from *circa* 1880 but, by the late 1920s, there had been a convergence of thought on the way ahead. The favoured grouping of apparatus comprised indirect, or flying spot, scanning (using a Nipkow disc) of the object, and a neon lamp source and Nipkow disc at the receiver. Such systems were in use in the US, the UK, Germany and France by the end of the decade.

In the GE demonstration, which was supervised by Alexanderson, the picture signals were broadcast from station 2XAF on 37.8 m and the sound signals were radiated from station WGY on 379.5 m. The flying spot scanner generated vision signals corresponding to 48 lines per picture, 16 pictures per second. These were received and displayed using the special neon lamp devised by D MacFarlane Moore. The image synthesised was about 1½″ long by 1″ wide and was magnified with a lens by a factor of two. Only three television sets, in different parts of

Figure 9.12   Merrill Trainer, one of the assistants of Dr Alexanderson in television research and development, standing before the television scanner

Reproduced by permission of the Schaffer Library

Schenectady, received the image signals: these were in the homes of Dr Alexanderson and two GE executives, E W Allen and E W Rice. A report stated that the received images were as good as those viewed in the laboratory and could represent every detail in faces.

On 13 March 1928 Alexanderson reported on a demonstration[39], due to Kell, of the projection of a television image on a silver screen by means of the Karolus version of the Kerr cell. The light source was sufficiently intense to produce a 48-line image of about 18 inches square. It was witnessed by Allen and was described as being 'of fair quality giving a good likeness of the person'. Alexanderson stated that this was the first time, to his knowledge, that an image of this quality and size had been projected onto a screen.

Further tests[40] were conducted on 26 July 1928. These used GE 24-line equipment operating in 'ordinary daylight'. Crude reproductions of subjects such as the full figures of two men boxing were shown. The tests indicated that the apparatus had adequate light sensitivity and so later tests were given with a 48-line scanning disc. Obviously, the holes in this disc only transmitted one quarter of the light flux transmitted by the apertures of the 24-line disc. Nevertheless, 'we could clearly see the skyline of the factories and the smoke coming from the chimneys. When we focused on the window itself, we could see not only the silhouette of figures and the bars in the window, but we could also clearly see the cords used for pulling the shades'.

The experience gained from these private tests gave GE confidence to attempt the first outside television broadcast[41]. On 21 August 1928 listeners and viewers tuned to WGY or to the shortwave stations 2XAF and 2XAD were able to hear, and see the face of, Govenor A E Smith, the Democratic candidate for president, when he made his acceptance speech for the nomination, in Albany.

Three tripod-mounted items of equipment were employed to televise the event. These comprised two units containing photelectric cells which were placed within 18″ of Smith's face, the 1 kW light source and the 24-hole scanning disc unit which was situated in front of him, face high, about 3′ away. From the hall the signals were sent 18 miles by landline to station WGY.

'Much, very much remains to be done,' said one report[41], 'but the first appearance of television pick-up outside of the laboratory is [the] forerunner of the day when such apparatus will be as familiar as the present microphone, and it may sometime be expected to find its place at all great public functions, at athletic events, etc., carrying not a verbal description of the event but an actual picture of the event.' According to Kell, the rehearsals were a success but the received images of the actual event were marred when the powerful arc lights of the newsreel cameras were switched on.

WGY was the pioneer station in the broadcasting of television in the US. A regular schedule of 15 to 30 minute programmes had been in operation for several months and by September 1928 four television broadcasts were being made weekly[42]. These were provided primarily for the assistance of the engineers who were developing the system of transmission and reception but, in addition, they were sent out to stimulate amateur interest in the subject. Reports of reception were received from Los Angeles, Detroit and several places in Pennsylvania. The greatest success in television reception had been obtained from the experiments conducted over short waves, maintained Alexanderson, who also stated that 48 lines was 'sufficient to give good results'.

Further public awareness was engendered by the television exhibits at the Radio World's Fair[43] held in New York on 21 September 1928. The *New York Times* headlined its report on the occasion as 'Television thrills radio show crowd' and said: 'More than 40 thousand persons filed past the television exhibits at the show yesterday according to one estimate. On Wednesday 11 thousand saw the General Electric Company's display between 1.00 p.m. and 11.00 p.m.. The visitors are sent through the rooms about 50 at a time. More than 220 performances of 2 to 3 minutes each are given daily . . . At the booth of the Carter Company, in the Theatre of Wonders, the arrangements permit 3 thousand persons to view the images in one hour . . . Those who appeared before the "eye" were Fay Cusick and Lee Crowe who with Mr J Hartley Manvers gave sketches from his play "The Queen's Messenger".'

Figure 9.13   Interior view of GE's home television receiver. At the top is a neon lamp, a standard motor and a revolving disc. In the centre are the batteries and a short-wave receiver, and at the bottom a radio amplifier. The photograph shows Dr Alexanderson and Mr R D Kell

Reproduced by permission of the Schaffer Library

The Queen's Messenger, a one act drama, was the first play to be televised[44]. It was broadcast on 11 September 1928 from WGY. Picture signals were sent out on 379.5 m and 21.4 m, and the sound signal on 31.96 m. The studio equipment included three portable cameras (one for each of the two characters and one for visual effects), which were based on 12″ scanning discs, each having 24 apertures, driven by small synchronous motors. Two tripod-mounted banks of photoelectric cells and a 1 kW lamp completed the basic studio set up. The two actors worked in front of white screens, to enable their features to be sharply defined[45], and to relieve the monotony of the transmitted head images the third camera was focused on the actors hands and various props such as keys, a ring, a pistol, a bottle of wine and a glass, a dagger and so on. In this way added realism was given to the production.

The complete performance was seen by a group of newspapermen and scientists gathered in one of the buildings of the General Electric Company, at a short distance from the television transmitter. According to one account the three inch by three inch received images had to be viewed at a distance of ten feet for a clear image to be observed. It had a 'quite good' definition. Another article said the pictures were sometimes blurred and confused, were not always in the centre of the viewed area and were hard on the eyes because of the flicker.

When the AT&T Company gave its impressive demonstration of television in April 1927 image signals were transmitted by radio and line links. The company had, of course, an extensive network of telephone lines and envisaged utilising these for two-way visual telephone services, and for the visual equivalent of a public address system.

GE, too, seemed keen to exploit the facilities of the telephone network for on 2, 3 and 4 October 1928 the company endeavoured to transmit video signals over long-distance transmission lines[46]. The field test originated from the NBC studio at 411 Fifth Avenue, New York, and employed the 24-line GE camera equipment.

In one test, signals were received at the WGY studio in Schenectady. Very bad line reflections and four pronounced transients spoiled the reproduction of a face.

A second test in which signals were sent 3000 miles from New York to Chicago, back to New York and then to Schenectady was a total failure as no faces were recognisable. Even the simplest geometrical figures could not be delineated. The trials, conducted by Kell and M Trainer, verified that video signals could not be propagated over a long unequalised telephone line.

Nonetheless, Alexanderson's group was undaunted and next attempted to send images from Schenectady to Australia and back to Schenectady, and also from Schenectady to Oakland, California and back to Schenectady, via radio paths[47]. The image used in the tests of 18 February 1930 was a rectangle drawn in bold lines on white paper. Multiple images were received but there were moments when a single fairly sharp image could be seen. Both tests gave substantially the same pictures—the California test being somewhat the better of the two. The *New York Times* said that Dr Alexanderson was 'much enthused with the result'[48].

These poor images were due to the reception of signals which had travelled over different path lengths. Interference between the signals and constantly varying path lengths led to changing phase differences and gave rise to the observed defects. On the basis of the experiments, Alexanderson suggested a method whereby the television apparatus might be utilised for the adjustment of broadcasting circuits 'because the eye is so much superior to the ear for analysis'. Nothing came of this plan.

The company persevered with its long range trials and on 13 February 1931 GE reported that it had successfully transmitted images from Schenectady to Leipzig, Germany[49]. The raison d'etre for this activity seems to have been based on the prospect of topical news events being displayed in cinemas contemporaneously with the actuality itself. The alternative method of showing newsreels depended on film being transported by (in those days) slow moving aircraft. Delays of many hours could occur therefore between a news report on radio and the subsequent projection of a news film in a theatre[50].

Rather intriguingly, the Baird Television Development Company, the American Telephone and Telegraph Company, the Radio Corporation of America and the Marconi Wireless Telegraph Company all gave consideration to the propagation of television signals over very long distances.

References

1 ANON.: 'Charles Francis Jenkins', *The National Cyclopaedia of American Biography*, p. 246
2 JENKINS, C.F.: 'Transmitting pictures by electricity', *The Electrical Engineer*, 25 July 1894
3 ANON.: 'Électricité', *Cosmos*, NS tome XXIX, (502), 1894, p. 161
4 ANON.: 'Motion pictures by wireless', *Moving Picture News*, 27 September 1913, **VIII**
5 '"Radio vision" shown first time in history by Capital inventor', *Sunday Star*, 14 June 1925. Also: 'First motion pictures transmitted by radio are shown in Capital, Washington', 14 June 1925
6 JENKINS, C,F.: 'Radiomovies, radiovision, television' (Jenkins Laboratories, Washington, D.C., 1929) p.60
7 JENKINS, C.F.: 'Spiral mounted lens disk'. US patent 1679 086, 2 January 1925
8 JENKINS, C.F.: 'Continuous motion picture machines', *Trans. Soc. Motion Picture Engineers*, May 1920, **20**, pp. 97–102
9 CLAUDY, C.H.: 'Motion pictures by radio', *Sci. Am.* November 1922, **127**, p. 320
10 DINSDALE, A.: 'Television apparatus. A description of the Jenkins system', *Wireless World*, 5 May 1926, p.642
11 ABRAMSON, A.: 'Pioneers of television—Charles Francis Jenkins', *SMPTE J.*, February 1986, pp. 224–238
12 HERNDON, C.A.: '1000 printed words a minute by radio', *Popular Radio*, January 1925, **7**, pp. 11–15
13 DAVIS, W.: 'Seeing by radio', *Popular Radio*, April 1923, **3**, pp. 266–275
14 DAVIS, W.: 'The new radio movies', *Popular Radio*, December 1923, **4**, pp. 437–443
15 ANON.: 'Bell Labs test coaxial cable', *Electronics*, December 1937, pp. 18–19
16 JENKINS, C.F.: 'The drum scanner in radiomovies receivers', *Proc. IRE*, September 1929, **17**, (9) pp. 1576–1583
17 Report in *Nature*, 1 December 1928, p. 853
18 'Broadcasting pictures', *New York Times*, 6 May 1928, p. 3:2
19 UDELSON, J.H.: 'The great television race' (University of Alabama Press, 1982) p. 37
20 'Jenkins television demonstration—Case 52301'. Memorandum, 17 January 1930, AT&T Archives, Warren, New Jersey, USA
21 'Radio reception of the Jenkins television broadcast'. Memorandum 20 January 1930, AT&T Archives, Warren, New Jersey, USA
22 JENKINS, C.F.: 'Radiovision', *Proc. IRE*, 1927, **15**, pp. 958–964
23 ANON.: 'Radio mirror', *Daily News*, 30 December 1926
24 ANON.: 'Predicts vision across the ocean', *The New York Times*, 16 December 1926
25 SHAW, G.B.: 'Back to Methuselah' (Constable, London, 1924)
26 ALEXANDERSON, E.F.W.: 'Radio photography and television', *General Electric Review*, February 1927, **30**, (2) pp. 78–84
27 ANON.: 'E F W Alexanderson', *The National Cyclopaedia of American Biography*
28 ALEXANDERSON, E.F.W.: Memorandum, 23 October 1925, MSS A379, Schaffer Library, Union College, Schenectady, USA
29 ALEXANDERSON, E.F.W.: Memorandum to H E Dunham on 'Television projector', 31 October 1925, MSS A379, Schaffer Library, Union College, Schenectady, USA
30 ALEXANDERSON, E.F.W.: Memorandum to C O Howland on 'Television', 18 September 1926, MSS A379, Schaffer Library, Union College, Schenectady, USA

31 DINSDALE, A.: 'Phototelegraphy and television', *Wireless World*, 20 April 1927, pp. 476–480
32 ALEXANDERSON, E.F.W.: 'Electrical transmission of pictures'. US patent 1 694 301, 19 October 1926
33 Mills Novelty Company: 'Improved process of an apparatus for telegraphically reproducing pictures and the like'. British patent 200 643, 16 May 1922
34 ALEXANDERSON, E.F.W.: Memorandum to C O Howland on 'Television projector', 24 December 1926, MSS A379, Schaffer Library, Union College, Schenectady, USA
35 BURNS, R.W.: 'The contribution of the Bell Telephone Laboratories to the early development of television' (Mansell, London, 1991) pp. 181–213
36 ALEXANDERSON, E.F.W.: Memorandum to A D Lunt, 21 May 1927, MSS A379, Schaffer Library, Union College, Schenectady, USA
37 ALEXANDERSON, E.F.W.: Memorandum to A D Lunt on 'Synchronisation of radio picture receivers', 26 April 1927. Also, memorandum to H E Dunham, 11 May 1927, MSS A379, Schaffer Library, Union College, Schenectady, USA
38 ANON.: 'Radio television to home receivers', *New York Times*, 14 January 1928, p. 1:6
39 ALEXANDERSON, E.F.W.: Memorandum to H E Dunham, 17 March 1928, MSS A379, Schaffer Library, Union College, Schenectady, USA
40 ALEXANDERSON, E.F.W.: Memorandum to H E Dunham, 2 August 1928, MSS A379, Schaffer Library, Union College, Schenectady, USA
41 ANON.: Memorandum, 23 August 1928, AT&T Archives, Warren, New Jersey, USA
42 *The New York Times*, 6 September 1928, p. 22:2
43 'Television thrills radio show crowd', *The New York Times*, 21 September 1928, p. 24:1
44 HERTZBERG, R.: 'Television makes the radio drama possible', *Radio News*, December 1928, pp. 524–527, 587
45 'Drama via television', *Science and Invention*, December 1928, pp. 694 and 762
46 TRAINER, M.A.: Memorandum, 8 October 1928, MSS A379, Schaffer Library, Union College, Schenectady, USA
47 ALEXANDERSON, E.F.W.: a memorandum to H E Dunham, 19th February 1930, MSS A379, Schaffer Library, Union College, Schenectady, USA
48 Report in *The New York Times*, 19 February 1930, p. 1:6
49 'Schenectady to Leipzig television a success', *The New York Times*, 13 February 1931, p. 15:3
50 'Television in the theater', *Electronics*, June 1930, pp. 113 and 147

Chapter 10

Excellence in low-definition television (1925–1930)

Of all the demonstrations of television that were given in the 1920s, none surpassed in technical excellence those mounted by the Bell Telephone Laboratories of the American Telephone and Telegraph Company[1].

The results that were obtained by the Laboratories in 1927 are particularly important historically because they were the best that could be expected with the technology as it existed at that time. With its vast resources in finance and equipment, and in staff expertise and experience, the Laboratories was uniquely able to demonstrate what could be engineered in the field of television. Subsequently, colour television and two-way television systems were realised in 1929 and 1930 respectively, all at great cost. As noted previously, from 1925 to 1930 (inclusive) the American Telephone and Telegraph Company approved the expenditure of $308 100 on low-definition television—a sum far in excess of anything available to any of the lone television workers of whom Baird, Jenkins, Mihaly, Belin and Karolus were in the vanguard.

After 1930 the Laboratories continued its work on television, but without achieving successes of the type which were being manifested contemporaneously by RCA of the US and EMI of the UK, despite the allocation of $592 400 to the work from 1931 to 1935 (inclusive). Thereafter, television research and development declined and ceased, sometime in 1940, to be part of the Laboratories interests.

Although the Laboratories initiated its television project in the mid-1920s, there is some evidence that the Bell System had an interest in television during the Great War and in 1921.

In November 1912, A McL Nicolson[2], a British subject living in New York, joined the research branch of the engineering department of the Western Electric Company (now Bell Telephone Laboratories) and soon after began active work on vacuum valve research in association with Dr H D Arnold and Dr H J Van der Bijl. Nicolson's first important contribution was the development in 1913 of the 'unipotential' (or indirectly heated) cathode; an innovation which achieved worldwide usage. His coated filament manufacturing process was employed in the production of all telephone and radio valves made by the company prior to 1917.

Later, from 1917, Nicolson undertook a major study of piezoelectricity and was the first person to discover that the mechanical vibrations of a crystal such as Rochelle salt could be coupled to an electric circuit by means of suitable electrodes, and that the resulting electromechanical vibrating system performed exactly like an electrically tuned circuit. He successfully used such crystals in loudspeakers, microphones, gramophone pick ups and valve oscillators. His patent on crystal-controlled oscillators was filed in April 1918, and after numerous interferences, was issued in August 1940: it antedated all others in this field.

Nicolson was a prolific inventor throughout the early years of radio and television, and during his life applied for about 180 patents. On 13 October 1916 he forwarded, to E H Colpitts, the head of the research branch, a 28 page paper[3] on television transmission and reception and mentioned in his covering letter: 'This completes my work on the subject as first submitted to you July 23rd, 1915.' At the end of the paper Nicolson referred to his original manuscript of 95 pages (including 170 figures) dating from 20 April 1916 to 7 October 1916. Regrettably, his paper gives no indication of the stimulus which led to his study of television, but it does list 11 features of his scheme that were considered to be patentable. Subsequently, on 7 December 1917, Nicolson sought patent protection for his system of television and this was granted on 16 October 1923: he assigned the patent to the Western Electric Company of New York. The company evidently felt that the patent had some worth for it obtained patent safeguards not only in the US but also in France and the UK[4].

Nicolson's scheme incorporated an electromechanical transmitter and cathode-ray tube receiver and was one of the first to include thermionic valve circuits operating as amplifiers and oscillators. Furthermore, the patent clearly illustrated wireless broadcasting and reception of the vision signals.

Figure 10.1 depicts his design. The scanner consisted of a mirror, 15, mounted at the intersection of two supporting wires, 80, 81, at right angles to one another, and was capable of moving in a magnetic field established between the poles of two electromagnets. By the application of appropriate currents to these electromagnets, the mirror was able to scan the object in a spiral path similar to that shown in Figure 1 or Figure 2 of the patent. As it did so, the light from it was focused by the lens, 21, onto the cathode of a photocell, 17. These photocurrents modulated a carrier wave which was additionally modulated by signals dependent on the position of the scanning spot. For this purpose Nicolson used a special photocell, 51, provided with two anodes, 53 and 54, and a cathode, 50. Light from an auxiliary constant source, 45, was reflected by the mirror, through a lens, 52, onto the cell's cathode so that as the radial position of the focused image on the cathode varied from an auxiliary electrode, 56 (maintained at a constant potential difference with respect to the cathode), the two anode currents altered as a function of the position of the scanning point on the object.

At the receiver the two low-frequency positional signal currents were filtered out from the transmitted modulated carrier wave and used to control the deflection of an electron beam in a Braun tube. Similarly, the picture signal currents were

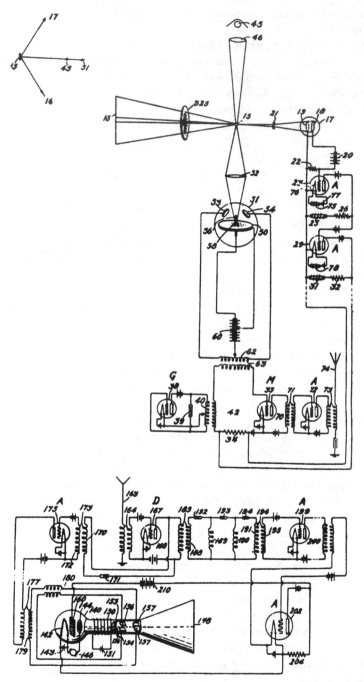

Figure 10.1   The television system of A McL Nicolson was based on an oscillating mirror transmitter scanner and a cathode-ray display tube

Source: British patent 288961, 7 September 1923

Figure 10.2   Dr H E Ives (1882–1953)

Reproduced by permission of AT&T Bell Laboratories

extracted and utilised to change the potential of the grid of the Braun tube in accordance with the photocell's output current.

Seventy-three claims were made by Nicolson in his US patent and thirteen in the equivalent British patents, but of these the most important from the point of view of the development and history of television concerned his use of a grid, in the cathode-ray tube, to control the intensity of the electron stream.

Prior to Nicolson's 1917 patent application, inventors such as Dieckmann and Glage[5], Rosing[6] and Campbell Swinton[7], had incorporated post anode control of the intensity of the electron beam, but now a new and much more effective control was available for this purpose.

F Skaupy[8] in 1919 also advocated the utilisation of grid control, in telephoto-graphic systems, and the method was mentioned by P Villard in his 1908 book on 'Les Rayons Cathodiques'. Zworykin[9] employed the same method in his US patent of 13 July 1925, discussed in Chapter 16, and it is now a standard feature of all modern television receiving tubes.

Another aspect of Nicolson's patent which is of some importance concerns his use of a single carrier wave for propagating both the picture and synchronising signals. Previously, the transmission link between the transmitter and receiver had been a transmission line: now, with the rapid advances which were taking place in the field of communications, a new mode of wireless transmission was available. Zworykin's 1925 patent likewise included a wireless link but, as the Western Electric Company had filed a prior claim, Zworykin necessarily had to use a different transmitting arrangement. His solution was to employ two carrier signals, of different frequencies, one of which was modulated by the picture signals and the other by the synchronising signals, both being radiated from a common antenna.

The remaining aspects of Nicolson's patent hardly advanced the state of the art. Oscillating mirrors were never popular with the large industrial organisations which were working in the field of television, and the simple versatile, robust and cheap Nipkow disc, and the more costly mirror drum, were never to be supplanted by oscillating devices, which perforce tended to be costly, fragile and limited in their performance as scanners.

Apart from Nicolson's 1915 and 1916 report and patent the only other document which relates to early (pre-1920) television interest in a Bell company is the memorandum[10] written in January 1894 by G K Thompson of the Mechanical Department of the Boston Laboratory of the American Bell Telephone Company. Thompson advocated the employment of Nipkow disc scanning analysers and synthesisers, an electric shutter type of light valve at the receiver and a light-sensitive cell comprising a large number of glass bulbs containing burnt cork (sic) at the transmitter. He thought the subject was 'worthy of being considered by [the] Company'.

Following the birth of broadcasting, *circa* 1920, interest in television was renewed. On 6 September 1921 PM, (probably Pierre Mertz), wrote an eight-page memorandum[11] 'to note down some preliminary considerations regarding a possible system of television'. Whether this was a personal initiative or was requested by the company was not stated. His paper was read and very briefly annotated, by a person having the initials AW, on 1 July 1923.

PM examined the prospect of transmitting picture signals by means of multi-channel communication methods. For a picture 5 × 5 inches, and using half-tone printing screens capable of printing 150, 100, 60 and 16 dots per inch, PM calculated the number of dots which would have to be sent every second (the signalling speed) for a televised image to be synthesised. Then, for a number of signal channels equal to the number of dots in a row of the picture, he determined the signalling speed per channel. His results are given in the table opposite.

These requirements are extremely severe noted PM. His observation was similar to that obtained, in 1908, by Shelford Bidwell who, by a congruous method of calculation, found that 160 000 synchronised operations per second would be needed to transmit a 2 × 2 inch image. This was 'widely impractable', he opined[12]. His view led to Campbell Swinton's all-electronic television notions of 1908 and 1911.

No. of dots per inch (n)	Total no. of dots ($25n^2=N$, say)	Signalling speed for 1 channel ($=16\times N=S$, say)	Multichannel signalling	
			No. of channels ($=5\times n=$ no. of rows)	Signalling speed per channel $=S/5n$
150	562 000	9 000 000	750	12000
100	250 000	4 000 000	500	8000
60	90 000	1450 000	300	4800
16	6400	102 400	80	1280

There appears to be no evidence to show that either PM's or Nicolson's ideas were ever implemented. Rather, AT&T commenced its experimental study of the television problem when 'it began to be evident that scientific knowledge was advancing to the point where television was shortly to be within the realm of the possible'[13]. The company was of the opinion that television would have a real place in worldwide communications and that it would be closely associated with telephony. It was certainly well placed to advance television, not only because of the extensive facilities of the newly formed (1925) Bell Telephone Laboratories but also because of the experience which had been acquired in the research and development work that had made transcontinental and transoceanic telephony and telephotography possible.

In January 1925 development work under the direction of Dr H E Ives had been completed on a system for sending images over telephone lines and so research resources and expertise existed for a new scientific venture. Dr Ives and Dr Arnold, the director of research, agreed that the next problem to be undertaken was television. 'At Arnold's request,' wrote Ives[14], 'I prepared and submitted to him on 23 January 1925, a memorandum[15] surveying the problem and proposing a programme of research.'

Ives was eminently well qualified to lead a television project team. His experience and erudition at this time (1925) had been founded on work and investigations on colour photography, phosphorescence, illumination, colour measurement, intermittent vision, photometry, photoengraving, photoelectricity and picture transmission, *inter alia*. His standing in his chosen fields had been recognised by the award (in 1906, 1915 and 1918) of three Longstreth Medals by the Franklin Institute of Philadelphia. Later he was to receive three more medals—in 1927 the John Scott Medal, in 1937 the Frederick Ives Medal of the Optical Society of America and, after World War II, the Medal for Merit, the highest civilian award of the US Government, for his war work[16].

Ives's memorandum discussed the characteristic difficulties of securing the requisite sensitiveness of the pick-up apparatus, the wide bandwidths which from his experience of picture transmission were indicated as necessary for television, the problem of producing enough modulated light in the received image to make it satisfactorily visible, and the problem of synchronising apparatus at the sending and at the receiving ends of the transmission link. The memorandum concluded with a proposal for 'a very modest attack' on the problem capable, however, of 'material expansion as new developments and inventions materialised'.

Ives felt that these difficulties could be examined by utilising a mechanically linked transmitter and receiver, each incorporating a Nipkow disc scanner operating on a 50 lines per picture, 15 pictures per second standard. A photographic transparency, later to be superseded by a motion picture film, would be used at the sending end, together with a photoelectric cell and a carbon arc lamp. At the receiving end Ives proposed the use of a crater-type gaseous glow lamp. His plan was thus based on the transmission of light through the object rather than on the reflection of light from an opaque body: the latter problem was found by experimenters to be much more difficult. $15 000 was approved for the project.

By May 1925 the apparatus for Ives's design had been constructed and was in operation. A memorandum[17] of 14 May 1925, by J G Roberts, a patent attorney, records: 'I witnessed today a demonstration of Mr Ives's system of television. He has constructed and put into operation substantially the system he described in his memorandum of 23 January 1925, to Mr Arnold. In viewing the picture at the receiving end, I could distinguished with fair definition the features of a man's face like that of a picture at the transmitting end and also observed that, when the picture at the transmitting end was moved forward or backward, or up or down, the picture at the receiving end followed these motions exactly.' (Figure 10.3.)

Figure 10.3   Research apparatus used in the mid-1920s at Bell Telephone Laboratories. The scanning discs at the transmitter and receiver are mechanically coupled to ensure synchronism of the two discs. Dr F Gray is standing behind the box which contains the photocell. J R Hefele is observing the synthesised image

With this initial success behind them Ives's group, Dr F Gray, J R Hefele, R C Mathes, R V L Hartley, *et al.* next tackled the problem of synchronisation when the two Nipkow discs were uncoupled. H M Stoller was given responsibility for this particular phase of the project and, by December 1925, the group was able to show motion pictures from a projector driven in synchronism with the discs.

Another stage in the research programme was passed on 10 March 1926 when, at the conclusion of the ceremonies to mark the fiftieth anniversary of the invention of the telephone by Alexander Graham Bell, F B Jewett, president, and E B Craft, executive vice-president, talked over a telephone circuit in the telephone laboratory and were able to see the face of the speaker at the far end of the line.

According to Ives,[14] the group forebore to announce their achievement because, from the beginning of their investigations, it had been considered that only when vision signals could be sent over large distances—to parallel 'what had been done for voice signals'—would their apparatus be worthy of the appellation 'television system'. 'It would be television when the laboratory experiment was expanded to cover distances beyond any the eye could reach.'

By 7 April 1927 the system was ready to be demonstrated. It has been estimated that over one hundred engineers, scientists and technicians contributed to the success of the project[18], although some reports mention a figure of one thousand.

The demonstration[19], using a wire link, consisted of the transmission of images from Washington, DC, to the auditorium of the Bell Telephone Laboratories in New York, a distance of over 250 miles. During the radio demonstration images were sent from the Bell Laboratories experimental station 3XN at Whippany, New Jersey, to New York City, a distance of 22 miles. Reception was by means of two forms of receiver. One receiver produced a small image of approximately 2.0in by 2.5 in, which was suitable for viewing by one person. The other receiver gave a large image of nearly 24 in by 30 in for viewing by an audience of considerable size, Figure 10.4.

Ives and his colleagues used a Nipkow disc, having 50 apertures, for scanning purposes. They arrived at this figure by taking as a criterion of acceptable image quality the standard of reproduction of the half tone engraving process in which it was known that the human face can be satisfactorily reproduced by a 50-line screen. Thus, assuming equal definition in both scanning directions, 40 000 elements per second had to be transmitted for a rate of picture transmission of 16 pictures per second. The frequency range needed to transmit this number of elements per second was calculated to be 20 kHz.

A spotlight scanning method[20] was adopted to illuminate the subject, the beam of light being obtained from a 40A Sperry arc. Three photoelectric cells of the potassium hydride, gas-filled type were specially constructed and utilised to receive the reflected light from the subject. These were probably the largest cells that had ever been made by that time and presented an aperture of 120 in², Figure 10.5.

For reception a disc similar to that at the sending end was used together with a neon glow lamp, Figure 10.6. The disc had a diameter of 36 inches and synthesised the 2.0 by 2.5 inch image. Another form of receiving apparatus comprised a single, long, neon-filled tube bent back and forth to give a series of fifty parallel sections of tubing. The tube had one interior electrode and 2500 exterior electrodes cemented

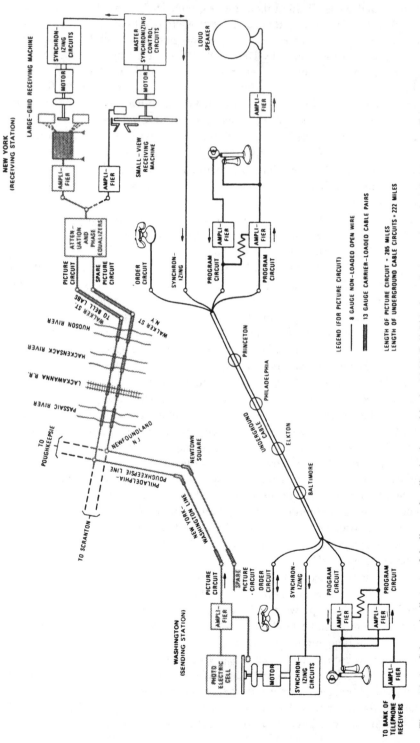

Figure 10.4   Schematic diagram of the line and radio circuits used in the 1927 Bell Telephone Laboratories' television demonstration

Reproduced by permission of AT&T Bell Laboratories

Figure 10.5   The photocells used in the April 1927 television demonstration were, at that time, the largest ever fabricated

Reproduced by permission of AT&T Bell Laboratories

along its rear wall. A high frequency voltage applied to the interior electrode and one of the exterior electrodes caused the tube to glow in the region of that particular electrode. The high frequency modulated voltage was switched to the electrodes in sequence from 2500 bars on a distributor with a brush rotating synchronously with the disc at the transmitting end. Consequently, a spot of light moved rapidly and repeatedly across the grid in a series of parallel lines, one after the other, and in synchronism with the scanning beam. With a constant exciting voltage the grid appeared uniformly illuminated, but when the high frequency voltage was modulated by the vision signals, an image of the distant subject was created, Figures 10.7 and 10.8.

To transmit the vision, sound and synchronising signals three carrier waves were employed; 1575 kHz for the image signals, 1450 kHz for the sound signals and 185 kHz for the synchronisation controls[22].

According to Ives, success of the system was due to the 'chief novel features' listed below[14]:

1   choice of image size and structure such that the resultant signals fell within the transmission frequency range of the available transmission channel;
2   scanning by means of a projected moving beam of light;
3   transmission only of the a.c. components of the image;
4   use of self-luminous surfaces of high intrinsic brilliancy for the reconstruction of the image;
5   utilisation of high frequency synchronisation.

*Figure 10.6   Close up of the transmitting apparatus installed at Washington for the 1927
demonstration. Mr E F Kingsbury of Bell Telephone Laboratories is sitting in front
of three large photoelectric cells, and a microphone*

Reproduced by permission of AT&T Bell Laboratories

The first demonstration consisted of the transmission of an image of, and an address
by, Herbert Hoover, Secretary of Commerce, from Washington to New York over
telephone lines. The second demonstration by radio comprised three events: first, an
address by E L Nelson, a Bell Laboratories engineer; secondly, a vaudeville act
featuring 'a stage Irishman, with side whiskers and a broken pipe, . . . [who] did a

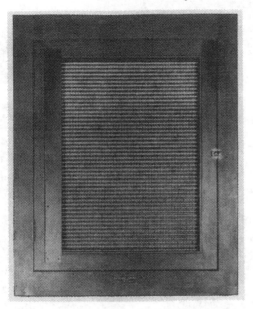

Figure 10.7   *The large-screen display consisted of a single, long, neon-filled tube bent back and forth to give a series of 50 parallel sections of tubing. The tube had one interior electrode and 2500 exterior electrodes cemented along its rear wall*

Reproduced by permission of AT&T Bell Laboratories

monologue in brogue' and then, after a quick change, returned with a blackened face and made a few quips in a negro dialect; and, finally, a short humorous dialect talk'[23].

The received images were subject to some fading and ghosting and occasionally appeared in the negative, but in general they impressed the audience. 'It was as if a photograph had suddenly come to life and began to talk, smile, nod its head and look this way and that,' said one observer.

Colonel Angwin, the deputy chief engineer of the General Post Office, UK, witnessed a demonstration of the Bell Laboratories television system some time after the public demonstration[24]. In his report he mentioned: 'This system reproduces a clear and undistorted picture and the results obtained are undoubtedly far in advance of those claimed for by the Baird system. The American system is a very costly and elaborate piece of mechanism and requires a special circuit for line transmission and exceptionally stable conditions for wireless transmission.' Angwin's visit occurred about one year later than the April 1927 trial, but during that time nothing had been done by the company to exploit the system commercially and Angwin was given to understand that it had no intention of proceeding further at that time.

Dr Dauvillier, the eminent French physicist, observed the wireless transmission and, in a historical review article on television published in the *Revue Générale de L'Électricité*, wrote: 'Finally, the Bell Telephone Company recently succeeded in transmitting to a considerable distance the human face, using (without acknowledgement) the Baird system.'[25]

Figure 10.8   A photograph of the commutator used to switch the received, amplified television signals to the 2500 electrodes sequentially

Reproduced by permission of AT&T Bell Laboratories

The April 1927 demonstrations were the finest that had been given anywhere, even though no especially novel features had been incorporated into the various systems. They established standards from which further progress could be measured. Moreover, the publication in October 1927, in the *Bell System Technical Journal* (**6**, pp. 551–653), of five detailed papers on the factors which led to Ives's group success enabled other workers to ponder on whether their own ideas and practices were likely to lead to similar favourable outcomes.

Certainly, the Bell Laboratories equipment could be further developed. The large-screen grid display was 'very much inferior' to that of the gaseous discharge lamp, the person being televised had to sit in a semidarkened room, there was a need to dispense with the separate synchronising channel and there was a requirement for more detailed images. In addition to these considerations, the policy of AT&T towards television advancement in the Bell System had to be defined.

Ives felt there were three principal projects to be tackled[26]. First, the introduction of a two-way appointment service between New York and Washington 'as a means of keeping the Bell System on the map in connection with the onward course of television, while at the same time securing information as to its possible uses and problems'. Secondly, transatlantic television would be the 'supreme achievement' in television and would have 'an appeal to the imagination of all ranks of humanity which would be unsurpassable'. Thirdly, the development of the public address television apparatus, and its possible exercise in televising a presidential inauguration, would find its justification if it were the 'policy of the Telephone Company to either provide a service of this sort or through its subsidiaries to manufacture apparatus from the sale of which, or from the use of which, income could be expected'.

Of these proposals, approval was given for projects one and three. Project two was thought to be a 'publicity affair'[27]. F B Jewett, president of Bell Telephone Laboratories, believed that Ives's group should proceed as vigorously as possible with the preliminary work of the latter project, without there being any definite commitment. In addition, the Laboratories should carry on, as adequately as possible, whatever fundamental work would be necessary to safeguard the company's position and advance the art along lines that were likely to be of interest to the company.

This mandate gave Ives ample scope to investigate a quite wide range of television problems. He seized the opportunities made accessible to him and during the next three years daylight television, large-screen television, television recording, colour television and two-way television were all subject to his group's scrutiny and engineering prowess.

The televising of objects illuminated by natural daylight by the method of direct scanning was demonstrated on 10 May 1928. Studies of the optical conditions peculiar to television had brought out the simple fact that the light gathering power of a lens and television scanning disc could be increased by enlarging the physical dimensions of the whole scanning system[28]. Other work in the Laboratories had led to the evolvement of photoelectric cells of greatly increased sensitiveness. With these cells, of the thalofide type, a 50 hole, 36 in diameter scanning disc and a lens system of aperture f/2.5 it was practicable to transmit images of a full length human figure when the apparatus was taken out of doors and set up on the roof of the Bell Laboratories building in New York. A press show was given on 12 July 1928[29]. This contained scenes of a sparring match, a golf exhibit and other movements.

The evolution of equipment for public address television proceeded along several paths.

First, in the grid display receiver of 1927 the form of brush used and the method of making contact with the individual electrodes on the grid were much improved: Ives could 'guarantee', in February 1928, that a face could be reproduced so as to be 'very satisfactorily recognisable'[30].

Secondly, in 1927, Hartley and Ives patented means for projecting televised images by photographing the received image with a cine camera and developing the film images with the minimum of delay[31]. Also, they patented the generation of television signals from rapidly processed cine film. The advantages of using motion picture film at the sending end of a television system was known to Ives in 1925. In his January 1925 memorandum[15] to Arnold, Ives referred to measurements which he had made of the brightness of the image of a sunlit landscape as projected onto a photoelectric cell by means of a wide aperture lens. The measurements revealed that the magnitude of the light flux which could be concentrated on the cell was about 1/500th of that employed in picture transmission. Hence, the degree of sensitivity of the photocell and the degree of amplification necessary in the proposed television system would be far greater than those that had been acceptable in connection with picture transmission.

Ives advocated[32] using transmitted light (from a film) rather than reflected light (from a scene) to ease the solution of the problem. He noted: 'It may be pointed out that the use of a moving picture film as the original moving object is the equivalent to a very great amplification of the original illumination brought about by the photochemical amplification process involved in the production and subsequent development of the photographic latent image.'

This intermediate film system of television was employed in the 1930s by Fernseh AG and Telefunken in Germany, and by Baird Television Ltd at the London Television Station, Alexandra Palace.

Thirdly, Gray devised a method of projecting television images based on the optical projection of a small, illuminated and moving section of a slot cut in a rotating disc and placed in front of a capillary light source[33]. This lamp could operate either as a glow discharge through mercury vapour or as a mercury arc. In February 1929 Gray could exhibit 50-line, projected 21 by 32 inch television images for viewing in a darkened room and 11 by 14 inch screen images for viewing in a room only partially darkened. A modification of this apparatus allowed 50-line television images at 18 images per second to be recorded on 35 mm motion picture film[34].

When Ives was admitted in 1905 to John Hopkins University, as a PhD student, it was to study colour photography under the supervision of R W Wood, then the leading authority on optics in the United States. Ives's choice of subject had possibly been influenced by his father's contributions to the art of photography and the science of optics. Dr F E Ives had invented a trichromatic camera, various processes of colour photography and a 'device for optically reproducing objects in both full modelling and natural colours; *inter alia*.

H E Ives's first two papers (both published in *The Physical Review*) relate to improvements in methods of colour photography; the earlier paper[35] (1906) pertains to Wood's device and the later paper[36] (1907) to Lippmann's scheme. The papers were written before Ives had completed his doctoral thesis (1908) which had the title 'An experimental study of the Lippman colour photograph'.

With such a background it was perhaps inevitable that Ives and his coworkers would want to attempt to create coloured televised images. On 27 June 1929 colour

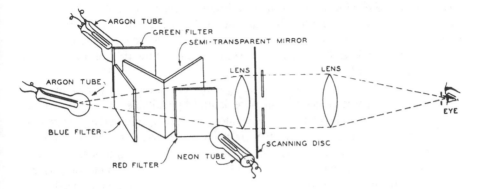

Figure 10.9   The schematic layout of the lamps and filters used in Bell Telephone Laboratories'
colour television demonstration of 1929

Reproduced by permission of AT&T Bell Laboratories

television was shown by Bell Telephone Laboratories to an invited gathering of scientists and journalists[37].

Ives utilised three signal channels so that the three colour signals could be sent simultaneously from the transmitter to the receiver[38]. An advantage of this arrange-

Figure 10.10 Side view of Bell Telephone Laboratories' colour television equipment. With the
exception of the photoelectric cabinet at the left, the apparatus was identical with that
used for the April 1927 demonstration of monochromatic television

Reproduced by permission of AT&T Bell Laboratories

ment was that the same scanning discs and motors, synchronising equipment and light sources and the same type of circuit and method of amplification were used as in the monochrome scheme. The only new features were the form and disposition of the specially devised photocells at the sending end and the type and grouping of the neon and argon lamps at the receiving end, Figures 10.9 and 10.10.

A neon glow lamp gave the desired red light but for the sources of green and blue light 'nothing nearly so efficient as the neon lamp was available'. Two argon lamps, one with a green filter and one with a blue filter, were finally adopted for the demonstration; however, various expedients were needed to increase their effective luminous intensity. Special lamps with long, narrow and hollow cathodes cooled by water were utilised and these were observed end on so that the thin glowing layer of gas was greatly foreshortened and the apparent brightness thereby increased.

To render the correct tone of coloured objects it was essential to obtain photoelectric cells which would be sensitive throughout the visible spectrum. A R Olpin and G R Stilwell constructed a new kind of cell which used sodium in place of potassium. Its active surface was sensitised by a complicated process involving sulphur vapour and oxygen instead of by a glow discharge of hydrogen as with the former class of cell.

An account of the demonstration, in which the transmission was over lines, was published in the *Telephony* journal for 6 July 1929[39]. The display 'opened with the American flag fluttering on a screen about the size of a postage stamp. The observer saw it through a peep hole in a darkened room. The colours reproduced perfectly.

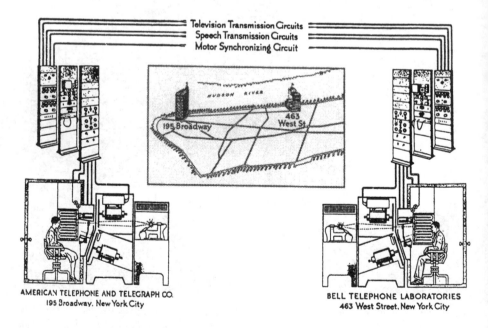

AMERICAN TELEPHONE AND TELEGRAPH CO.
195 Broadway. New York City

BELL TELEPHONE LABORATORIES
463 West Street. New York City

Figure 10.11 Diagrammatic layout of Bell Telephone Laboratories—two-way television

Reproduced by permission of AT&T Bell Laboratories

Then the Union Jack was flashed on the screen and was easily recognised by its coloured bars.

'The man at the transmitter picked up a piece of watermelon, and there could be no mistake in identifying what he was eating. The red of the melon, the black seeds and the green rind were true to nature, as were the red of his lips, the natural colour of his skin and his black hair . . . '

As previously noted, of the two projects approved by Jewett, one pertained to two-way television, Figure 10.11. This was established between the main offices of the AT&T Company at 195 Broadway, New York and the Bell Telephone Laboratories at 463 West Street, New York and was demonstrated on 9 April 1930[40]. It consisted essentially of two complete television transmitting and receiving sets of the kind employed in the 1927 one-way television scheme. Spot light scanning, Nipkow discs and neon lamps were still incorporated but with several improvements. Two discs, each containing 72 holes to give double the image detail of the 50-hole discs of the 1927 apparatus, were utilised at each end of the line links, one was for image analysis and the other for image synthesis. In addition to the photoelectric cell and neon lamp each 'ikonophone' booth had a concealed microphone and loudspeaker. Special precautions had to be taken with the telephone circuit to prevent singing due to the closeness of the two electroacoustic transducers. Also, the increased band-width of the system led to problems of amplitude and phase equalisation which were more difficult than those encountered in the earlier tests[41].

SOMETHING MUST BE DONE ABOUT THIS TELEVISION MENACE

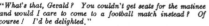
"*What's that, Gerald? You couldn't get seats for the matinee and would I care to come to a football match instead? Of course! I'd be delighted.*"

"*Is that you, Sir? I can't possibly come down to the office to-day I think I'm sickening for 'flu or something.*"

Reproduced by courtesy of "The Humorist"

Figure 10.12 Something must be done about this television menace

Source: *Television*, **2**, 1929–1930

On the optical side the principal problem[42] was that of regulating the intensity of the scanning light and of the received image so that 'the eye [was] not annoyed by the scanning beam or the neon lamp image rendered difficult of observation'. Ives and his colleagues, Gray and Baldwin, solved this difficulty by using a scanning light colour to which the eye is relatively insensitive, but to which the photoelectric cells could be made highly sensitive.

When the two-way system was withdrawn from service it had been seen by more than 17 000 people. A novel application was observed when two deaf persons carried on a telephone conversation by reading each others lips[43]. The cost of providing the service, by the New York Telephone Company, was estimated to be $15 350 per year, excluding the cost of the technical operation and maintenance which was borne by Bell Telephone Laboratories.

Much publicity was given to the demonstrations: for the first six months of 1931 more than 700 column inches of news print were devoted to reports in the city papers[44]. Rather oddly, perhaps, the desirability of this type of publicity was questioned by the administration department of the AT&T Company. 'Our exhibit emphasises to me that commercial use of television is a long way off,' wrote a member of the department, in March 1931[45] (see Figure 10.12).

Following the completion of the two-way link, Ives undertook an important appraisal of the progress which had been made by his group and endeavoured to define the course of action that had to be implemented for the future advancement of television. His prognosis was gloomy in outlook[46]. For Ives the statement of the problem that had to solved was simple:

> 'An electrically transmitted photograph 5 in by 7 in in size, having 100 scanning strips per inch, has a field of view and a degree of definition of detail which, experience shows, are adequate (although with little margin) for the majority of news events pictures. It is undoubtedly a picture of this sort that the television enthusiast has in the back of his mind when he predicts carrying the stage and the motion picture screen into the house over electrical communications channels.'

The difficulty of achieving this desirable result was readily apparent. In the photograph the number of picture elements is 350 000, and at a reception speed of 20 per second (24 per second had now become standard with sound films) this meant the transmission of 7000 000 picture elements per second and a bandwidth of 3.5 MHz for the system on a single sideband basis. Ives compared the criteria for high-definition television and the results which had been obtained in America and observed: 'All parts of the television system are already having serious difficulty in handling the 4000-element image.' (This was the number of image elements used in the 72-line picture of the two-way television link.)

The obstacles that had to be overcome before a high-definition system could be implemented concerned the use of the scanning discs at the transmitter and receiver, the photoelectric cells, the amplifying systems, the transmission channels and the receiving lamps. Ives noted that the disc, although quite the simplest means for scanning images of few elements, was entirely impractical when really large numbers of image elements were in question and wrote: 'As yet, however, no practical substitute for the disc of essentially different character has appeared.'

Turning next to photocells there were, in 1930, two types of cell which could be utilised for television; a gas-filled cell which had a good sensitivity but poor frequency response, and the vacuum cell, which was much less sensitive than the gas-filled cell, although it was free from its failing. The self capacity of the cells and the associated wiring and amplifier caused the high frequencies to be attenuated relative to the lower frequencies and consequently equalising circuits, with their attendant problems of phase adjustment, together with more amplification, were needed. But amplifiers capable of handling frequency bands extending from low frequencies up to 100 000 Hz or more gave serious problems, observed Ives.

The communication channels, either radio or wire, also posed grave difficulties for high-definition television and its related bandwidth specification:

> 'In radio, fading, different at different frequencies, and various forms of interference stand in the way of securing a wide frequency channel of uniform efficiency. In wire, progressive attenuation at higher frequencies, shift of phase, and cross-induction between circuits offer serious obstacles. Transformers and intermediate amplifiers or repeaters capable of handling the wide frequency bands here in question also present serious problems.
>
> '[Finally at the receiving end of the system the neon glow could not] follow satisfactorily television signals well below 40 000 Hz, and, in the case of the 4000-element image, the neon had to be assisted by a frequently renewed admixture of hydrogen, which again [could not] be expected to increase the frequency range indefinitely. In the receiver disc, as at the sending end, increasing the number of image elements, rapidly reduced the amount of light in the image. With a plate glow lamp of given brightness, the apparent brightness of the image is inversely as the number of image elements.'

These considerations led Ives to one clear conclusion: 'The existing situation is that if a many-element television image is called for today, it is not available, and one of the chief obstacles is the difficulty of generating, transmitting and receiving signals over wide frequency bands.' A partial solution was to employ multiple scanning and multiple channel transmission. The beginnings of such an approach had been given by P Mertz in his 1921 memorandum[11], now (1930) a multichannel set up was to be investigated. This is discussed in Chapter 14.

References

1 BURNS, R.W.: 'The contributions of the Bell Telephone Laboratories to the early development of television' (Mansell, London, 1991) pp. 181–213
2 HEISING, R.A.: 'Alexander McLean Nicolson', *Bell Laboratories Record*, May 1950,

p. 221. Also: ANON., 'A McL Nicolson, video pioneer, 69', *New York Times*, 4 February 1950

3 NICOLSON, A. McL.: 'Television transmission and reception'. Memorandum and report to E H Colpitts, 13 October 1916, pp. 1–28, correspondence folder 200670, **B** (1/1/16 to 11/29/16), AT&T Archives, Warren, New Jersey

4 NICOLSON, A.McL.: 'Television'. US patent 1 470 696, 7 December 1917. British patents 228 961 and 230 401, 7 September 1923

5 DIECKMANN, M., and GLAGE, G.: 'Verfahren zur Uebertragung von Schriftzeichen und Strichzeichnungen unter Benützung der Kathodestrahlrohre'. German patent 190 102, 12 September 1906

6 ROSING, B.: 'New or improved methods of electrically transmitting to a distance real optical images and apparatus therefor'. British patent 27 570, 13 December 1907

7 CAMPBELL SWINTON, A.A.: 'Presidential address', *J. Roentgen Soc.*, 1912, **8**, pp. 1–5

8 SKAUPY, F.: 'Braunsche Rohre mit Gluhkathode, inbesondere fur die Zwecke der Elektrischen Bildübertragung'. German patent 349 838, 28 November 1919

9 ZWORYKIN, V.K.: 'Television system'. US patent 1 691 324, 13 July 1925

10 THOMPSON, G.K.: 'Telephotography'. Memorandum for file, January 1884, Case 37014, correspondence folder 459 (2/18/86 to 9/30/97), Boston file, Radiophony, AT&T Archives, Warren, New Jersey

11 P.M.: 'Television'. Memorandum, 6 September 1921, Television folder, file 6.035.3,8, AT&T Archives Warren, New Jersey

12 BURNS, R.W.: 'Seeing by electricity', IEE Proc. A, 1985, **133**, pp. 27–37

13 'Remarks by Frank B. Jewett at the television demonstration', *Bell Laboratory Record*, May 1927, p. 298

14 IVES, H.E.: 'Television: 20th anniversary', *Bell Laboratory Record*, May 1947, **25**, pp. 190–193

15 IVES, H.E.: 'Television'. Memorandum to H D Arnold, 23 January 1925, case file 33089, **A**, (8), AT&T Archives, Warren, New Jersey

16 FINDLEY, P.B.: 'Biography of Herbert E. Ives'. Internal report, date unknown, pp. 1–12, AT&T Archives, Warren, New Jersey

17 ROBERTS, J.G.: 'Invention of Mr H.E.Ives—system of television'. Memorandum for file, 14 May 1925, case file 33089, **A**, (1), AT&T Archives, Warren, New Jersey

18 ANON.: 'Television—a group achievement', *Bell Laboratory Record*, May 1927, pp. 316–323

19 IVES, H.E.: 'Television', *Bell Syst. Tech. J.*, October 1927, pp. 551–559

20 GRAY, F., HORTON, J.W., and MATHES, R.C.: 'The production and utilisation of television signals', *Bell Syst. Tech. J.*, October 1927, pp. 560–581

21 See also Ref.47, chapter 8

22 NELSON, E.L.: 'Radio transmission for television', *Bell Syst. Tech. J.*, October 1927, pp. 633–653

23 ANON.: 'Far off speakers seen as well as heard here in a test of television', *New York Times*, 8 April 1927, p. 1

24 ANGWIN, A.S.: memorandum to the secretary of the GPO, minute 51/1929, file 9, 7 June 1928, Post Office Records Office

25 DAUVILLIER, A.: 'La télévision électrique', *Revue Générale de l' Électricité*, 7 January 1928, pp. 5–23

26 IVES, H.E.: 'Development program for television'. Memorandum to H D Arnold, 4 May 1927, case file 33089, **A**, pp. 1–9, AT&T Archives, Warren, New Jersey

27 H.S.R.: 'Television'. Memorandum for file, 18 March 1966, case book 1538, case file 20348 and case file 19350, pp. 1–12, AT&T Archives, Warren, New Jersey

28 GRAY, F., and IVES, H.E.: 'Optical conditions for direct scanning in television', *J. Opt. Soc. Am.*, 1928, **17**, pp. 423–434

29 ANON.: 'Television shows panoramic scene carried by sunlight', *New York Times*, 13 July 1928, p.4

30 IVES, H.E.: Memorandum to H D Arnold, 18 February 1928, case file 33089, pp. 1–5, AT&T Archives, Warren, New Jersey

31 HARTLEY, R.V.L., and IVES, H.E.: 'Improvements in or relating to television'. British patent 290 078, application date (UK) 19 March 1928, (convention date, US, 14 September 1927), issued 19 June 1927

32 IVES, H.E.: 'Television'. Memorandum for file, 10 July 1925, case file 33089, **A**, pp. 1–2, AT&T Archives, Warren, New Jersey
33 GRAY, F.: 'The projection of television images'. Memorandum for file, 13 February 1929, case file 33089, pp. 1–5, AT&T Archives, Warren, New Jersey
34 GRAY, F.: 'Recording television images on a movie film at television speeds'. Memorandum for file, 11 February 1929, case file 33089, pp.1–4, AT&T Archives, Warren, New Jersey
35 IVES, H.E.: 'Improvements in the diffraction of colour photography', *Phys. Rev.*, 1906, **22**, p. 339
36 IVES, H.E.: 'Three colour interference pictures', *Phys. Rev.*, 1907, **24**, p. 103
37 ANON.: 'Television in colour shown first time', *New York Times*, 28 June 1929, p.25
38 IVES, H.E.: 'Television in color', *Bell Laboratory Record*, **7**, July 1929, pp. 439–444
39 ANON.: 'Television in colour successfully shown', *Telephony*, 6 July 1929, **97**, pp. 23–25
40 ANON.: '2-way television in phoning tested', *New York Times*, 4 April 1930
41 IVES, H.E., GRAY, F., and BALDWIN, M.W.: 'Image transmission system for two-way television', *Bell Syst. Tech. J.* 1930, **9**, pp. 448–469
42 IVES, H.E.: 'Some optical features in two-way television', *J. Opt. Soc. Am*, 1931, **21**, pp. 101–108
43 FARRELL, W.E.F.: Memorandum to J Mills, 23 June 1930, ref. 592–C–WCFF–EO, p.1, AT&T Archives, Warren, New Jersey
44 SARGENT, W.D.: Letter to L S O'Roark, 9 July 1931, ref. 0603–2, p.1, AT&T Archives, Warren, New Jersey
45 W.J.O'C.: Memorandum to A W Page, 23 March 1931, ref. 0603–2, p.1, AT&T Archives, Warren, New Jersey
46 IVES, H.E.: 'A multi-channel television apparatus', *J. Opt. Soc. Am.*, 1931, **21**, pp. 8–19

Chapter 11

German and French developments (the 1920s and early 1930s)

As in Britain and the United States practical television development in Germany, after the First World War, initially stemmed from the work of one person. That person was Demes von Mihaly, a Hungarian by birth.

Little seems to be known about Mihaly's early life and even the year when he commenced working on television is uncertain. According to Dr Eugen Nespen[1]: 'In the spring of 1918 Denes von Mihaly told me of his television project. His aim was to realise the centuries old dream of humanity to see into the far distance: a device,

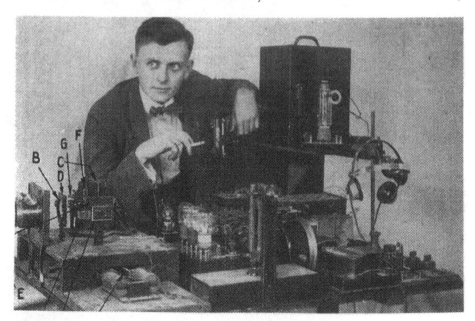

Figure 11.1   Denes von Mihaly with some of his experimental apparatus

Source: *The Wireless World and Radio Review*, 19 March 1924, p. 763

in fact, which should be complementary to wire or radio telephony. Further development of this idea had to be postponed until peace reigned again, and even then the unfavourable conditions did not allow a practical realisation of his ideas. It is hoped that they will soon become reality.'

Nicholas Langer[2], on the other hand, has stated that Mihaly began his experiments in 1916 and succeeded in 1919 in transmitting simple pictures of size 10 × 10 cm. As Langer was Mihaly's assistant from about 1919, his statement might be supposed to be definitive, but Nesper's account was given in his foreword to Mihaly's book on television (published in 1923). From this and other information it appears highly unlikely that the inventor transmitted even simple pictures in 1919.

Mihaly's improved transmitter[2] (1924) is shown in Figure 11.2. In this an image of an object or scene was projected, by means of lenses A and B, upon the small oscillating mirror, D, (of oscillograph C) which was caused to oscillate at 500 Hz by an appropriate current. Additionally, the mirror, D, was capable of a motion in a direction perpendicular to the former movement so that any point in or near to the plane of the object could be incident on the selenium cell, SE. This motion was achieved by pivoting the frame, F, holding the platinum wire loops and mirror, about an axis at right angles to the loop and connecting the frame through a lever to an eccentric point on the phonic motor rotor, G, which rotated at five revolutions per second.

The motor, of La Cours' type, comprised a hollow drum G, made of a nonmagnetic material, on the outside of which 20 equidistant thin iron bars had been placed, and an electromagnet, N. The air gap was approximately 0.5 mm and the stator coil, N, was placed in series with a self-oscillating tuning fork which determined the frequency of oscillation, 5 Hz, of the frame, F. A small commutator, K, consisting of one hundred insulated segments mounted on the rotor of the motor, together with a battery, T, and transformer, M, provided the necessary 500 Hz current.

The selenium cell used had an effective area, determined by the aperture, E, of 1.0 mm^2 and a dark resistance of the order of 100 kΩ.

Mihaly's receiver[2], Figure 11.3, employed an identical scanning arrangement to that used in the transmitter, and a light valve Q. This consisted of a highly sensitive bifilar oscillograph and diaphragm which allowed light from an arc lamp, A, to be incident on the screen, R, so that the intensity of illumination at any point on R was dependent on the magnitude of the received picture current flowing at that instant. The use of an oscillograph for this purpose was, as mentioned previously, due to Brillouin in 1891, and Mihaly's scanning mirror system was essentially an adaptation of an idea originally put forward by Le Blanc in 1880, Figure 11.4.

To ensure synchronism between the transmitter and receiver the inventor placed in front of the lens, B, a glass plate marked with three dots, X, Y and Z. At the receiver three selenium cells were placed on the screen in identical relative positions X,Y,Z to the dots so that if the two apparatuses were synchronised the resistances of the cells would be very high. A departure from synchronism therefore resulted in changes of resistance and, by means of 'sensitive relays, magnetic couplings, brakes, etc.', the changes resulted in synchronism being restored.

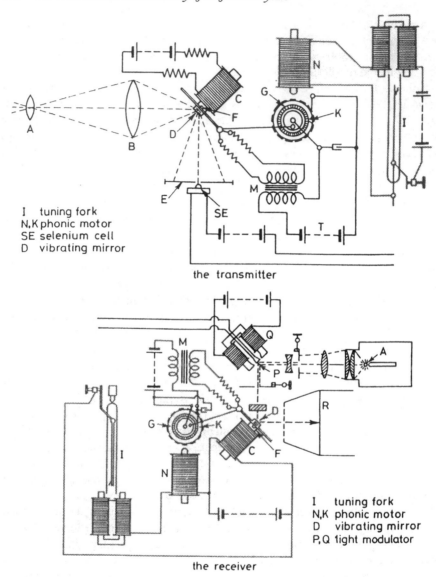

I   tuning fork
N,K phonic motor
SE selenium cell
D   vibrating mirror

the transmitter

I   tuning fork
N,K phonic motor
D   vibrating mirror
P,Q light modulator

the receiver

Figure 11.2 & 3 D von Mihaly's television transmitter and receiver utilised vibrating mirrors for scanning and phonic motors and tuning forks for synchronisation

Source: *The Wireless World and Radio Review*, 19 March 1924, p. 762

In his 1924 account of the work of Mihaly, Langer gave no indication of the success or otherwise which Mihaly obtained with this equipment but ended his article by stating: 'The experiments were partly conducted under difficult conditions especially during the Great War and in some of the subsequent years, when the materials of many kinds necessary for experimental purposes were scarcely obtainable, so much so that at one time the experimenters were obliged to undertake the

Figure 11.4   Photograph of von Mihaly's receiver when arranged for wireless reception. The arc lamp, O, the light relay oscillograph, Q, the mirror of the light relay, P, the screen, R, the diaphragm, E, the oscillograph, C, and the electromagnet of the phonic drum, H, are indicated

Source: *The Wireless World and Radio Review* 26 March 1924, p. 794

making of the amplifier valves themselves. Material difficulties interrupted the experiments in the middle of 1923, but it is hoped to be able to continue them in the near future, and by using more elaborate arrangements and some improvements, to obtain perfectly satisfactory results.'[4] This statement perhaps suggests that all was not well with the equipment, a view which is supported by comments made by Mihaly in his book 'Das Elektrische Fernsehen und das Telchor' (1923)—the earliest book to be published which dealt exclusively with television.

Although the book contains very full descriptions of the apparatus which he invented, and named the telehor, there is a notable lack of narrative on the successes achieved. Mihaly mentions in two places that with the first model he succeeded in transmitting some simple geometrical figures, drawn white on a black background, artificially illuminated, but the reception was not quite satisfactory and he was depressed by the poor results ('Missgestimmt durch die schlechten Resultate . . . ').

He ascribed this lack of a favourable outcome to instrumental deficiencies, which he hoped to improve, but there is some firm evidence that this was due to an oversight on the part of the inventor in designing the optical arrangement of his system.

The oversight arose because Mihaly's lens configuration focused a very small real image of the object onto the minute mirror of the oscillograph so that after reflection an out of focus patch of light appeared on the apertured screen, E, covering the selenium cell. This patch of light, naturally, did not represent a true image of the object and, of course, if the transmitted picture was faulty good reception could not be expected. The error occurs three times in the text, albeit expressed differently on each occasion.

The explanation given above would be invalidated if either the mirror were not plane or the image were focused on the screen. But the description[5] of the mirror implied that it was plane ('a thin glass plate, silvered, 3 mm × 3 mm, area 8 mm²), and, if the second alternative were correct, the minute mirror could only intercept a very small portion of the wide beam necessary to give a fair-sized picture and most of the light would pass straight on instead of being reflected to the cell. It is significant to note that no lens is shown between the mirror and screen, which would have enabled an image to be focused onto the screen.

If Mihaly had utilised a concave mirror, some form of enlarged image could have been produced on the screen by focusing the image, using the lens system, not onto the mirror (as Mihaly's account states) but at a position some distance in front of it, outside its focus. A concave mirror, however, gives a very unsatisfactory picture of an extended object and although with a fixed mirror compensatory errors might be introduced into the accompanying lens system, a moving mirror would upset the calculations.

In addition to the above fundamental error in design, the use of a very small mirror posed grave practical problems. The normal function of a galvometer or oscillograph is simply to reflect a small spot of light onto a scale; quite a different matter from that in Mihaly's scheme where the mirror was required to transmit detailed picture information. For this purpose an optically-worked reflecting surface would have been necessary. Whether the inventor's 3 × 3 mm mirror, made from a microscope slide cover glass, was so worked is not known: the preparation of an optical surface on such a fragile object would certainly have taxed his skill.

The letter[3], in *Experimental Wireless and the Wireless Trader*, by Miss A Everett, which pointed out Mihaly's mistake, was published in September 1927. It is likely that Mihaly saw it for he had himself submitted to the same journal, in April 1927, a contribution[6] to the discussion of a paper by Baird which was printed in December 1926. Mihaly does not appear to have commented on Miss Everett's observations but at the Berlin Radio Exhibition held in 1928 he demonstrated the transmission of shadowgraphs using a disc scanner operating on a standard of 30 lines scanned at 10.0 frames per second[7]—similar to the standard employed by Baird at that time.

Mihaly used an alkali metal photocell at the transmitter and a glow lamp, of the type devised by D McFarlan Mooore, at the receiver. The 1928 demonstration consisted of the reproduction of moving silhouettes and transparencies on a screen of size 4 × 4 cm. The Hungarian inventor had previously considered the Nipkow disc but had concluded[1]: 'Dieses Projekt wurde zwar eine sehr einfache Lösung ergeben, aber leider ist dasselbe praktisch undurchführbar.'

Later, in 1931, Mihaly attributed his abandonment of oscillographic scanners to the opinion that 'it was not worth the trouble to utilise oscillographic image dissectors for the transmission of only a few thousand picture elements'.

The foregoing narration perhaps places in perspective the statement of Dr Nesper who said[1] that on 7 July 1919 Mihaly was able, 'to instantaneously reproduce blurred, coarsely scanned pictures from a distance by electrical means'. Nevertheless, even though his early efforts were unrewarding and he was a self publicist, Mihaly was one of the first to stimulate the evolution of television in Germany[8].

At the Berlin Radio Exhibition of 1929 Mihaly showed two types of receiver. The smaller of the two utilised a light ebonite Nipkow disc, having 12 holes and a small neon tube. The disc was driven by a phonic motor, the stator current of which was supplied from the a.c. mains, and the television transmission consisted of the reproduction of the image of a lantern slide only[9]. On the authority of one reporter who was at the exhibition, 'the received picture was of very poor quality' due to the small number of image elements.

The larger instrument also employed a Nipkow disc, driven by a phonic motor, but used a standard of 30 lines. Again, a neon lamp was utilised and the picture had a square format, of size about 25 mm square. This was enlarged to approximately 75×75 mm by a magnifying lens.

In both receivers Mihaly used the phonic motor driving scheme of his earlier 1924 oscillograph transmitters and receivers and, because these were not self-starting, means had to be provided to enable the user to run them up to speed by hand. For correct phasing of the received image Mihaly employed the method described by Baird in 1924, the rotation of the motor stator.

A report of the 1929 Berlin Radio Exhibition noted: 'The brilliancy of the image of the larger instrument was good, as was also the detail. Synchronism was assured of course, owing to the fact that the disc was driven by a synchronous motor working off the same supply as that used for the transmitter driving motor.'

Mihaly did not demonstrate direct television but employed a form of telecinema for the exhibition and so his received pictures were obtained by using transmitted light at the transmitter rather than by reflected light, Figure 11.5. The explanation for this is contained in a paper by Dr A Gradenwitz[10]: 'Others who have had a better chance to investigate the Mihaly system told me that he had to show at the present moment mainly telecinema, that as a matter of fact he had done actual television, but that this was yet of a very crude description, though the light used at the transmitting end in scanning the person to be televised—namely, powerful Jupiter lamps—was rather unpleasant both by its blinding effects and the enormous heat developed.'

Mihaly's telecinema images were broadcast daily from the Berlin-Witzleben transmitter operating on a wavelength of 475.7m[11]. By the summer of 1929 he felt the time had come to show the British public his television system. For this purpose he had to obtain the cooperation of the General Post Office and the British Broadcasting Corporation.

In June 1929 a Mr Lynes of Wireless Pictures (1928) Ltd wrote to the GPO on behalf of Mihaly and asked the Post Office to sanction a test of his system[12]. One of the points he put forward in support of the request was the cost of the receiver—50 Reichmarks (£2.50) in Berlin. Phillips of the GPO informed Reith of the BBC and asked for the corporation's observations on the subject[13]. The BBC replied that no application for experiments with the system had been submitted to them and they did not have any information on the method of working, although their chief engineer and a Mr Hayes had seen the system at the 1928 Berlin Radio Exhibition; reproducing a 'moving dark silhouette against a light background'. The BBC's conclusion was not enthusiastic[14]: 'The corporation is not anxious to embark on any experiments with this system.'

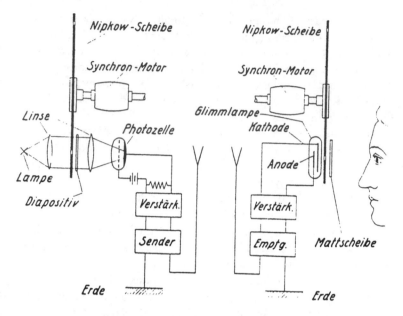

Figure 11.5   *Von Mihaly's telecinema images were broadcast daily, in 1929, from the Berlin-Witzleben transmitter operating on a wavelength of 475.7 m. The diagram shows the transmitter and receiver scanning discs, the film transparency and photocell, the neon lamp, and the synchronous motors*

Source: *Archiv für das Post-und Fernmeldewesen*, August 1985, p. 193

A few days later Mihaly called at Angwin's office accompanied by a representative of Wireless Pictures and a Mr L H Jackson of the Beaconsfield Trust Ltd, to explain his method of television, and to show Angwin two types of his receivers (prices £5.00 and £2.50). The reception given to Mihaly and his colleagues must have contrasted markedly with the enthusiasm of the German Post Office, for Angwin could offer no immediate support for the system. He told Mihaly that the question of broadcasting was a matter for the BBC in the first instance and also that no opinion could be formed of his scheme based on an examination of the receivers shown[15].

Mihaly did not at once give up his desire to introduce his system into the United Kingdom, and so his company, Messrs Telehor AG of Berlin—the proprietors of the Mihaly television and telecinematography system—appointed S Sagall as their duly accredited representative in the UK[16]. Sagall's terms of reference were to enter into negotiations with the BBC with a view to obtaining from them the necessary facilities for a demonstration before the BBC and GPO.

Solomon Sagall, who later became the managing director of Scophony Limited (1931) was, like Zworykin of RCA and Shoenberg of EMI, born in Russia[17]. His father Yakov Meir Sagalowitsch was at first a rabbi in Tver, Russia, but in 1923 became the chief orthodox rabbi in the Free State of Danzig. In 1933 he was elected chief rabbi of Brussels. When the German armies invaded Belgium in 1940 Sagall's parents were again uprooted and escaped, via France, to Lisbon, Portugal.

Sagall had a deep commitment to the Zionist cause and helped to organise the emigration of Jews from the Ukraine to Palestine, the US and the UK in the early 1920s. These relief activities were somewhat short lived and Sagall felt obliged to leave the USSR. He made his way to Berlin, where he continued to work for the Zionist movement as secretary to the Central Committee of Russian Zionists outside Russia. At the same time he pursued his educational studies in Berlin and at the University of Jena.

In 1925 Sagall again emigrated, this time to London, where he stayed until 1940. His academic work was furthered at the London School of Economics and he obtained just enough money to live on from freelance journalism and some part time employment at the London headquarters of the World Zionist Organisation. By now he was fluent in Russian, German, Hebrew, Yiddish and English.

His first business venture, to revitalise a small silent cinema in North London, was not successful. He then conceived the notion of exporting made to measure men's suits (tailored in Leeds) to Germany and published advertisements in the leading newspapers of Berlin. This entrepreneurial speculation also was unprofitable and the scheme was discontinued.

In March 1929 Sagall read, in the *Observer*, an account of a demonstration of the Telehor system, which had been given by Mihaly. The report caught Sagall's imagination and he returned to Berlin where he met the inventor at the Telehor Laboratories. Here he was introduced to Gustav von Wikkenhauser and Ferenc von Okoliczanyi—both Hungarians like Mihaly—who were to become leading engineers in the Scophony company, and who were the principal engineers of Telehor Ltd.

Although Mihaly had received some encouragement and financial support from the Reichspostzentralamt, it was a wealthy German wine merchant, Paul Kressman, who provided the finance for the development of the laboratories. Kressman's backing in March 1929 had allowed Mihaly to give the public demonstration of rudimentary television broadcasting from the Witzleben transmitter. Sagall remained in Berlin for two weeks and negotiated with Kressman for the British rights of Telehor AG. Kressman gave him the exclusive rights to the Telehor system in exchange for a promise from Sagall that he would use his best endeavours to promote the system in Britain.

Sagall's action might, with or without hindsight, seem reckless. He had no engineering background, he had not previously seen any other trial of television and he had not, in 1929, established himself as a successful businessman. Moreover, the demonstration which he had witnessed was of an inferior quality compared to those which had been given by AT&T, Baird and others.

Moseley, a shrewd judge of people, was not taken in by the inventor's claims: 'Mihaly represents a different type from John Baird. He is inclined to be more self assertive, and by this means has managed to impress the wireless world of Berlin of capacities which he does not altogether possess,'[18] Moseley wrote in a report published in *Television*. He said also: 'Mihaly is clever enough, however, to surround himself even after all these years with an air of mystery.'

Sagall, on his arrival back in London, was soon in touch with the BBC and paid the corporation a visit[19] on 12 August 1929 to discuss the feasibility of obtaining the necessary facilities for a demonstration of the Telehor television system.

Carpendale, the BBC's controller, passed the issue back to the Post Office by asking for the Postmaster General's consent and when this was not forthcoming the matter rested—for a few months. At this time both the BBC and the General Post Office were suffering a hostile press campaign in support of the Baird scheme and in Phillip's view it seemed doubtful 'judging by experience' whether any reliance could be placed on any undertaking (from television promoters) to avoid publicity. For this reason, he thought that it would be inadvisable to arrange a demonstration of the Mihaly system just at the present, although he considered the issue could be raised later[20].

Sagall returned to his brief[21] in December and on 3 January 1930 a meeting was arranged at the Post Office[22]. He was seen by Phillips and Weston and during the discussion the following points were noted by the Post Office[23]:

1   the German administration had given to the Baird company the same facilities as to the Mihaly interests, and it was only natural to expect that the British administration would reciprocate as regards the Mihaly system;
2   receiving sets were not yet on sale to the public in Germany;
3   the prices previously quoted in the press of £2.50 and £5.00 for Mihaly's small and large receivers were now altogether too low and £12 to £15 would be nearer the mark;
4   the Mihaly system could transmit film only for the present;
5   synchronisation of the Mihaly set was simpler than that of the Baird system, Sagall claimed.

For the Post Office, Phillips mentioned that they were most anxious to explore every system of television but that as tests of the Mihaly system were being conducted in Germany, no useful object would be served so far as the Post Office was concerned by having a demonstration from a BBC station. And in any case, as they had given the Baird company facilities for broadcast transmission in this country they wished to give that system a fair chance. Sagall was clearly disappointed by this lack of enthusiasm and hinted darkly that 'although he was reluctant to adopt means at his disposal of bringing pressure to bear on the Post Office, he feared in the circumstances he was left with no alternative'.

Nothing further was heard by the BBC or the GPO from the Mihaly organisation until after the publication of the Television Committee's report in 1935. The plain facts were that the company could not show direct television, it had no sets available for sale and, finally, it had ample facilities in Germany for experimental purposes.

On the question of the cost of his sets Mihaly had seriously underestimated the retail price. His figures of £2.50 and £5.00 were later revised so that at the October 1929 Berlin Radio Exhibition a representative of Kramolin and Company of Berlin (a firm of wireless manufacturers having the manufacturing and sales monopoly of Mihaly's receivers) said that the price might be between 200 and 300 Reichmarks

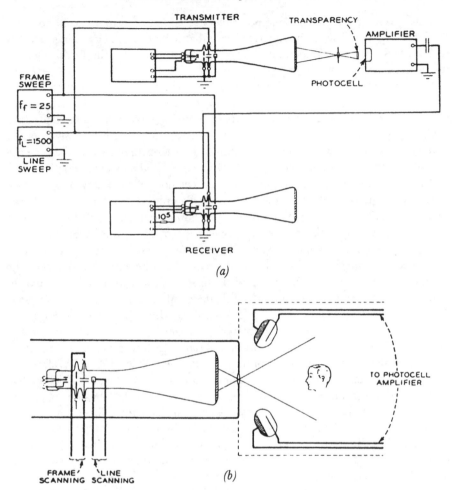

FRAME
SWEEP

$f_f = 25$

$f_L = 1500$

LINE
SWEEP

TRANSMITTER

TRANSPARENCY

AMPLIFIER

PHOTOCELL

10^5

RECEIVER

(a)

FRAME   LINE
SCANNING SCANNING

TO PHOTOCELL
AMPLIFIER

(b)

Figure 11.6   *The use of film, for picture generation, enabled transmitted light rather than reflected light to be used. Adequate signal strength could be obtained thereby, and several inventors and companies (including HMV and EMI) utilised film during their development work.*

Manfred von Ardenne carried out much experimentation on cathode-ray tube television.

The diagrams show his 1930 proposals for film scanning (a), and for live pick-up (b), using c.r.t spot scanning

Reproduced by permission of the Royal Television Scoiety

($£10$ to $£15$). The Karolus receiver was likely to cost about 700 to 800 marks and the Fernseh product had had its price fixed at 350 marks.

In the months following his visits to the BBC and the GPO, Sagall attempted to raise money to create a company of his own. Unfortunately for him, the world's business community was distressed by a massive depression. Sagall's proposed venture was ill-timed and he found no backers. A further blow to his aspirations was the information, given to him by Kressman, that Mihaly was negotiating privately with a Mr Sweeney, a London financier, and a Mr Stafford for the British rights to the Telehor system.

Sagall returned to Berlin and there he was acquainted with the deterioration of the association between Kressman and Mihaly. It seems that Mihaly, a handsome man who enjoyed the good life and was a frequenter of night clubs, had taken advantage of Kressman's considerable generosity and was not vigorously applying himself to the advancement of his inventions. From Kressman's viewpoint the original agreement between himself and Mihaly had been in the inventor's favour, for it had given Mihaly sufficient voting power in Telehor to block any action which Kressman might take, although the business and administrative functions had been vested in him. He decided that he would not continue to finance Telehor Ltd unless he procured Mihaly's shares. However, Sweeney's presence on the business stage upset this plan because Mihaly looked to Sweeney as a new and independent source of finance. Sweeney was informed by Kressman that if he wished to proceed with the purchase of the British rights he would have to deal with Sagall who had the rights, but Sagall had no intention of selling them. After staying about five weeks in Berlin, Sagall succeeded in arranging for the sale of Mihaly's shares to Kressman.

On his return to London Sagall formed and registered, in 1929, a company called British Telehor Ltd. The press announcement did not attract much interest because of the prevalent widespread unemployment but it was seen by G W Walton. He wrote to Sagall and described the novel optical-mechanical scheme[24] which he had devised and which he thought would replace the Nipkow disc. Walton impressed Sagall, who submitted the inventor's concept to Kressman for evaluation by Wikkenhauser and Okoliczanyi, and suggested that Kressman should invite Walton to Berlin. This Kressman arranged and Walton spent several months in the Telehor Laboratories with Wikkenhauser and Okoliczanyi. Walton's stay was highly successful and he formed a lasting friendship with the two Hungarians.

Later, when Sagall returned to Berlin with Walton, a small German company was formed and registered with the aid of financial help from Kressman. Walton, Sagall and Kiessman were the sole stockholders and Walton put forward the name of Scophony for the company. A British registered subsidiary of the Berlin registered company Scophony GmbH was established in London in November 1930. It, too, had the title of Scophony Limited. Chart 11.1 shows the development of the Scophony companies[25].

Although television's progress in Germany was not so advanced as that in the UK or the US in the late 1920s, the Deutsche Reichspost (DRP), much to its credit, had taken an interest in the subject from July 1927, when Dr Fritz Banneitz, the head of the department for high speed wireless telegraphy and facsimile transmission at the National Centre for Telegraph Techniques (Leiter des Referats für drahtlose Schnelltelegraphie und Bildübertragung beim Telegraphentechnischen Reichsamt) was given the task of 'concerning himself with television'[26]. This was so that his department could 'remain up to date regarding the state of development and, by means of suggestions and regular transmissions, promote progress'. The DRP was

Chart 11.1 Development of the Scophony companies

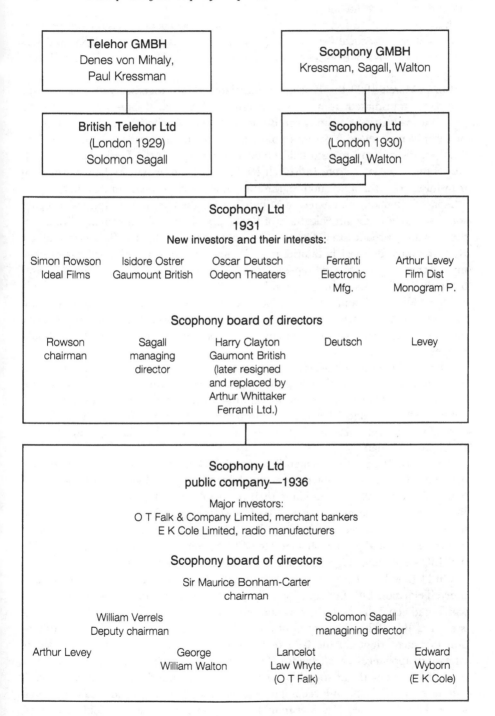

the first government department, anywhere, to encourage officially the advancement of the new technology.

Banneitz and his assistants investigated various aspects of television and profited from Mihaly's practical experience. Mihaly was sponsored by the DRP and was allowed to demonstrate his system on the Reichpost stand at the fifth German Radio Exhibition held in Berlin in 1928[27].

On 9 March 1929 the Reichpostzentralamt (RPZ), together with the Telehor company, transmitted, using a wavelength of 475.4 m, 30-line images from the 1.5 kW broadcasting transmitter in Berlin-Witzleben for the first time. 'The moving pictures were received at various points in the town and were, without exception, faultless, distinct and undistorted,' noted one reporter in the Berlin 12-Uhr Blatt. About one week later, from 14 March 1929, the RPZ took part in the experimental transmissions—which had no programme value—with its own film scanner[28].

Apart from taking an interest in Mihaly's work, the German authorities also paid some attention to the investigations of Baird. Dr Bredow, the managing director of the German Broadcasting Corporation, accompanied by Dr Banneitz and Dr Reisser, visited Baird's laboratories in 1928 and, according to J L Baird, Dr Bredow invited the company to send representatives over to Germany to install a transmitter in the Berlin broadcasting station[29].

Dr E C Rassbach and Mr K M Wild of Robert Bosch GmbH, Stuttgart, had also seen, on another occasion, Baird's equipment and had been duly impressed[30]. They had found the demonstrations convincing and suggested that the two firms should cooperate.

The Baird Company's television transmitter was installed in the VOX-Haus and was operated by some of the company's staff. Live experimental transmissions commenced on 15 May 1929 and continued until 13 July 1929. They took place daily between 9.00 a.m. and 10.00 a.m. and sometimes between 1.00 p.m. and 2.00 p.m. and after midnight. From 13 July the Baird equipment was operated in the Reichpostzentralamt.

The cooperation of the German Broadcasting Corporation came at a most opportune time for Baird and his companies. After many months of opposition by the British Broadcasting Corporation to their proposals, the adventitious attention of Dr Bredow and his colleagues must have been a welcome relief. Dr Bredow's helpful attitude contrasted markedly with that of Reith (who rarely attended a Baird television demonstration) and possibly shamed the BBC. Subsequently, on 30 September 1929, the BBC's experimental, regular television service was inaugurated.

On 11 June 1929 Fernseh AG was founded[31]. It comprised Robert Bosch GmbH, Baird Television Ltd, London, Zeiss-Ikon, Dresden and DS Loewe, Berlin, all of which had equal shares in the new company. Fernseh was entered in the trade register on 3 July 1929. 'The object of the company [was] the acquisition and utilisation of patent rights in the field of television and the manufacture and sale of television appliances of all kinds.' The rationale on which the creation of this company was based was sound and well considered: Baird Television would contribute expertise in picture scanning and electrooptical conversion, Loewe would supply knowledge and practical manufacturing experience in the field of wireless broadcasting, Zeiss-Ikon would specialise in the solution of optical and photographic problems and Bosch would add its considerable skills in precision engineering and experience of measurement techniques.

This collaborative venture lasted until 1935 when, following Hitler's directive, Baird Television was obliged to withdraw from the partnership. Later, David S Loewe, chairman of the company which bore his name and a Jew, also had to revoke his company's interest in Fernseh AG. The shares of these former partners were taken over by Bosch and Zeiss and the board of directors comprised Dr E C Rassbach and Karl Martell Wild of Bosch and Alexander Ernemann and Alfred Simader of Zeiss-Ikon. In 1939 the latter company left the undertaking, leaving Bosch as the sole proprietor of Fernseh AG. This state of affairs persisted until 1973 when the company became the video equipment division of Robert Bosch GmbH.

The initial capital investment of the Fernseh company was a modest 100 000 Reichmarks and Dr Ing Paul Goerz was entrusted with its management. Dr Banneitz was appointed as a consultant. Two rooms on the top floor of the Goertz works[32], in Berlin-Zehlendorf, which belonged to Zeiss-Ikon, were made available to Fernseh and work began with three employees. Soon after the company was established Dr R Moller, who had just completed his doctorate thesis in physics under Professor H G Moller of Hamburg, became responsible for developments in the field of physics and high vacuum technology. Shortly afterwards Dr-Ing G Schubert, who had studied under Professor Barkhausen in Dresden, was assigned to advance work in appliance and high frequency technology. They were both made deputy board members.

Within three or four years from its foundation the directors had made provision for the company to engage in all branches of television engineering. A well equipped general workshop and a glass blowing shop had been set up and precision fabrication facilities for the manufacture and assembly of electrodes for hard-vacuum tubes had been furnished. In addition, the high vacuum department was able to develop its own materials for hot cathodes and fluorescent screens and for the vapour deposition of photocathodes. Laboratories existed for the design, construction and testing of sweep generators, high voltage power supplies, electron optic devices, wideband amplifiers and receivers.

Although the formation of Fernseh was timely from the viewpoint of television progress, the world-wide recession imposed a major constraint on the independent financial viability of the company. Even in 1935 the annual report for the year contained the statement: 'The research work was continuing according to plan . . . no earnings are to be anticipated in the foreseeable future.'[32]

It seems that K M Wild (mentioned above) played a decisive role in the continuing existence of the company. According to Rudert, Wild 'believed with unshakeable optimism in the future of the new technology and was able, again and again, to obtain the financial support from the partners, particularly from Bosch, that was necessary for the research and development work until the company first began to make a profit'.

Soon after the establishment of the company and the active participation of the DRP in television transmission, the Deutsche Reichpost considered that the time

Figure 11.7   In 1929, at the Berlin Funkausstellung, the Reichspostzentralamt demonstrated two-way television. The photograph shows one of the two-way television-telephone call boxes

Reproduced by permission of the Deutsches Museum

had arrived when a common scanning standard ought to be adopted for television reproduction. Accordingly, a conference to discuss this matter was arranged and held in Berlin on 12 July 1929[33]. President Kruckow, of the RPZ, Professor Goldberg, managing director of the Zeiss-Ikon Company, Dr Loewe, of the Loewe Radio company, J L Baird, D von Mihaly and representatives from the Telefunken and Bosch companies attended. Standardisation of the methods of synchronisation and transmission and reception was clearly of assistance to mass production methods as manufacturers would not be required to design and produce many different models.

The German conference did not immediately adopt a common standard because President Kruckow had decided that enquiries should be made in the US (to Bell Telephone Laboratories) and in Britain (to the GPO) about their picture formats with a view to the possibility of a universal agreement being reached on the matter[34]. Whether these enquiries were initiated by Baird Television as a ploy to force the GPO's hand is not known but, in any case, the GPO thought that the art of television had not yet advanced sufficiently to justify international standardisation and that such standardisation might in effect cramp development[35].

Subsequently, on 20 July 1929, the German Administration adopted the ensuing specification for its television system[36]:

1   number of lines: 30, horizontally scanned
2   number of pictures per second: 12.5
3   aspect ratio (vertical/horizontal): 3/4

The first two of these standards were based on the frequency bandwidth available, namely, 9 kHz; the third was put forward partly because of the German plan to broadcast cinematograph films, in the first instance, during the initial experimental period.

These standards were slightly different to those which had been chosen by Baird, for he used vertical scanning and a 7:3 aspect ratio. He had selected vertical scanning because the tendency of the reproducing scanner to hunt was less noticeable than with horizontal scanning. This point seems to depend on the fact that most movements in everyday life take place in a horizontal plane and the eye and brain are therefore more sensitive to such movements rather than to vertical movements. On the other hand, when walking, the head moves slightly up and down to oppose the small vertical motions.

Fernseh showed, for the first time, one of its products at the sixth German Radio Exhibition which was held at the Funkturm in Berlin in the autumn of 1929. The Funkausstellung, which had been founded in 1924, was the premier occasion for the display of radio and television goods and technologies. Television originally made an appearance at the exhibition in 1928 when Mihaly and Karolus presented their receivers[26]. Then, in 1929, both the Reichpost and Fernseh, in addition to Mihaly and Karolus, had stands of their own[37].

Receivers of the Baird type, for the reception of 30-line images, generated by a Nipkow disc film scanner, were demonstrated by the new exhibitors. Neither organisation highlighted transmitters for televising live scenes, although the RPZ had such a transmitter on its stand. This transmitter was one of the German Post Office's earliest attempts to construct for themselves such an apparatus, but it had not proved very successful. Prior to the exhibition, in the summer of 1929, the RPZ had developed[26] a transmitter which only produced silhouettes even though the object had been illuminated with 50 000 lux of lighting and even though eight stages of picture signal amplification had been employed.

The approach and policy of the German broadcasting and Post Office authorities during the formative period of German television was one of great support and stimulation[38]. They made a noteworthy contribution to its promotion: Mihaly had received some much needed sponsorship, Fernseh AG had been formed, a television standards conference had been set up, transmitting facilities had been provided and encouragement had been given to the Funkausstellungen. The attitude of the German administrations was beyond reproach and contrasted strikingly with some views held elsewhere. Indeed, whereas low-definition television had been the subject of a deal of derision and scorn in several sectors of opinion in the UK, in Germany it was felt that progress could only be achieved by the undertaking of systematic research into image analysis and synthesis, synchronisation control and glow discharge processes in gas-filled tubes. These aspects of the television problem (as well as low-definition television, generally) were studied by Dr Banneitz and his assistants at the Reichpostzentralamt (RPZ). Regular experimental transmissions were a feature of their efforts, and the RPZ's standardisation work was always in accord with the state of the art.

At the 1929 Berlin Funkausstellung the RPZ installed two television-telephone boxes, at opposite ends of their stand, to allow members of the public to see and hear each other simultaneously when making a call. A single 30-hole Nipkow scanning disc running at 12.5 revolutions per second was used at each end and the analysing and synthesising operations were carried out by the scanning apertures situated near the bottom and the top of the discs, respectively. According to one report[9]: 'The received image, however, was not at all good, and, owing to the spotlight playing on one's face, was difficult to see.' Furthermore, since the two scanning regions of each disc were positioned a disc diameter apart, the received image showed a person looking away from the direction of the analysing beam. In the Bell Telephone Laboratories' scheme the two regions were only about six inches apart.

The 1920–30 decade heralded not only important advances in television engineering in the United Kingdom, the United States, France, Germany and elsewhere, but also progress in the related field of picture telegraphy/facsimile reproduction. Although Bain and Bakewell had put forward essentially practical ideas for this in the 1840s and services had been implemented for brief periods in the 1860s and the 1900s, it was not until the 1920s that regular, commercial, line and wireless phototelegraphy services were initiated.

As noted previously, numerous systems appeared within a few years of each other, as with television, and by 1928 the following schemes were in use: Ranger-Marconi,

Figure 11.8   The 1924 apparatus of Professor A Karolus in the University of Leipzig. The transmitter and receiver Nipkow scanners are mechanically coupled. The number of lines per picture was 45

Reproduced by permission of the Deutsches Museum

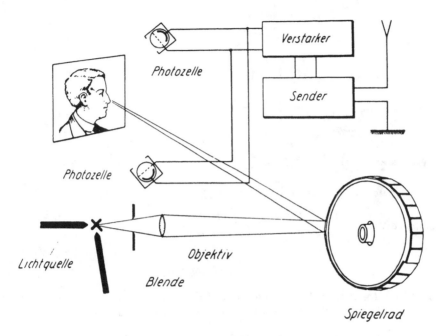

Figure 11.9   Schematic drawing of the mirror drum transmitting apparatus of Karolus; 45 lines per picture, 25 pictures per second

Reproduced by permission of AEG Aktiengesellschaft

Western Electric, Thorne-Baker, Belin, Siemens-Karolus-Telefunken. Prior to the commencement of the Second World War the systems most commonly used in Europe were those of Belin and Siemens-Karolus-Telefunken. London was linked to the capital cities of most European countries and the UK GPO had provided a public service which operated through the medium of the ordinary trunk telephone network. In addition, the Karolus apparatus was employed in Japan by two newspaper companies and also by the Australian Post Office which had a service between Melbourne and Sydney. The service between Peking and Mukden utilised the Belin method of phototelegraphy.

In all these systems the transmitter incorporated a rotating cylinder on which was mounted an image of the object/scene/manuscript to be transmitted. This took the form of either a photograph or a transparency, in those instruments in which the image was scanned by a moving spot of light, or a specially prepared image when electrical scanning was adopted. Pictures sent by the Karolus system did not require any special preparation and if they were of a suitable size could be attached directly to the transmitting cylinder. A further advantage of this equipment was the high speed of transmission and reception with good quality. These two features were made possible by the use of a ring-shaped photoelectric cell, designed by Dr Schroter, of the Telefunken Company, and a Kerr cell developed by Professor A Karolus of Leipzig University. As believed by one authority, these

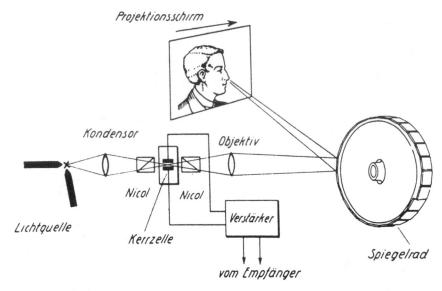

Figure 11.10 The diagram shows the disposition of Karolus's mirror drum scanner and Kerr cell light valve. This receiver operated with the transmitter of Figure 11.8

Reproduced by permission of AEG Aktiengesellschaft

devices enabled the Siemens-Karolus-Telefunken method to be superior to others (in 1926).

Notwithstanding the above claim, the system used two components, the photocell and the Kerr cell, which also featured in some television schemes. August Karolus started his television research towards the end on 1923 when he was an assistant professor at the Physical Institute of Leipzig University[26]. For the analysis and synthesis of his images Karolus employed two Nipkow discs, each about a metre in diameter, coupled together and rotated at a speed of 600 revolutions per second. Figure 11.8. His mesh-type photcell had a hydrogenated potassium cathode and was filled with an argon-hydrogen mixture. At the receiver Karolus used a Kerr cell to control the light flux from a 30 A arc lamp. Until this time (1924), Kerr cells had been considered quite unsuitable for television purposes[39]. Karolus's main contribution to the new art was his work on the adaptation of these cells for use as almost inertialess light modulators[40]. By shortening the optical cell length, reducing the electrode spacing, increasing the optical cross section and using nitrobenzol as the dielectric, Karolus succeeded in designing a workable light control valve. A four stage d.c. amplifier provided the appropriate picture signals for modulating the Kerr Cell. 'The resulting picture quality and the picture brightness . . . were quite good'[41] in the case of diapositive images which were projected onto the Nipkow disc by means of a 30 A arc lamp. Later, Karolus replaced the d.c. amplifier with an amplifier using resistance-capacitance coupling, and with the aid of a bridge circuit, introduced a carrier signal so that the low frequency signals could be amplified, Figures 11.9, 11.10 and 11.11.

Figure 11.11 The 96 lines per picture mirror drum scanner shown at the 1928 Berlin Funkausstellung

Reproduced by permission of AEG Aktiengesellschaft

Karolus was led to consider the Kerr cell because of the limitations of the existing methods 'as regards their usefulness and efficiency'. These constraints he identified as: the lag of the light controlling device, the small intensity of the modulated light source, the poor effectiveness of control of the light source, the great loss of energy and the impossibility of obtaining sharp impressions of the source of light when employing glow lamps. With his invention Karolus hoped to provide 'a means for quantitative image tone control of practically any amount of light, with negligible losses, up to frequencies of over 10 MHz'.

The utilisation of the Kerr cell did not negate all the five defects listed by Karolus: the cell gave rise to a picture of relatively low intensity and then only with a considerable loss of light energy.

It has been stated[42] that the light losses in an optical system of the Kerr cell type can amount to approximately 80 %, made up as follows:

1   50 % loss due to the total reflection of the ordinary ray;
2   10% loss due to the reflection from the inclined surface of the Nicol prism upon which the light is incident—a loss which occurs in both the polariser and analyser;
3   20 % loss due to the limited opening angle of the prism which is about 25%.

A report[36] on the 1929 Berlin Radio Exhibition, at which the Karolus system was exhibited, noted that, although the Nicol prisms associated with the cells were water cooled, owing to their proximity to the intense arc light and the detail of the 48-line picture was very good, the image was 'extremely difficult to see, as the illumination of the screen was very poor indeed'. Karolus ascribed this to the adverse conditions prevailing at the exhibition even though the television section was lit 'by very subdued lights'. The size of the reproduced image was 30 cm by 30 cm and was synthesised by a mirror drum having fifty small mirrors.

A disadvantage of the above method of scanning was the high cost of a television set using the method compared with that of a set utilising, for example, a Nipkow disc. Fernseh fixed the 1929 price of its sets at Dm 350 but the cost of the Telefunken receiver was between Dm 700 and Dm 800, and this was for the small set which employed a crater neon lamp and gave a projected picture of approximately four inches square. The cost of the large receiver using the Kerr cell was not given. At the exhibition spectators were told that for the present the Telefunken Company did not intend to market instruments. This reluctance may have been due to a factor other than cost.

Both Dr Schroter and Dr Karolus were of the opinion that the Reichspostzentralamt's television standards were unsatisfactory because they believed greater detail was essential to ensure the success of television. To further this opinion the Telefunken Company in 1929 commenced experiments on television broadcasting using wavelengths of 30 m and 80 m. These wavelengths would allow wider bandwidths and hence a higher definition to be to be utilised.

The existing (1929) German broadcasts on 418 m (716 kHz) were radiated from the 1.5 kW Witzleben station daily from 9 a.m. to 9.30 a.m. and from 1.00 p.m. to 1.30 p.m., except Tuesdays and Thursdays when the times were devoted to instructional broadcasts to schools[43]. These transmissions, which had a range of about 100 km, were concerned exclusively with telecinema and were made available free of cost to developers and experimenters.

A further television station was opened in the second half of 1930 at Königswusterhausen[44] and, in 1931, the Doberitz station was commissioned[45] to cater for the growing curiosity of the German public in the new form of communication.

In 1930 also, Fernseh AG was successful in supplying, against some competition, the Reichspostzentralamt with a spotlight scanner[46]. This was shown at the Funkaussetellung in August 1930 and was based on Baird's work on indirect scanning and his use of infrared radiation for this purpose. The scanner employed a 900 W projection lamp, with associated parabolic mirror and projection lens, and used adjustably mounted photoelectric cells (housed in screened enclosures to reduce the pick-up of spurious interference) having a sensitive surface area of

450 mm by 600 mm (to suit the aspect ratio of 3:4). These cells were highly sensitive to infrared radiation and, by using an infrared filter in the path of the incident radiation, a subject could be televised in comfort.

Apart from Fernseh AG, the only other company which displayed exhibits in 1930 was Telehor AG[47]. This company showed commercial type apparatus (based on the use of the Nipkow disc and phonic motor) which employed a synchroniser, operated independently of the a.c. mains, which permitted 'a remarkable stability to be obtained, any fluctuation of images being eliminated'.

Karolus and Telefunken were not alone in concluding that the RPZ's 1929 television standard was inadequate[48] for entertainment by television and in 1931 all the exhibitors at the Berlin Radio Exhibition showed equipment capable of working at up to 100 lines per picture[49]. Details of the equipments are given in Table 12.2[50]. The scanners were still of the Nipkow disc, lens disc and mirror wheel types, but in addition cathode-ray tube displays were shown by four exhibitors. Nothing of a startlingly original nature was disclosed; it seems that German developments were proceeding along methodical conventional lines.

The most noteworthy German television contribution in 1931 was not that connected with a piece of hardware. In an issue of the periodical *Fernsehen* (which was first published in 1930 and which was the first German television journal), Moller[51] and Kirschtein[52], in two separate papers showed how the optical efficiencies of scanners could be calculated according to the principles of optics and photometry.

One of the most extraordinary features of the evolution of practical systems of television is that for approximately 50 years, from the first 1878 suggestions for seeing by electricity to 1930, no-one anywhere had sought to compare the efficien-

Figure 11.12 Marconi low-definition television, 1931/32. A type of scanning apparatus of high optical efficiency, (see Table 11.2). Horizontal movement of the light spot was caused by the aperture or lens drum, and vertical movement of the spot was by means of the mirror drum

Reproduced by permission of the Marconi Company Ltd.

cies of the various mechanical scanners. Indeed, the technical literature for this period is notable for the almost complete absence of mathematical analyses of the merits or otherwise of the different methods. This was a period of empiricism in television matters.

As a consequence, just about every conceivable type of scanner for analysing or reconstituting an image was suggested prior to the inauguration of the world's first all electronic television service in 1936. There were vibrating mirrors, rocking mirrors, rotating mirrors, mirror polyhedra, mirror drums, mirror screws and mirror discs; there were lens discs, lens drums, circles of lenses, lenticular slices, reciprocating lenses, lens cascades and eccentrically rotating lenses; there were rocking prisms, sliding prisms, lens prisms, electric prisms, lens prism and rotating prism pairs; there were aperture discs, aperture bands, vibrating apertures, intersecting slots, multispiral apertures and ancillary slotted discs; there were cell banks, lamp banks, rotary cell discs, neon discs, scintillating studs, corona discs and convolute neon tubes and tubes with bubbles in them; there were cathode-ray tubes, Lenard tubes, X-ray tubes with fluorescent screens, gas screens, photoelectric matrices, secondary emitting surfaces, electroscope screens and Schlieren screens[53] (see Appendix 2).

And yet, as Moller and Kirschtein showed, an analysis of the optical performance of a scanner is a relatively straightforward matter based on the elementary principles of illumination and physics, and simple mathematics. A few years after their papers were published Myers in his book[42] 'Television optics' compared the characteristics of many mechanical-optical systems, see Appendix 3.[54] He determined the relative efficiencies of the apertured disc (Nipkow, 1884) and the mirror wheel (Weiller, 1888) and concluded: ' . . . the aperture disc should be used at the transmitting end for low-definition systems with about 40 lines and upwards. Below 40 lines it is advisable and more efficient to employ the mirror drum. The advisability rests on economic grounds. This is the policy which was adopted by the BBC in their low-definition 30-line transmitter.' Actually, Baird seems to have arrived at this result from purely experimental investigations. Although he used aperture discs, of various types, in his early work, by 1930 he was utilising a mirror drum scanner for his 30-line broadcasts (see also Figure 11.12).

Of all the multifarious mechanical scanners mentioned above there was one whose use contributed more to the advancement of television than any other. This was the aperture disc scanner invented by Paul Nipkow[55] in 1884. With it Baird first demonstrated the crude reproduction of a half-tone image of a human face by reflected light on 2 October 1925. It was used by the American Telephone and Telegraph Company in their ambitious and well engineered field trial of April 1927, and during the succeeding years the Nipkow disc formed an essential component of the television schemes of inventors/engineers in the UK, the US the USSR, France and Germany. The disc was successfully adapted to show the world's first broadcast colour television images and the first demonstration of stereoscopic television. It was employed for several years in telecine equipment and enabled films to be televised with a degree of clarity and brilliance which rivalled and probably, initially, surpassed the early iconoscope teleciné cameras[56].

Nipkow's invention was greatly developed: the scanner was used in 30, 50, 60, 90, 120, 180 and 240-line television apparatus; it was run in air and in vacuo; it was constructed with a single spiral, multiturn spirals and spiral segments; it was made with round holes, square holes and hexagonal holes; it was designed to produce orthodox scanning, discontiguous scanning, graduated scanning, overlapping scanning and, when pairs of discs were used, interlaced scanning and bilateral scanning.

Considering the many different mechanical scanners which were patented before 1926, the Nipkow disc was probably the most robust, the simplest and the cheapest. It could be made from cardboard or metal, in a workshop or a home, by the unskilled or the skilled, for a few pence or many pounds. It was used by amateurs and professionals alike.

Nevertheless, the Nipkow scanner had a number of inherent defects:

1   the scanned area was in general truncate-sector shaped due to the fact that if the angular separation of the holes was constant, the linear spacing between consecutive holes decreased as the distance of the holes from the centre of the disc decreased;
2   the scanning strips were necessarily arcuate in shape;
3   the size of the traversed area decreased as the number of scanning lines increased;
4   the size of the scanning apertures decreased as their number increased and consequently difficulties could be caused by dirt entering the holes.

Furthermore, as Myers proved[57], 'the aperture disc cannot satisfactorily be embodied in the receiving system' for reasons of optical efficiency. Also: 'combinations of aperture disc and mirror drum have been tried out [in receiving systems], apparently without previous analytical investigation, the aperture disc functioning as the line scanner and the mirror drum as the frame scanner; but such arrangements have, of course, not met with any success.'

The 1931 analyses of Moller and Kirschtein appear to have had an influence on the developing German television industry. At the 1932 Berlin Radio Exhibition[58] all of the transmitter scanners were of the Nipkow disc type, and half of the mechanical receiver scanners were of the mirror screw pattern, see Table 11.1[59]. The decline in the use of the Nipkow disc and the increasing importance of mirror screws and cathode-ray tubes in receivers is highlighted in the Table.

On the other hand, the supremacy of the Nipkow disc as an analyser is apparent from the data given in Table 11.2[59].

The mirror screw, due to H Hatzinger[60], was first shown at the 1931 Berlin Radio Exhibition. Essentially, the scanning element comprised a number of mirrors, equal to the number of scanned lines, having the form of flat microscope slips, arranged like a spiral staircase. For the 84-line images displayed in 1931 the mirror screw used 84 mirrors, each 10.0 cm long and 1.0 mm deep to produce a picture 8.4 by 10.0 cm in size[61].

At the 1932 Funkausstellung[58] the smaller screws were constructed from chromium plated brass and the edges of the mirrors polished to give good reflecting surfaces. With this form of scanner a light source having a luminous area equal to the product of the height (8.4 cm) of the image and the width of the scanning spot

Table 11.1   Types of receiver scanner employed at Berlin Funkausstellungen

Receiver scanner type	Number demonstrated				
	1931	1932	1933	1934	1935
Nipkow disc	4	3	1	1	–
lens disc	1	–	–	–	–
mirror drum/wheel	1	1	1	1	–
mirror screw	1	7	4	3	3
mirror ring	–	–	1	–	–
cathode-ray tube	4	3	7	7	12
intermediate film	–	–	–	–	1

Table 11.2   Types of transmitter scanner employed at the Berlin Funkausstellungen

Transmitter scanner type	Number demonstrated				
	1931	1932	1933	1934	1935
Nipkow disc	only discs	10	8	9	9
lens disc	used	–	–	–	1
mirror drum		–	1	1	1
cathode-ray tube		–	1	–	–

(1.0 mm) was required. A special neon tube with a long, thin cathode was employed for this purpose.

The advantages which seemed at first sight to make the mirror screw an attractive scanner were: (1) the simpler and more accurate construction, compared with a mirror drum; (2) the absence of viewing lenses; (3) the small power consumption of the driving motor; (4) the compactness and equality in sizes of the viewed image and of the scanner; and (5) the low cost (estimated in 1932 to be £2 for a fully adjusted mirror screw 13 cm by 15 cm).

Considerable development of the screw was carried out by von Okolicsanyi of the Tekade Company. The screw was constructed with a single helix and with a double helix, with flat mirrors and with mirrors shaped to give a concave curve to the surface of the mirror group; and it could be utilised in either a virtual image or a real image mode[62].

With the former of these modes a definite relationship exists between the source and viewing distances. Herein lies the disadvantages of the mirror screw for high-definition television since the viewing distance becomes too great for the eye to resolve satisfactorily the available detail in the received picture, and the improvement afforded by the increase in the number of scanning lines is not apparent to the observer. Even the improvement embodying the spherical mirror screw cannot overcome this limitation, since, when too large a curvature is imparted to the surface of the screw, the picture becomes rapidly distorted.

For the production of the real image only a small portion of the flux from the light source is available for the production of the light spot on the screen. Myers, *op.cit.*, has shown that the mirror screw, when operated in this mode, has only π/n of the efficiency of the simple mirror drum (n being the number of lines per picture). When n is more than c.30 the system becomes extremely inefficient.

Thus, although the mirror screw achieved some popularity for a few years, particularly in Germany, it did not possess the potential for high-definition television and could not compete with the cathode-ray tube for this purpose.

Table 11.3 gives the principal parameters associated with the transmitters and receivers of the 1932 Funkausstellung and indicates the considerable improvements which had been made since the previous exhibition. These included an increase in definition from 60 to 90 lines per picture, an enhancement of the general level of image brightness, a substantial enlargement of image size, a move towards the general adoption of the cinematograph film aspect ratio of 3 (vertical) to 4 (horizontal), and the first appearance of the intermediate film process.

The most important television event in 1932, in Germany, was the inauguration in August of the 15 kW, 43 MHz ultra short wave transmitter situated in the broadcasting tower at Witzleben. Manufactured by Telefunken, the transmitter used the video signals which were generated in the nearby Berlin broadcasting House. From the top of the 420′ antenna tower, the highest point in the city, the simple 6′ long metal rod antenna radiated electromagnetic waves over an area of c. 50 miles radius.

During the 1932 Berlin Radio Exhibition the Fernseh 90-line studio apparatus in Broadcasting House provided daily televised film images between 9.00 and 10.00 a.m. and between 1.30 and 3.00 p.m.. These were intended to be received and displayed by all the six exhibitors but only the Reichspostzentralamt—and on one occasion the Loewe company—demonstrated television reception from the Witzleben transmitter. This was the first time in Europe when ultra short waves had been employed for the public demonstration of 90-line images. A report on the exhibition mentions: 'The general impression of the quality of the images was that 90-line images . . . can give real entertainment value of almost any type of subject.'

Fernseh also supplied, for the Reichs Rundfunk Gesellschaft, a 120-line head and shoulders Nipkow disc spotlight scanner. The holes in the disc were just 0.08 mm in diameter.

Until 1932 no demonstrations of direct television, i.e. the televising of scenes by television cameras, had been given at the Berlin Radio Exhibition. The images shown at the several Funkaustellungen prior to 1932 had all been obtained by telecine means—the scanning of a ciné film by a Nipkow disc. This position changed in 1932 when Fernseh showed its intermediate film scanner (see Chapter 13).

Electronic cameras of the image dissector and iconoscope types were not demonstrated until the 1936 exhibition when Fernseh AG and Telefunken showed their versions of Farnsworth's and Zworykin's cameras respectively. In both cases the basic research and development of the electronic camera tubes had been undertaken elsewhere, in the US, and it seems that no fundamental work on purely electronic scene analysis was carried out in Germany contemporaneously with that being executed by Farnsworth, Zworykin and Shoenberg's teams.

This early lack of interest in all-electronic television may have stemmed from the views of Fritz Schröter who was director of research at Telefunken Gmbh. In 1932

Table 11.3 Details of the various television systems exhibited at the 1932 Berlin Funkausstellung

Stand	Telefunken		Tekade			RPZ (German Post Office)			HHI	Fernseh AG				Loewe
Exhibit no.	a	b	a	b	c	a	b	c	a	a	b	c	d	a
Type	talkie	talkie	*Transmitter* talkie	talkie		cine	cine	cine	*Receiver*	talkie	spot light	inter-mediary film		cine
Scanner	disc	disc	disc	disc		disc	disc	disc		disc	disc	disc		disc
Lines	48	90				90	90	90		120	90	90		90
Points	5760	10 800	9720	9720		10 800	10 800	10 800		19 200	6075	10 800		10 800
Images/p./sec.	25	25	25	25		25	25	25		25	25	25		25
Ratio	4:5	3:4	5:6	5:6		3:4	3:4	3:4		3:4	4:3	3:4		3:4
Scanner	mirror drum	cathode ray tube	*Transmitter* mirror screw	mirror screw	mirror screw	mirror screw	disc	cathode ray tube	*Receiver* mirror screw	disc	mirror screw	disc	mirror screw	cathode ray tube
Light source	kerr cell		neon lamp	neon lamp	neon lamp	sodium lamp	sodium lamp		HF mercury argon lamp	sodium lamp	neon lamp	sodium lamp	sodium lamp	
Lines	43	90	90	90	90	90	90	90	90	120	90	90	120	90
Points	5760	10 800	9720	9720	9720	9720	10 800	10 800	9720	19 200	6075	10 800	19 200	10 800
Images/p./sec	25	2	25	25	25	25	25	25	25	25	25	25	25	25
Ratio	4:5	3:4	5:6	5:6	5:6	5:6	3:4	3:4 up to	5:6	3:4	4:3	3:4	3:4	3:4
Size (cm)	40 × 50	9 × 12	6 × 7	13 ×15	30 × 36	13 × 15	12 × 16	15 × 20	13 × 15	7 × 9	16 × 12	12 × 16	12 × 16	9 × 12
Colour	white	pale green	pink	pink	pink	yellow	pale yellow	blue	blue	pale yellow	pink	pale yellow	not shown working	blue
Detail	v. good	excellent	v. good	v. good	v. good	v. good	v. good	v. good	v. good	excellent	excellent	v. good		excellent
Intensity	v. good	excellent	good	good	good	v. good	v. good	v. good	excellent	v. good	good	v. good		v. good
Flicker	slight	none	none	none	none	none	none	none	none	none	none	none		none
F. max	72 000	135 000	121 500	121 500	121 500	121 500	135 000	135 000	121 500	240 000	76 000	135 000	240 000	13 000
Channel						USW&	USW&							
Synchronising	wire mains	wire mains	wire mains	wire mains	wire mains	wire mains	wire mains	wire mains or sawtooth impulse	wire mains	wire mains	wire mains	wire mains		wire picture current

Note: HHI exhibit a — "working off the Tekade transmitter".

he edited[64] the 'Handbuch der Bildtelegraphie und des Fernsehens'. This contained an excellent survey of the various mechanical scanning systems and of the state of the art of television. However, one section, by Schroter and F Banneitz, was quite critical of the efforts of Campbell Swinton and others to evolve an electronic scanning tube. On Campbell Swinton's ideas they wrote:

'This proposal though not free from electrical or quantum theoretical objections, and moreover never put into practice, was later followed by similar proposals from Schoultz, Seguin, Zworykin, Sabbah, von Codelli, von Tihanyi and others.

'The idea common to all these was to project the image as a whole onto the photoelectric surface and there to convert it into a uniform distribution of electrical quantities such as conductivity or charge potential. These variations in conductivity or potential are sampled point by point by the scanning beam and converted into internal resistance fluctuations which after amplification, modulate the carrier frequency.'

In the next paragraph Schröter and Banneitz described the first proposal of an image dissector by Dieckmann and Hell[65], later taken up and developed by Farnsworth, and identified the inherent limitations of the method and observed: 'None of the above described scanning systems has much chance of being developed in the foreseeable future . . . '

Schroter and Banneitz's pessimism was based on the very appreciable vacuum technology difficulties which existed at that time. Moreover, they felt that neither the potassium nor the selenium camera screens could be produced, with the large number of individual cells necessary, in a homogeneous and sensitive way. However:

'Those arrangements that provide effective integration of the effect of the light with the help of the capacity of individual cells . . . present the only solution achievable in the future for the transmission with fine detail of pictures of normal brightness.'

Finally, they concluded:

'The raster of photoelectric cells with charge integration and a cathode-ray beam as commutator opens up a new way ahead which probably must one day be followed if it does not happen that the other methods can be made practicable by the new discoveries which improve their photoelectric efficiency.'

In 1932 it was known that Telefunken was receiving the RCA patent applications and licensee bulletin. Also, Farnsworth's work had become public knowledge. Nevertheless, Telefunken did not engage in all-electronic television development until c. 1936 and it would appear that Schroter's gloomy outlook had an inhibiting effect on German all-electronic television's progress in the early 1930s.

Elsewhere, in the US and the UK, much effort had been expended, from the mid-1920s, and was continuing to be expended to develop an all-electronic television camera. The successful outcome of one of these endeavours led to the evolution in 1936 of the world's first, regular, public, all-electronic, high-definition television service. However, before this is narrated and discussed, some consideration must be given to the measures which were being taken in France to resolve the television problem.

Experimental work on low-definition television began in France in the early 1920s and hence was contemporary with similar activities being undertaken in the United Kingdom, the United States of America and Germany. And just as some of these activities, for example those of Jenkins, GE, AT&T and RCA in the US, and of Karolus in Germany, followed on from developments in the field of still picture transmission, so in France also the earliest practical enquiry into television in the 1920–1930 decade succeeded efforts to establish phototelegraphic and phototelephonic services.

Unlike the position in Great Britain where effectively only one person, J L Baird, engaged in practical invention and innovation in the field of seeing by electricity, in France several enquiries proceeded independently during the decade. Among the principal investigators mention must be made of E Belin, G Valensi and A Dauvillier. However, none of these investigators was as adept as Baird in demonstrating the basic principles of television, and none of the French demonstrations of the 1920s matched those given by either the AT&T Company or the General Electric Company.

In the US the endeavours of Jenkins, AT&T, GE and others were accorded much publicity in newspapers, semipopular magazines and learned society journals; in Germany the annual Berlin Funkausstellung (from 1928) enabled the press and general public to view the efforts of Mihaly, Karolus, Fernseh A G, Telefunken and RPZ; and in Britain the work of Baird and the results he achieved were widely reported and commented upon. This was not the situation with French developments in the 1920s. Indeed, there is a dearth of first-hand independent eye witness reports of the televised images which were reproduced by the French workers. Essentially, their successes lagged behind those being accomplished elsewhere and were not particularly newsworthy.

The progress of Baird, Jenkins, AT&T, GE, among others, stemmed from their adoption of rotary scanners of the apertured/lensed disc or mirror drum types. Some of the French television pioneers utilised vibrating mirrors, attached to galvanometers or tuning forks, as scanners but these were never viable for practical studio television. By the end of the decade all the mechanical scanners used for analysis and synthesis in the USA, the UK and Germany were of the disc or drum forms. A further impediment to early success by the French workers was the utilisation, by some of them, of the cathode-ray tube. This needed much development before it could display satisfactory images. Elsewhere the accomplished exponents of the television art were using either directly modulated discharge tubes/lamps, or lamps/arcs and an associated light valve—usually a Kerr cell. Baird's shrewd selection of components (from the considerable range available) for his television schemes highlights his considerable talent for reducing a problem to its simplest form and then finding an effective and simple solution. His system of analysing and synthesising apparatus was either independently arrived at by a few investigators, for example those at AT&T, or followed by others after the demonstration of 1926 and 1927 by Baird and AT&T. The early French workers were not so fortunate or so perspicacious in their choice of subsystems.

A further determinant in Baird's case is that he had not worked in the field of phototelegraphy or phototelephony and was not constrained to attempt to apply or to develop the devices employed to the field of television. In France Edouard Belin had acquired much experience of the technology of still picture transmission and initially tried to adopt some of this for television purposes. This approach was not rewarding.

In 1907 and 1908 Belin[66,67] conducted various experiments using an apparatus, which he called a télestéréograph, to send still images over the long distance telephone line between Paris and Lyon. Belin used the gelatin-bichromate relief process, devised by Amstutz[68] in 1891, to form his photographic images and made his first transmission in 1907 with the sending and receiving instruments coupled together in Paris, the signal current being sent from Paris to Lyons and then back to Paris, a distance of c. 400 miles.

As stated earlier, a photograph in relief can be obtained by printing from a negative onto a gelatine coated paper which has been sensitised to light by the addition of bichromate. When exposed and developed in hot water the gelatine washes away from the unexposed areas but remains insoluble where the exposure to light has been intense. In the half-tones the gelatine dissolves to an extent dependent on the strength of the incident light flux. Thus, a relief print is obtained in which the thickness of the surface coating is a function of the optical density of the associated photographic negative.

In the télestéréograph the relief picture was wrapped around a cylinder which rotated and moved laterally, parallel to its axis, so that a fixed sapphire stylus pressing on the relief surface scanned a helical path on the print. Movement of the stylus altered, by means of a mechanical linkage and sliding contact, the resistance in series with the telephone line and so enabled a variable current to be transmitted to the receiving instrument. This comprised an aperiodic mirror galvanometer, to which the signal was applied, of the Blondel type, a constant intensity light source, either a wedge-shaped aperture or a graduated grey scale glass wedge, and a scanning cylinder similar to that utilised in the transmitter and synchronised with it. Light reflected from the galvanometer mirror passed through the aperture or wedge and was concentrated by a lens onto the photographic film wrapped around the receiving cylinder. Movement of the light beam altered the emergent light flux from the aperture/wedge and hence the exposure of the film.

Some success was achieved with this equipment although the pressure exerted by the stylus on the relief picture caused the surface to be damaged. To obviate this defect Belin conceived the idea of making the stylus bear against the diaphragm of a microphone, so that the force exerted on it would vary in accordance with the relief of the photographic print. Experimentation led to the use of a carbon microphone in which displacement of the diaphragm caused the resistance of three carbon balls to change.

Belin maintained his interest in picture transmission for many years and in 1921 he sent the first visual message[69] across the Atlantic ocean. One year later, on 1 December 1922, at a meeting of Les Société des Électriciens held at the Sorbonne, Paris, Belin gave a demonstration of long distance sight by wireless. The report in the *New York Times*[70] stated: 'Flashes of light were directed on a selenium element which through another instrument produced sound waves. These waves were then taken up by a wireless apparatus that reproduced the flashes of light on a mirror. This was offered as proof that the general principle of projecting a stationary scene has been solved.'

Belin applied for his first television patent[71] on 27 December 1922. His scheme included at the transmitter two mirror scanners, oscillating about orthogonal axes, and a photoelectric cell; and at the receiver a source of light, a mirror galvanometer oscillograph with a graduated symmetrical neutral filter and lens, which in combination were intended to function as a light valve (as in his téléstéréograph), and two mirror scanners synchronised to the transmitter scanners. It is not known whether any success was obtained with this system.

On 19 December 1925 the *New York Times*[72] reported that Belin had informed the members of the Society of French Photography that his efforts to find a solution to the problem of television had finally succeeded. 'The announcement created a profound impression since a veritable race is going on between the inventors of France, America, Germany and England to reach the goal which M Belin now claims to have achieved,' noted the reporter.

It seems that the inventor had exhibited a steel disc, having 20 silver facets, which in addition to its rotation at 4000 r.p.m. was capable of 'an ingenious movement backwards and forwards' which permitted a pencil beam of light to scan an image. No other details of the invention were revealed. But: ' "Television" by radio is on the point of being realised, M Belin told the Society.' The next month J L Baird demonstrated rudimentary television for the first time anywhere to 40 members of the Royal Institution, London.

Belin gave an important demonstration[73], on 28 July 1926, of his new receiver system to General G A Ferrie, head of the French Military Telegraph, Professor C Fabrie of the Sorbonne and R Mesny of the French Academy of Sciences. The inventor was now working with F Holweck[74,75], who had been chief of staff of the Madam Curie Radium Institute. He was a noted expert on vacuum technology and the inventor of a high vacuum pump which had been named after him. The demonstration, which was given at the Madam Curie Radium Institute and not at Belin's laboratory at Malmaison, near Paris, was supervised by G N Ogloblinsky, who had constructed the apparatus and was chief engineer of the Laboratoire des Etablissements Edouard Belin. When and why Belin sought collaboration with Holweck is unknown. Possibly he, Belin, had concluded that television progress could not be made with a light valve of the type first used in his still picture transmission experiments of 1907–1908 and that the approach of Campbell Swinton—which was described in its modern form in 1924 in *Wireless World*—was the way forward.

Figure. 11.13 shows the layout of the Belin and Holweck sending apparatus. The novel feature concerned the 10 Hz mirror scanner which was linked to a microphone diaphragm by a light metallic bar. At the end of each oscillation movement of the diaphragm generated a pulse of current which was used for synchronisation purposes. Here, again, Belin was adopting an aspect of his picture equipment to his television apparatus. Both mirrors were driven from the same motor to maintain the constancy of the ratio of their oscillation frequencies (500 Hz and 10 Hz).

The principal innovation of the system as a whole was the utilisation of a cathode-ray tube to display the received images. Figure 11.14 illustrates the electron gun of the tube which was continuously evacuated by a molecular pump. Both the pump and the cathode-ray tube had been designed by Holweck. Magnetic deflection was used for the line and frame scanning, and collimation of the electron beam on the screen was accomplished by the apertured grid, G, and anode, F, electrodes, and by the 'concentrating coil', D.

During the demonstration[76] the 'image of a moving finger as well as of several small objects moved before the sending apparatus' were transmitted from one end of the laboratory to the other. The images were 'only jet black and white' and no half-tones were displayed.

A soft (i.e. gas-filled) cathode-ray tube was included in the system of G Valensi, the chief engineer of the French Postes et Télégraphes. His apparatus was based on patents[77,78] dated 29 December 1922 and 3 January 1923 and was described in several journals[79] in 1927. Valensi's work was carried out in collaboration with M Johannes of the Établissements Gaiffe-Gallot-Pilon. Figure 11.15 shows a diagrammatic arrangement of the signal generating equipment. Two scanning discs, of an unusual type, analysed the object in a manner 'continue et rigoureusement uniforme', unlike the Lissajous type scanning achieved with two oscillating mirrors. At the receiver it was intended that a cathode-ray tube (having a construction very similar to that of Belin and Holweck) should synthesise the received image. However, no results were described in the 1927 articles. In a paper, written by Valensi, the author mentioned that from 1925 he had not pursued the solution of the television problem by means of cathode-ray tubes because 'elles semblent comporter des limitation inévitables'. Given an image of 22 500 elements (150 lines per picture of 1:1 aspect ratio) scanned in 0.1 s, each element of the fluorescent screen would be bombarded by the electron beam for 1/225 000 of a second. Valensi felt that to secure a visual impression the electrons would have to strike the screen with such a high velocity that the fluorescent screen material would probably very quickly suffer 'une fatigue' which would render its operation unreliable.

At the end of his 1927 paper the chief engineer opined that an intense light source and Kerr cell would appear to lead to the best results and that discharge lamps would give an acceptable and economic solution. These were the devices being vigorously developed in the US, the UK and Germany.

M A Dauvillier was another French worker who pursued the aim of television image display by means of cathode-ray tubes[80,81]. He was the head of the physical research laboratory for X-rays founded by L de Broglie and had been working in his organisation from 1921. Dauvillier's interest in television dates from 1923.

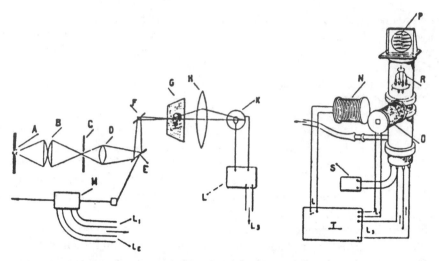

Figure 11.13 Schematic diagram of the apparatus of Belin and Holweck. A, arc; B, condenser lens; C, diaphragm; D, lens; E, F, vibrating mirrors; G, screen; H, lens; K, photoelectric cell; L, amplifier; L_1, L_2 transmission lines for synchronising signals; L_3, signal line; M, driving motor for E and F; N, O, electromagnets; P, fluorescent screen; R, electrodes of the cathode-ray tube

Source: *Television*, Dunod, 1939, p. 195

A diagrammatic representation of his apparatus is given in Figure 11.16. Again, orthogonal oscillating mirrors were used to analyse the object and a cathode-ray tube was employed to reconstitute the image. Table 11.4 summarises the main features of the sending and receiving scanning systems of the Belin and Holweck, Valensi and Dauvillier equipments[82] and compares the forms of their cathode-ray tubes with the general purpose Western Electric c.r.t which was described in 1922 and which became available to researchers in the 1920s.

Much experimental work, from 1924, on the reproduction of images was undertaken by Dauvillier using a modified Western Electric Type 224 c.r.t. He noted in a paper[83] published in 1926 that the construction of a tube sensitive to changes in the intensity and deflection of the electron beam had presented 'sérieuses difficultés'. By 1925 Dauvillier had succeeded in transmitting the image of a very luminous point but observed in his 1926 paper that an increase in the sensitivity of the apparatus of the order of 1000 times would be necessary to render the television system usable in practice.

Belin and Holweck, Valensi, and Dauvillier seem to have become discouraged by their efforts[84,85] and from c. 1928 little was heard of their investigations. Belin and Holweck also seem to have discontinued their television investigations, Oglobinsky emigrated to the United States and worked for the Westinghouse Company and Dauvillier turned his attentions to the field of cosmic ray physics. Valensi applied for some patents in the 1930s but does not appear to have engaged in any further practical work on television. Only Barthélémy and de France were pursuing enquiries at the end of the 1920s decade[86].

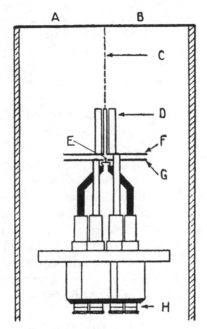

Figure 11.14 The Holweck cathode-ray tube. The diagram shows the fluorescent screen, A–B; the electron beam, C; the concentrating coil, D; the filament, E; the anode, F; the grid, G; and the terminals, H

Source: *Radio News*, **8**, December 1926, p. 739

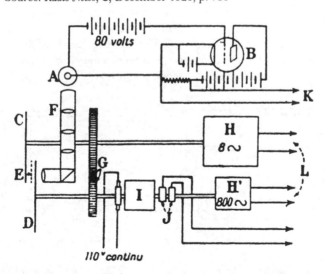

Figure 11.15 Diagram illustrating Valensi's signal generating apparatus. A, photoelectric cell; B, amplifying valve; C, D, scanning discs; E, image of the film to be transmitted; G, gears; H, H', alternators; I, motor; J, slip rings; K, signal lines to the receiver

Source: *Television*, Dunod, 1937, p. 188

In France R Barthélémy, Figure 11.17, is known as 'le père de la télévision Francaise'. He became associated with television development in the mid-1920s following work, before and during the First World War, under General Ferrie, on wireless.

In 1928 the director general, M E Chamon, of La Compagnie pour la Fabrication des Compteurs, à Montrouge, visited London and learnt of Baird's endeavours. Convinced that television would progress Chamon asked his assistant M le Duc to establish a radiovision research laboratory. Barthélémy was appointed director of the laboratory in 1928[87,88].

His first crude 30-line images[80,90] of the five fingers of a hand were reproduced in September 1929. Later, on 14 April 1931, under the auspices of the Société Francaise des Électriciens of the Comité Central des Sociétés de TSF and of the Société Francaise de Télévision, the first public demonstration of television was given. This was held at the Ecole Supérieure d'Electricité à Malakoff, Montrouge before about 800 persons[87]. The demonstration was repeated for the benefit of the Academie des Sciences in November 1931. Barthélémy's approach to his task was based essentially on the use of Nipkow disc scanners, following Baird, although the method of synchronisation used incorporated Barthélémy's ideas. 30-line images were projected onto a ground glass screen.

Of the other television researchers in France, Henri de France in August 1931 at Le Havre gave an official demonstration to the Mayor, M Leon-Meyer. Shortly afterwards, in February 1932, de France was given facilities to continue his experiments at Radio Normandy[91].

From January 1932 the Paris Postes, Télégraphes et Téléphones (PTT) began regular television transmissions[87,92,93], with Baird equipment, using a 441 m, 0.8 kW transmitter. The transmissions were sent out at variable times each day (except Sundays) but usually from 8.15 a.m. or 2.15 p.m. and lasted for 30 to 45 minutes. The broadcasts were intended primarily for amateurs eager to overcome the various difficulties associated with television reception, including those presented by the

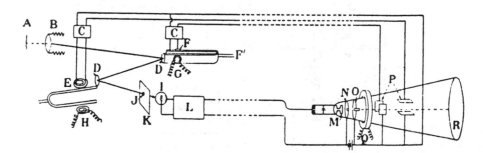

Figure 11.16 Line diagram of Dauvillier's apparatus. A, object; B, lens; C, amplifiers; D, mirrors;
E, F, coils for generating the synchronising signals; F', tuning fork (800 Hz); G,
H, driving electromagnets; I, photoelectric cell; J, aperture; K screen; L, amplifier;
M, filament; N, grid; O, anode; P, deflecting plates; Q, focusing coil; R, fluorescent
screen

Source: *Television*, Dunod, 1937, p. 192

Table 11.4   *The television schemes of Belin & Holweck, Valensi and Dauvillier*

Investigator	Type	Sending end scanner Driving device	Scanning	Frequencies
Belin & Holweck 1922–	two orthogonal oscillating mirrors	two galvanometers	Lissajous figure	500 Hz, 10 Hz
Valensi 1922–	two scanning mirrors	an electric motor and gears	linear	800 Hz, 8 Hz
Dauvillier 1923–	two orthogonal oscillating mirrors	two tuning forks	Lissajous figure	800 Hz, 10 Hz

Investigator	Type	Intensity control	Receiving end cathode-ray tube Cathode type	Collimation means	Defelction means	Screen material	Results
Belin & Holweck	hard, pumped	grid-cathode voltage	thermionic	apertured disc (1)	magnetic		crude images, no half tones
Valensi	soft, sealed off	"	"	"		magnetic	none publishd
Dauvillier	soft, sealed off	"	"	"	electrostatic	Willemite	none published
Western Electric Type 224 (described in 1922)	soft, sealed off	none	"	" (2)	electrostatic	Willemite (3)	not applicable

(1) with tubular anode and coil, (2) with tubular anode, (3) plus calcium tungstate

diversity of the a.c. and d.c. domestic power supplies. One account gives the number of these 'bricoleurs' (potterers) as 30.

At this time the state had no intention of organising an official public television service and the policy of La Compagnie pour la Fabrication des Compteurs (CDC) was to proceed with caution. Displays of television were given from time to time and in September 1932 at the Grand Palais, Paris a notable public exhibition was organised with the patronage of the 'industries radioéléctriques'. Picture signals were transmitted from the studio of CDC and were shown on five receivers. Among the artistes who participated were 'le chansonnier Devilliers' and his troup, the clown Bilboquet, Mme Ninon Guedald of l'Opera Comique and Fredo Gardoni and his orchestra[94].

From May 1933 Barthélémy, and Roger R Cahen, who later collaborated with de France, worked from the small studio of the PTT situated at 97, Rue de Grenelle. They still used a definition of 30 lines per picture. The only real progress during the year was due to de France who, in July 1933, demonstrated 120-line television[95].

In 1934, following the setting-up of the UK's Television Committee, to consider and report on the various systems of television, letters of enquiry were sent to several communication administrations in Germany, France and the US. Subsequently,

Figure 11.17 Monsieur R Barthélémy

Reproduced by permission of Mr Bruno Ruiz

Figure 11.18 The mirror drum television scanner which was used during some of the transmissions from the PTT station in Paris, (1932)

Reproduced by permission of Radio Rentals Ltd.

visits were made to Germany and the US where demonstrations were given, and confidential talks were held on the problems concerned with the inauguration of a public television service. These visits were of some assistance to Lord Selsdon's committee. No visit was made to view French television, presumably because it was felt that such a visit would not be worthwhile. Television had advanced rapidly in the US, the UK and Germany during the early 1930s and much progress was being achieved in the US and the UK on the development of an all-electronic system. These efforts were being pursued by RCA, Farnsworth Television and EMI with most talented research staffs working in well equipped and resourced laboratories. A comparable French industrial research programme did not exist. Indeed, French television work during the early years of the 1930–1940 decade was unremarkable in its achievements. It lacked the excellence of the corresponding mechanically scanned television schemes being developed in Germany and did not advance the principles, or the engineering implementation of the principles, on which television is based[96].

References

1 MIHALY, D.: 'Das Elektrische Fernsehen und das Telehor' (M.Krayn Verlag, 1923) foreword
2 LANGER, N.: 'Television, an account of the work of D. Mihaly', *The Wireless World and Radio Review*, 26 March, p. 796
3 EVERETT, A.: letter to the editor, *Experimental Wireless and Wireless Engineer*, September 1927, pp. 580–581
4 Ref.2, p. 796
5 Ref.1, p. 162
6' VON MIHALY, D.: contribution to a discussion on a paper by J L Baird, *Experimental Wireless and Wireless Engineer*, April 1927, pp. 239–240
7 GRADENWITZ, A.: 'Television at the Berlin Radio Exhibition', *Television*, October 1930, p. 317
8 GOEBEL, G.: 'Das Fernsehen in Deutschland bis zum Jahre 1945', *Arch. Post-& Fernmeldwes.*, (5) August 1953
9 The Editor: 'Television at the Berlin Radio Exhibition', *Television*, October 1929, p. 386
10 GRADENWITZ, A.: 'Mihaly's telecinema', *Television*, April 1929, p. 59
11 ANGWIN, A.S.: Memorandum, 24 June 1929, 33/5141, file 2, Post Office Records Office, UK
12 LYNES, W.: Letter to the GPO, 13 June 1929, 33/5141, file 1, Post Office Records Office, UK
13 PHILLIPS, F.W.: Letter to the BBC, 19 June 1929, Post Office Records Office, UK
14 CARPENDALE, C.: Letter to the GPO, 24 June 1929, 33/5141, file 2, Post Office Records Office
15 ANGWIN, A.S.: Memorandum to the secretary, 24 June 1929 33/5141, file 2, Post Office Records Office, UK
16 TELEHOR, A.G.: Letter to S Sagall, 26 July 1929, 33/5141, file 2, Post Office Records Office, UK
17 SINGLETON, T.: 'The story of Scophony', (Royal Television Society, London, 1988)
18 'S A Moseley writes from Berlin', *Television*, June 1929, p. 175
19 CARPENDALE, A.: Letter to the secretary, GPO, 12 August 1929, 33/5141, file 2, Post Office Records Office, UK
20 PHILLIPS, F.W.: Letter to Carpendale, 14 August 1929, 33/5141, file 2, Post Office Records Office, UK
21 SAGALL, S.: Letter to the GPO, 24 December 1929, 33/5141, file 2, Post Office Records Office, UK
22 Letter to Sagall from GPO, 31 December 1929, 33/5141, file 2, Post Office Records Office, UK
23 PHILLIPS, F.W.: Report, 3 January 1929, 33/5141, file 2, Post Office Records Office, UK
24 Ref.17, pp. 15, 152
25 Ref.17, p. 135
26 GOEBEL, G.: 'From the history of television—the first fifty years', *Bosch Tech. Ber.*, 1979, **6**, pp. 3–27
27 GOEBEL, G.: 'Der Fernseh Start in Deutschland', *Funkschau*, 1978, **19**, pp. 906–909
28 Ref.26, p. 23
29 BURNS, R.W.: 'British television, the formative years' (Peter Peregrinus Ltd, London, 1986) pp. 157–158
30 Ref.26, p. 25
31 RUDERT, F.: '50 years of "Fernseh", 1929–1979', *Bosch Tech. Ber.*, 1979, **6**, p. 28
32 'A visit to the Goerz works', *Television*, November 1929, pp. 430–431
33 HUTCHINSON, O.G.: Letter to the PMG, 12 July 1929, minute 51/1929, file 43, Post Office Records Office, UK

34 HUTCHINSON, O.G.: Letter to the PMG, 28 July 1929, minute 51/1929, file 43. Also: President Kruckow, letter to Col. Purvis, 27 July 1929, minute 51/1929, file 43, Post Office Records Office, UK
35 Purvis: letter to President Kruckow, 2 August 1929. Also: KRUCKOW; Letter to Col. Purvis, 9 August 1929; letter to Baird Television Ltd, 5 September 1929, minute 51/1929, file 43, Post Office Records Office, UK
36 *Television*, October 1929, pp. 379–389
37 The Editor.: 'Television at the Berlin Radio Exhibition', *Television*, October 1929, pp. 379–389
38 MOSELEY, S.A.: 'Sydney A Moseley writes from Berlin', *Television*, July 1929, pp. 244–246
39 KORN, A., and GLATZEL, B.: 'Handbuch der Phototelegraphie und Telautographie' (Nemnich, Leipzig, 1911) p. 446
40 GRADENWITZ, A.: 'Television progress in Germany', *Television*, June 1929, pp. 165–166
41 ILBERG, W.: 'Ein Jahrzehnt Bildtelegraphie und Fernsehen, Telefunken-Karolus', *TZ*, 1933, **65**, pp. 5–26
42 MYERS, L.M.: 'Television optics' (Pitman, London, 1936)
43 RUSSELL, A.V.F.V.: 'Television in Germany', *Television*, March 1930, pp. 10–11
44 *The Engineer*, 4 July 1930, **150**, p. 11
45 ROSEN, H.: 'Television on short waves at Doberitz', *Television*, August 1931, pp. 214–215
46 'The exhibit of the Fernseh A.G. as shown at the Berlin Radio Exhibition', *Television*, October 1930, pp. 338–339
47 GRADENWITZ, A.: 'Television at the Berlin Radio Exhibition', *Television*, October 1930, pp.327–340
48 IVES, H.E.: 'Visit of Dr Karolus'. Memorandum for file, case 33 089, 10 October 1930, pp. 1–3, AT&T Archives, Warren, New Jersey
49 GRADENWITZ, A.: 'Television at the Berlin Radio Exhibition', *Television*, October 1931, pp. 310–312, 318
50 TRAUB, E.H.: 'Television at the 1931 Berlin Radio Exhibition', *J. Television Society*, 1931, pp. 100–103
51 MOLLER, R.: 'Das Weillersche Spiegelrad', *Fernsehen*, 1931, **2**, pp. 80–97
52 KIRSCHESTEIN, F.: 'Nipkowscheibe oder Spiegelrad', *Fernsehen*, 1931, **2**, pp. 98–104
53 WILSON, J.C.: 'Television engineering' (Pitman, London, 1937)
54 Ref.42, p. 234
55 NIPKOW, P.: 'Elektrisches telescop'. German patent 30 105, 6 January 1884
56 Ref.29, p. 428
57 Ref.42, p. 187
58 TRAUB, E.H.: 'Television at the Berlin Radio Exhibition', *J. Television Society*, 1932, pp. 155–166
59 See the reports on the Berlin Radio Exhibition in *J. Television Society* for 1931, 1932, 1933, 1934, and 1935
60 HATZINGER, H.: 'Improvements in or relating to devices for transmitting and receiving stationary and animated images'. British patent 358 411, 25 March 1930
61 Ref.50, pp. 102–103
62 Ref.42, pp. 201–228
63 Ref.58., pp. 165–166
64 SCHROTER, F.: 'Handbuch der Bildtelegraphie und des Fernsehens' (Julius Springer, Berlin, 1932) p. 61
65 DIECKMANN, M., and HELL, R.: 'Lichtelektrische Bildzerlegerrohre für Fernseher'. DRP patent 450 187, 5 April 1925
66 BAKER, T.T.: 'The telegraphic transmission of photographs' (Constable, London, 1910) Chapter 6, pp. 116–126
67 BAKER, T.T.: 'Wireless pictures and television' (Constable, London, 1926) pp. 104–105
68 Ref.1, pp. 8–9
69 BELIN, E.: 'Sur la transmission télégraphique des photographies, dessins et écritures', *C. R. Seances Acad. Sci.*, 6 March 1922, **174**, pp. 678–680
70 ANON: 'Belin shows television', *New York Times*, 2 December 1922, p. 15:3

71 BELIN, E.: 'Procédé et appareillage par réaliser par T.S.F. la télévision'. French patent 571 785, 27 December 1922, addition 29 259, 5 January 1924
72 ANON.: 'Reports television an accomplished fact', *New York Times*, 19 December 1925, p. 10:2
73 ANON.: 'Moving images sent by wire or wireless by Professor Belin before Paris experts', *New York Times*, 29 July 1926, p. 1
74 FOURNIER, L.: 'New television apparatus', *Radio News*, December 1926, **8**, pp. 626–627, 739
75 ANON.: 'Television developments of Edmund Belin', *Science and Invention*, September 1927, p. 410
76 BELIN, E., and HOLWECK, F.: 'Sur la télévision. Premiers résultats dans la transmission des images animées', *C. R. Seances Acad. Sci.*', 1927, **184**, pp. 518–520
77 VALENSI, G.: 'Dispositif récepteur pour la télévision'. French patent 577 762, 29 December 1922, and additions 28 926, 9 April 1923, 28 927, 24 April 1923, 28 929, 30 July 1923
78 VALENSI, G.: 'Dispositif de transmission et de synchronisation pour la télévision'. French patent 572 716, 3 January 1923
79 ANON.: 'New European television scheme', *Science and Invention*, July 1927, pp. 204–205
80 DAUVILLIER, A.: 'Procédé et dispositifs permettant de réaliser la télévision'. French patent 592 162, 29 November 1923, and addition 29 653, 14 February 1924
81 FOURNIER, L.: 'Television by new French system', *Science and Invention*, March 1927, pp. 988, 1066
82 HEMARDINQUER, P.: 'Télévision' (Dunod, 1937) Chapter 6, p. 194
83 DAUVILLIER, A.: 'Sur la Téléphot, appareil de télévision par tubes à vide; Resultats experimentaux preliminaires', *C. R. Seances Acad. Sci.* August 1926, **183**, pp. 352–354
84 VALENSI, G.: 'L'état actuel du problème de la télévision', *Annales des Postes et Telegraphes*, November 1927, **6**, pp. 1047–1067
85 Ref.81, Chapter 6, 'La télévision cathodique ses principes ses debuts', pp. 182–201
86 PAUCHON, B.: '50 ans de télévision en France', *Revue de l'uer technique*, (220) December 1986, pp. 3–12
87 BARTHÉLEMY, R.: 'French progress in television', *Television and Short-wave World*, April 1939, pp. 196–197
88 BARTHÉLEMY, R.: 'Système de télévision comportant, en particulier, un dispositif de synchronisation et de mise en phase automatique', *C. R. Séances Acad. Sci.*, **191**, December 1930, pp. 1051–1053
89 BARTÉLEMY, R.: 'La reception en télévision', *L' Onde Électrique*, July–August 1931, pp. 3–12, 52–59, 338–346
90 BARTÉLEMY, R.: 'L'emission en télévision', *L' Onde Électrique*, January 1931, (10), pp. 5–35
91 ANON.: 'Are the Eiffel tower transmissions a failure?', *Television and Short-wave World*, May 1936, p. 297
92 ANON.: 'La television Francaise' BBC report T8/27/1, undated, unsigned, BBC Written Archives Centre, Caversham, UK
93 ASSAILLY, E.: 'La télévision en France'. BBC report E/726, 18 April 1945, BBC Written Archives Centre, Caversham, UK
94 Ref.87, p. 8
95 Ref.87, p. 12
96 HEMARDINQUER, P.: 'Survey of television progress in France', *in* WEST, A.G.D. (Ed.). 'Television today: practice and principles explained' (Newnes, London, 1935) **2**, pp. 748–756

Chapter 12

Some low-definition television broadcasting services, c. 1930

The period 1928 to 1934 saw both the rise and the fall of low-definition television broadcasting not only in the United Kingdom but also in the United States, Germany and France. Initially, the impetous to establish the new form of entertainment came from the lone inventors. J L Baird applied to the British General Post Office for a licence on 4 January 1926[1], C F Jenkins[2] was issued with the Federal Radio Communication's first television licence for station W3XK in 1927 and in 1929 the German Post Office allowed D von Mihaly the use of the Witzleben transmitter for experimental transmissions with his system[3].

In these three cases companies were formed to advance the interests of the inventors: television development was costly, patents had to be obtained, equipment purchased, staff engaged and valuable publicity sought. Unlike the large corporations, the lone inventors at first had no source of funding to sustain their television activities. Thus, company formation was a *sine qua non* to the furtherance of their aspirations. The creation of companies required people who would invest in what were, at the time, somewhat speculative ventures. Fortunately for Baird, and others, prior to the Wall Street crash, there was no shortage of investors willing to take a risk, in the expectation that quick and handsome profits would result[4].

The need for profits for their shareholders influenced the business strategies of the early pioneers. In the short term profits could arise from the sale of patent rights and the sale of receivers: for both an imperative factor was the construction of television transmitters, and the broadcasting of vision programmes.

Television flourished in the United States in 1928[5]. An early list of stations published[6] in the *New York Times* in July 1928 named nine stations. By October three more stations had been included in a radio broadcasting survey, and a November Federal Radio Commission communication added another six. In 1929 22 stations were authorised[7] by the FRC to transmit visual images, and over the next 15 years no fewer than 104 stations were granted construction permits and licences.

The early station planners had a wide latitude in their choice of frequency. In 1927, the first year of the FRC's existence, television broadcasting was permitted in the medium waveband (550 kHz to 1500 kHz) with a bandwidth restricted to 10 kHz[8], if no interference was caused to other services. Television experimentation was also sanctioned within the band 1500 kHz to 2000 kHz[7]. And during the formative period radiated powers were allowed to increase; W3XK initially operated, in 1928, with a transmitter power of 250 W but by 1931 this had increased to 5000 kW[9].

Many of the experimenters were inexperienced, however, and brought about some damage to the development of television by their use of scanners, having large numbers of holes, which were rotated at high speeds to reduce flicker. Consequently, interference was produced because of the wide frequency bands employed[10]. The

Table 12.1   Stations licensed to transmit television experimentally in the United States (1931)

Call letters	Company and location	Power (watts)
2000–2100 kHz		
W3XK	Jenkins Laboratories, Wheaton, Md	5000
W2XCR	Jenkins Television Corporation, Jersey City, NJ.	5000
W2XAP	Jenkins Television Corp Portable	250
W2XCD	De Forest Radio Co, Passaic, NJ	5000
W9XAO	Western Television Corp, Chicago, Ill	500
W2XBU	Harold E Smith, North Beacon, New York (school)	100
W10XU	Jenkins Laboratories, aboard cabin monoplane	10
W1XY	Pilot Electrical & Manufacturing Co, Springfield, Mass	250
W1XAE	Westinghouse Elec & Manufacturing Co Springfield, Mass	20000
2100–2200 kHz		
W3XAK	National Broadcasting Co, Bound Brook, NJ	5000
W3XAD	RCA-Victor Co, Camden, NJ	500
W2XBS	National Broadcasting Co, New York	5000
W2XCW	General Electric Company, South Schenectady, NY	20000
W9XAP	Chicago Daily News, Chicago, Ill	1000
W2XR	Radio Pictures Inc, Long Island City, NY	500
2750–2850 kHz		
W2XBO	United Research Corporation, Long Island City, NY	20000
W8XAA	Chicago Federation of Labour, Chicago, Ill	1000
W9XG	Purdue University, West Lafayette, Ind	1500
W2XBA	WAAM Inc, Newark, NJ (broadcasting station)	500
W2XAB	Columbia Broadcasting System, New York	500
2850–2950 kHz		
W1XAV	Shortwave and Television Labs, Boston, Mass	500
W9XR	Great Lakes Broadcasting Co, Downers Grove, Ill	5000

belief was held that listeners would be antagonised by the annoying signals radiated by their loudspeakers to the exclusion of the wanted, regular broadcast programmes.

With the proliferation of television stations in 1928 it was obvious that some control had to be exercised by the FRC. The Federal Radio Commission performed three roles during this formative period of American television[11]. First, it determined the frequencies and bandwidths for the nascent services, secondly, it established procedures and standards for assigning these frequencies to transmitting stations and, thirdly, it regulated the services made accessible to the public. As a result of its powers the FRC had to decide whether licences should be issued exclusively for experimental work or whether some commercialisation should be approved. Again, the FRC had to protect public interests, including those of the listener as well as those of the television viewer, and it had to undertake these tasks in a way that would not restrict advancement of the new medium.

The immediate issues which had to be tackled by the FRC in 1928 concerned interference and bandwidth. The commission permitted television broadcasting to continue in the medium waveband but limited the broadcasts to one hour per day and banned them totally from 6.00 p.m. to 11.00 p.m. to prevent conflict with radio users[12]. Later in 1929 these transmissions were further constrained to the hours of 1.00 a.m. to 6.00 a.m.[13], and were finally terminated in 1931 with the cessation of broadcasting, on 660 kHz, by Westinghouse's station W8XT[14].

Practical experience soon emphasised the severe restraint imposed on television programme makers by the need to operate within a 10 kHz bandwidth[15]. The FRC referred to this matter in its second annual report[12], 1928, and quoted from an engineering report submitted to it by Dr A N Goldsmith (of RCA):

> 'A 5-kilocycle bandwidth permits the television broadcast of a crude image of a head, with comparatively little detail. A 20-kilocycle bandwidth will permit the broadcasting of the heads and shoulders with more detail. An 80-kilocycle band will permit transmission of the picture of two or three actors in fairly acceptable detail.
>
> The allocation of bands of 100 kilocycles wide for television is strongly advocated, since this is clearly the minimum basis for a true television service of permanent interest to the public.'

100 kHz bandwidth signal channels could not be accommodated in the medium waveband and so there was a need to allocate higher frequency wavebands to television.

In January 1929 a North American conference held in Ottawa decided that radiation of television signals was to be limited to a bandwidth of 100 kHz, within the frequency bands 2000–2100, 2100–2200, 2200–2500, 2750–2850 and 2850–2950 kHz[16], Figure 12.1. The powers subsequently employed varied from 10 W to 20 kW, with the majority of stations operating at 5 kW.

One of the effects of the FRC's order was the restriction it placed on definition standards. Weinberger, Smith and Rodwin of the Radio Corporation of America showed[17] that a standard based on 60 lines per picture, 20 pictures per second and

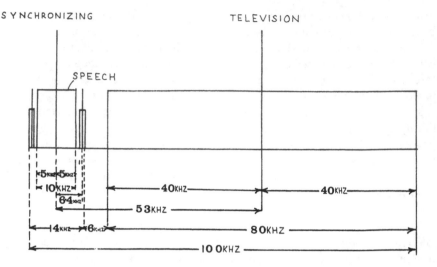

Figure 12.1   Diagram illustrating how the video, speech and synchronising signals were to be accommodated within the 100 kHz bandwidth, for television transmissions, agreed at the January 1929 conference in Ottawa

Source: *Proc. IRE*, **17**, (9), September 1929, pp. 1584–1590

on an aspect ratio of 5:6 (height to width) would be realistic and apposite for the television bandwidth specified by the FRC. These suggested television parameters became the norm for television transmissions in the US, although a committee appointed by the Radio Manufacturers Association of the United States had recommended[18] that all radiovision pictures then being broadcast should be standardised on the Jenkins system, namely 48 lines per picture, 15 pictures per second. Presumably the RCA proposals prevailed because they enabled a better definition to be realised: the RCA and the RMA picture standards were based on 2500 and 2304 picture elements respectively (Baird used a 2100 element image).

Another result which stemmed from the FRC's ruling was that, as none of the commercial receivers then on the market could receive signals in the 100–150 m wavelength range, an entirely new and distinct receiver was required for television reception. The existing broadcast receivers which covered the band of 200–600 m had been developed to a considerable degree of excellence, but now manufacturers were faced with the prospect of designing and constructing sets for a frequency band which had not been previously investigated for commercial use. Somewhat naturally, a number of firms felt uneasy about embarking on large scale manufacturing programmes for television sets when the art was still in the experimental stage.

Many different types of scanning were employed during the relatively brief period (1928–1934) of low-definition television, but the most popular was the flying spot method which had been patented in the UK[19] in 1926 and in the US[20] in 1927.

Receivers were usually of the Nipkow disc, gas-filled lamp type and synchronising was based on the use of synchronous motors. Several resistance-capacitance coupled

Figure 12.2   Front view of a 1928 Baird 30-line televisor. The cover has been removed

Reproduced by permission of the BBC

amplifier stages were normally necessary to produce an adequate signal to excite the glow discharge tube. This basic system was used not only in the UK and the US but also in Germany, France, the USSR and elsewhere. Essentially, it conformed to the scheme which Baird and Bell Telephone Laboratories had demonstrated in January 1926 and April 1927 respectively.

The BBC transmitted its first experimental television programme[21] on 30 September 1929 even though the general public (in the UK) could not purchase a receiver until March 1930. 'At last,' said *Wireless World* on 12 March 1930, 'a Baird receiver built for sale to the public has arrived.' Hutchinson's remark of 1926 had come true. 'High class workmanship, but with a non too pleasing external appearance, owing to the use of a light metal cabinet and poorly devised controls, are one's first observations on acquaintance with the instrument,' said the writer, but it did 'give reception of images with sufficient definition to be readily intelligible,' he concluded[22].

The receiver (Figures 12.2, 12.3 and 12.4) consisted of a thin, 20 inch diameter aluminium scanning disc having 30 apertures, the majority of which were square (about six near the ends were rectangular), an electric motor which carried the disc and a synchronising mechanism, a neon tube and a voltage regulating resistance. A width/length ratio of the image area of 1:2.5 was produced by the spiral design so that the resulting picture was suited to accommodate the head and shoulders of a

Figure 12.3   Rear view of the 1928 Baird 30–line televisor

Reproduced by permission of Radio Rentals Ltd

person speaking. The universal motor could be operated from either a d.c. or a.c. supply and synchronising was effected by means of a toothed wheel running between the poles of an electromagnet. According to the company, a power output, from the last stage of an amplifier, of 1.5 W was sufficient to actuate both the neon lamp and the synchronising gear (which were in series), but the *Wireless World* reviewer found that the process of synchronising became easier if the L55 output valve used in the amplifier was changed to a L56A valve as this had a rated output of 5 W. 'When once correctly set up and a little practice gained in the operation of the speed regulating control, reception became reliable,' he said. 'For quite long intervals the picture remained steady, though in the case of head and shoulders images the lighting effect was far from perfect, and it was not possible to glean the significance of the movements, though if accompanied by speech the effect might have been different.'

The television receiving set was produced to give an image only and not sound. It was assumed that the purchaser of the set already possessed a suitable receiver and aerial installation for the reception of both the vision and sound signals.

The first dual UK transmission—of sound and vision signals simultaneously—took place on 31 March 1930[23]. Prior to this date the test transmission announcements had been given in the form of wording running across the aperture. According to the author of the aforementioned review 'the capital letters forming

Figure 12.4   Top view of the Baird televisor showing the scanning disc, the synchronising gear, the neon lamp, and the enlarging lens

Reproduced by permission of Radio Rentals Ltd

the words were clearly defined and easy to read, while a clock face could be read to the nearest half minute'. And on the achievement the reviewer said: 'Such results will interest the enthusiast, and these have become possible since the adoption of the signal controlled toothed-wheel method of synchronising first introduced towards the end of last year.' It was thought that the price of the set would be about £20.

German television transmissions, using Baird equipment, commenced on 15 May 1929 and terminated on 13 July 1929. Details of German television receivers, Figure 12.5, are given in Table 12.2 and Chapter 11. None were on sale to the public until 1939[24].

The need for kits which would allow amateurs to build their own television sets at reduced cost was soon recognised. Jenkins (in the US) explained his purpose[25] in making these accessible to the FRC in 1930:

Figure 12.5   A receiver of the Ferseh company, which was based on the Baird system

Reproduced by permission of Radio Rentals Ltd

'We do offer the radio amateur kit parts for the construction of an excellent receiver for our broadcast movie stories. The kit includes (except the motor) every essential in the construction of a really excellent receiver, i.e. a superior neon lamp specially made for this kit, a lamp holder, a 12-inch scanning disc (die perforated), a shaft and cast frame mounting therefor: a motor hub and driving disc therefor; and a synchronising screw.

'All these cost the amateur but $7.50 packed and postage paid—less than the cost to us—and the pictures of this assembled receiver are good pictures, the equal, the superior of some, of those obtained with any other 12-inch disc receiver at any price. But we can afford this loss because by making these receiver parts available at this low cost we enlist the cooperation of the amateurs of the country in helping us improve our methods, our mechanisms, and our broadcasts.'

A year later (1931) a Jenkins Universal Television Receiver cost $69.50, plus an extra $13.45 for the valves, and the Jenkin Radiovisor kit was $42.50 plus $5.00 for the magnifying lens[26].

Figure 12.6 shows the Baird Television Company's advertisement for their kit.

Table 12.2 Details of the television systems on view at the 1931 Berlin Funkausstellung

Stand	RPZ (German Post Office)				Fernseh AG				Tekade		Loewe
Exhibit	a	b	c	d	a	b	c	d	a	b	a
Scanning Method	mirror wheel	Nipkow disc	cathode ray tube	lens disc	Nipkow disc	Nipkow disc	Nipkow disc	cathode ray tube	mirror screw	cathode ray tube	cathode ray tube
Light source	crater neon	positive column		Kerr cell	positive column	positive column	positive column		plate neon		
Lines	48	48	60	100	90	60	80	†	84	†	100
Points	3000	3000	4800	13 000	10 800	4800	4800	†	8400	†	6-8000
Revs./p./sec.	25	25	25	25	25	25	16⅔	†	25	†	16-20
Picture ratio	3:4	3:4	3:4	3:4	3:4	3:4	4:3	–	5:6	–	3:4
Picture size	3½" × 4½"	3" × 4"	2" × 3"	4½" × 6"	3½" × 4½"	3½" × 4½"	4½" × 3½"	†	3½" × 4"	5 × 5cm	3" × 4"
Colour	red	sepia	green or blue	white	white	white	white	†	red	blue	green or blue
Detail	fair	fair	good	v.good	excellent	†	excellent	†	v. good	†	v. good
Intensity	good	fair	fair	good	excellent	†	excellent	†	v. good	†	v. good
Flicker	none	none	none	none	none	†	very slight	†	none	†	slight
Cycles (max.)	37 500	37 500	60 000	160 000	135 000	†	40 000	†	105 000	†	150 000
Subject	film	film	films	talkies	talkies	†	semi-close-up	†	talkies	†	films
Synchronising	mains	mains	generated impulses	mains	mains	†	mains	–	mains	–	plates in parallel
Transmitter	Nipkow disc				Nipkow disc				42 hole disc		cathode ray

† not working

BAIRD "JUNIOR" KIT

PRICE LIST OF PARTS COMPRISING THE BAIRD "JUNIOR" KIT

	£	s.	d.
MOTOR AND SYNCHRONISING GEAR	5	5	0
DISC	1	1	0
NEON LAMP	1	5	0
NEON LAMP HOLDER		1	6

PRICE COMPLETE, with Two Blueprints . £7 . 12 . 6

The following are other parts not included in the kits of parts :

	£	s.	d.
MAIN CASING	3	12	6
WELL CASING		15	0
NEON CASING		2	6
TERMINAL CASING		3	6
WOOD BASE		15	0
FEET FOR BASE (FOUR)		11	0
SCREWS, ETC. (VARIOUS)		4	0
FLEXIBLE LEAD AND ADAPTOR		2	6

ANY SINGLE COMPONENT
MAY BE PURCHASED SEPARATELY

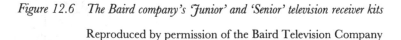

Figure 12.6   The Baird company's 'Junior' and 'Senior' television receiver kits

Reproduced by permission of the Baird Television Company

Several of the major American television stations of the late 1930s commenced their activities during the low-definition era[27]. The General Electric station in Schenectady operated on a 60 lines per picture, 20 pictures per second standard,

5

BAIRD "SENIOR" KIT

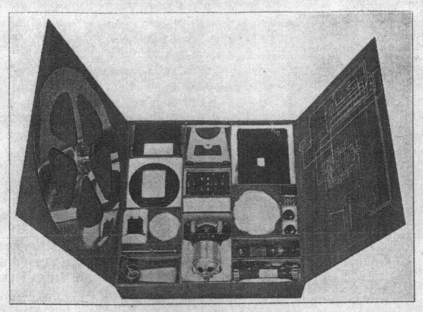

PRICE LIST OF PARTS COMPRISING THE BAIRD "SENIOR" KIT

	£	s.	d.		£	s.	d.
MOTOR & SYNCHRONISING				DISC	1	1	0
GEAR		5 5	0	ZENITE RESISTANCE AND			
MOTOR BRACKET, CLAMPING				SUPPORTS		17	0
BOLTS & NUTS	1	0	0	RESISTANCE, WITH BRACKET			
LENS—LARGE		13	6	AND CONTROL KNOB		6	3
LENS—SMALL		7	6	SPINDLE ASSEMBLY, WITH			
LENS—BOX ASSEMBLY, WITH				CONTROL KNOB & SCREWS		5	9
VIEWING TUNNEL (6 pieces)	1	0	0	CONDENSER		2	0
NEON LAMP		1 5	0	TERMINAL BOARD		7	6
NEON LAMP HOLDER		1	6				

PRICE COMPLETE, Packed in attractive Carton with Blueprint,   £12. 12. 0

Figure 12.6   continued

and a 20 kW power output, from 1929 to 1932. The Columbia Broadcasting System station in New York started in July 1931 and transmitted pictures, having the above mentioned standard, for a total of 2500 hours in the period ending February 1933. A typical day's programme is shown in Figure 12.7[28] (see also Figure 12.8).

Columbia Broadcasting System, Experimental Television Programme – 3, Tuesday, December 29th 1931, W2XAB and W2XE New York

2.00–6.00 p.m.	*Experimental sight programmes.* Card station announcements, and drawings of radio celebrities.
8.00 p.m.	*Hemstreet Quartet.* All-girl novelty group with Helen Andrews, soloist. Long shot group picture. White and silver backdrop curtains.
8.15 p.m.	*Grace Voss* – Pantomimes. Long shot and close-up. Silver backdrop curtain.
8.30 p.m.	*Senorita Soledad Espinal and her Pamperos* in a half-hour programme of Spanish and Latin American music and songs. Guitar Sextette and Mezzo Soprano. Group projection with various backdrop screens.
9.00 p.m.	*The Television Ghost.* Mystery character in weird costume enacts the murder mysteries in the character of one risen from the grave . . . the murdered!!
9.15 p.m.	*Hazel Dudley.* Song recital. Series of close-up pictures to be scanned.
9.30 p.m.	*Three-round exhibition boxing bout.* An experimental television demonstration of what the flying spot can do at a fight. Minature ring will be used. Blow by blow description by Bill Schudt. Dark backdrops. Long shot pick-up. Scanner will follow boxers around ring.
9.45 p.m.	*Gladys Shaw Erskine and Major Ivan Firth present.* Television novelties with visual illustrations. Alternate backdrops will be used.
10.00 p.m.	*'Tashamira'* introduces new German modernistic dances and technique. Extreme long shot focus. Close-up with varied backdrop curtains.
10.15 p.m.	*One man novelty band* with *Vincent Mondi.* Close-up against white backdrop.
10.30 p.m.	*Eliene Kazanova.* Violinist.
10.45 p.m.	*Grace Yeager*, song recital. Close-up shot of an artist singing semi-classic favourites.
11.00 p.m.	*Sign off*

Figure 12.7   Columbia Broadcasting System's experimental television programme for 29 December 1931. The programmes were radiated by stations W2XAB and W2XE, New York

Source: *Journal of the Television Society*, 1932, pp. 148–149

Of all the large companies in the US which made contributions to the furtherance of television, the Radio Corporation of America was pre-eminent. RCA com-

THE FUTURE OF TELEVISION . . . . . How the cartoonists saw it in the 'thirties

This peep into the future was drawn by FITZ for the *Daily Mirror* in June 1932. 'Will it prove to be a help or hindrance to life?' he asked.

Figure 12.8   *'The future of television . . . How the cartoonists saw it in the 'thirties'*

Source: *The Daily Mirror,* June 1932

menced[29] its activities in 1927 when the technical and test department's laboratory group at Van Cortland Park carried out some experimental investigations on mechanically scanned television systems. The approach and work were similar to those which had been undertaken by E F W Alexanderson at General Electric in Schenectady and, by April 1928, the group had advanced its developments to the trial stage. It applied for and was granted in April 1928 a permit for the first RCA television station, known as W2XBS, and the initial broadcasts took place from the Van Cortland site.

T J Buzalski of the NBC's commercial television outlet in New York has described these early transmissions[30].

'The programme's material usually consisted of (a) posters with black images upon a white background, such as W2XBS-New York-USA . . ., (b) photographs, (c) moving objects such as Felix the cat . . . revolving on a phono turntable, and (d) human talent. Communication from observers indicated that the service area of the station was quite large and, under favourable conditions, good television reception was reported at distances of several miles, the farthest point being Kansas . . . Many interesting tests were conducted between W2XBS at Times Square and the RCA laboratory at 411 Fifth Avenue.'

Figure 12.9   Comedy sketch by Bill and Elsa Newell (October 1932) from the television studio in the basement of Broadcasting House, London

Reproduced by permission of the BBC

The apparatus used the 60 lines per picture, 20 pictures per second standard and the received image was almost 1.5 inch by 2.0 inch in size. Regular transmissions took place from 7.00 to 11.00 p.m. daily, but these were for experimental purposes only as it was thought that the state of development of television was not sufficiently advanced to entertain and maintain the interest of the general public.

Of the early television stations in the US, three were operated by higher education establishments, *viz.*: the State University of Iowa, Purdue University and the Kansas state College of Agriculture and Applied Science in Manhattan[31].

Purdue University, by March 1932, was broadcasting a regular two-hour weekly schedule of motion picture film, mainly news reels, whereas the programmes of the State University of Iowa's station were based on live television. Among the studio productions were: 'Elementary art lessons', 'Home planning', 'Personal shorthand', 'Iowa wild life', 'Spring birds' and 'Portrait sketching'. Each series was presented by a member of the university. Students could also benefit from studies of the generation, transmission and reception of video signals.

In 1931 it was apparent that progress could not be made on the restricted channels of the 2 MHz band. The FRC in its fifth annual report (1931) noted: 'The consensus of engineering opinion indicates that in order to transmit a picture having satisfactory detail the bandwidth required will be many times that now available in this [2 MHz] frequency range[32].'

And so the trend towards the utilisation of the higher frequencies commenced. The FRC authorised three additional television band allocations: (43.0 to 46.0) MHz, (48.5 to 50.3) MHz and (60.0 to 80.0) MHz. No limitations were placed on the bandwidths of the transmitters using these frequencies.

One of the earliest to apply for permission to use the higher frequencies was the Don Lee Broadcasting System of Los Angeles, California. In December 1931 the licence of station W6AO was granted to this organisation. Permission was also given to several other stations, including NBC's W2XF and W2XBT in New York, Jenkins' W3XC in Wheaton, Maryland, W1XG in Boston and W8XF in Pontiac, Michigan. Additional VHF stations were licensed in 1933 and 1934. From 1936 television activity in the 2 MHz band ceased and all VHF stations were given frequencies in the (42–56) MHz and (60–86) MHz bands[2].

From the beginning of television broadcasting the FRC refused to authorise any commercialisation of television. It held that any other policy would violate its regulating responsibilities by seeming to encourage public investment in apparatus which could quickly become obsolete. This policy was not reversed until 1941 when it was thought that acceptable quality had been achieved[33].

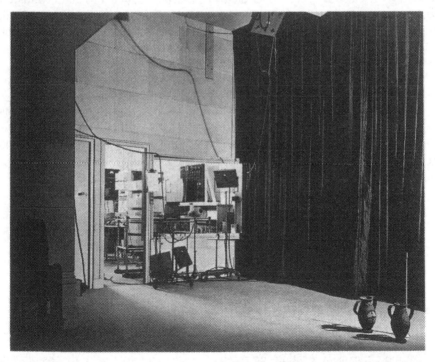

Figure 12.10 Television studio at 16, Portland Place looking towards the control room, which can be seen through the projection window. The large black curtain screens the orchestra and their lights from the photocells. One bank of cell has been placed high in the ceiling for top lighting

Reproduced by permission of the BBC

The immediate effect of the FRC's approach was the limitation of licence permits to those prospective licensees who could fulfil the commission's conditions.

The FRC's principal objectives were to determine the experiences of the staff to be associated with the running of the new station, the existence of a viable plan of research, the financial resources of the applicant and the public interest to be served by the granting of the particular licence. Moreover, even after a licence had been issued a licensee was required to file regular reports indicating the station's hours of operation, the general results of its broadcasting and the technical studies being undertaken.

The commission explicitly rejected any considerations of public entertainment, the lack of a local outlet and commercial advantages as relevant criteria for deciding the worth of licence applications. In the FRC's opinion, television broadcasting could only be approved if it remained strictly experimental and furthered the progress of the new media. Only when sufficient technical advances had been made could entertainment be a proper issue for consideration and commercial promotion of television a legitimate activity.

Apart from the efforts of the early experimenters to enhance their images, no organisation until 1929 had systematically investigated the many parameters which define a good television picture. RCA devoted much thought to this matter and published a series of papers on its findings in the Proceedings of the Institute of Radio Engineers from 1929 to 1936. The corporation's conclusions very considerably helped to rationalise the standards of the television industry for both low and high-definition systems.

The work of Weinberger *et al.* did not represent a definitive study on picture standards (Engstrom later inquired into the problem much more fully), but nevertheless their results provided a basis on which low-definition systems could be satisfactorily designed and they allowed manufacturers to produce sets which were able to receive radiations from many different television transmitters using these standards. The RCA group especially had in mind criteria for a commercial television service. Fortunately, a definition of what constituted commercial television had been given in 1929 by a special committee of the National Electrical Manufacturers Association appointed to deal with the subject and this is quoted[17] in the following:

> 'Commercial television is the radio transmission and reception of visual images of moving subjects comprising a sufficient proportion of the field of view of the human eyes to include large and small objects, persons and groups of persons, the reproduction of which at the receiving point is of such size and fidelity as to possess genuine educational and entertainment value and accomplished so as to give the impression of smooth motion, by an instrument requiring no special skill in operation, having simple means of locating the received image and automatic means of maintaining its framing.'

So far as is known, this was the first formal, explicit definition of television ever advanced. It posed the question of the meaning to be given to the phrase 'genuine entertainment and educational value' and, in an effort to clarify it, Weinberg's group consulted various persons engaged in the motion picture and theatrical fields and came to the conclusion that genuine entertainment value would be achieved if a clear reproduction of a semiclose-up of two persons was obtainable. From the group's discussions it appeared that the majority of dramatically interesting situations reduce to two, or at most three persons, and in motion picture practice it was customary to photograph 80 or 90 % of a picture in close-up or semiclose-ups. Consequently, the RCA research team felt that, if it chose a degree of picture detail which would acceptably render two persons in a semiclose-up view, the entertainment requirement would be met, at least in the early stages of the art.

Baird's images in 1928 were restricted to head and shoulders and so did not satisfy the above criterion. Eckersley was correct in his view of what constituted entertainment television.

Weinberger, Smith and Rodwin found 'acceptable detail would just be obtained with an approximately 60-line scanning system and, although greater detail would naturally be obtainable with a greater number of lines, such a system would meet the entertainment requirement'. This conclusion was based on the results of a series of experiments using convential half-tone pictures of various degrees of fineness of detail. After a correlation between the number of half-tone lines per inch and corresponding television scanning lines had been obtained, a number of half-tone letters and photographs had been prepared and their appearance compared with the television image, on a 48-line system, of the same original until apparent equality of detail had been obtained.

The RCA workers noted in their paper the agreement of their conclusion with the unpublished view of Dr F Conrad of the Westinghouse Electric and Manufacturing Company.

Two years later Gannett, of the American Telephone and Telegraph Company, gave the results of his researches on the quality of television images. In his paper[34] he illustrated the appearance of such images when reproduced with 625, 1250, 6250 and 12 500 picture elements. For this purpose Gannett used facsimile equipment, in which the received picture element was 1/100 of an inch square, and photographs, which were reduced in size until they contained, in their reduced form, the appropriate number of elements. Subsequent enlargement to a given size thereby enabled photographs to be obtained having the number of elements stated above.

Gannett illustrated his work with pictures of a head and shoulders portrait, a footballer (full length) and a boxing ring (containing three people). He arrived at no firm conclusion for the optimum number of lines required in a televised picture but noted: 'It appears that when several objects are being observed, some psychological factor causes one to expect and be satisfied with fewer details in each subject. Another matter of interest is that in television a somewhat more pleasing effect is obtained than can be represented by still pictures because the eye, in following the movements in the television scene, is less aware of the streaks caused by dividing the image into rows of elements.'

Gannett's work was considerably extended by V H Wenstrom[35], of the US Signal Corp, in 1933. He chose four photographs to 'represent approximately equal gradations of scene comprehensiveness from the lowest to the highest', namely:

1   a single face;
2   a small group (four people in three-fourth's view);
3   a musical comedy stage;
4   a general view of a football game.

These photographs were transmitted on a facsimile machine so as to represent the performance of 60, 120 and 200-line television systems. Approximate resemblances of 400-line television were obtained by means of half-tone screens.

Wenstrom's paper was the first to mention specifically 400-line images. The author probably used this figure because he had previously determined that the definition of home movies corresponded to approximately a 400-line picture. This was an important finding for it was likely that high-definition television would be compared with home cine films. Wenstrom was well aware that the method which he had employed had a serious disadvantage that might lead to an underestimation of the possibilities of television; namely, the apparent clarity of a scene in motion compared with a still picture owing to the integrating effect of 'eye on mind'. Nevertheless, notwithstanding this defect and the possible enhancement of televised images by the use of sound accompaniment, Wenstrom came to the conclusions shown in Table 12.3.

From these results and from observations of existing television systems he arrived at certain general deductions— 'which, if verified, may have the force of basic laws governing the performance of television:

1   'The degree of definition required in any given television application is conditioned chiefly by the comprehensiveness of the scene to be portrayed.
2   'The higher the contrast in a given scene, the lower the order of definition required to portray it.
3   'While a lower order of definition may suffice for the momentary presentation of certain scenes, a higher order of definition is required if the observer's continued attention is to be held.'

Wenstrom's analysis thus showed that for continuous home television involving full theatre stage productions and outdoor spectacles a 400-line standard was needed.

All of this supported the FRC's contention that low-definition television could never provide an acceptable service of entertainment. The commission strictly adhered to its policy for more than a decade, despite strong protests from the

Table 12.3

Type of picture	Number of lines			
	60	120	200	400
face	fair	good	excellent	perfect
group	poor	fair	good	excellent
stage	poor	poor	fair	good
game	worthless	poor	fair	good

Figure 12.11 A studio set up, for a 30-line television production, at 16, Portland Place

Reproduced by permission of the BBC

industry. During the formative years the smaller firms which were endeavouring to obtain a stake in the new industry were especially aggrieved. Of the construction permits and licences granted to no fewer than 104 stations from 1929 to 1944, 21 never got beyond the construction permit stage and 48 were licenced but ceased transmission. Only 35 remained in operation in 1944: nine of these stations were licenced for regular transmissions to the public and could accept programmes from commercial sponsors. The rest were experimental stations, many of which rendered occasional public service[36].

As stated previously, by 1931 more than 20 stations had been authorised by the Federal Radio Commission to broadcast visual images; many of these were in New York and New Jersey. Possibly as a consequence, the *New York Times* was led to seek its readers' opinions on the words which ought to be used to describe various aspects of television. A report[37], 'Search continues for the right word', published in the 19 April issue, contained some varied and bizarre word formations. 'We see with our eyes, we hear with our ears. Why not combine the two and make the word "eyear" or "earyer" whichever is more euphonic?' wrote a listener of New Brunswick, New Jersey.

'I suggest the word "tellser" as a fitting name for one who sees by television,' said a citizen of Hartford. Again, '"sightener" is easy to pronounce,' noted a New Yorker, 'and is not outlandishly shocking to the ear.'

A reader from Florida had several suggestions to offer: 'For an easy sounding and clear meaning designation in television, why not adopt the following derivatives? Audivise for operating, audiviser for the name of the receiver, and audiovision for the science of seeing by radio?'

One correspondent was not too sanguine about the prospects for the future of television since he commented: 'I suggest noisivision because the vision will be noisy.'

Other words proffered were:

for the owner of a set—*teleceiver*
for viewers—*radiospects* (taken from the word spectacles)
for the apparatus—*raduo*
for a performer—*raduolist*

Of the notable figures in industry who took part in the word naming exercise the following synonyms for the present day (1996) viewer are of interest:

Dr Alexanderson (RCA)—radio spectator
Mr Aylesworth (NBC)—radio audience
Dr de Forest—televiewer or teleseer
Dr Goldsmith (RCA)—lookstener
Mr La Fount (FRC)—observer
President (RMA)—viseur
Dr Pupin—televisioner
Mr Repogle (Jenkins Television Corp)—looker-in
Mr Horton (General Radio Co)—viewer

Similar word forming activities were to be conducted in 1935 following the publication of the UK Television Committee's report on television[38].

In the UK, the position concerning television broadcasting was simpler than that in the US. Only the Baird companies were authorised to radiate television signals, from one of the BBC's transmitters, and the proliferation of low-definition TV stations which existed in America was not a feature of the British scene.

Mention has been made of Baird's efforts to commence TV broadcasting. On 30 September 1929 the BBC transmitted its first experimental television signals[39]. 'A great day for Baird and all of us,' wrote Moseley. The honour of opening the proceedings at 11.04 a.m. fell to Moseley who announced: 'Ladies and Gentlemen: you are about to witness the first official test of television in this country from the studio of the Baird Television Development Company and transmitted from 2LO, the London station of the British Broadcasting Corporation.' Then followed a number of short messages from the Rt Hon William Graham, PC MA LLD MP, Sir Ambrose Fleming and Professor E N da C Andrade. Graham took as his theme one which had been raised on a number of occasions previously by Hutchinson—the establishment of a new industry which would provide employment for large numbers of our people, and which would prove the prestige of British creative energy. This point was further mentioned by Fleming who acknowledged that the creation of the new industry owed so much to the genius of Mr Baird. Andrade compared the event to that:

'On which the records of the early phonograph were publicly tried. The voices that then issued from the horn were not of the clarity which we now expect, and the faces that you will see today, by Mr Baird's ingenious aid, are pioneer faces, which will no doubt be surpassed in beauty and sharpness of outline as the technique of television is developed. One face, however, is as good as another for the purpose of today's demonstration and I offer mine for public experiment in this first television broadcast.'

The second half of the proceedings consisted of very short solo performances:

11.16 a.m. Sydney Howard: televised for two minutes
11.18 a.m. Sydney Howard: gave a comedy monologue
11.20 a.m. Miss Lulu Stanley: televised for two minutes
11.22 a.m. Miss Lulu Stanley: sang 'He's tall, and dark, and handsome' and 'Grandma's proverbs'
11.24 a.m. Miss C King: televised for two minutes
11.26 a.m. Miss C King: sang 'Mighty like a rose'

Only one transmitter had been allotted for the transmission, and so each artist had to be televised for two minutes and then repeat the act before a microphone.

The programme was not without its difficulties—the most serious of which was the reproduction of a negative image instead of a positive image. This was rapidly corrected and fairly clear images received, said *The Times*[40]. *The Daily Herald*[41] thought that Sydney Howard and a woman artist were 'quite recognisable—looking like the earliest photographs in the daily paper' but 'the image jerked up and down like a film when the operator is having serious trouble with his machine'.

Baird told a *Manchester Guardian* reporter that 'he was satisfied with the demonstration but he hoped to obtain much better results as the experiments continued. He pointed out that there had been very little time indeed for tuning-in, a most important operation'[42]. *Amateur Wireless*[43] also referred to the hunting but mentioned that the results were of good quality:

'One sees the image through a wide lens about eight inches in diameter and the general effect is similar to that of looking into an automatic picture machine as installed in amusement halls. The image appears as a 'soft tone' photograph illuminated by a reddish light.'

The general impression, the writer stated, was that the present situation had reached the stage of development of the early flickering cinematograph. He thought there was 'much, very much, yet to be done, but the present stage is highly creditable and the fact that public broadcasts are now being given will undoubtedly hasten progress'.

Unfortunately, very few people witnessed this historic event. Asked by an *Evening News* reporter immediately after the broadcast how many people he thought had been able to receive the transmissions, Baird himself put the total at under 30:

'There is one receiving set at my home at Box Hill, and I believe the BBC and the Post Office each have one. That makes three and I should say there are half a dozen other sets in the country. Add to these the receivers which clever amateurs have built for themselves from our directions and you might count another twenty. That makes 29 in all.'

The experimental service has been described in the author's book[15], ànd is not repeated here.

Later, on 22 August 1932, the experimental service was replaced by the BBC's first, public, low-definition TV service[44]. From this date the production of the programmes and the transmission of the picture signals were carried out by the BBC's staff. The corporation was particularly fortunate in having two first-clcss enthusiasts for the posts of television research engineer and studio producer. Birkinshaw, the ex-Cambridge science graduate, and Robb, the ex-Guards officer, achieved some very considerable improvements in both programme and transmission quality. Robb dealt with some of these in a very interesting internal BBC report[45] (dated 10 May 1933). Although this is fully considered in Burns, *op. cit.*, several points are quoted here since they are evocative of the struggles of the nascent service:

1   'I think I may claim that this invention of proper make-up to overcome the peculiar reactions to photoelectric cells and certain limitations of frequencies has succeeded in crispening the radio picture received.
2   'The innovation of visual announcements and illustrated subtitles.
3   'The first use of suitable scenery and sets.
4   'The introduction through the cooperation of the television engineers of continuous movements from extended to close-up position, thereby obtaining a perspective which was impossible at the outset.
5   'The special study of costume and its modification for television use: a subject of extreme importance, as the wrong use of colour will nullify the picture.
6   'Valuable work had been done bringing representatives of different schools of dancing to broadcasting and adopting their art to the exigencies of television thus paving the way for the eventual expression of this art to the great radio public.
7   'The transmission of art exhibitions by television is the beginning of an era when the public will be taught to appreciate great works of art, seeing them in their homes and at the same time that the finer points are demonstrated by an expert lecturer—in other words illustrated talks.
8   'Animals, trick-cycling balancing acts, roller skating, all of which have been done, are useful for the light entertainment programme of the future as being the means by which the ear will be relieved of the intolerable strain of concentration by the eye.'

Initially, the BBC's television equipment was housed in the basement of Broadcasting House, London. The studio was larger than that at Baird's Long Acre site and next to it, and on the same level, was the control room separated only by a glass window. The control desk was manned by three engineers, Birkinshaw,

Bridgewater and Campbell, who controlled the sound, lighting and vision. It was soon noticed that the initials of the surnames of these engineers formed BBC and this was felt to be a good omen for the future of the 30-line broadcasts.

In 1934 the BBC's television activities were moved to 16 Portland Place. The new studio was a converted first floor drawing room, 28' wide instead of the previous 18', and the new control room was larger too, 24' by 14', with a large window into the studio. Campbell, with more space and larger and more sensitive photocells, was able to refine his lighting techniques to the stage where he could create or eliminate a shadow on a performer's nose. He also made a caption scanner, a circular drum with a dozen flat facets into which could be slipped a postcard-size caption. This is shown in Figure 12.12.

The spotlight projector beamed into the studio through the glass which partitioned the control room from the studio. Bridgewater has written[46]:

'[The projector] was on rails. There was a tubular steel framework with two cross-bars front and back and the projector itself was on wheels which ran along these two cross-bars so that you could pull the whole thing sideways, one way or the other, and therefore, by doing that you

Figure 12.12 The BBC's control room at its 16, Portland Place, London television centre. The drum in the foreground was used to make various announcements. The mirror drum transmitter camera is shown on the right hand side of the photograph and is pointed into the studio which is adjacent to the control room

Reproduced by permission of the Science Museum

adjusted to the subject rather than pushing the subject all the time into your field of view. It could focus from extreme close-ups to as far as the back of the studio—25 or 30 feet. And you could see whether it was in focus or not by just peering into the studio, because you could see the scanning strips and you could focus quickly. And the final point of flexibility was that you could tilt the scan up or down so that if somebody wasn't just the right height, you could adjust to them—not the other way round.'

But did the BBC's productions have entertainment value? Robb was in no doubt about the answer.

'Entertainment value does exist in the present television programmes, in spite of all that is said about their merely holding scientific interest . . . We are able to infuse considerable variety and attraction into the programmes. And topicality is always uppermost in our minds . . . Television is building up a technique absolutely its own. While we like to put over stage items as near to their theatre performances as possible, all programmes are developed specially for television.

References

1 HUTCHINSON, O.G.: Letter to the Postmaster General, 4 January 1926, minute 4004/33, file 1, Post Office Records Office, UK
2 MURRAY, G.: Memorandum to the Director General (BBC), 19 October 1928, BBC file T16/42
3 ANGWIN, A.S.: Memorandum, 24 June 1929, post 33/5141, file 2, Post Office Records Office, UK
4 MOSELEY, S.A.: 'John Baird' (Odhams, London, 1952)
5 UDELSON, J.H.: 'The great television race' (University of Alabama Press, 1982) p. 29
6 ANON.: 'Stations licensed for television', *New York Times*, 21 July 1926, p. 16
7 FINK, D.G.: 'Television broadcasting practice in America—1927 to 1944', *J. IEE*, 1945, **92**, part III, pp. 145–160
8 JENKINS, C.F.: 'Radiomovies, radiovision, television' (Jenkins Laboratories, Washington, 1929) pp. 10–11
9 ANON.: Report, *Science*, (supplement), 1931, p. xiv
10 ANON.: 'Some impressions of the New York Radio Show, September 23–29, 1929', *Television*, November 1929, pp. 360–463
11 Ref.5, p. 40
12 US Federal Radio Commission, second annual report, 1928, pp. 252–253
13 US Federal Radio Commission, third annual report, 1929, p. 2
14 Ref.5, p. 41
15 BURNS, R.W.: 'British television , the formative years' (Peter Peregrinus Ltd, London, 1986)
16 DINSDALE, A.: 'Television in America today', *J. Television Society*, 1932, pp. 137–149
17 WEINBERGER, J., SMITH, T.A., and RODWIN, G., 'The selection of standards for commercial radio television', *Proc. IRE*, September 1929, **17**, (9), pp. 1584–1594
18 ANON.: Report, *Nature*, 1 December 1928, p. 853
19 BAIRD, J.L., and Television Ltd.: 'Apparatus for the transmission of views, scenes or images to a distance'. British patent 269 658, 20 January 1926
20 Electrical Research Products Inc. (assignees of F. Gray): 'Improvements in television

systems'. British patent 288 238, 18 January 1928 (UK)
21   Ref.15, Chapter 6, pp. 132–147
22   ANON.: Report, *Wireless World*, 12 March 1930, p. 277
23   ANON.: 'First sound television broadcast', *Evening Standard*, 31 March 1930
24   GOEBEL, G.: 'From the history of television—the first fifty years', *Bosch Tech. Ber.*, 1979, **6**, p. 25
25   Quoted in ref.15, pp. 52–53
26   Advertisement, *Television News*, **1**, (1) March/April 1931, p.67
27   Ref.7, p. 146
28   Ref.16, Appendix 2, pp. 148–149
29   BITTINGS, R.C.: 'Creating an industry', *J. SMPTE*, November 1965, **74**, pp. 1015–1023
30   BUZALSKI, T.J.: 'Experimental television station W2XBS'. Development Group Engineering report 95, NBC, 1 March 1933
31   Ref.5, pp. 73–76
32   US Federal Radio Commission, fifth annual report, 1931, p. 54
33   Ref.7, p. 147
34   GANNETT, D.K.: 'Quality of television images', *Bell Lab. Rec.*, April 1931, **8**, pp. 358–362
35   WENSTROM, W.H.: 'Notes on television definition', *Proc. IRE*, September 1933, **21**, (9), pp. 1317–1327
36   Ref.7, p. 145
37   ANON.: 'Search continues for the right word', *New York Times*, 19 April 1931, section IX, p. 10:1
38   Ref.15, pp. 353–354
39   Ref.15, Chapter 7, pp. 148–175
40   ANON.: 'Television. First experimental broadcast', *The Times*, 1 October 1929
41   ANON.: 'Television humour. Machine breaks down in first day broadcast. Mugs Monday', *Daily Herald*, 1 October 1929
42   ANON.: 'Broadcast of television begins', *Manchester Guardian*, 1 October 1929
43   Editorial, *Amateur Wireless*, October 1929
44   Ref.15, Chapter 11, pp. 240–273
45   ROBB, E.: Internal report to DP, DIP and CE, 10 May 1933, BBC file T16/214
46   NORMAN, B.: 'Here's looking at you' (BBC and Royal Television Society, London, 1984) p. 85

Chapter 13

Large-screen television
(1930–1935)

Mention has been made of Alexanderson's reference to the play 'Back to Methusala' and his early ideas on large-screen projection television using multiple light sources. Although this approach was abandoned, GE continued throughout the late 1920s to be active in the field of theatre television. The company was not alone in pursuing this objective. During the late 1920s and the 1930s several influential organisations engaged in the development of large-screen equipment, including Baird Television Ltd, Electric and Musical Industries Ltd and Scophony Ltd, in the United Kingdom; Fernseh AG and Karolus-Telefunken in Germany; and de Forest, the Radio Corporation of America and General Radio, in the United States. There seemed to be a view that cinema television was an essential application of the new media form.

Langer, who had worked with Mihaly in Germany, wrote in 1922[1]:

> 'Personally I look forward with confidence to the time when we shall not only speak, but also see those with whom we carry on telephone or wireless telephone conversation, and the distribution of cinematograph films will be superseded by the direct transmission from a central cinema.'

And in 1924 the *Hastings and St Leonards Observer* reported[2]:

> 'A Scotsman has come south, in fact he has come to Hastings, and this particular Scotsman is now engaged upon perfecting an invention which at some not very distant date may enable people to sit in a cinema and see on the screen the finish of the Derby at the same moment as the horses are passing the post, or maybe of the Carpentier-Dempsey fight.'

A number of factors probably influenced the growth of cinema television: first, the excitement of watching sporting and other events while they were actually taking place rather than viewing the same scenes on cinema newsreels several hours or days after the events, and when the results were already known; secondly, the expectation that domestic television receivers would be expensive and beyond the reach of the masses—particularly during the depression of the late 1920s and early 1930s; thirdly, the popularity of cinemas as a cheap form of entertainment; fourthly, the very considerable time that would be required to manufacture the millions of television sets needed for home use even if these could be made at low cost to give satisfactory image quality; and fifthly, the possibility that cinema television receivers could be much more complex than domestic receivers and hence more capable of producing a better image. Also, of course, good cinema television might encourage people to purchase domestic sets when they became available.

Alexanderson exhibited[3] his television projector at RKO's Proctor's Theatre, Schenectady, on 22 May 1930. The screen was 6' by 7' in size. A standard 175 A arc, Karolus cell, and 48-hole scanning disc enabled an image to be synthesised which was about half as bright as that obtained from normal motion picture film. Figure 13.1, 13.2, 13.3 and 13.4 illustrate the layout of the apparatus.

The programme produced comprised an announcement by Mr Trainer, a series of vaudeville acts and the conductor of the theatre's orchestra who directed it from the studio, one mile away. Synchronisation was effected without difficulty, and all that marred the performance was the tendency of the 48 lines per picture, 16 pictures per second images to sway slightly. There was an opinion that these images

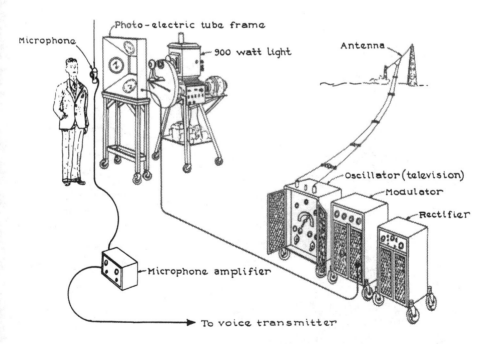

Figure 13.1 Schematic diagram showing the image signal generation equipment used in General Electric's theatre television project. Spot-light scanning was used

Source: Electronics, June 1930, p. 147

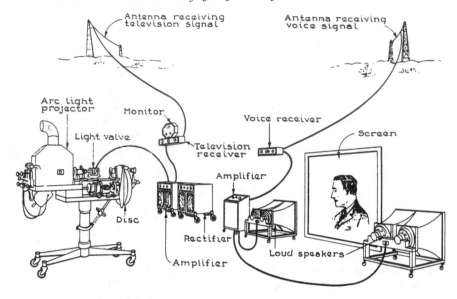

Figure 13.2   GE's theatre television receiver employed a Nipkow disc for image synthesis, and a Karolus cell light valve for modulating the light flux, from the arc lamp, in accordance with the vision signals

Source: *Electronics*, June 1930, p. 147

Projector Lens System

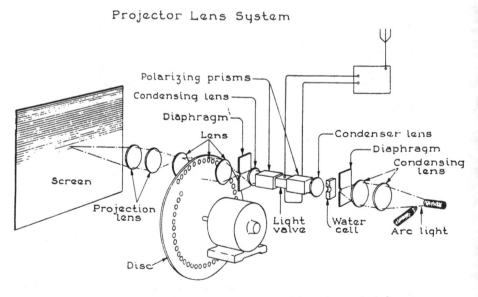

Figure 13.3   A detailed diagram showing the arrangement of the various optical elements

Source: *Electronics*, June 1930, p. 147

Figure 13.4   GE's theatre television equipment installed on the stage of the Proctor's Theatre, Schenectady (May 1930)

Reproduced by permission of Schaffer Library

were much superior, in lighting and in their comparative freedom from flicker, to the early cinema pictures.

The first public demonstration[4] of Baird's large-screen receiver was given at the London Coliseum, where Sir Oswald Stoll had arranged with Baird to allocate part of each performance, for a period of two weeks from 28 July 1930, to showing the reception of television to a large audience. Public interest was unequivocally great[5] and on many occasions the house full notice had to be displayed. Baird Television made a profit of £1500.

Subsequently, the equipment was taken to the Scala Theatre, Berlin, where performances were given from 18 September to 30 September 1930; then to the Olympic Cinema, Paris and finally to the Kvarn Cinema in Stockholm. According to Moseley, the demonstrations were a huge success and amazing enthusiasm prevailed[6].

The size of the screen was 5′ by 2′ and consisted of 2100 very small metal filament lamps, each lamp being placed in a cubicle so that the screen had the appearance of a large honeycomb. The front of the screen was covered by a ground glass sheet to soften the harsh light intensity of the lamps. Each lamp was connected to a separate bar of a commutator, Figure 13.6, which switched on one lamp at a time, the various

THE WORLD'S FIRST
PUBLIC PERFORMANCE OF
TELEVISION IN A THEATRE

BAIRD
TELEVISION
at the
LONDON
COLISEUM
commencing
JULY 28th, 1930

LIVING CELEBRITIES AND ARTISTES
TELEVISED THREE TIMES DAILY
BY THIS MARVELLOUS INVENTION

Figure 13.5   The first demonstration of J L Baird's large screen television receiver was given at the London Coliseum on 28 July 1930

Reproduced by permission of Radio Rentals Ltd.

lamps being excited in succession. Because the speed of the selector arm was 750 r.p.m. the individual lamps were energised every twelfth of a second.

By 1930 it was obvious that the existing methods of television (e.g. Figure 13.7 and 13.8) would not provide the quality of image production necessary for a high-grade service for theatre reception. There had to be a move towards high-definition television.

Dr L de Forest, the inventor of the audion, has related in his book[7] 'Father of radio' how he came to propose a new solution to the problem of theatre television:

'I had witnessed in Jersey City pathetic attempts by Repogle and Hoffman to turn the Jenkins system[8] [then owned by de Forest Radio] into something merchantable. In 1929 Allan du Mont and I had journeyed to Schenectady to see Dr Alexanderson's giant lens-studded scanning disc throw on a 6 ft by 6 ft screen in an absolutely dark theatre a

Figure 13.6 *The screen used at the Coliseum comprised 2100 very small metal filament lamps arranged in a rectangular array. Each lamp had to be connected in sequence to the output of the vision receiver. This was achieved by means of a commutator*

Reproduced by permission of Radio Rentals Ltd

much too faint picture from a great water-cooled glow lamp, amplifier-modulated by a television signal from a small scanning disc pick-up a quarter of a mile distant. I realised then the utter futility of theatre television by such methods. Obviously, adequate screen illumination would require a theatre projector arc lamp or its equivalent.'

De Forest, who was then (1930–31) living on Hollywood Boulevard and had been associated with the film industry, on studio sound, came to the conclusion that if a

Figure 13.7   Marconi low-definition television: optical system, based on an arc lamp and Kerr cell, for large screen projection (1932)

Reproduced by permission of the Marconi Company Ltd

television image could be formed on 35 mm film stock, a conventional film projector could be used to give a brilliantly illuminated picture on a cinema screen.

In his method a blank 35 mm film was coated on one side with a thin layer of pure metallic silver and then run at 24 frames per second over a brass anvil block located directly on the theatre projector about three inches from its lens. Above the anvil there was a high speed rotating drum carrying on its periphery a series of fine needles spaced 27 mm apart and of such a length that their points passed 0.001″ above the silvered film surface transversely to the film. The drum diameter and speed (3600 r.p.m.) enabled 90 needle traverses per picture frame to be obtained, thus giving a 90 line picture. The needle drum carried a commutator so that each needle, as it swept across the film, was 'brush connected' to the output of a small high frequency generator, excited by the video output of a television receiver. Consequently, a very fine modulated spark train traversed the film and eroded the deposited silver in proportion to the strength of the television signals. The film then passed in intermittent motion behind the projector lens and caused a brilliant picture to be imaged on the theatre screen—approximately two and a half seconds after the erosion of the silver.

This simple idea depended on silvered 35 mm film stock for its success, but the production of this material proved to be 'an unexpectedly difficult problem, or succession of tough problems—in fact, the bottleneck of the entire invention'. Professor A. Beckmann (later of pH-meter fame) of California Institute of Technology was responsible for the design of the chemical process for depositing the minutely thin silver coating on the film. But as de Forest noted: 'The coatings were too light, too heavy, or not of uniform density from foot to foot, as the scrupulously clean film passed through the ester and silver nitrate troughs to the distilled water laundry, the

Figure 13.8   Marconi 100-line large-screen television projector (April 1935)

Reproduced by permission of the Marconi Company Ltd

dryer and to the wind-up reel. Dr Beckmann toiled skillfully, ingeniously and zeal-ously.' Results were obtained and at times these were 'amazingly good'. 'Bright, life-size pictures of almost photographic quality were projected on the large screen', although to observers sitting close to it a Venetian-blind effect was seen with the 90-line picture. Designs were therefore advanced for a 150-line picture employing film material on which a silver layer was to be deposited by a vacuum sputtering process, but the pilot plant for that process was never completed.

Dr Forest and his colleagues encountered a problem common to inventors, which proved intractable of solution—shortage of money. 'We all felt that, given six months more and another 100 000 dollars, we would be ready for actual theatre television demonstrations, at which time the system would begin to earn its way. This was late in 1932. Then the Kalamazoo, and a thousand other banks closed; no more money for the experiment was in sight.'

And so ended an interesting and novel line of development. So far as is known, de Forest's application of video signals to electric spark erosion was the only one, of several suggestions, to use sparks in television reception which reached the stage of practical implementation.

The first proposal to employ sparks in television receiving apparatus seems to be that contained in von Jaworski and Frankenstein's German patent[9] in 1904. These inventors visualised a controlled spark as a light source. This was to consist of a vibrating wire and a platinum point, arranged as a spark gap. The sparks were to be modulated by the picture signals.

Later, in a British patent[10], Baird put forward the use of 'an image receiving screen formed of material capable of being heated to a degree that would cause emission of visible radiations . . . '. An exploring disc was to be provided with a specially arranged series of spikes (the points of which were to pass close to the back of the thin metallic foil screen when the disc was in rotation), and as each spike approached the screen an arc was to be initiated by the received television signals, thereby causing the screen to be heated in a series of bands. The purpose which Baird had in mind for this idea is not mentioned in the patent, but as it refers to apparatus of comparatively large dimensions, for example, 'the disc might be as large as 8 feet in diameter' it is obvious that a nondomestic viewing situation was envisaged. Baird was considering, at this time (1928–1929), the possibility of large-screen televisors being installed in cinemas and theatres and so it appears possible that the above described equipment—if it could be made to work—was intended for the same application as that of de Forest.

In both de Forest's and Baird's inventions the spark or arc would cause heating and erosion of the thin metallic film or foil, but whereas de Forest designed his equipment to enable the erosion characteristic to be dominant, Baird stressed the heating effect of the arc.

During the month, January 1931, when Baird Television, HMV and Bell Telephone Laboratories released details of their multichannel television systems, Baird disclosed[11] for the first time the use of direct arc modulation as a means of increasing the brilliance of the image received on a large screen[12]. The demonstration was given before representatives of the scientific press, including *Nature*, and the technical representatives of *The Times*. In this demonstration the video signals from the Baird transmitter were applied after suitable processing to a specially adapted arc. 'The detail and definition of the received image was comparable to that received on the standard commercial "Televisor" receiver, and the brilliance of illumination was remarkable. This demonstration of the successful modulation of the arc with television signals appears to open up considerable possibilities, and the television arc would appear to have a useful future,' commented *Nature*.

The first public exhibition of the modulated arc did not take place until the British Association Meeting, in September 1931, at the French Institute, Cromwell Gardens, when it was shown in the section devoted to mechanical aids to learning. Here, the new form of light source for video signals was used to project a picture onto a screen approximately four feet by two and a half feet.

The most noticeable effect for a reporter of the *Manchester Guardian*[13] was: 'The light was no longer orangey (as with the neon tube) but white. This put it on a level more comparable with photography and the cinema, whereas the black and orange picture received on the ordinary televisors reminds one of the early flickering films and so emphasises the distance television has yet to travel. A black and white picture seems to mark a definite milestone.'

Although the directly modulated high intensity arc had not previously been used for distant vision, its properties and susceptibility to modulation had been known for some time.

The discovery of the speaking arc was made by Bell and Hayes in the US and independently by Simon in Germany in 1897 and numerous arrangements were devised to control the arc by speech signals[14].

Monasch in his book[15], 'Alternating and direct current electric arcs', of 1904, referes to the speaking arc: 'With a ten ampere direct current arc of between three and five millimetres, between either solid or cored carbons, a clearly audible sound was produced even when an alternating current of one milliampere was superimposed on the direct current, and having a periodicity of 50 to 5000 Hz. The sound became inaudible only when the frequency was raised to 30 000 Hz.'

Ernest Ruhmer, in 1901, seems to have been the first person to have used the luminous properties of the speaking arc in a system for transmitting speech over an optical path[16]. In his apparatus the fluctuating light beam was rendered parallel by a parabolic mirror at the transmitter and after propagation was allowed to fall onto a selenium cell placed at the focus of a similar mirror at the receiver. The consequential changes in the cell's resistance, due to the action of the variable light flux incident on it, produced corresponding changes in the receiver circuit containing the telephone, so that the whole apparatus comprised a speech communication system, rather similar to Bell and Tainter's photophone.

Many experiments were carried out in photophony in the first quarter of the 20th century, by Ruhmer, Schukert, Blake and others. Ruhmer and Schukert were particularly successful in their work on photophony on the Wansee and the Havel, near Berlin, and in 1902 two permanent stations were erected, one at the works of Messrs Siemens, Schukert and Company in Berlin, and the other at the parish school in the Baumschulweg, two and a half kilometres away. Trials showed that speech could be transmitted by day or by night with good clarity, and even in wet weather transmission was possible, although with poorer fidelity of reproduction[14].

Another application of the directly modulated arc was made by Bernouchi who sent phototelegraphic signals over a beam of light. The first proposal[17] for the utilisation of the speaking arc for television was made by E W Whiston[18] in 1921. Other experiments were conducted by Blake in 1925[19].

In the same year Baird applied for his first patent on the use of arcs in television[20]. At that time the commercial arcs, when modulated, gave a light output which consisted of an alternating component superimposed on a direct component. This was unsatisfactory for television reception since the steady component would lead to a decrease of the contrast range of the reproduced picture. For photophony, the direct component would be unresponsive in producing an audible effect.

The Baird company carried out numerous investigations on speaking arcs. These showed that the light variations were dependent upon the luminosity of the gaseous envelope partly surrounding and partly within the crater of the negative electrode. Later, arc electrodes were made of metal which permitted the heat generated at the electrode tip to be rapidly conducted away and allowed the crater gases to contribute to the major portion of the light from the arc. This discovery[21] greatly increased the contrast of the reproduced picture. An additional improvement stemmed from the coring of the negative electrode with a refractory material, such as cerium oxide mixed with certain other salts of the metals of the alkaline earths, to increase the brightness and improve the colour of the light given out by the flame of the arc.

Considerable power was needed to fully modulate the arc. For an arc capable of passing a current of 10 A and dissipating 300 W an amplifier having an output of 100 W was required[21].

Experiments by the Baird company showed that its arcs had a frequency response which was essentially flat from 10 Hz to 10 000 Hz. Since the 30-line transmitter bandwidth did not have to exceed 10 000 Hz, the arc was entirely suitable for low-definition telecinema-type applications, and Baird Television successfully demonstrated cinema television on screens measuring 3 × 7 feet by its use.

During the period when this work was in progress, other members of the Baird company were investigating the use of the Weiller mirror drum scanner as a means of augmenting the optical efficiency of the system.

Baird first demonstrated[22] the mirror drum transmitter on 8 May 1931. This was to be a prelude to the televising of the Derby and its subsequent showing, live, in a theatre in 1932. For his purpose Baird equipped a van with a transmitter of the drum type. Natural light was used during the May test but although there were appreciable variations in the quality of the reception due to the varying cloud cover, reporters saw, in a room in Long Acre, the images of people passing along the street outside[23].

The *Daily Telegraph* noted[24]: 'The televising of great national events, Mr Baird considers, is now well within the bounds of possibility, though so far no definite arrangements for such broadcasting have yet been made.' Moseley, the radio critic of the *Daily Herald*, observed on the same day, however: 'Mr Baird said that the fact that one was able to pick up the street scene showed that the idea of televising the Derby or the cricketers at Lords was not so fantastic as some imagined.'[25]

Following this latest trial Moseley[26] wrote to the BBC and enquired whether the BBC could give Baird Television five or ten minutes on Derby Day for a television broadcast during the race. Mosely pointed out that his company had willingly given up its half hour transmissions when asked to do so and consequently such a request would more than make amends.

Figure 13.9   Baird's large television screen mounted on the stage of the Metropole Cinema, London

Reproduced by permission of Radio Rentals Ltd

The BBC's reply[27] stated that although Baird Television could not have the London regional wavelength for the vision transmitter it might be possible 'to arrange for the London national wavelength [of] 261 m to be placed at [its] disposal for the television signals from approximately 2.45 to 3.15 p.m. on Wednesday, 3rd June [provided] the following conditions were fulfilled:

1   'that the speech accompaniment, which of course would not be broadcast, would be on a telephone line quite separate from any telephone line rented by the BBC;

2   'that the BBC engineers would be satisfied in a preliminary test that nothing involved in this television transmission should in any way interfere with the normal service transmission of the running commentary'.

So, on Derby Day 1931, Baird Television's outside broadcast television van was taken to Epsom and parked opposite the winning post. It was connected by telephone line direct to the company's London studio, from where the signals were sent to the Brookman's Park transmitter[27].

The parade of the horses was seen, although the horses were not individually recognisable, together with shadowy images of moving people. 'As the moment of the finish of the race approached, interference became worse and the screen, seen through its enlarging lens, at times dissolved into a blurred mass of flickering lines.

Notwithstanding this, however, the horses were plainly seen as they flashed past the post.'[28] The *Daily Telegraph*[29] was more charitable: 'Fifteen miles from the course, in the company's studio at Long Acre, all the Derby scenes were easily discernible—the parade of the horses, the enormous crowd, and the dramatic flash past at the winning post.'

After the transmission Baird said he was quite satisfied with the experiment. 'This marks the entry of television into the outdoor field,' he noted, 'and should be the prelude to televising outdoor topical events.'

Many years later Moseley[30] wrote: 'The first Derby broadcast captured the imagination of the general public so strongly and so fired the enthusiasm of amateur experimenters as to impress even the BBC.'

The following year, 1932, Baird again televised the Derby from Epsom[4]. The broadcast was made through the BBC's transmitter and in addition there was a special transmission by landline to the Metropole Cinema, London. Baird wrote:

> 'I used the same van as the previous year for a much more ambitious experiment, and fitted up a large screen, nine feet by six, at the Metropole Cinema, Victoria. The transmitter was the same as that used the previous year and consisted of a large revolving drum. The picture sent out by the BBC was narrow and upright in shape, seven feet high and only three wide. To give a large picture at the Metropole I had three pairs of telephone wires from Epsom and sent out three pictures side by side. The three pictures thus formed one big picture on the screen at the Metropole seven feet high and nine feet wide.
>
> 'The demonstration was one of the most nerve racking experiences in all my work with television, second only in anxiety to the Parliamentary Committee. The night before the show we were up all night putting the finishing touches to the apparatus, and when the great moment drew near I remember literally sweating with anxiety. The perspiration was dripping off my nose.
>
> 'A vast audience had gathered in the cinema; even the passages were packed, and the entrance hall and the street outside were filled with a disappointed crowd, unable to get in. If the show had been a failure the audience would have brought the house down and I should have been a laughing stock.
>
> 'All went well. The horses were seen as they paraded past the grandstand. When the winner, April the Fifth, owned and trained by Tom Walls, flashed past the post, followed by Daster and Miracle, the demonstration ended with thunderous applause. I was hustled to the platform to say a few words but was too thrilled to say more than "Thank you".'

The 1932 Derby transmissions were undoubtedly much better than those of 1931. Two factors contributed to this improved performance; first, the use of zone television, and secondly, the much greater signal level used on the actual lines. The first

Derby broadcast had been much impaired by interference and so Campbell, one of the Baird engineers, had unofficially increased the permitted line voltage of 3 V to 30 V.

In Germany large-screen television development effectively began in 1931 when Ferseh AG confirmed Ives' view that television scanning in the open or by artificial light was possible in principle but required a very high level of illumination of the scene to be televised[31].

Early in 1931 the company sought to overcome this limitation and began work on the intermediate film method of television. It was based on standard motion film practice and used film cameras, film material and film developing and fixing processes which had been evolved over a period of many years. Fernseh hoped to concentrate the successive steps of exposure, development, fixing, drying and negative picture scanning in a short time period so that an almost live transmission would be possible.

After 18 months' effort, Fernseh in August 1932 had a system working to a 90 lines per picture standard. Much painstaking work had been necessary. A film with a high gamma and thin emulsion had been produced in the film works of Zeiss-Ikon, developing baths of the highest possible concentration and a special fixing solution had been formulated, and the optimum temperatures of the individual processing baths had been determined. By these means the time interval from taking the picture to its reproduction had been reduced to 15 seconds, although this necessitated the film being scanned while it was still wet. Alternatively, the film could be dried using hot air, and scanned 85 seconds after exposure. The film could then be stored and subsequently copied when required. The sound track was recorded on the film according to the standard techniques of the trade so that the delay was of no consequence.

Figure 13.10 shows the 1932 equipment. Apparatus of this type was built and installed in a 3.5 ton television camera truck for the Reichsrundfunkgesellschaft in 1934 (see Figures 13.11, 13.12 and 13.13).

Fernseh's equipment was shown at the 1932 Funkausstellung[32] but not at the 1933 Berlin Radio Exhibition. A report on the latter exhibition stated[33]:

> 'The system is technically highly interesting, but the results were photographically "thin" denoting underexposure or under-development of the film . . . Unfortunately splashes and bubbles sometimes appeared on the film, due to insufficient wiping. The defects of the system at present are therefore mainly chemical.'

Furthermore, the intermediate film method had the serious disadvantage of very expensive running costs: the cost of the film was approximately £45 per hour.

In order to reduce this expenditure a new method was developed which used a continuous film loop, about 70 m long. From the technical point of view the new apparatus was 'nothing more or less than the combination of a film factory, a film processing laboratory and a complicated television receiver'[34]. Referring to Figure 13.10, the blank film was passed through a chamber, 1, where a photographic

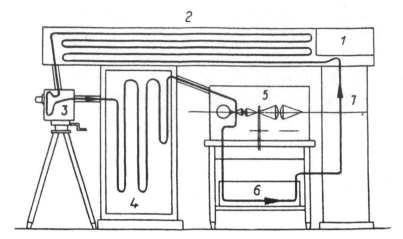

SCHEMATIC VIEW OF THE CONTINUOUSLY WORKING INTER-
MEDIATE FILM SCANNER

*Figure 13.10 Fernseh's intermediate film scanner. It comprised the emulsioner, 1, the drier 2, the
ciné camera, 3, the developer and fixer, 4, the scanner, 5, and tank 6 'where the
emulsion was washed off the film' and the drying chamber, 7*

Source: *Television Today*, Newnes, London, 1935, pp. 252–255

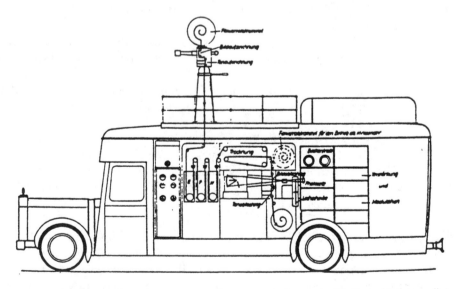

*Figure 13.11 Diagram of the German Post Office's transmission truck with Fernseh's intermediate
film equipment (1936)*

Source: *Television Today*, Newnes, London, 1935, pp. 252–255

Figure 13.12 Interior view of Fernseh's intermediate film apparatus

Reproduced by permission of the BBC

emulsion was deposited on the film, then into tank, 2, where it was hardened and dried. Following exposure in the camera, 3, the film was developed, fixed and scanned as in the previous system. After scanning the emulsion was removed in tank, 6, and then dried in compartment, 7, ready again to receive a new emulsion.

With the use of a fixed length of film the material cost was reduced to c. £2 per hour.

In 1932 Fernseh's engineers conceived the idea of utilising the same process for displaying large screen television images. The received and amplified picture signals would be applied to a Kerr cell which would modulate the light from an arc lamp and the variations in light flux would then be synthesised, by means of a scanning Nipkow disc, and recorded on film. Subsequently, the film would be projected using a standard cine projector and the film reproduced as described above[35].

Apparatus working on the above principles was constructed but the results generally were not of high quality. In 1934 a report on the Berlin Radio Exhibition noted[36]: 'The results, although improved since last year, are still far from satisfactory as regards detail and modulation. Although this method seems the ideal one to overcome the optical difficulties associated with big screen projection, the difficulties in other directions are very great and it will probably take some considerable time before the system is perfected.'

One year later (1935) Fernseh AG again exhibited at the Funkausstellung an intermediate film projection receiver. It differed from previous models in two respects.

THE FERNSH A.G. INTERMEDIATE FILM CAMERA AND ITS ASSOCIATED APPARATUS
MOUNTED ON A MOBILE VAN
The film of the scene to be transmitted is photographed, rapidly developed and fixed, and then scanned
in an orthodox film scanner.

Figure 13.13 The Reichs Rundfunk intermediate film processing truck with roof mounted cine camera

Reproduced by permission of the BBC

First, a hard vacuum cathode-ray tube replaced the scanning disc and Kerr cell arrangement; secondly, new film stock supplanted the continuous film loop. 'The results shown [by] this intermediate film projector are still disappointing, and the improvement from year to year is only very small,' observed one commentator. 'The difficulties associated with this process appear to be so great that many experts express great doubt whether this system can ever be perfected,' he noted[37].

An intermediate film transmitter, but not receiver projector, was also developed by the Agfa company in collaboration with Karolus. The transmitter used new film stock rather than a continuous loop of film and standard developers and fixers which were maintained at c. 65°C to ensure speedy processing. A 1933 report stated[38] that 'the transmitter worked extremely well and reliably', but later annual reports of the Berlin Radio Exhibition failed to comment on any demonstrations/displays of the apparatus.

Also at this exhibition, Karolus showed a four-channel large-screen television system in which each of the channels transmitted a 24-line video signal[39]. These were interlaced to give a 96-line image by arranging that channel 1 transmitted image detail associated with lines 1, 5, 9, 13, etc., channel 2 with lines 2, 6, 10, 14, etc., and similarly for channels 3 and 4. A mirror drum was used in the film

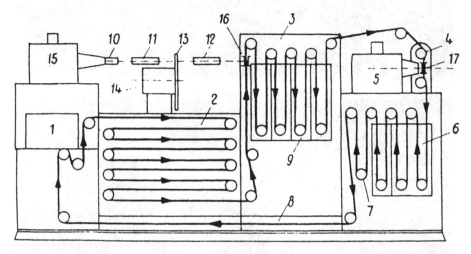

Figure 13.14 Schematic diagram of Fernseh's continuously working intermediate film, large-screen projection receiver. It consisted of the emulsioner, 1, the drier, 2, the developing tank 3, the fixing tank, 9, the projector lamp housing, 5, the Maltese cross, 4, the projection window, 17, emulsion removing tank, 6, the compensating loop, 7, the drying chamber, 8, the arc lamp, 15, the Kerr cell, 10, the optical system, 11, 12 & 16, the scanning motor, 14 and the scanning disc, 13

Source: *Televeision Today*, Newnes, London, 1935

transmitter together with four photocells. At the receiver a single arc lamp and a special Kerr cell (a Karolus multielectrode cell) enabled the image to be reconstituted. 'Although only a 15 A arc lamp was used, the picture brightness [20 lux] was good. Detail and modulation were also very pleasing. The system is capable of further improvement in all directions, the limit having not nearly yet been attained.' An image size of one metre square was produced.

Karolus's four-channel system seems to have been the precursor of a much more ambitious multi-channel large-screen television set. In 1935 he demonstrated a 100-channel scheme, which employed nine valves per channel, in conjunction with a 10 000 element lamp screen to synthesise an image of 100 lines per frame. The screen was two metres square, so that each picture element (pixel) was two centimetres square. 'Standing about 40 yards away, the cell formation was not apparent and the brightness was very remarkable. The picture had a brilliance of some 1000 lux—some 6 to 8 times the brilliance of a good cine screen. 50 pictures per second were used and there was no flicker. The brightness was actually such that when the windows in the hall were open, the screen could be watched in perfect comfort. The colour was quite a pleasant fleshy tint.'[40]

Karolus's lamp screen was similar to that of Baird which had been displayed in 1930 but the two schemes differed in the number of channels which were utilised. Baird used one channel only and a commutator at the receiver to distribute the line signals to the individual lamps. Karolus required 100 channels so that the wide

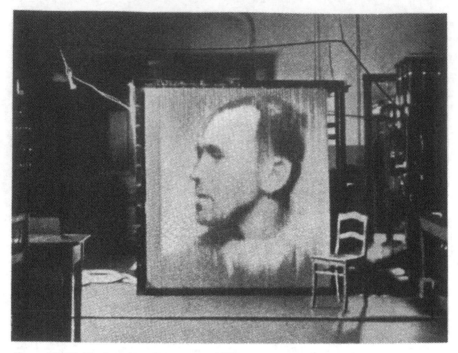

Figure 13.15 Karolus's large lamp screen, 1937

Reproduced by permission of the Deutsches Museum

bandwidth of 250 kHz needed for a 100 line per frame image could be divided into 100 parts each of 2.5 kHz bandwidth. In this way his generated picture signals could be allocated to 100 ordinary telephone lines each of which would handle signals having a maximum frequency of 2.5 kHz.

The stimulus for Karolus's work stemmed from a change of government in Germany. On 30 January 1933 Adolf Hitler was appointed Chancellor of the Third Reich, and soon afterwards Dr P J Goebbels became Minister of Propaganda and Public Enlightenment. His propaganda policy of calumny and bias necessitated the banning of all forms of dissent and free expression and the furtherance of means which would allow his messages to reach a wide audience. Shortly after Hitler's rise to power, Goebbels proscribed all amateur radio transmissions as part of this policy, and sought to impose his views on all matters relating to the advancement of Hitler's authority[41]. Essentially, Karolus's large-screen television receiver was the television equivalent of a public sound address system, and was intended to be used by the government for propaganda work. Ives, in 1927, had referred to this application of television when the AT&T Company had demonstrated its television equipment.

A layout of Telefunken's public address television system was shown at the 1937 Berlin Radio Exhibition[42]. Instead of the 10 000 lamp screen, the apparatus employed a new form of spotlight scanner, Figure 13.17.

Figure 13.16 The large lamp screen of Telefunken, 1937

Reproduced by permission of AEG Aktiengesellschaft

None of the experimental large-screen television projectors developed by Jenkins, Baird, GE, the AT&T Company, Telefunken, Fernseh AG and others achieved any long-term success. During the formative period of cinema/theatre television much work was being undertaken in the UK, the US and Germany on electron optics, on fluorescent screen materials and on the physics of materials and of electron emission. This effort led to the development not only of electronic camera tubes of the iconoscope/emitron and image dissector types and domestic cathode-ray tube receivers but also of high intensity projection tubes. From about the middle of the 1930–40 decade firms including *inter alia* Electric and Musical Industries Ltd, the Radio Corporation of America, Baird Television Ltd and Telefunken engineered large-screen television equipment based on nonmechanical methods. Of the mechanical scanners for this purpose only those of the Scophony Company achieved any substantial fame.

Electronic camera tubes will be dealt with in Part III which is concerned with high-definition television.

Figure 13.17 Intermediate film image generation equipment of Telefunken (1936)

Reproduced by permission of the Deutsches Museum

References

1 LANGER, N.: 'A development in the problem of television', *The Wireless World and Radio Review*, 11 November 1922, pp. 197–210

2 Report in *The Hastings and St Leonards Observer*, January 1924

3 'Television in the theater', *Electronics*, June 1930, pp. 113 and 147

4 BAIRD, M.: 'Television Baird' (HAUM, South Africa, 1974) Chapter IX, pp. 102–105

5 A report in the *Sphere* quoted *in* MOSELEY, S.A.: 'John Baird' (Odhams, London, 1952)

6 MOSELEY, S.A., and BARTON CHAPPLE, H.J.: 'Television today and tomorrow' (Pitman, London, 1933)

7 DE FOREST, L.: 'Father of radio' (Wilcox and Follett, 1950) pp. 418–422

8 ABRAMSON, A.: 'Pioneers of television—Charles Francis Jenkins', *SMPTE J.*, February 1986, pp. 224–238

9 VON JAWORSKI, W., and Frankenstein, A.: 'Verfahren und Vorrichtung Fernsichtbarmachung von Bilden und Gegenstanden mittels Selenzellen, Dreifarbenfilter und Zerlegung des Bildes Punktgruppen durch Spiegel'. German patent 172 376, 20 August 1904

10 BAIRD, J.L.: 'Improvements in or relating to television and like apparatus'. British patent 323 817, 10 October 1928

11 Ref.6, p. 152

12 ANON.: Report in *Nature*, 10 January 1931

13 ANON.: 'A stage nearer cinema television', *Manchester Guardian*, 26 September 1931

14 BARNARD, G.P.: 'The selenium cell: its properties and applications' (Constable, London, 1930)
15 MONASCH, A.: 'Direct and alternating current arcs' (Julius Springer, 1904)
16 RUHMER, E.: 'Neue Sende und Empfangsanordnung fur drahtlose telephonie', *Phys. Zeit.*, 1901, **2**, pp. 339–340
17 BAKER, T.T.: 'Wireless pictures and television' (Pitman, London, 1926)
18 WHISTON, E.W.: 'A mode of and/or means of transmitting photographs, messages, views and like devices by wire or wireless telegraphy'. British patent 185 463, 4 May 1921
19 BLAKE, G.G.: 'Communications on wavelengths other than those in general use', *Experiemental Wireless*, 1925, **2**, pp. 561–572
20 BAIRD, J.L.: 'Improvements in or relating to television and to apparatus for use in transmitting views, scenes, images, pictures, and other objects of an animated or inanimate nature to a distance'. British patent 269 219, 21 October 1925
21 BANKS, G.B., and WILSON, J.C.: 'Modulating the arc', *Television*, September 1934, pp. 397–398
22 ANON.: 'Street television. Test of new apparatus in London', *Daily Mail*, 9 May 1931
23 ANON.: 'Daylight demonstration', *The Times*, 9 May 1931
24 ANON.: 'Television by daylight. Watching the man in the street. National events to be brought near. First open air apparatus', *Daily Telegraph*, 9 May 1931
25 ANON.: 'Televising the Derby', *Daily Herald*, 9 May 1931
26 MOSELEY, S.A.: Letter to G Murray, 19 May 1931, quoted in 'John Baird', *op.cit.*, p. 150, see Ref. 5
27 MURRAY, G.: Letter to S A Moseley, May 1931, quoted in 'John Baird', *op.cit.*, p.151, see Ref.5
28 ANON.: 'Television scenes. Horses seen flashing past post', *Daily Mail*, 4 June 1931
29 ANON.: 'Race at home. Television success', *Daily Telegraph*, 4 June 1931
30 MOSELEY, S.A.: 'John Baird', *op.cit.*, p. 153, see Ref.5
31 RUDERT, F.: '50 years of "Fernseh", 1929–1979', *Bosch Tech. Ber.*, 1979, **6**, p. 32
32 TRAUB, E.H.: 'Television at the 1932 Berlin Radio Exhibition', *J. Television Society*, 1932, p. 165
33 TRAUB, E.H.: 'Television at the 1933 Berlin Radio Exhibition', *J. Television Society*, 1933, p. 276
34 Quoted in Ref.31, p. 33
35 WEST, A.G.D., *et al.*: 'Television today: practice and principles explained' (George Newnes, London, 1935) **1**, pp. 252–261
36 TRAUB, E.H.: 'How far has Germany progressed?', *Television*, October 1934, p. 454
37 ANON.: 'Television progress in Germany', *Television and Short-Wave World*, October 1935, pp. 565–566
38 Ref.33, p. 281
39 Ref.33, p. 280
40 TRAUB, E.H.: 'Television at the Berlin Radio Exhibition', *J. Television Society*, 1935, **2**, part III, p. 56
41 ANON.: 'German Government and television', *Electronics*, January 1935, pp. 10–11, 30
42 TRAUB, E.H.: 'Television at the Berlin Radio Exhibition, 1937', *J. Television Society*, 1937, pp. 292–294

Part III

The era of pre-war high-definition television 1934 to 1939

Chapter 14

Between low and high-definition television, (1930–1931)

Until January 1931 it seemed to the general public of Great Britain that J L Baird had no effective competition from any other company in the UK. He had been active in television matters from the winter of 1922/23, had established, with others, several companies (Television Ltd, Baird Television Development Company and Baird International Company), had given many demonstrations of television based on his basic 30-line scanning system, had devised numerous variations of this scheme to show rudimentary colour television, stereoscopic television, noctovision, daylight television, long distance television, *inter alia* and had secured much needed publicity and finance thereby. But then, in January 1931, the Gramophone Company, or HMV (His Master's Voice) as the company was popularly called, showed equipment[1] at the Physical and Optical Societies Exhibition, London which was to herald an ominous situation for Baird and his collaborators.

At the exhibition the Gramophone Company displayed 150-line images by combining the signals from five separate television channels. Prior to 1931 only 30-line pictures of a head and shoulders format had been produced by Baird. Now, with the emergence of HMV, the progress of television in the UK would be on a more competitive basis.

The demonstrations created great interest and queues of people had to wait outside the small theatre, which the Gramophone Company had set up, to view the new system. G A Atkinson, the film critic of the *Daily Telegraph*, referred to the notable gain of the system and wrote: 'It marks a considerable technical advance on any system yet demonstrated, especially in the direction of bringing television rapidly into use for entertainment purposes'[2]. The *Evening Standard* noted that television seemed now to be within sight of realisation[3].

During the exhibition the Gramophone Company showed images from ordinary cinematograph films projected by means of its apparatus onto a screen measuring 24 by 20 inches. No longer were head and shoulders shown but instead the audience saw images of buildings, soldiers marching, cricketers walking on and off the field and so on. 'Everything was easily recognisable. An LCC tramcar showed up so clearly that its number on the front was decipherable without difficulty.' 'The pictures were steady on the screen. They were in good focus at very short range. They would probably stand enlargement up to four or five times the size of the screen actually used . . . ' 'The general effect was that of looking at a performance of miniature films lacking full illumination.'[2]

HMV's success obviously pointed the way to the favourable adoption of a public television broadcasting system. Baird Television now had a rival in the UK and considerable animosity was to be engendered by this fact. J L Baird wrote about the gigantic Radio Trust of America spending vast sums of money on the perfection of television 'not only in America but also in England where their subsidiary the HMV Company [had] recently [given] a demonstration of most elaborate apparatus'[4].

This reference to HMV (and later EMI) being a subsidiary of RCA, with the implication that a great deal of the knowhow of television was being imported from the US was to be a feature of the Baird Company's writings to the BBC and to the Post Office for the next few years. Baird was most anxious at this time about the emergence of a competitor in Britain which could compound the troubles that the company was experiencing with its financial affairs and the introduction of a satisfactory low-definition service.

Unfortunately for the Baird Television Company but fortunately for the prestige of the UK, the work on television carried out in the laboratories of the HMV, and later EMI, companies was undertaken by British technicians and engineers and all the equipment was designed and constructed in British research laboratories and workshops.

The link between HMV and RCA must be explained because it was this relationship which caused Baird himself so much unease.

The Gramophone Company was founded[5] in 1898 in London as a private company with purely British capital of £15 000. Its only business was recording music, selling duplicated pressings of this music made by an associated company in Hanover (which became Deutsche Gramophon) and vending gramophones manufactured in the US by the Victor Talking Music Company. In 1900 the Gramophone Company became a public company with a capital of £600 000 and engaged in the business of pressing its own records and making its own reproducers.

A further change took place in 1920 when the Victor Company obtained a 50 % interest in the British company—although the management remained undisturbed. Nine years later the Radio Corporation of America procured all the shares of the Victor Company and so became the beneficial owner of the shares held by Victor in the Gramophone Company Ltd.

The association of the Victor Company with the British company came about after the 1914–1918 war when the latter company needed capital for expansion. Both companies had similar interests: they had exchanged artists, matrices, trade marks and patents relating to talking machines over a period of many years, following a contract negotiated in 1902–1903. Hence, when the directors of the Gramophone Company decided to expand further they felt that it would be desirable to approach the American company rather than a company having different sympathies. They offered the Victor Company a half interest in the British business

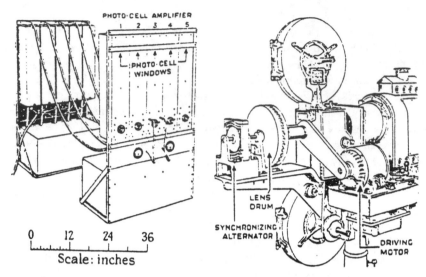

Figure 14.1   Disposition of the scanning equipment of HMV's five-channel zone television system

Source: *Journal of the IEE*, **70**, 1932, pp. 340–349

for £1 250 000. As a consequence the Victor Company had two directors on the board of the Gramophone Company, which comprised nine members, but the two rarely attended meetings and there was no agreement or understanding of any kind as regards policy. Essentially, though, the Victor Company operated in the western hemisphere and the Gramophone Company functioned in the rest of the world[6].

Although the British company competed very successfully, its field of business was restricted and this provoked some concern for the directors. They endeavoured to add typewriter manufacture to the range of products, but without success. However, wireless broadcasting had commenced in the UK in 1922 and in 1925 the Western Electric Company had developed a system of electrical recording which enabled recorded sounds to be reproduced with a very marked improvement of quality compared with those obtainable from acoustical recordings. The Gramophone Company quickly followed the example of Western Electric and embarked on a programme of development of electrical gramophone pick ups, turntables, amplifiers and loudspeakers.

By the end of 1929 the boom in gramophone records had subsided and the recording industry was in a somewhat parlous state. Certain record manufacturers, notably the Gramophone Company, turned their attention to the radio industry. Initially, HMV had sold receivers imported from Germany but later they manufactured sets to their own designs before acquiring in 1928, from the Marconi Company, the Marconiphone Company and the entertainment rights in all the Marconi Company's patents[7]. Even though the broadcasting of music and its reception by wireless sets was seen as a rival to the sale of gramophone records, nevertheless, the broadcasting medium effectively acted as a publicity medium for recorded sound.

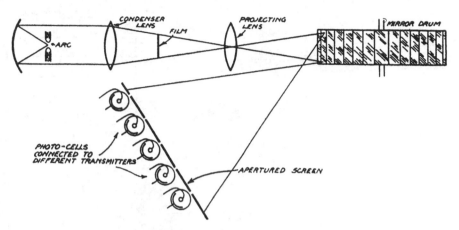

Figure 14.2   Schematic diagram of the HMV company's five-zone television transmitter. Effectively five 30-line television systems were combined to give a 150-line image

Source: *Journal of the IEE*, **70**, 1932, pp. 340–349

Further stimulus was given to the recording industry in 1928 when the talkies arrived. HMV became active in producing sound tracks using both disc and film material. Discs were usually manufactured to a 16" diameter, 33⅓ r.p.m. standard so that the playing time of a disc equalled that of one reel of film.

Two of HMV's engineers, W F Tedham and C O Browne, who were later to play an important role in the development of television, devised a sound-on-film system, and for a short period the company was involved in film production. Only one film, the Ben Travers farce 'Rookery Nook', was made, although the sound tracks for other films were recorded[8].

Contemporaneously with this activity the company made an attempt to devise a recording process which would allow audiovisual duplicated recordings to be sold to the domestic market. The ordinary silver image celluloid kinematograph film had two severe disadvantages; namely, the high cost of its production and its inflammability. It was thought essential to the ultimate commercial viability of home movies for a cheap, non inflammable storage medium to be developed. Photomechanically printed paper film was felt to be a solution. This would have the further advantage that the production of a paper film in colours would be an easy matter whereas the coloured kinematograoh films of c. 1930 could only be prepared by very laborious processes.

F Ellington was assigned to work on the problem[9]. Following design and experimental studies he proved the validity of the method and estimated the cost of 1000' of 16 mm paper film as 1.25 p against the cost of £1.00 for the same length of film. With this paper film a continuous motion projector of the episcope type was needed. Although such a projector was evolved which showed 'exceptional promise' in the experimental stage, it was decided by the company that it would be too expensive for the domestic market. Thus, this further attempt to diversify into the mass consumer market failed.

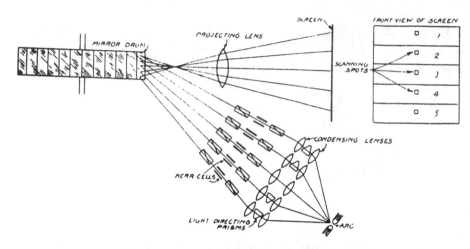

Figure 14.3   The diagram illustrates the disposition of the elements of the five-zone television receiver. The five picture signals were applied to five Kerr cells; these modulated the light flux from the arc and enabled the image to be synthesised. Mirror drum scanning was used at both the transmitter and receiver

Source: *Journal of the IEE*, **70**, 1932, pp. 340–349

HMV's interest in television may have been stimulated by the first demonstration, in January 1926, of rudimentary television by J L Baird since in that year a Mr Hallawell, at the direction of Mr B Mittel, the production director of the company, compiled a list of patents of film recording, picture telegraphy and television[10].

Nothing further was undertaken until October 1929 when consideration of the television problem was renewed by the preparation of a case opening report on the subject, written by M Bowman-Manifold. In this, mechanical scanning was regarded as the more practical method but the advantages of cathode-ray systems in both transmission and reception were noted[10].

A few months later, in January 1930, a television section was set up in the research department and an allotment of £800 was given for work, which was on mechanical scanning lines, for the period January to June 1930.

In April 1930 A Whitaker, of the advanced development department, visited RCA, which had just taken over the Camden works (see Chapter 17), and had absorbed Zworykin's section from Westinghouse. Demonstrations of 60-line television using cathode-ray tube displays were given to Whitaker and although the definition of these was still not satisfactory for the commercial market, nevertheless the brightness of the cathode-ray tubes used was 'so remarkably in advance of anything previously known', that he returned with the belief that this method of reception was worthy of progression. Indeed, 'the prospects of this line of attack seemed so bright' that in the second half of the year, July to December 1930, a further allotment of £800 was again requested. An additional sum of £1200 for work on the transmission system was also sought. This followed a report written by W J Brown which recommended the utilisation of the 3 m to 5 m short wave band for the system. Prior

Figure 14.4   Photograph of HMV's five-channel television receiving apparatus

Reproduced by permission of Thorn EMI Central Research Laboratories

to Brown's report the General Post Office had been approached and asked to waive some of the restrictions which had been applied to experimental transmissions[10].

During the second half of the year (1930) some investigations were carried out on cathode-ray tube reception. The correctness of this course was confirmed when G E Condliffe, of HMV, visited RCA's Camden works in November 1930. These visits by Whitaker and Condliffe to RCA entirely changed their views regarding television receiving apparatus. Before April 1930 HMV had adjudged cathode-ray tube reception as 'an interesting theoretical scheme but probably impracticable owing to the instability and low luminosity of the tubes available at that time'. After their visits and the demonstrations of the new tubes which had enhanced brilliance, focus and efficiency they were converted to the opinion that the RCA way was the right one. As Whitaker noted: 'Our decision to change our line of attack was caused by these RCA demonstrations . . . and as we had no wish to start repeating all the work which [had] been so excellently carried out in America, we decided to investigate the possibility of using a very considerably greater number of picture elements than [had] been possible,'[10,11] (Whitaker to Dr Goldsmith, vice president of RCA, 22 August 1930.)

Such a plan necessitated a wide bandwidth and therefore a very high radio frequency. Whitaker, in his letter of 22 August, informed Goldsmith that HMV was 'at present' working with wavelengths from 2 m to 5 m, although the radio section had recently been experimenting as far down as 30 cm. Goldsmith replied on 4 September 1930: ' . . . I am indeed glad to learn that your research group is working along somewhat different lines from those which are being carried forward in

Camden. The field is so new that duplication of effort would be uneconomical . . . .
I note with some astonishment that you propose to use 250 kilocycle sidebands.'[12]

The Gramophone Company's attitude to the television problem was cautious and
sensible. After a watching brief from 1926 the company naturally was keen to avoid
some of the pitfalls which had been encountered by others. Apart from the visits to
RCA, representatives of the company had, as noted in Chapter 8, been to the Baird
Television Company's studio in Long Acre, London, and had had discussions with
O G Hutchinson about current and future trends. Moreover, HMV had subjected
each Baird patent specification to a careful examination for claims which might subsequently 'be awkward in operating a television system' (see Appendix 4).

Again, the company had sought confirmation, from both the General Post Office
and BBC, of statements made by Baird Television concerning future licences and
wavelength allocations[13].

On the practical, side HMV wished to advance from a position where a reasonable knowledge of the essential components of a television system had been accumulated by carefully conducted experimental studies. The photoelectric cell and the
light valve were two of these components, and hence work was initiated on investigations of photoemissivity and the Kerr effect. These were pursued by W F Tedham
and W D Wright respectively and reports were submitted by the researchers in
September and November 1930. It is clear from Tedham's report[14] 'An investigation of the factors affecting emission in photoelectric cells' that HMV had in mind
the manufacture of such cells since an appendix deals with the costs involved in
producing 1000 cells in a six-month period.

Tedham achieved some success. This is shown by the sensitivity figures which he
quoted for various cells:

Cathode	Sensitivity ($\mu A/L$)
potassium (bulk)	1.0
Tungsram sodium	1.0
Weston sodium	1.8
Zworykin's cell	2.0
RCA cells (7)	4.0 to 8.0
Osram cells (caesium)	12.0 (catalogue)
one Osram cell	25.0
HMV (normal process)	15.0
HMV (special process)	30.0

He had difficulty obtaining some repeatable results 'presumably' as he said 'because
quantitative methods are not easily applied'. He investigated the sensitivity of cathodes of caesium on silver oxide and was able to produce consistently good cells. The
effect of gas filling was researched but Tedham concluded: 'There is . . . considerable advantage in using vacuum cells at 30 $\mu A/L$ for television rather than gas-filled
cells at say 60 $\mu A/L$ and compensating for the serious high frequency loss [of the
signal].' An additional conclusion was: 'Cell manufacture needs more skill than
valve manufacture.'

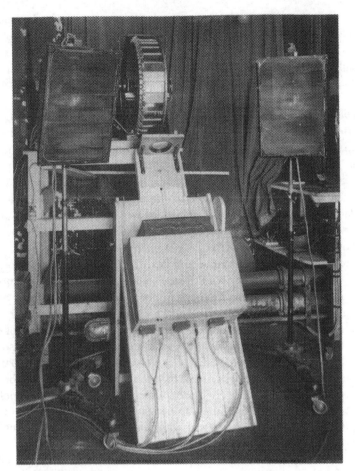

Figure 14.5   Photograph of the triple zone receiving equipment used at the Metropole cinema to back project a three-zone televised image onto a screen for public viewing

Reproduced by permission of Radio Rentals Ltd

Tedham's first hand practical experience in producing these photosensitive surfaces was to stand him in good stead in the early 1930s when his research group embarked on the fabrication of an all-electronic camera tube of the iconoscope type.

Meanwhile, C O Browne, a colleague of Tedham, was giving some thought to a 'proposed system for five channel television'. Browne's report[15], dated 20 September 1930, indicates that the Gramophone Company was anxious to become a competitor in the infant but growing television industry for he wrote: 'In the proposed system of television as far as possible known results and data are utilised with a view to producing a workable system in a short time with a reasonable chance of success. It is for this reason, largely, that a number of transmission channels are to be used.' He succeeded in his task and his plan was engineered in the remarkably short time of approximately four months.

Browne's work was aided, not only by the experience which the Gramophone Company had acquired in its venture into talking films and Tedham's researches on photoelectric cells, but also by the enquiries of his colleague, W D Wright, on Kerr cells. Wright's report[16] (dated 26 November 1930) on this subject states: 'The research and development on Kerr cells was undertaken with a view to their use as light valves for projection of television images . . . The two main requirements were maximum light transmission and minimum biasing and modulating potentials.'

Browne presented a paper[17], on his multichannel TV system, to the Institution of Electrical Engineers in January 1931. He referred to the principles which had to be observed in any new scheme of television: 'Either we must realise the practical limitations imposed by the very limited frequency band it is possible to transmit along one channel from the transmitter to the receiver and design apparatus to accommodate only this range of frequencies, or we must disregard this practical limitation and attempt to produce correspondingly better results at the expense of a number of transmission channels.'

Browne adopted the latter alternative and described the advantages which would be obtained. 'The total frequency band necessary to transmit a given picture may, in the case of a multichannel television system, be divided into a number of channels, each of which accommodates a frequency band given by the total frequency range divided by the number of channels. On this account the difficulties of design, not only of the apparatus situated at the transmitter and receiver but also of the transmission line between the two stations, are considerably reduced. Apart from this advantage of the multichannel system, the amount of light available for illuminating the receiver screen is increased in proportion to the number of channels used. Further, the velocity with which the scanning spots travel over the surface of the picture to be transmitted is decreased so that the accuracy necessary for synchronising is reduced as the number of channels is increased.'

Five channels were chosen and each picture was scanned at a rate of 12.5 per second (this figure being adopted because then the 50 Hz mains could be used without the inclusion of awkward gear ratios between the synchronous driving motors and scanning devices at the transmitter and receiver).

The aspect ratio of each picture was three (width) by two (height) and the amount of detail which it was decided to transmit was that corresponding to 15 000 picture points. From these factors the frequency band could be calculated to be 117 000 Hz.

The HMV television apparatus did not represent true television in which reflected light is received from an object and allowed to fall onto a photoelectric cell but was more in keeping with the early systems of television of Jenkins and Baird in which a powerful light source was situated behind the object or scene to be televised. There was an important difference, however: whereas the early workers had used opaque objects Browne employed cinematograph film bearing the subject to be transmitted. Half tones were thus taken into account as in true television. Film was used mainly because it was plentiful and it enabled the conditions existing for any particular transmission to be repeated with comparative accuracy on any subsequent occasion. Browne, like the engineers of the Bell Telephone Laboratories who were responsible for the 1927 demonstration of television between Washington and

New York, was particularly interested in investigating the problems associated with the transmitter and receiver and examining the electrical and optical conditions necessary for the picture channels in order to secure good results. Repeatability was thus an important point to be considered and the employment of film allowed this to be achieved.

The film was scanned, while stationary (using the intermittent motion provided by a Maltese cross shutter), by a revolving lens drum containing thirty eight lenses, and the light passed over five equidistant photoelectric cell apertures which distributed the light simultaneously into five vertical section. Although thirty lenses were used in the lens drum for scanning purposes, actually thirty eight were included— eight of these being operative during picture changing. Each photoelectric cell, and associated amplifier, was required to handle frequencies up to 23 400 Hz.

At the receiver the modulated image signals were further amplified by a bank of five amplifiers and the outputs impressed on a bank of five Kerr cells, complete with crossed Nicol prisms. These cells caused the light flux from a powerful arc lamp to be modulated. Reconstruction of the picture from the five fluctuating pencils of light was accomplished by a revolving drum fitted with polished steel mirrors corresponding to the arrangement of the lenses at the transmitter.

Synchronisation was achieved by electrically coupling a 1200 Hz generator of the phonic wheel type, mounted on the lens drum spindle of the transmitter, to a motor of similar design on the receiver mirror drum. Phase adjustment was made by viewing a predetermined mark on the mirror drum in the light of a neon lamp, which was excited once every revolution of the transmitter lens drum.

Great attention was paid by Browne to a consideration of the various kinds of distortion which could be produced in the received picture and to ways of overcoming or minimising their effects. Effort was also expended in achieving a high standard of workmanship and technique, both optical and electrical, in the apparatus, although the Gramophone Company had no intention, at the time of the Physical and Optical Society Exhibition, of manufacturing television apparatus on a commercial basis. Even so, the reproduction of the images was not without some defects: 'The effect was, however, marred by a series of five wavebands which kept travelling constantly across the picture, cutting it with travelling light and dark areas. The general effect can best be described as a small poorly lit photograph seen through a disc of concentric circles constantly travelling to the centre. Nevertheless, the picture was in great, but poorly lit, detail and if the scanning circles could have been eliminated might have passed for a good film inadequately projected.'[1]

For a public used to 30-line head and shoulders images the HMV Company's display of medium-definition (150-line) television created quite a stir. Newspaper reporters were quick to realise the potentialities of such a system and for some it opened up the near prospect of a new entertainment era, apart from the other almost infinite uses of television. 'For the first time in history a stage or film producer will be able to take his show to the public instead of waiting for the public to come to his show.'[2]

The demonstration of telecinema at the Imperial College of Science and Technology was interesting not only from the technical point of view but also from a personality point of view. In the front row at one of the shows sat John Baird. He was recognised by His Master's Voice demonstrators and later was allowed to examine the apparatus. When asked for his opinion he replied darkly 'Wait a while.'[18]

Subsequently, Baird Television Ltd brought an action against the Gramophone Company and, as mentioned earlier, alleged that 'it had manufactured, exhibited and used at the Imperial College of Science and Technology, Kensington, a certain apparatus in January 1931—which presumably infringed a patent of the plaintiff company. However, in a hearing before Mr Justice Clauson, Chancery Division, on 15 March 1932—by which date Baird Television was out of time in prosecuting the action—the defence advocate stated that the defence was that the apparatus was experimental, of purely scientific interest and of no commercial interest and that no such apparatus had ever been offered for sale. Baird Television did not proceed with the case and costs were awarded against the company[19].

The HMV Company had not been alone in working on multichannel television in 1930. In the same month that they had given a demonstration of their apparatus to audiences at the Physical and Optical Societies Exhibition, Baird Television had shown its zone television system to newspaper reporters.

Todays Cinema referred to the new equipment which had been displayed to the press in the Baird laboratories on 2 January as 'an amazing new development in the Baird television process, which makes it possible to project pictures onto an ordinary full sized cinema screen, televise people and objects illuminated only by arc lighting or daylight instead of an intensive exploring beam and show an unlimited amount of detail in the picture'[20]. The technical editor in a special editorial note wrote: 'It is my firm belief that Baird has at last hit on the very method which will bring television into the cinema. It is a bold statement, but I make it in all seriousness and when I saw yesterday's demonstration I could see beyond the tiny screen shown to a visitor to a new revolution in our industry.'

In pursuing development work in this field Baird Television probably had in mind two important points: first, the restriction of its 30-line television system to head and shoulders images and secondly the enthusiasm engendered by the London Coliseum screen television shows. These had been very successful both in the UK and on the continent, the press had reported favourably on them and there had been suggestions that a new form of entertainment had been founded—telecinema. But the system of lamps was somewhat crude and cumbersome and did not lend itself to higher definition images due to the complexity of the commutator. Also, the use of the 30-line system was completely unsuited for showing outdoor scenes such as cricket matches, processions and, in general, any scene having a number of artists or performers.

Baird's solution was to use three channels in a similar way to that adopted by the HMV Co. But whereas Browne had restricted his equipment to show films only, Baird's apparatus had the advantage that it could be used to televise objects or subjects both indoors and outside. Unfortunately, John Baird never emulated Browne's openness in giving exhaustive technical details of his equipment, and information about it has to be gleaned from newspaper and other reports.

The Times[21] of 5 January 1931 devoted several paragraphs to the new system and stated:

> 'The chief difficulty in broadcasting large images by television is that of the scarcity of available wavelengths. When the communication is by means of telephone lines, however, the difficulty is overcome by using several pairs of lines, each line being, as it were, responsible for a portion of the picture.
>
> 'A demonstration of such zone television was given by the Baird Company when Mr H Strudwick, the England and Surrey cricketer, was "televised" in action as a batsman. His movements and those of the wicket-keeper could be clearly seen.
>
> 'In this latest apparatus the scene to be transmitted is not scanned by a rapidly moving spot of light, but is illuminated by ordinary floodlighting, such as is used in theatres, and ordinary daylight is equally suitable. The picture was shown, made up of three sections, transmitted side by side. The transmitter consisted of a large mirror drum with 30 mirrors, which revolving rapidly, caused a succession of images to be moved over three different apertures admitting light to three photoelectric cells. These cells controlled the light emitted by three neon lamps, which produced the final image on a ground glass screen.'

Baird's three-zone system[22] was later employed to televise the Derby of 1932 but in the meantime the engineers of Bell Telephone Laboratories had been exploring the potentialities of multichannel television. They, too, expounded the *modus operandi* of their system in January 1931[23].

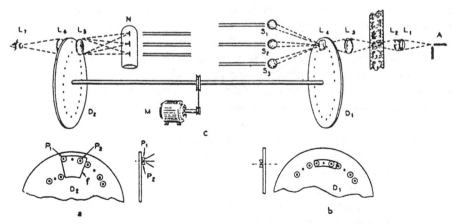

Figure 14.6   *Schematic diagram of Bell Telephone Laboratories three-zone television system: (a) receiving end disc with spiral of holes provided with prisms, (b) sending end disc with circle of holes provided with prisms, (c) general arrangement of apparatus*

Source: *Journal of the Optical Society of America*, **21**, January–June 1931, pp. 8–19

That three quite independent companies should publish descriptions of their multichannel apparatuses in the same month indicates the state of the art at that time and the convergence of ideas on the way forward.

Figure 14.6 depicts the principles of the Bell Telephone Laboratories apparatus. It utilised scanning discs over the apertures of which were placed small prisms so that the beams of light from the successive apertures were diverted to three photo-electric cells. At the receiving end, the prisms enabled the beams of light from three lamps to be deflected to a common direction.

In Figure 14.6 the disc's holes are shown disposed in a spiral path. The angular separations are such that three holes are always included in the frame, f. Over the first hole of a set of three there is positioned a prism, P_1, which refracts the normally incident light beam upwards, the second hole is left clear and the third is covered by prism P_2 so arranged that it refracts the incident light beam in a downwards direction.

This disposition led to apparatus of manageable proportions and permitted the signals generated by each photoelectric cell to be continuous. The number of holes was 108, the aspect ratio of the picture was the same as that of the sound motion picture, and the repetition frequency was 18 pictures per second.

At the receiving end a similar spiral of holes was used and the three sets of signals applied to the three electrodes of a special lamp.

The use of motion picture film permitted a simplification in the design of the transmitting disc since the scanning holes could be set out in circle rather than a spiral by utilising the longitudinal motion of the film. Browne[17] had considered this possibility but had rejected it because the method would have required the use of perfect gearing between the film and the scanning device so that any errors in the positions of the scanning lines did not exceed the width of a line, which was considered to be the limit of resolution of the eye at the receiver. 'It is doubtful,' wrote Browne, 'whether after a limited number of traversals of one length of film through the apparatus the accuracy of the perforations would be good enough to satisfy this conditions. In the case of the intermittent motion, however, which is provided by a Maltese cross, the film is stationary during scanning and, moreover, even with a defective mechanism the film obtains identical framing in the gate at every fourth picture.'

Ives found that his three-channel apparatus yielded results strictly in agreement with the theory underlying its conception and observed that the 13 000 element image was a marked advance over the single channel 4000 element image:

> 'Even so, the experience of running a collection of motion picture film of all types is disappointing, in that the number of subjects rendered adequately by even this number of image elements is small. "Close-ups" and scenes showing a great deal of action are reproduced with considerable satisfaction, but scenes containing a number of full length figures, where the nature of the story is such that the facial expression should be watched are very far from satisfactory. On the whole the general opinion . . . is that an enormously greater number of elements is required for a television image for general news or entertainment purposes.'

This point had been made by Bidwell in 1908 and an idea for the solution of the problem had been outlined in the same year by Campbell Swinton and further elaborated by him in his presidential address to the Roentgen Society in 1911.

In the 1920s several inventors gave some thought to the means—based on the use of cathode rays—which would allow an all-electronic television system to be implemented. Pre-eminent among them were V K Zworykin and P T Farnsworth. Their early schemes were to be much developed and were to be the only viable all-electronic notions, of those put forward in the 1920s, which were utilised in the 1930s. Consideration must now be given to their endeavours.

References

1 ANON.: 'Television nears the cinema. Great advances by HMV. New system demonstrated', *The Bioscope*, 7 January 1931
2 ANON.: 'Great advance in television tests. Nearing practical success. Broadcast of film and plays. When all may see the Derby', *Daily Telegraph*, 6 January 1931
3 ANON.: 'Success of new tests today. Film of everyday life projected. Ride on a bus. Everything clearly recognisable', *Evening Standard*, 6 January 1931
4 BAIRD, J.L.: Statement, 28 February 1931, BBC file T16/42
5 Memorandum of evidence before the Television Committee, 14 June 1934, minute 33/4682
6 Notes of a meeting, held at the GPO, 27 June 1934, of the Television Committee, p. 7, minute 33/4682
7 STURMEY, S.G.: 'The economic development of radio' (Duckworth, London, 1958) p. 166
8 LODGE, J.A.: 'The early days of television in HMV and EMI', *IEE Conf. Publ.*, (271), 1986, pp. 17–20
9 ELLINGTON, F.: 'The photomechanical reproduction of kinematograph films'. Report GB1, 18 September 1930, pp. 1–21, EMI Archives
10 WHITAKER, A.: 'Television development of HMV', 17 February 1938, EMI Archives
11 WHITAKER, A.: Letter to Dr A N Goldsmith, 22 August 1930, EMI Archives
12 GOLDSMITH, A.N.: Letter to A Whitaker, 4 September 1930, EMI Archives
13 History sheet on Baird television, Baird II, BM22, p. 1–3, EMI Archives
14 TEDHAM, W.F.: 'Report on an investigation of the factors affecting emission in photoelectric cells'. Report GD2, 17 September 1930, EMI Archives
15 BROWNE, C.O.: 'Proposed system for five channel television'. Report GC1, 20 September 1930, EMI Archives
16 WRIGHT, W.D.: 'Report on Kerr cells'. Report GC5, 26 November 1930, EMI Archives
17 BROWNE, C.O.: 'Multi-channel television', *J. IEE*, 1932 **70**, pp. 340–349
18 ANON.: 'Rival inventors', *Daily Herald*, 9 January 1931
19 ANON.: 'Baird television v. gramophone', *Financial Times*, 16 March 1931
20 ANON.: Report, *Today's Cinema*, 2 January 1931
21 ANON.: Report, *The Times*, 5 January 1931
22 BAIRD, J.L.: 'Improvememnts in or relating to television and like apparatus'. British patent 360 942, application dates 6 August and 8 December 1930
23 IVES, H.E.: 'A multi-channel television apparatus', *JOSA*, Jan.–June 1931, **21**, pp. 8–19

Chapter 15

Early electronic camera tubes, and the work of Farnsworth, (c. 1920–1935)

An all-electronic television system uses, of course, special types of cathode-ray tubes as camera tubes and as display tubes at the transmitting and receiving ends of the system respectively. The manufacture of these tubes could only be accomplished after much research and development activity had been expended on thermionic and secondary emission of electrons, electron optics, photosensitive and fluorescent materials, and vacuum practice.

In the 1890s several scientists had employed cathode rays to determine the ratio of the charge to the mass of the electron, and this work had led to the construction of the first commercial cathode-ray oscilloscope, known as the Braun tube. The original form of the Braun tube required a steady potential difference of 10 kV to 50 kV to accelerate the electrons—Campbell Swinton had shown a 100 kV (sic) supply in his 1911 scheme. Such a voltage source was not only potentially dangerous but would, in the 1920s, have been bulky and very expensive. High voltages were needed because cold cathodes were utilised in tubes of the Braun type.

In 1904 Professor A Wehnelt, of Erlangen, reported[1] that electron emission from pure metals was enormously increased by coating them with an oxide of one of the alkaline earth metals, preferably barium or strontium, or a combination of these. His first observations had been made with calcium oxide. Pure metals such as tungsten, platinum, tantalum and molybdenum, were good emitters but only when heated to more than c. 2600°F. The process of emission was studied by Richardson (1901), Langmuir (1925) and Dushman (1923) of General Electric and others, and the relationship between electron emission current per square centimetre of heated surface and temperature was well known[2]. Wehnelt's discovery had the potential for large savings in heating power since the oxide coated surfaces had to be raised to just c. 1100°F.

Furthermore, there was the possibility of a longer life because advantage could be taken of the much smaller work function of the coating.

Subsequently, many researchers in universities, and later in industry, were led to study thermionic emission from oxides under various conditions, and the literature on the subject for the first quarter of the 20th century is voluminous.

The first application of the hot cathode Braun oscilloscope was described in April 1919. Sir J J Thompson, of the Cavendish Laboratory, Cambridge, was studying explosion phenomena[3] using the piezoelectric effect to measure pressure and employed the oscilloscope with an a.c. time base. Also in 1919, F Skaupy, of Berlin, applied for a patent[4] for a Braun tube having a hot cathode for the purpose of sending images electrically. In the United States, Western Electric had assembled a powerful team which included W Wilson, A M Nicolson and H J van der Bijl, working under E H Colpitts and H D Arnold, to investigate the most fundamental problems of electron emission[5]. This work resulted in J B Johnson, of the Engineering Department of Western Electric, designing in the early 1920s a cathode-ray tube, the type 224[6], which needed an accelerating voltage of only 300 V to 400 V. He accomplished this by using a Wehnelt cathode, consisting of an oxide coated platinum ribbon, of the same type that had been developed by the company for long-life repeater valves.

With a copious supply of electrons available, the necessary accelerating voltage was established primarily by the energy which the electrons had to possess to bombard the fluorescent material of the screen. The low voltage electron beam gave rise to a focusing defect but Johnson's coworker, van der Bijl, solved[7] this ingeniously by introducing a small amount of inert gas into the tube. The electron beam was diffuse for two reasons, first because it was originally divergent from the cathode, and secondly because of the electrostatic repulsion that exists between electrons. The presence of the gas causes ionisation by collision and hence the production of heavy positive ions. A column of positive ionisation extends down the length of the electron beam with a negative space charge surrounding it. This produces a radial electrostatic field which tends to deflect the paths of the outer electrons of the beam inward towards the centre and so counteract the spreading due to mutual repulsion within the beam.

These techniques made the cathode ray oscilloscope a practical laboratory instrument and Western Electric assigned many of its tubes to experimenters and to manufactures. Figure 15.1 illustrates the electron gun and the two pairs of deflecting plates.

Soft (gas-filled) cathode-ray tubes (c.r.t.s) of the Western type have two disadvantages: first, they suffer at all deflection frequencies from origin distortion which is a

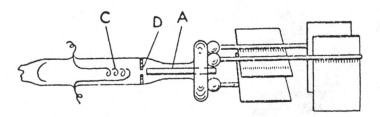

Figure 15.1   Diagram of the electron gun of the Western Electric cathode-ray oscilloscope type 224. The cathode, C, apertured disc, D and anode, A, are indicated. Gas focusing was employed. The two pairs of deflecting plates are also shown

Source: *Journal of the Franklin Institute*, **212**, (6), December 1931, pp. 687–717

lack of sensitivity near the undeflected position of the electron beam; and secondly, and more seriously, there is a loss of focus when the beam is intensity modulated at frequencies above about 0.2 MHz. Now in a television system the highest modulation frequency depends on the line standard, that is the number of lines per picture. For a 405-line picture the highest modulation frequency is c. 2.5 MHz. Consequently, for high-definition television in which an image is reproduced on the screen of a c.r.t., by means of intensity modulation of the electron beam, soft tubes are wholly unsatisfactory. With hard (vacuum) television tubes, suitably designed, defocusing does not occur. Such tubes are usually operated so that the electron beam scans the image raster with a constant velocity.

In the 1920s television workers who wished to utilise c.r.t.s in their receivers either had to design and construct their own hard tubes, or perforce had to purchase the more easily accessible soft tubes. These factors led to several television schemes based on soft tubes in which the tubes were used in a constant intensity, variable velocity mode, rather than in the variable intensity, constant velocity mode of the later hard tube television systems.

The concept of the velocity modulation principle was due to Boris Rosing who in 1911 obtained a patent[8] which embraced the principle. From that date velocity modulation seems to have fallen into oblivion until it was considered by R Thun, of Germany, and R H George, of the US, who applied for patents[9] in 1929. The principle was investigated by George in 1929, and by M von Ardenne,[10] of Germany, in 1931, and in 1932 L H Bedford and O S Puckle of A C Cossor Ltd, a British company, commenced development on a velocity modulation television scheme[11]. This was demonstrated in August 1933. When the Cossor workers began their investigation they were in ignorance of the earlier publications of Rosing, Thun and von Ardenne on the subject and were led to the principle 'in search for a means of circumventing the difficulty of satisfactorily modulating the intensity of the cathode ray'.

The principles of 'variable intensity, constant velocity' and 'constant intensity, variable velocity' scanning[12] are illustrated in Figure 15.2*a,b,c*[11] Figure 15.2*a* shows

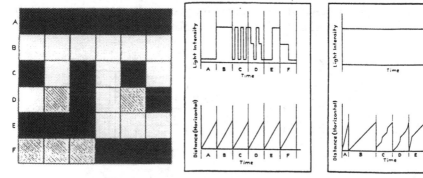

Figure 15.2   (a) Image with three degrees of shade to be transmitted, (b) Constant velocity scanning leads to a variable light intensity, (c) A constant light intensity output requires variable velocity scanning

Source: *Electronics*, November 1935, pp. 30–32

a simple image comprising squares having three degrees of shading. If this is reproduced on the screen of a c.r.t. by scanning the electron beam from left to right, top to bottom, at a constant velocity (Figure 15.2*b*) the relationship between beam intensity and time is as given in Figure 15.2*b*. However, if the beam intensity remains constant, Figure 15.2*c*, the velocity of scanning must vary as shown in Figure 15.2*c*. Dark areas must be scanned rapidly and light areas slowly to give the required effects and the synthesis of the image must be completed in a period of time within the persistence of vision. The essential principle is that the apparent brightness of the image is inversely proportional to the local instantaneous scanning velocity of the electron beam when the persistence of vision condition applies (see also Appendix 5).

One drawback to the use of a c.r.t. receiver based on velocity modulation is the fundamental impossibility of reproducing a velocity modulated image from a unformly scanned object: the scanning at the transmitter must also be velocity modulated. Consequently, as Bedford and Puckle noted, the subject matter has on 'the grounds of scanning light economy to be restricted to film material, at least so long as anything comparable with the ordinary low-voltage oscillograph was employed as a scanner'.

Figure 15.3*a* illustrates Thun's early (1929) apparatus[9]. An oscillatory mirror line scanner reflects a beam of light back and forth across the film: the transmitted light falls onto a photocell and following amplification its output modulates the common speed of both the transmitter and receiver mirror scanners.

In a later (1932) patent Thun disclosed a similar application of the velocity modulation principle using cathode-ray tubes at the transmitter and receiver, Figure 15.3*b*.

Thun's 1929 patent was basically an adaptation for television of a photo-telegraphic apparatus patented by G Sellers[13] of the US in 1908. His apparatus comprised a rotating transparent cylinder upon which a photographic film was placed. Within the cylinder a selenium cell received the light transmitted through the film from an external source. An electric motor, controlled in speed by the output from the cell, caused the cylinder to rotate and the light source selenium cell assembly to move longitudinally with respect to the cylinder. Thus, when the transparent portions of the film were being scanned, and the cell was strongly illuminated, the scanning speed was high and *vice versa*. The corresponding receiving apparatus operated in synchronism with the transmitter and permitted a negative image to be obtained on a sheet of photographic film wrapped around the receiver's cylinder, Figure 15.3*c*.

During the 1920s many proposals for electronic camera tubes were advanced, see Table 15.1. It seems reasonable to suppose that Western Electric's work on the type 224 tube had some influence on the initiation of some of these proposals. Apart from those of L and A Sequin, and of A and W Zeitline, the camera and display tubes incorporated thermionic cathodes; and most of the tubes, whether of the hard or soft type, used apertured diaphragms to give a collimated electron beam. Both electric and magnetic deflecting fields, producing scanned rasters of the Lissajous, spiral or parallel line forms on the photosensitive surfaces, were a feature of the divers suggestions.

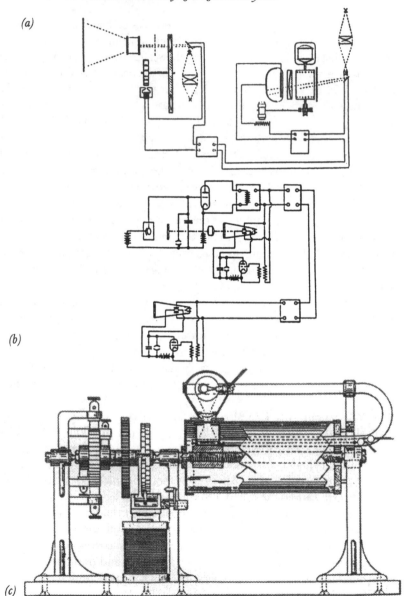

Figure 15.3   (a) Diagram from R Thun's British patent of 1931. An electrically oscillated mirror scanner reflected a beam of light back and forth across a cine film, and the output from the photocell modulated the velocity of the mirror's motion accordingly. The same output controlled the velocity of a similar scanner at the receiver, (b) Thun's 1932 patent described a velocity modulation scheme based on the use of c.r.ts. (c) in 1909 G Sellers patented an apparatus, for electrically transmitting still images, in which the speed of rotation of the transmitter and receiver cylinders was controlled by the output from the selenium cell at the transmitter

Source: *Electronics*, November 1935, pp. 30–32

Table 15.1 Summary of the proposals made during the period 1911–1930 for electronic television cameras using cathode-ray tubes

Date	Name	Type of cathode	Line scanning	Frame scanning	Type of raster	Focusing means	Light sensitive target	Remarks
07.11.1911 (lecture) Jan. 1912 (paper)	A A Campbell Swinton UK	cold	magnetic field	magnetic field	Lissajous	use of aperture to give a collimated beam	mosaic of photosensitive cubes (e.g. of rubidium) in contact with sodium vapour; the target is double sided, i.e. the electrons strike one side and the photons the opposite side	'... it must be distinctly understood that my plan is an idea only, and that the apparatus has never been constructed.'
23.08.1921 (28.06.1922) French patent	E G Schoultz[15] France	thermionic	not applicable	not applicable	spiral by means of coils 20 Hz	not described	the light sensitive plate has a coating of potassium, or thallium sulphate, or selenium; the tube is double sided, i.e. the photons and electrons have opposite directions	the first patent on electron camera tubes, a parabolic mirror, mounted in the tube, reflects an image of the object onto the p.e. surface; a small hole in the mirror allows the electron beam to scan the surface; no record of experimentation
29.12.1923 (20.12.1938) US patent	V K Zworykin[16] US	thermionic	electric field 1 kHz	magnetic field 16 kHz	Lissajous	use of aperture to to give a collimated beam	a double sided p.e. plate; a thin sheet of aluminium foil is coated with aluminium oxide; on this potassium hydride is deposited in the form of small globules; the overall thickness 'need not exede half mill'	a grid collects the photo-emission which may be intensified by the use of argon vapour and ionisation; (cf. Cambell Swinton) claims 1, 13, 14, 15, 16, 17, & 18 refer to storing elements; curiously the patent description makes no mention of these
08.02.1924 (06.09.1924) French patent	L and A Seguin[17] France	cold	electric or magnetic field	electric or magnetic field	substantially parallel lines or spiral, means not described	use of aperture to give a collimated beam	a thin selenium plate	a small mirror mounted in the tube reflects an image of the object onto the selenium plate; a single-sided tube; no evidence of experimentation

Table 15.1 continued

Date	Name	Type of cathode	Line scanning	Frame scanning	Type of raster	Focussing means	Light sensitive target	Remarks
28.02.1924 (28.05.1925) UK patent	G J Blake and H J Spooner[18] UK	thermionic	electric or magnetic field	electric or magnetic field	probably Lissajous—not explicitly stated	use of aperture to give a collimated beam	a plate having a very thin coating of selenium	a single sided tube; no evidence of experimentation
18.03.1924 (17.07.1930)	A and W Zeitline[19] Germany	cold	electric or magnetic field	electric or magnetic field	probably Lissajous	not stated explicitly	a plate having a coating of potassium or other suitable substance	the patent describes a double sided tube; no evidence of experimentation
German patent 10.04.1924 (03.09.1935) US patent	H J McCreary[20]	thermionic	electric field	electric field	Lissajous	use of aperture to give a collimated beam	a double sided photoelectric plate; the insulated plate has a large number of minute pins imbedded in it; embedded on the faces of the plate are conducting grids; the image side is coated with potassium hydride or selenium	the patent mentions that both sides of the plate may be coated with potassium hydride, selenium or other suitable substance; also three camera tubes and appropriate filters may be combined to enable colour television to be produced
09.04.1924 & 16.04.1924 23.04.1924 (papers)	A A Cambell Swinton[21,31] UK	thermionic	electric or magnetic field	electric or magnetic field	Lissajous or parallel lines	use of aperture to give a collimated beam	a mosaic of photosensitive cubes in contact with sodium vapour	basically the same as his 1911 proposal but updated to take account of current (1924) prcatice; no evidence of experimentation
11.07.1924 (28.02.1928) US patent	K C Randall[22] US	thermionic	electric field 1 kHz	magnetic field 16 Hz	Lissajous	use of aperture to give a collimated beam	a double sided photoelectric plate; a thin sheet of aluminium foil is coated with aluminium oxide; on this an alkali metal (e.g. potassium hydroxide) is deposited	the camera tube is identical to that shown in Zworykin's 1923 patent; the object of Randall's patent was to monitor the operation of substations from a central control station

Author / Patent	Electron source	Deflection (1)	Deflection (2)	Scanning	Beam	Photoelectric surface	Remarks
M Dieckmann[23] and R Hell German patent 05.04.1925 (15.09.1927)	not applicable	magnetic field 500 Hz	magnetic field 10 Hz	probably Lissajous	not applicable	cathode coated with potassium or rubidium	an image dissector tube; Hell claimed a tube was constructed but operation was not possible
C A Sabbah[24] US 27.05.1925 (11.12.1928) US patent	thermionic	not applicable	not applicable	spiral, by use of two pairs of deflecting plates 16Hz	use of aperture to give a collimated beam	a double sided photoelectric plate; the transparent plate is coated on one side with a thin semi-transparent film of photoelectric material (caesium, sodium or potassium)	Lissajous raster also mentioned; no evidence of experimentation
V K Zworykin[25] 13.06.1925 (13.11.1928) US patent							An adaptation of Zworykin's 1923 patent[16] for colour television. The system uses Paget colour filters
T W Case[26] US 25.08.1925 (03.02.1931) US patent	thermionic	electric or magnetic field 6 kHz	electric or magnetic field 10 Hz	probably Lissajous	use of aperture to give a collimated beam	the inner surface of the end of the cathode-ray tube is coated with a semitransparent or opaque layer of some conducting photoelectric material (e.g. potassium)	both double-sided and single-sided operation mentioned; no evidence of experimentation
F W Reynolds[27] US 04.12.1926 (04.11.1930) US patent	thermionic	electric or magnetic field 100 Hz	electric or magnetic field	substantially parallel lines or spiral	use of beam focusing anode	a multiple unit photoelectric element comprising a large number of closely compacted relatively fine and long glass or quartz tubes; a film of rubidium or potassium is deposited on the inner walls of the tubes	the photoelectric elements permit the passage of the cathode-ray beam in proportion to the light activation; the discrete space charges of the photo-electrons successively present a varying impedance to the cathode-ray beam; no evidence of experimentation

Table 15.1   continued

Date	Name	Type of cathode	Line scanning	Frame scanning	Type of raster	Focussing means	Light sensitive target	Remarks
07.01.1927 (26.08.1930) US patent	P T Farnsworth[28] US	not applicable	electric field 5 kHz	electric field 10 Hz	substantially parallel lines 500 lines per raster	not stated	the cathode is a fine mesh screen covered with a light sensitive material (sodium, potassium or rubidium)	an image dissector tube
10.07.1928 (British patent void)	K Tihany[29] Hungary	thermionic	electric field 3.2 kHz	electric field 8 Hz	substantially parallel lines	use of apertures and magnetic field to give a uniform collimated beam	the photosensitive layer of the image plate could be of (1) a photoelectric material (e.g. an alkali metal, (2) a photo-conductive substance (e.g. selenium, (3) crystals in which a change of state occurs	the British patent had 127 claims and described many different types of image carrier; however, the complete specification was not accepted and the patent became void, double-sided tubes were described; no evidence of experimentation
22.06.1928 (05.09.1929) UK patent	C E C Roberts[30] UK	not applicable	electric or magnetic field	electric or magnetic field	probably Lissajous	not stated	the cathode may be formed of one or more different photo-electric materials	a single-sided tube of the image dissector type; several arrangements shown; several tubes were constructed but could not be made to operate satisfactorily
24.08.1928 (31.10.1929) UK patent	D N Sharma[51] UK	not stated explicitly	electric or magnetic field	electric or magnetic field	substantially parallel lines or spiral	use of aperture to give a collimated beam	a small opaque region is produced in a transparent or semitransparent 'composite' plate when the region is struck by the electron beam	an image of the object is formed on the surface of the plate and as the opaque spot traverses the image it is reflected spot by spot onto a photoelectric cell; no evidence of experimentation

Date	Inventor						
26.11.1928 (21.04.1936) US patent	P T Farnsworth[32] US	not applicable	magnetic field	magnetic field	substantially uniform parallel lines field		the cathode is a photosensitive film deposited on a metallised surface supported on a glass plate	an image dissector tube having two apertures (one large and one small) to produce a signal current which comprises a low frequency component and a component covering the entire desired frequency range; object: to improve sensitivity and detail
01.06.1928 (02.09.1929) US patent	Associated Telephone and Telegraph Co[33]	thermionic	elecric field	electric field	Lissajous	use of aperture to give a collimated beam	the double sided image plate comprises a large number of short, parallel, insulated conductors; two metallic grids coated with e.g. potassium hydride or selenium make contact with the two faces of the plate	the wires are arranged to be orthogonal to the faces of the plate; the British patent hints at some attempt at experimental practice but apart from this there is no other evidence of experimentation
26.11.1929 (14.07.1931) US patent	W J Hitchcock[34] US							A photoelectric control element is scanned by an electron beam from one of two cathodes
01.05.1930 (17.06.1941) US patent	V K Zworykin[35] US	thermionic	electric field	electric field	parallel lines	use of aperture to give a collimated beam		

Of the tubes listed in Table 15.1 only those of V K Zworykin and of P T Farnsworth were extensively developed. Campbell Swinton[36], Takayanagi[37], Hell and Roberts[37] claimed to have constructed tubes based on their ideas but without achieving any successes. The others presumably took out patents in the expectation that some of their claims might prove to be valuable and lead to the payment of royalties by a manufacturer. Case, McCreary, Reynolds and Sabbah, among others, all instituted actions for infringement of their claims against Zworykin's 1923 patent. These delayed its publication until 1938.

Great credit is due to Zworykin and Farnsworth for the skill and perseverance which they presented in overcoming the many problems encountered in the realisation of their inventions. The two experimenters were of entirely different backgrounds; their working environments were totally dissimilar; and their camera tubes bore almost no affinity in operating principles. Zworykin possessed formal academic qualifications (including a PhD), unlike Farnsworth who never graduated; Zworykin was employed by large electrical engineering companies, and had access to excellent workshops and laboratory facilities, staff and financial resources; whereas Farnsworth was a lone inventor who at first worked in a room of his domestic residence with financial support that was for several years in a somewhat parlous state; and the iconoscope of Zworykin utilised the very important charge storage principle which did not feature in Farnsworth's image dissector tube.

Of all the persons who characterise the history of television, Farnsworth is probably the most remarkable. Although scientifically untrained (in a formal sense), and just 20 years of age when he embarked on his early life's work, Farnsworth succeeded with some of his innovative ideas to such an extent that the mighty Radio Corporation of America, in 1938, paid royalties to his company for the use of some of his patents. The corporation had never previously been in a position which required this action.

Philo Taylor Farnsworth[38,39], the son of Lewis Edwin and Serena Bastian Farnsworth, was born, on 19 August 1906, on a farm near Buckhorn, Utah. He was the eldest of five children. Both the Farnsworth and the Bastian families had been Mormon pioneers and Philo's paternal grandfather had been one of Brigham Young's lieutenants.

From the Spring of 1919 the family lived at Bungalow Ranch (close to Rigby, Idaho), which was shared with his uncle. Philo attended Rigby High School and there he met J Tolman, the school's superintendent and chemistry teacher. Tolman became his mentor and it was to Tolman that the young Farnsworth confided, sometime in March/April 1922, his first thoughts on an all-electronic television system. From an early age Philo had shown an interest in science and invention and at the age of 12 he had won a $25 first prize for inventing a thief proof car lock.

In 1923 the Farnsworth family moved to Provo, Utah, where Philo attended the local high school. He left after one year to obtain employment to support the family following his father's death in January 1924. Joining the Navy, he became a member of the Naval Academy Group, and after an admission test, in which he was ranked second in the nation, gained entry to Annapolis. Farnsworth was sent to San Diego for training and there he learnt that his notions on television would become government property if developed while he was in the Navy. This did not accord with his plans and so, with the assistance of a Navy chaplain and a US Senator from Utah, he gained release from the Navy after only three months of service.

He then, with the help of the Mormon Church, attended some courses in mathematics and electronics at Brigham Young University. His job as a part-time janitor at the university provided much needed finances.

A further move was made in the Spring of 1926 when Philo went to Salt Lake City. Here, he established a radio service shop in partnership with a friend, Clifford Gardner, the brother of his future wife. The business soon failed and once again Farnsworth was seeking work, registering with the employment agency at the University of Utah. Fortunately, the university's employment office referred him to George Everson and Leslie Gorrell, two fund raisers from California who had arrived in Salt Lake City to organise the city's community chest drive[40].

Farnswoth's introduction to Everson was to have a most profound effect on the career of the television pioneer; Everson subsequently became Philo's biographer. Gorrell was a graduate, in mining engineering, of Stamford University.

Soon, Farnsworth was working for the charity drive together with Gardner and Gardner's sister Elma, whom Philo married on 27 May 1926[41]. He quickly showed, to his new employers, his penchant for invention and inspired in them such confidence in his abilities that Everson and Gorrell formed a partnership with him to develop an all-electronic system. He was just 20 years of age at that time.

Under the terms of the agreement, Everson advanced $6000 and he and Gorrell had a quarter share each in the partnership: Farnsworth undertook to develop his ideas for the remaining half[42]. He received $150 per month for living expenses[41], as part of the contract, and with his wife lived at 1339 N New Hampshire Street, Hollywood. The dining room of their four-room furnished apartment became Philo's laboratory and work on television began there in May, 1926.

Everson has mentioned in his biography that Farnsworth's objective was not merely to devise a television system but also to use any patent rights to acquire finance which would allow him to pursue research in other fields. However, he soon encountered difficulties not uncommon to many inventors. The $6000 was spent in three months and the naive belief that a scheme could be perfected in a short time was quickly shattered[43].

Luckily for Farnsworth, Everson was a capable entrepreneur in matters of finance and he set about organising an additional group of financial backers, which was to comprise executives of the Crocker First National Bank in San Francisco. The new investors, W W Crocker, J J Fagan and the vice president of the bank, J B McCrager, acquired 60 % of the interests in Farnsworth's endeavours in return for their acting as trustees[44]. The remaining 40 % was retained by Everson, Farnsworth and Gorrell. (McCrager was chairman of the fund raising campaign for California, Inc which Everson had managed.)

From the end of 1926 Farnsworth carried out his inventive work in a loft at the Crocker Laboratories, 202, Green Street, San Francisco. His allowance was $1000 per month, which included his salary of $200 per month[44] and rent for the loft of $75 per month. The total capital put forward was $25 000.

The immediate problems which the inventor faced, and which delayed the implementation of his concepts, arose from the lack of availability of the apparatus needed
for his work. He was forced, as were most of the early television pioneers of the
period, to construct, with assistance, his own equipment and to acquire the necessary skills in radio technology, the deposition of photoelectric surfaces and the fabrication of cells. Farnsworth was aided in his efforts by his brother-in-law, Clifford
Gardner, who set up a glass blowing laboratory[45] with the help of W Cummings, a
consultant at the University of California, and additionally, by H Metcalf, a radio
engineer and physicist, who had had some experience on cathode-ray tube development and photcells while at the University of Illinois. Gardner's total training for
his task was a high school education. Two radio technicians were employed to work
on the design of the transmission and receiving circuits but they were not trained
engineers.

Notwithstanding his difficulties, Farnsworth was so confident that his plan was
feasible, he filed his first patent[28], Figure 15.4, on 7 January 1927, at the age of 20.
Later that year, on 7 September, at the Green Street laboratory, Farnsworth transmitted his first image. Everson says that a glass slide with a black triangle was used
and that the image was a 'fuzzy, blurry, but wholly recognisable image' of a
triangle[46].

Farnsworth's novel conception of a television camera tube was based on the
implementation of an electrical analogue of the Nipkow disc-photocell picture generator. In this a stationary optical image of the object being televised was scanned
sequentially by the moving aperture, of a mechanical scanner, and the photons
which pass through were collected by a photocell. In the image dissector, as
Farnsworth called his apparatus, a moving electronic image of the object was electrically swept sequentially past a stationary aperture and the electrons which pass
through were collected by an electrode.

Figure 15.5 illustrates Farnsworth's image dissector camera tube. Basically, the
high vacuum cylindrical tube contained a flat photosensitive surface onto which an
image of the object to be televised was focused by a lens. Electrons emitted from the
surface under the influence of, and proportional in number to, the incident light,
were accelerated by an anode which consisted of a flat mesh plate through which the
electrons passed into an equipotential space. Too orthogonal electric fields produced
by two pairs of deflecting plates, to which appropriate sweep voltages were applied,
then caused the electron image to pass systematically before a small aperture in a
target structure to enable the image to be analysed (or dissected) into picture elements. In effect, the aperture scanned the electron image. An anode in the structure
collected the electrons associated with the picture elements. The first image dissector tube was constructed by Metcalf and Cummings[47].

Unknown to Farnsworth when he filed his patent, two German workers, Dr M
Dieckmann and Dipl Ing R Hell had patented[23] a similar camera tube on 5 April
1925. The patent was made public on 15 September 1927, about eight months after
Farnsworth's application, and so the two patents were independent of each other.
The principal difference between the operation of the camera tubes concerned the
method of scanning: Farnsworth used electric fields, Dieckmann and Hell employed

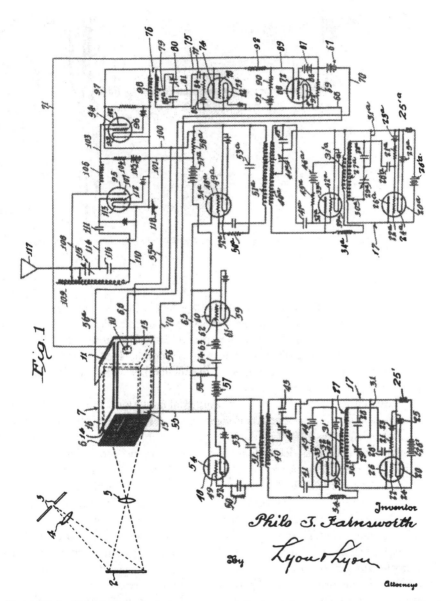

Aug. 26, 1930.   P. T. FARNSWORTH   1,773,980

TELEVISION SYSTEM

Filed Jan. 7, 1927   4 Sheets—Sheet 1

Figure 15.4   *Page 1 of Farnsworth's first patent. Farnsworth was just 20 years of age when he applied for the patent*

Source: US patent 1773 980, 7 January 1927

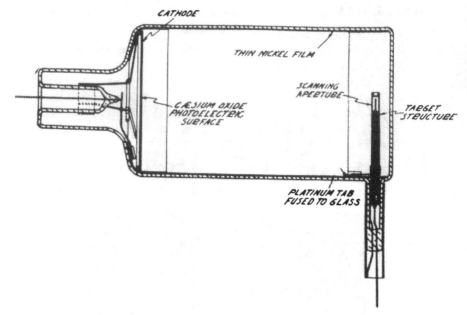

Figure 15.5   Image dissector tube, showing the various elements of the structure. A later version of the tube is illustrated

Source: US patent 1773980, 7 January 1927

magnetic fields. Generally, camera and display tubes have a simpler internal electrode configuration when magnetic deflection is utilised since the coils are necessarily external to the tube. The photoemissive cathode surfaces of both tubes consisted of a coating of potassium, or rubidum, or sodium. Hell claimed in 1951 that he had made a tube but could not get it to function because of the inadequate knowledge (in 1925) of electron optics.

These early image dissectors had two fundamental limitations. First, neither tube had any means of focusing the electron image, which was formed immediately adjacent to the photoemissive surface, in the plane of the aperture. Secondly, the signals obtained from the collecting electrode gave rise to an unsatisfactory signal to noise ratio unless the televised scenes were extremely well illuminated. A typical image dissector tube (of the 1930s) had a length of 20 to 25 cm and a diameter of 75 to 80 mm so that an area of the photocathode of 40 mm × 50 mm could be used. In such a case, for a 400-line image, the aperture in front of the collecting electrode had to be only 0.1 mm square. Thus, only electrons from one picture point can enter the aperture at a given time, all the rest are wasted. Assuming similar optical systems and photoelectric sensitivities it follows that the early image dissectors had an efficiency of the same order as the Nipkow disc-photocell arrangement.

In April 1928 Farnsworth applied for a patent[48] for an improved image dissector. The new device incorporated a long solenoid, in which the tube was placed, so as to establish a uniform longitudinal magnetic field along the axis of the image dissector. It is easy to show that an appropriate magnetic field enables the electronic image formed in the plane of the cathode to be focused in the plane of the aperture. Although the magnetic field configuration does not constitute an electron lens— since it can neither focus a parallel beam of electrons nor produce a magnified or diminished image of an extended source of electrons—it does lead to sharper images. This follows because electrons which leave a given point on the photosensitive surface at different divergence angles are brought to a common focus provided the divergence angle (the angle between the electron's path and the axis of the tube) is small.

The augmented finances available and the transfer to the Green Street laboratories allowed Farnsworth to increase his staff. By 1929 12 persons were employed. Besides Farnsworth, Metcalf and Gardner, there was A B Mann, the managing director, A Brolly, H Lubcke, R Rutherford, H J Lyman, R Varian, P Tait and the secretary Miss D Hagerty.

With improvements being made to his system Farnsworth decided during the summer of 1928 to demonstrate it to the public. Prudently, he gave a private demonstration to R Bishop (acting for the trustees) and the associates before the public showing. In addition, a demonstration[49] was given to representatives of the Pacific Telephone and Telegraph Company on 24 August, approximately one week before the public exhibition on 2 September. The latter was the first public display anywhere in the world of an all-electronic television system.

At the demonstration on 24 August the transmitting apparatus was located in one room and the receiving equipment, placed about 10 m away, in an adjacent room. The demonstration consisted of the transmission of silhouettes and films but not of television in which light reflected from an object was used.

The received images were displayed on a cathode-ray tube screen and were 1.25 × 1.50 inches in size. They had to be viewed preferably in a darkened room because of their faint brightness. In one demonstration, of a person, the chief engineer, G H Senger, of the Pacific Telephone and Telegraph Company reported:

> 'It could be easily identified as a picture of a man but the detail was hardly good enough to identify it as a particular man. This was partly due to the fact that the light intensity on the receiving screen is not uniform. As the film moved about at the transmitting end, the motion was followed exactly at the receiving end.
>
> 'The words "Radio News" in black letters about 0.5″ high on clear glass when moved about in the transmitter could be easily read at the receiving end.'

Senger had previously witnessed other demonstrations of Farnsworth's system and was 'particularly impressed', on 24 August, with the progress which had been made 'in the past three or four months'. 'The demonstration as given was far from what one might term reasonably commercial transmission. However, anyone at all familiar with the difficulties of television could readily overlook certain crudities.' Senger, who communicated his views to E B Craft, the executive vice president of Bell Telephone Laboratories, attended the demonstration following an invitation from

Farnsworth to the Laboratories. Whether Farnsworth hoped to secure some financial backing from the giant AT&T Company is not known, but he did agree to supply the Laboratories with photographs, drawings and explanatory notes covering the operation of his apparatus.

Several aspects of the design and operation of the tube marred the quality of the received image. First, the use of sinusoidal line and frame scanning produced a ghosting effect because signals were generated in both the forward and the return scanning directions and it was not possible for the two sets of signals to be perfectly in phase; secondly, the nonuniform scanning velocity of the electron image gave rise to a black smudge down the centre of the picture (a velocity modulation effect); and, thirdly, the use of potassium hydride for the photosensitive surface was unsatisfactory because of the low sensitivity achieved.

Farnsworth soon realised that the utilisation of saw-tooth scanning instead of sinusoidal scanning would prevent the first two difficulties although he seems to have had some problem designing the circuits. But when H L Lubcke[50], a young electrical engineer from the University of California, was engaged on 15 February 1929, the problem was soon solved. Lubcke's principal project, initially, was to develop the scanning generators: by July 1929 the necessary circuits had been designed.

A study of the photosensitivity of some photoemissive materials was carried out by another new member of staff, R Varian[51]. He later achieved considerable fame for his work on klystrons, but when he was employed by Farnsworth he showed that caesium oxide, with a sensitivity of 20 μA/1, was one of the most sensitive materials that could be used for the photoelectric surface.

Meanwhile, changes were taking place in the financing of the company. Gorrell, who married in late 1926, initially sold a[52] tenth of his ten per cent stake for $5000, and then, later, his total holding: Farnsworth also sold small amounts to sustain his work. By the summer of 1928 $60 000[53] (well beyond the original allocation of $25 000) had been spent on the project but, as Everson has related, at first it was 'not difficult to find men who were willing to take a flier in this glamorous speculation when we needed money to meet expenses'. The principal backers at this time were W W Crocker and R N Bishop.

Nevertheless, there was a pressing need for substantial further financial backing. Bishop, who was the engineer for the group which financed the laboratory, suggested that, because of rising costs, one of the large electrical companies should be invited to share the expenditure of the research and development activities. On 22 May 1928 he wrote to C E Tullar of the patent department of General Electric and tried to interest him in the Farnsworth scheme. This followed a demonstration[54] on 28 March which had been given to Dr L F Fuller and Mr J Cranston of GE. However, 'the demonstration was not satisfactory'. 'The disector[sic] coil system heats up very badly. In fact one tube was spoiled by the potassium becoming hot and distilling off the cathode. It was decided to write a report disclosing our system completely as regards to our tube . . . and submit this to General Electric Laboratory.'

It seems that A G Davies, vice president of G E, was not particularly impressed but thought that Farnsworth had carried out some interesting scientific work with limited means. Davis indicated that General Electric would be pleased to have

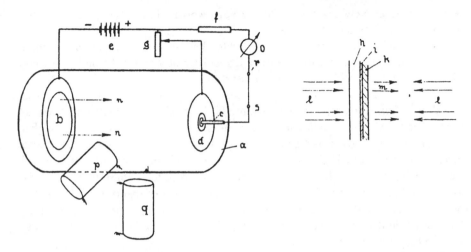

Figure 15.6   Diagram from Dieckmann and Hell's 1925 camera tube patent. The photoelectric surface, b, scanning aperture, c and coils, p and g, for magnetically deflecting the photo electrons, are indicated

Source: German patent 450187, 15 September 1927

Farnsworth on its staff, and pay for his inventions up to the time of the appointment, but that subsequent to this 'whatever he invents while in our employ comes to us under the regular engineering contract'. Such a situation was not congruent with Farnsworth's aspirations.

Farnsworth, like Baird, was anxious to derive profits from the sale of his equipment. Following the public showing of 2 September Farnsworth had enthusiastically stated his view that his sets would soon be available commercially at a price of $100. Like many inventors he was keen to secure a monetary return from his efforts to ensure the financial viability of his ventures. But, as with many inventors, he seriously misjudged the length of time needed to develop a system which would provide real entertainment value for the general public. Even when Baird Television Ltd reached an agreement with the Farnsworth company, in 1935, to utilise the image dissector camera, its sensitivity was so low that it required 94.4 kW of studio lighting, compared with the 24 kW of lighting necessary for the operation of the emitron (iconoscope type) camera, and it could not compete satisfactorily.

Farnsworth's hope was not advanced by the report[55] of the public demonstration which was printed in the San Francisco Chronicle for 3 September 1928. The image, it reported, was 'a queer looking little image in bluish light now, that frequently smudges and blurs'. (The image comprised 8000 picture elements and was transmitted 20 times per second.)

Still, the 1928 demonstration was an important landmark in the progress of all-electronic television. Moreover, it gave Farnsworth some much needed publicity. A report of his work was sent by a wire service and national media attention followed. *Radio News*, in January 1929, described the system, and a photograph of an image produced by it was published in the December 1929 issue of *Radio*. The apparatus was displayed in New York, in 1930, by Farnsworth. He claimed it 'makes commercial television practical at once'.

This publicity was supported by private demonstrations to several famous persons. Among the visitors to the Green Street laboratory were the film stars Mary Pickford, Douglas Fairbanks, his brother and the film producer Joe Schenck[56].

Actually, the necessity to give private and public demonstrations did not arise solely from Farnsworth's desire to seek fame for himself; rather, it stemmed from an urgent need to obtain extra finance for his endeavours. From the start of his work in 1926 to 1929 his project had cost $140 000. It could not be sustained by private investors alone and in 1929 it was felt that some public involvement was required. Accordingly, on 27 March, Farnsworth's backers agreed to establish their enterprise as Television Laboratories, Incorporated with an authorised capital of 20 000 shares of no par stock. 10 000 of these were issued to the original partnership of Everson, Farnsworth and Gorrell and to the trustees[57]. By September 1931 the remainder were selling at $125 each.

1930 was a difficult year for speculative business ventures generally. The effects of the recession were having an adverse influence on financial returns, and since the Farnsworth system still appeared to need much further development before profits could be realised some of the supporters of the Farnsworth Television Laboratories were keen to sell out. The firm of Carroll W Knowles Company had obtained an option to buy the stock of the laboratories and wished to have a technical assessment of its worth.

It sent an invitation—without Farnsworth's knowledge—to RCA to inspect the laboratories and its patents. As a consequence Dr V K Zworykin spent several days from 16 to 18 April 1930 appraising the apparatus[58]. He was astonished at Gardner's success in sealing a disc of optical glass to an image dissector tube. Zworykin had been told by Westinghouse and RCA engineers that such discs could not be fused onto glassware.

He was shown the Farnsworth equipment in operation. The inventor was not averse to the visit taking place because he recognised that if television were to be a commercial success RCA would eventually be involved in its advancement.

At the initial meeting in the laboratory with McCrager, Lippincott and Everson, Zworykin picked up an image dissector and said: 'This is a beautiful instrument. I wish that I might have invented it.'[58]

After the visit Zworykin wrote a report on what he had seen. It was read by Dr E F W Alexanderson, the chief engineer of RCA. He opined[59]:

> 'Farnsworth has evidently done some very clever work but I do not think that television is going to develop along these lines. However, this is a question that can be settled only by competitive experimentation and I think that Farnsworth can do greater service as a competitor to the Radio Corporation group by settling this provided that he has financial backing. If he should be right, the Radio Corporation can afford to pay much more for his patent than we can justify now, whereas, if we buy his patents now it involves a moral obligation to bring this situation to a

conclusion by experimentation at a high rate of expenditure. I feel that we can use our experimental funds to better advantage and that we should not assume such responsibility.'

Several days after the visit A Mann, managing director of the Farnsworth laboratory, sent a letter to Farnsworth requesting a report. This was to include[58]:

'all time spent with Zworykin, both at the laboratory and elsewhere, mentioning subjects discussed generally and the tests and observations made at the laboratory. We want your signature witnessed by a notary. If any data or photographs were furnished him, attach copies to your report; if any apparatus is given him include description of such apparatus.'

Farnsworth complied with the request. It appears that at least one of the backers was concerned about the visit and what Zworykin had seen. There was the possibility that Zworykin could have acquired ideas which would be helpful in his work on an all-electronic system[60]. Subsequently, several Farnsworth and Zworykin patents were to be the subject of a protracted interference action.

Following his visit Zworykin asked Dr E D Wilson of Westinghouse to construct several image dissectors for experimental purposes. Wilson had collaborated with Zworykin on the development of the caesium-magnesium photocell and so the cathodes of the experimental tubes were coated with caesium-magnesium rather than the potassium hydride which Farnsworth used. The new tubes were more sensitive than those of Farnsworth[61]. It may be that the tubes were superior to the crude double-sided experimental camera tubes which were being investigated by Zworykin. The problems associated with the latter were not resolved for many years and it was not utilised for television broadcasting until 1939.

Zworykin's visit, report, and work on the Farnsworth tube led to RCA sending its patent attorney, T Goldsborough, and A F Murray, director of the advanced development division at RCA Victor, to the San Francisco laboratories to assess again the Farnsworth patent position. This interest may possibly have stimulated Zworykin to advance his work on the double-sided tube. On 1 May 1930 he filed a patent (his first since July 1925) for a two-sided tube.

Farnsworth's research and development expenses continued to mount and when, in 1931, the system was still not ready to be marketed more capital was needed. Some of his staff had already been laid off and the financial position of the company was causing concern. Farnsworth did not wish to sell out and so a sponsor in the form of a large manufacturing company was urgently required. News of the company's predicament led to approaches being made to the San Francisco laboratories by Sarnoff of RCA, and by the Philiadelphia Storage Battery Company of Philadelphia.

Sarnoff, himself, visited the Green Street laboratories in May 1931 and was shown around the establishment by Everson[62]. (Farnsworth, at the time, was with McCrager in Philadelphia negotiating a contract with the battery manufacturers.)

Sarnoff had several reasons for making the visit: first, to see the research and development work in progress and to learn what had been accomplished; secondly, to assess the patent position and to consider whether it posed a threat to RCA's activities; and thirdly, to purchase the system if it were felt to be a serious competitor to Zworykin's all-electronic system which was being developed at RCA. Previously, in April and July 1928, Farnsworth had lost two patent interference actions to Zworykin. Perhaps Sarnoff had been appraised by his patent department that claims in Farnsworth's 1927 patent, which had been issued in 1930, could be the basis for a costly interference case against Zworykin's 1923 patent which had not yet been published, and which was being opposed by several patent holders.

Sarnoff seems to have been quite impressed with all he saw and offered $100 000 for the business, including the service of Farnsworth. Of the picture generating and display apparatuses, Sarnoff was more interested in the former than the latter since he felt that Zworykin's kinescope receiver did not infringe any of Farnsworth's patents. However, with a contract about to be concluded with the Philadelphia company, such an offer could not be accepted. Sarnoff, on his departure, stated: 'There's nothing here we'll need.' This was to be a grossly mistaken view.

The Philadelphia Storage Battery Company was a manufacturer of the Philco radio and was RCA's principal rival in the radio industry. Philco, as the company was commonly called, was established in 1892 as the Helios Electric Company. It adopted the name Philadelphia Storage Battery Company in 1906 and when national broadcasting began in c. 1920, it built up a thriving country-wide business manufacturing battery eliminators for domestic radio sets. When valves with indirectly heated cathodes became readily available in the mid-1920s sales of the eliminators slumped. Philco's management decided to enter the domestic radio market but, since RCA was restricting (in 1927) the total number of licences it granted for the use of its patents, the only way Philco could enter the industry was to purchase an existing company which already had a licence. In February 1928, the firm of William J Murdock Co was acquired for $100 000 and its licence transferred to the Philadelphia company[63].

The original terms of the licence demanded a royalty of seven and a half per cent of the price of the radio set, including its cabinet. Later, in May 1929, RCA permitted its licensees to subtract the cost and profits on the cabinet from its rate base and to add $2.00 in lieu of the deduction for the useful value of the cabinet. Three years afterwards RCA again relaxed its terms by reducing the royalty rate to five per cent on receivers for home use. Even this was considered to be excessive by the majority of the licence holders and Philco conceived the idea of separating its manufacturing interests from those which were concerned with engineering and marketing. The latter became the responsibility of a new company, the Philco Radio and Television Company, while the parent firm confined its operations to the production of various components and their assembly onto a chassis. Under this grouping the cost of the cabinet and dials and their assemblage and packaging as a complete radio set was charged to marketing, as were all research and development engineering expenses. Philco argued that these expenses were necessary to improve its products and so should be free from royalty payments.

Naturally, RCA was unhappy with this reasoning and instituted a chancery court action in 1937. In this year the expenses of the new company amounted to $500 000. The difference between the royalty receipts under the pre-1932 and post-1932 arrangements was very substantial, and when in 1939 judgement was made in favour of Philco the difference was $450 000[63].

Philco's subterfuge led to RCA reducing its rates for the entire industry from five per cent to two and a quarter per cent of the net selling price: this had the effect of appeasing the industry. By their astute management Philco's directors had not only entered a field in which RCA was pre-eminent but had succeeded by 1940 in making and selling more radio sets than RCA.

In 1931 Philco was anxious about the vulnerability of its position if commercial television were launched as rapidly as some observers had predicted. The company was keen to be independent of RCA's patent holding in the new field of television. Since Zworykin worked for RCA, and the only alternative scheme of electronic television (in the US) was associated with Farnsworth, Philco decide to offer some support to the inventor. A contract was drawn up in June 1931 by McCragar and Everson[64]. Farnsworth's research expenditures were to be credited as prepaid royalties.

In the same year Philco applied to the FRC for authority to build an experimental television station[65,66].

Farnsworth would have preferred to have stayed in California, but the terms of the offer and the worrying financial position persuaded him to move to the Philco laboratories at the Ontario and C Street plant, Philadelphia. Here, Farnsworth was still in control of his inventions, and Philco received a nonexclusive license to manufacture receivers of his design. A staff of two remained at the Green Street laboratory[67].

The Philco contract was the first substantial recognition by a major radio or electrical company of the Farnsworth system. The arrangement was short lived. During the second year it became apparent that Farnsworth's aim, in establishing a broad patent base by means of advanced research, was not in keeping with the production programme of the Philco plant. $250 000 had been spent in two years and the Philco directors were disappointed that after this time and investment the image dissector tube was not in a perfected state. Consequently, the relationship between the company and the investor terminated[68].

Farnsworth re-established, during the summer of 1934, his own independent research facility in the Chestnut Hill area of Philadelphia and small amounts of the corporation's stock were sold privately to finance his endeavours[69]. Philco, as a licensee of RCA, knew of the efforts of RCA's television research staff in Camden and, being keen to pursue a free course of action in this field, set up its own television laboratory. A F Murray, who had been director of the advanced development division at RCA Victor, was engaged and became the engineer in charge of the Philco Television research department. He had extensive knowledge of the work being undertaken by Zworykin's research group and thus was a considerable asset to Philco's research staff. Furthermore, he attracted several important RCA persons from Camden to Philco[70].

An examination of the expenditures of Farnsworth Television Incorporated for the period 1929 to 1934 highlights the importance given by Farnsworth to a strong patent holding[71].

Period		Development expenses	Patent and legal fees	Total
27 March to 31 December 1929		15 484	1 561	156 804
original acquisition $139 759				
Year ending	31.12.30	16 229	3 104	19 333
	31.12.31	59 287	4 895	64 182
	31.12.32	93 183	6 951	100 134
	31.12.33	54 525	16 994	71 519
	31.12.34	47 184	14 364	61 548

This importance had been accepted by Everson and Gorrell from the outset of their championing of Farnsworth and in 1926 one of their first moves was to retain an able patent attorney, C S Evans, in San Francisco. D K Lippincott, a graduate in radio engineering of the University of California and a former chief engineer of the Magnavox company, handled the radio patent work in Evans' office. His friend and associate was H Metcalf who had assisted Farnsworth. Shortly after the retention of Evans, Lippincott set up his own patent office and the Farnsworth account was handed over to him and Metcalf.

As a consequence of this foresight the Farnsworth interests were able to defend successfully several interference actions involving the Zworykin and Farnsworth patents. When Zworykin commenced his television research in 1923 he did not appreciate the need for great care and attention to detail in the drafting of patents. His 1923 patent was subjected to interference actions and was only sealed 15 years later. He has written: 'I had considerable difficulty in securing patent protection . . . My experience in this regard has impressed me tremendously with the importance of a good patent lawyer in the process of invention.'

On 28 May 1932 an interference case was opened which involved the two inventors. The case centred on Farnsworth's patent of 7 January 1927 and Zworykin's still-pending patent application of 29 December 1923. There was only one point at issue, namely, whether Farnsworth could implement claim 15 of his patent:

> '15 An apparatus for television which comprises means for forming an electrical image and means for scanning each elementary area of the electrical image, and means for producing a train of electrical energy in accordance with the intensity of the electrical area of the electrical image being scanned.'

The Farnsworth interests sought to dissolve the interference on the grounds that Zworykin could not make this claim but this motion was denied by the examiner of interference. After a further preliminary motion was dismissed, the case began. By

16 April 1934 the testimonies of the experts for the two sides had been heard. Briefs were submitted to the US Patent Office, Washington, DC and on 24 April 1934 the final hearing commenced[72].

On 27 June 1935 the examiner ruled that Farnsworth should be awarded priority of invention for his system of television. He based his decision on the following points: Zworykin had no right to make the count:

1   'by virtue of the specific definition of the term "electrical image" given in the Farnsworth patent;
2   because it is not apparent that the device would operate to produce a scanned electrical image unless it has discrete globules capable of producing discrete space charges and the Zworykin application as filed does not disclose such a device;
3   even if the device originally disclosed operates in the manner now alleged by Zworykin because this alleged mode of operation does not produce an electrical image that is scanned to produce the television signals.'[73]

An appeal was made and heard and a decision—again in Farnsworth's favour—given on 6 March 1936. The case had now lasted more than three years and had cost $30 000.

RCA had recourse to three alternatives[74]:

1   to appeal to the civil courts against the verdict of the Board of Appeals;
2   to develop an electronic camera tube without violating the Farnsworth patent;
3   to negotiate with Farnsworth over the payment of royalties for the use of his patent.

Of these possibilities the first was the least attractive. RCA had already lost on two occasions in the presentation of its case, and a civil case might last several years which could seriously delay the introduction of domestic television broadcasting. The second alternative was complicated by the fact that other, conflicting claims existed regarding the priority of certain synchronising techniques employed in television which had been advanced by RCA and Farnsworth. Moreover, the contestants had counterclaims over the priority of the method used in interlaced scanning.

In reviewing these financial and legal issues, RCA came to the conclusion that a settlement with Farnsworth was necessary. Negotiations led to a contract, dated 15 September 1939, according to which RCA was granted a licence relating to the 'television receiving set'. The agreed amount of total payments would be $1 000 000 for a period of ten years[75]. This was the first time ever that the Radio Corporation of America had had to pay royalties for the use of any radio/TV patents.

In addition to this valuable contract, Farnsworth Television and the AT&T Company had signed, on 22 July 1937, a nonexclusive agreement whereby each company had the right to the use of the other company's patents.

Although the settlements with RCA and AT&T were considerable triumphs for Farnsworth and his patent policy they did not connote an unconditional approval of the image dissector principle. The plain fact was that Farnsworth's electronic camera tube had an inherent deficiency, namely its inadequate sensitivity. The inventor referred to this defect in 1930 when, in a patent application, he noted[76]:

> ' . . . that only a relatively small portion of the electrons emitted from the photosensitive surface are used at any given instant, and therefore extremely photosensitive screens and amplifiers are necessary in order to transmit satisfactory pictures.
> 'When it is attempted to amplify the picture currents above a certain level, background noise, "Schottky effect" and other ordinarily negligible factors come in to make the amplified picture currents unsatisfactory and distorted.'

During his development work Farnsworth used photosensitive cathodes four inches in diameter. Thus, for a definition corresponding to 240 lines per picture, the diameter of the scanning aperture had to be less than 4/240 inches (=0.016″). Farnsworth regularly used, in the early 1930s, an aperture of just 0.015″ diameter and, as he noted[77] in 1934, this was 'not adequate for scanning directly from [a] subject, and if [the tube was] used in the form shown [in Figure 15.5] very intense light [was] required for illuminating the subject'. Such a tube could only be utilised to televise images from ciné film.

To overcome this limitation Farnsworth developed several types of electron multipliers. When an electron beam is incident on a suitably sensitised surface secondary emission takes place from that surface. The number of secondary electrons emitted may exceed in number those of the primary beam if the primary electrons have a sufficient velocity. This fact can be employed to amplify an electron flow. Typically, for electron multipliers of the 1930s, the secondary emission ratio, S of the sensitised surface was about 2.6 (i.e. on average 2.6 secondary electrons were released for every incident primary electron) when the primary electrons had been accelerated by a potential difference of c. 60 V. Thus, if the process of electron multiplication was allowed to occur n times, the overall gain of the multiplier would be 2.6^n (or S^n more generally): an eleven-stage multiplier would have a gain of c. 36 500.

Figure 15.7 illustrates the dissector-multiplier tube and camera which Farnsworth used in c. 1934. In an important paper published in that year he observed: 'Addition of the multiplier to the dissector in the form shown [in figure. 15.7] has increased the sensitivity of the latter by 3 or 4 thousand times with no observable increase of noise level. The sensitivity is now adequate for outdoor scanning even when there is no direct sunlight.'[78]

This point was publicly verified when, during the summer of 1934, Farnsworth received an invitation from the Franklin Institute of Philadelphia to demonstrate his television system. For ten days, commencing 25 August 1934, his company Television Laboratories gave daily exhibitions of its equipment to the general public. An improvised studio had been set up on the roof of the Institute and from 10.00 a.m. each day 15 minute programmes were produced and displayed on a 12 by 13 inch screen. The line standard was 220 lines per picture, 30 pictures per second. Various items of entertainment were televised including vaudeville talent

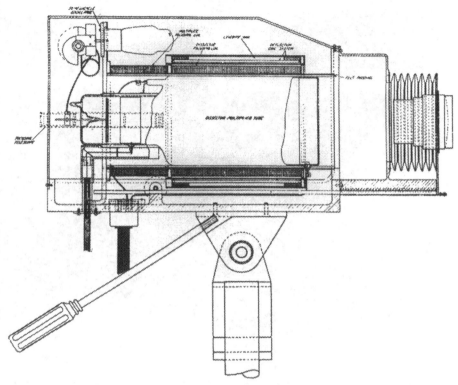

Figure 15.7   Detailed drawing of the image dissector camera. The focusing coil is clearly shown. The camera illustrated dates from the mid-1930s

Source: US patent 1986330, 17 April 1928

and athletic and sporting events such as tennis. When the camera was used out-doors, the displayed images showed 'moving automobiles and passengers leaving their cars. The swaying leaves of nearby trees could easily be distinguished'[79]. Farnsworth claimed that this test showed that it was now possible for football, base-ball and other games and sports to be satisfactorily televised. The demonstration was the first to be given anywhere of all-electronic television.

Farnsworth and his backers were, of course, keen to see a start to sponsored tele-vision. As with all the companies founded to develop the ideas of the lone inventors, Television Laboratories Ltd had had to carry the financial burden of its research and development work: it could not do so indefinitely without a return on its invest-ment. The situation which faced Farnsworth and his associates in the early 1930s was similar to that which was experienced by Baird and his supporters. Both Television Laboratories Ltd and Baird Television Ltd wished to commence televi-sion broadcasting at the earliest opportunity so that revenue could be earned from the sale of patent rights, television receivers and transmitters and programmes. For McCrager, the president of Television Laboratories Ltd, 1933 was an appropriate year for the birth of the new medium. 'There is no logical reason why commercial

Figure 15.8   *Farnsworth image dissector camera. The image tube is on the right inside the deflection coils*

Reproduced by permission of the Science Museum

television should not be launched in the near future,' he advised. He had in mind the organisation of television on a regional basis (using the ultra short wave channels and relay stations spaced 25 to 50 miles apart), if 'business conditions continue to improve'. However, there were many factors and interests to be pondered by the Radio Manufacturers Association and the Federal Communications Commission before commercial television was inaugurated in the US, in July 1941.

In May 1934 in the United Kingdom the Television Committee (sometimes known as the Selsdon Committee after its chairman Lord Selsdon) was established by the Postmaster General, to advise him on the desirability of commencing television broadcasting. The principal contenders for the provision of transmitting and studio equipment in the event that television broadcasting would be initiated were Marconi EMI Television Ltd and Baird Television Ltd. Electronic scanning was the central feature of the former company's research and development endeavours, and the latter company had concentrated its efforts on mechanical scanning. There was a view that high-definition television could only be achieved by all-electronic means and so Baird Television could be at a disadvantage in any trial of the two systems.

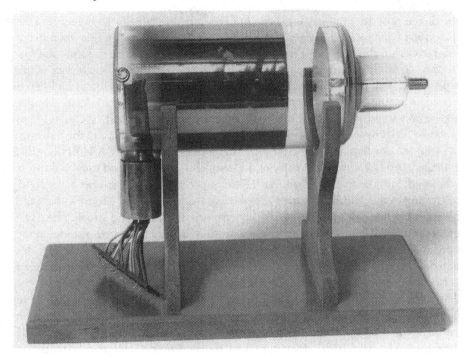

Figure 15.9   The image dissector tube of Figure 15.8

Reproduced by permission of the Science Museum

Following an approach by Baird Television Ltd, Farnsworth travelled to the UK in the autumn of 1934 and demonstrated his equipment to officials of the company. It seems that Baird was favourably impressed, for subsequently a licensing agreement was signed on 19 June 1935 between the two television companies. This allowed Baird Television Ltd to use the Farnsworth camera at the London television station, Alexandra Palace[80].

The Baird company's association with Fernseh provided the means by which a cross-licensing arrangement between the US and German firms was negotiated—Farnsworth having demonstrated his apparatus in Berlin after his visit to Baird Television in London. In Germany the image dissector was known as the Farnsworth Bildsondenrohre. It was soon realised by Fernseh that the image dissector lacked the sensitivity, for studio work, of the iconoscope, which was being employed by their competitor Telefunken, and subsequently Fernseh adopted the image dissector for its intermediate film process. Under the terms of the agreement Farnsworth was permitted to employ the Fernseh teleciné scanner and the Fernseh intermediate film process in his television transmissions from Philadelphia.

On 30 July 1935 Farnsworth gave another demonstration[81] of his system to members of the press and radio editors. Possibly it was felt by his financial backers that a good demonstration to such an influential body would produce beneficial comment and publicity, and spur the onset of commercial television. Possibly, too, the directors of the company had been impressed by the January 1935 report of the Television Committee (UK) which contained the recommendation that 'high-definition television [had] reached such a standard of development as to justify the first steps being taken towards the early establishment of a public television service of this type'. But as other companies, based on the work of a lone inventor, had determined, sometimes a demonstration intended to highlight a particular advance served only to expose the limitations of the system being exhibited.

At the Television Laboratories Ltd, Chestnut Hill, Philadelphia site the audience of pressmen saw film and live television programmes presented by A H Brolly, chief engineer, and G Everson, secretary of the company. Both wire and radio links were employed, and the line standard was 240 lines per picture, 24 pictures per second. According to the *New York Times*[82]: 'The broadcasting of Mickey Mouse's voice and antics, as well as the dancing of the chorus, was from a sound movie film. The orchestra and William Eddy, the announcer, were in Mr Farnsworth's studio in the same building as the receiving apparatus.' On the quality of the images the reporter noted: 'The images, despite their relatively high clarity, could not be considered perfect because of an oscillation which sometimes distorted them.'

The imperfections of the image dissector camera were further exposed, in November–December 1936, following the inauguration of the London television station at Alexandra Palace. During the trial between the systems of Marconi-EMI Television Ltd and Baird Television Ltd the advantages and disadvantages of studio cameras based on the iconoscope and image dissector principles were readily apparent.

An unbiased view of Farnsworth's electron camera was given by G Cock, the BBC's director of television, in an important confidential report dated 9 December 1936. He concluded: 'The cameras are in a somewhat primitive stage of development . . . Electron cameras do not appear likely seriously to compete with emitrons [iconoscope type cameras] at any rate for a considerable time . . . At present their [electron cameras'] operation seems somewhat precarious'.[83]

References

1 WEHNELT, A.: 'Empfindlichkeitssteigerung der Braunschen Rohre durch Benutzung von Kathodenstrahlen geringer Geschwindigkeit', *Phys. Zeit.*, 1905, **6**, (2), pp. 732–733
2 JONES, T.J.: 'Thermionic emission' (Methuen, London, 1936) Chapter 1
3 KEYS, D.: 'A piezo-electric method of measuring explosion pressures', *Phil. Mag.*, October 1921, **42**, pp. 473–484
4 SKAUPY, F.: 'Braun'sche Rohre mit Glühkathode, insbesondere fur die Zwecke der elektrischen bildübertragung'. German patent 349 838, 28 November 1919
5 FAGEN, M.D.(Ed.): 'A history of engineering and science in the Bell System: the early years (1875–1925)' (Bell Telephone Labs., 1975) pp. 967–977
6 JOHNSON, J.B.: 'The cathode ray oscillograph', *J. Franklin Inst.* December 1931, **212**, (6) pp. 687–717
7 Ref.5, p. 322
8 ROSING, B.L.: 'Improvements relating to the transmission of light pictures in electrical telescopic and similar apparatus'. British patent 5486, 4 March 1911
9 THUN, R.: 'Method of and apparatus for transmitting pictures'. British patent 355 319, 15 May 1930

10 VON ARDENNE, M.: 'Practical development of the Thun velocity modulation', *Fernsehen*, October 1932
11 BEDFORD, L.H., and PUCKLE, O.S.: 'A velocity modulation television system', *J. IEE*, 1934, **75**, pp. 63–82
12 THUN, R.: 'Intensity modulation and velocity modulation', *Fernsehen*, July 1931
13 SELLERS, G.: 'Electrical transmission of graphic messages'. US patent 939 338, 18 July 1908
14 SWINTON, A.A.C.: 'Presidential Address', *J. Rontgen Society*, January 1912 **VIII**, (30)
15 SCHOULTZ, E.G.: 'Procédé et appareillage pour la transmission des images mobiles a distance'. French patent 539 613, 23 August 1921
16 ZWORYKIN, V.K.: 'Television system'. US patent 2 141 059, 29 December 1929
17 SEGUIN, L., and SEGUIN, A.: 'Méthode et appareils pour la télévision'. French patent 577 530, 8 February 1924
18 BLAKE, G.J., and SPOONER, H.J.: 'Improvements in or relating to apparatus for television'. British patent 234 882, 28 Fenruary 1924
19 ZEITLINE, A., and ZEITLINE W.: 'Elektrischer Fernseher'. German patent 503 899, 18 March 1924
20 McCREARY, H.J.: 'Television'. US patent 2 013 162, 10 April 1924
21 SWINTON, A.A.C.: 'The possibilities of television, with wire and wireless', *The Wireless World and Radio Review*, 9, 16 and 23 April 1924, pp. 51–56, 82–84, 114–118
22 RANDALL, K.C.: 'Signalling system'. US patent 1 660 886, 11 July 1924
23 DIECKMANN, M., and Hell, R.: 'Lichtelektrische Bildzerlegerrohre für Fernseher'. German patent 450 187, 15 September 1927
24 SABBAH, C.A.: 'Transmission of pictures and views'. US patent 1 694 982, 27 May 1925; and 1 706 185, 27 May 1925
25 ZWORYKIN, V.K.: 'Television system (color)'. US patent 1 691 324, 13 June 1925
26 CASE, T.W.: 'Method and apparatus for transmitting pictures'. US patent 1 790 898, 25 August 1925
27 REYNOLDS, F.W.: 'Electro-optical transmission'. US patent 1 780 364, 4 December 1926
28 FARNSWORTH, P.T.: 'Television system'. US patent 1 773 980, 7 January 1927
29 TIHANY, K.: 'Improvements in television apparatus'. British patent 315 362, 10 July 1929, complete not accepted
30 ROBERTS, C.E.C.: 'Improvements in television and telephotographic apparatus'. British patent no. 318 331, 22 June 1928
31 SHARMA, D.N.: 'Television system incorporating transmitting and receiving apparatus with cathode rays'. British patent 320 993, 24 August 1928
32 FARNSWORTH, P.T.: 'Method and apparatus for television'. US patent 2 037 711, 26 November 1928
33 ASSOCIATED TELEPHONE AND TELEGRAPH CO.: 'Improvements in systems and apparatus for television'. British patent 318 565, 1 June 1928
34 HITCHCOCK, W.J.: 'Improvements in and relating to photo-electric apparatus'. British patent no. 363 103, 25th November 1930 (UK)
35 ZWORYKIN, V.K.: 'Photoelectric mosaic'. US patent 2 246 283, 1 May 1930
36 SWINTON, A.A.C.: 'Television by cathode rays', *Modern Wireless*, June 1928, pp. 595–598
37 Quoted by Abramson, *op.cit.*, pp. 101–102, 121
38 EVERSON, G.: 'The story of television, the life of Philo T. Farnsworth' (Norton, Pennsylvania, 1949)
39 HOFER, S.F.: 'Philo Farnsworth: the quiet contributor to television'. PhD thesis, Bowling Green State University, US, 1977
40 Ref.38, p. 36
41 Ref.38, p. 47
42 Ref.38, p. 45
43 Ref.38, p. 57
44 Ref.38, p. 72
45 Ref.38, p. 76
46 Ref.38, p. 91
47 Ref.38, p. 77
48 Farnsworth, P.T.: 'Electrical discharge apparatus'. US patent 1 986 330, 17 April 1928

49 SENGER, G.H.: Letter to E B Craft, Bell Telephone Laboratories, 4 September 1928, file 552–8, AT&T Archives, Warren, New Jersey
50 Ref.38, p. 115
51 Ref.38, p. 110
52 Ref.38, p. 101
53 Ref.38, p. 114
54 Ref.39, p. 62
55 ANON.: 'S.F.man's invention to revolutionise television', *San Francisco Chronicle*, 3 September 1928
56 Ref.38, p. 122
57 Ref.38, p. 116
58 Ref. 39, p. 69
59 ALEXANDERSON, E.F.W.: Memorandum to H E Dunham, 4 June 1930, GE Archives, Schenectady, 175
60 Ref.39, p. 70
61 Ref.38, pp. 125–127
62 Ref.38, pp. 132–135, 199
63 MACLAURIN, W.R.: 'Invention and innovation in the radio industry' (Macmillan, New York, 1949) pp. 137–139
64 Ref.38, p. 132
65 ANON.: 'Philadelphia to look-in', *New York Times*, 20 December 1931, sect. IX, p. 10:8
66 See also report on 'Farnsworth claims for narrow television band', *Electronics*, September 1931, p. 119
67 Ref.38, p. 133
68 Ref.38, pp.192, 195
69 Ref.38, pp. 136, 198
70 ABRAMSON, A.: *op cit*, p. 196
71 Ref.63, pp.210–211
72 Patent interference 64 027, Philo T. Farnsworth v. Vladimir K. Zworykin, final hearing, Washington, DC, United States Patent Office, 24 April 1934
73 Ref.39, p. 77
74 UDELSON, J.H.: 'The great television race' (University of Alabama Press, 1982) pp. 112–113
75 Ref.39, p. 82
76 FARNSWORTH, P.T.: 'Electron image amplifier'. US patent 2 085 742, filed 14 June 1930, issued 6 July 1937
77 FARNSWORTH, P.T.: 'Television by electron image scanning', *J. Franklin Institute*, October 1934, pp. 411–444
78 Ref.77, p. 433
79 ANON.: 'An electron multiplier', *Electronics*, August 1934, **7**, (8), p. 243
80 Ref.38, p. 147
81 ANON.: 'Television transmitters planned', *Electronics*, September 1935, pp. 28–29
82 ANON.: 'At Television Laboratories Inc., P T Farnsworth demonstrates gains', *New York Times*, 31 July 1935, p. 15:4
83 COCK, G.: 'Report on Baird and Marconi-EMI systems at Alexandra Palace'. TAC paper 33, 9 December 1936

Chapter 16
Zworykin and the kinescope, (1923–1930)

Of all the contributions made by individuals towards the realisation of an all-electronic television system in the United States of America, none were of greater importance than those of Dr V K Zworykin. His invention and development of the iconoscope and development of the cathode-ray tube as a television display tube (which he called a kinescope) were outstanding in conception and execution. The tubes were essential components of RCA's high-definition television system of the 1930s, and the iconoscope was the forerunner of a family of electronic camera tubes manufactured by RCA.

Vladimir Kosma Zworykin[1] was born on 30 July 1889 at Murom, Russia, which is 220 miles east of Moscow. He was the youngest of seven children and became interested in electrical devices when he was only nine years of age. His father operated boats on the Oka River and from this age Zworykin began to spend his summer vacations aboard the craft and eagerly helped his father with the various electrical repairs which had to be undertaken from time to time.

After graduating from high school, Zworykin enrolled at the University of St Petersburg determined to become a physicist. This did not accord with his father's wishes because he felt that Russia's rising new industries offered a richer future in engineering than in physics, and so Zworykin was persuaded to transfer to the Imperial Institute of Technology. Here he remained for six years, from 1906–1912.

Zworykin loved the life of a student, even during the restlessness and repression which characterised the last years of the Czarist government. He was a keen and diligent student, unlike some of his fellow students who 'tried to evade their laboratory work', to the annoyance of Professor Rosing, and as a consequence Rosing became friendly with the young engineering student. Zworykin assisted Rosing with his work on distant vision which was conducted in Rosing's private laboratory, a little cubby hole in the basement of the artillery school, which was situated across the street from the Institute[2]. Their relationship developed into a close friendship and Zworykin found Rosing to be not only 'an exceptional scientist but [also] a highly educated and versatile person'.

Figure 16.1   *Dr Vladimir Kosma Zworykin (1889–1982) holding a kinescope, a cathode-ray display tube*

Reproduced by permission of RCA

During the 'glorious three years' (1910 to 1912) when he aided Rosing[3], much of the apparatus had to be manufactured by them, including the photocells and all the glass vessels needed for their work. Of these times, Zworykin has written[4]:

'At that time the photocells . . . were in their infancy and although potassium photocells were described in the literature, the only way to have them was to make them ourselves. Vacuum technique was very primitive and it required a tremendous amount of time to obtain the vacuum needed. The vacuum pumps which we had were manually operated and quite often we had to raise heavy bottles of mercury up and down for hours at a time in order to produce a vacuum. Electronic amplifying tubes had just been discovered by de Forest and our reconstruction was very inefficient. We were struggling to improve it ourselves. Even the glass for the bulbs was not suitable—it was very brittle and therefore difficult to work with; we had to learn to be glass blowers ourselves. Still, at the end of my association with Professor Rosing, he had a workable system consisting of rotating mirrors and a photocell on the pick-up end, and a cathode ray tube with partial vacuum which reproduced very crude images over the wire across the bench.'

Zworykin's love for physics took him to Paris, following his graduation in 1912 from the Institute, and there he worked on X-rays under the guidance of Paul Langevin.

When the Great War broke out Zworykin returned to Russia and was drafted immediately. He was sent, a few months later, to the Grodno fortress near the Polish frontier, but after a year and a half was transferred to the Officers Radio School[1]. There, he was commissioned and began teaching soldiers how to operate and repair electrical equipment. A further period of time was spent in the Russian Marconi factory, which was constructing radio equipment for the Russian Army and which was situated on the outskirts of St Petersburg. He was attached to the factory as an inspector of radio equipment[4].

In 1917 the Russian Revolution started and Zworykin, fearing that it would disrupt his scientific career, decided to leave his native land. At first he could not obtain the necessary permission and the United States refused him a visa. For months he wandered around Russia to avoid arrest during the chaos of the civil war between the Reds and the Whites. Then, when an Allied expedition landed in Archangel, in September 1918, to aid Russia's northern defences against the Germans, Zworykin decided to make his way to that town. Pleading his case with an American official, Zworykin told him something about the work he could do in advancing television[1]. He was given a visa and arrived in the US on 31 December 1918.

His first requirement, of course, was to seek employment. After several unsuccessful applications, a locomotive company called Baldwin recommended him to the head of research at Westinghouse's factory in East Pittsburg. Zworykin was hired at a salary of $200 per month and a promise that his salary would be increased the following year. When this year came there was an announcement that due to the hard times everyone's salary would be reduced by ten per cent.

His first task was to assemble vacuum tubes on a production line. 'The assembly was cumbersome and took a lot of time. There was a tremendous number of rejects—about 70 %. I did this for about three months and almost went crazy,' Zworykin has written[3]. He then devised an apparatus to make and test simultaneously 100 tube filaments for the WD11 tube which had an oxide coated cathode. During this work an explosion occurred which burnt his right hand and resulted in him spending some time in hospital—and filing a claim for damages. The episode had a fortunate outcome for Zworykin because he was given an opportunity to work on television. It seems Westinghouse felt responsible for the accident and allowed him to engage in his beloved field of endeavour in order to 'humour him'[2].

Zworykin subsequently spent a year developing a high vacuum cathode-ray tube, but in 1920 he left Westinghouse following a dispute over some patents relating to the WD11 valve[2]. He obtained employment with the C&C Development Company in Kansas City but soon received a most attractive proposal from Westinghouse. It seems that Zworykin had previously greatly impressed O S Schairer, the manager of the Westinghouse patent department; he persuaded the new manager of the research laboratory, S M Kintner, to extend an invitation to Zworykin to return to the company. According to Zworykin, he was offered a three-year contract at about three times his former salary[4]. Under the terms of the contract Zworykin would retain the rights to his prior inventions with Westinghouse although the firm would hold an exclusive option to purchase his patents at a later date. Zworykin accepted the offer and recommenced his employment with Westinghouse in February/March 1923.

On arrival he was asked by Kintner to suggest a suitable research project. Now at this time there was much interest, by the leading electrical companies, in picture/phototelegraphy, which is essentially a very slow speed form of 'seeing by electricity'. In addition, N Langer had written a paper[5] on 'A development in the problem of television' which had been published, in the widely read *Wireless World and Radio Review*, in November 1922. He refered to 'the problem of television [as having] been already partly solved by the methods adopted by Professor Korn' and concluded by stating that he had endeavoured to indicate the lines along which a solution to the problem could be found. 'I feel that it may be of interest to other experimenters to have these suggestions put forward. The solution of the problem from a wireless point of view must be looked for as a logical outcome of television by line wires.' Langer's paper was sanguine in outlook and may have influenced some inventors/experimenters to consider the subject of television as one which was ripe for investigation.

Also, M J Martin in his 1921 book[6] 'The electrical transmission of photographs' had included a full description, with a circuit diagram, of Campbell Swinton's 1911 scheme which embodied electronic camera and display tubes. Whether Zworykin was familiar with these works is unknown but his biographer has commented[7]: 'It is almost certain that the Westinghouse patent department purchased a copy of this book [Martin's] for the research library, since they were committed to keeping up with the state of the art in phototelegraphy and television.'

Whatever the stimulus for Zworykin's choice of research project, he was permitted by Kintner to engage in the field which had held his attention about a decade previously. Zworykin worked rapidly and on 8 October 1923 submitted a plan, Figure 16.2, to the patent department for an all-electronic television system. His proposed camera tube was not too dissimilar to that of Campbell Swinton. However, there was one major difference which indicates that Zworykin was not then acquainted with the very important charge storage principle (Appendix 1). Campbell Swinton's photoelectric target comprised a mosaic of rubidium cubes whereas Zworykin's target was based on a uniformly deposited layer of photoelectric material.

Zworykin applied for a patent[8] on his television system on 29 December 1923, but it was not granted until 20 December 1938. The patent which was based on Zworykin's electronic camera tube, a later version of which became known as the 'iconoscope', was the subject of a number of interference actions. A brief history of this saga is as follows.

When the patent (2 141 059) was examined, the patent examiner rejected most of the claims in the application because of prior art in previously issued US and

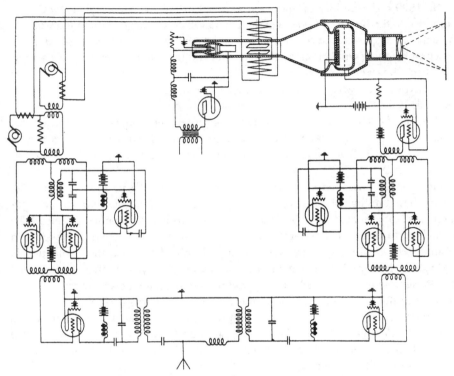

Figure 16.2   Circuit diagram included in Zworykin's 1923 patent 2141059

Reproduced by permission of the Science Museum

German patents. Zworykin's attorney tried to amend the application but to no avail since the examiner found even earlier work which would invalidate the application. He particularly stressed the system devised by Campbell Swinton in 1911 and described in M J Martin's book[6], 'Wireless transmission of photographs' of 1921. The attorney argued that the Campbell Swinton system was never constructed and would not be operative anyway because the photosensitive surface in the transmitter tube was specified to be rubidium which would explode in air. The examiner maintained that this element would not explode in a vacuum and that a rubidium mosaic could work well in a suitable tube. He again rejected the application (November 1926).

While Zworykin was preparing to appeal against this decision, six contenders appeared on the scene with television systems which incorporated one or more of his claims. Two basic interference actions ensued[9]. The first was Zworykin against Schneider against Reynolds against Case against Farnsworth against Sabbah against McCreary on the photochemistry and physics of the image plate and associated circuitry[10]. The seven-way interference underwent a complex series of legal manoeuvres in several separate proceedings in the Patent Office. By late 1928 all the parties with the exception of Zworykin and McCreary were adjudged to have abandoned their actions or to be ineligible for further consideration.

'In 1929 the examiner of interferences ruled in favour of Zworykin. McCreary appealed to the Board of Appeals of the Patent Office and argued that the Zworykin system was inoperative and based upon the scheme of Campbell Swinton, that the electron scanning would melt the image plate and that X-rays would pose a serious problem. However, the board of appeals ruled otherwise. Zworykin's case was again sustained in 1932 when McCreary appealed to the Court of Customs and Patent Appeals.'

The second basic interference action concerned the conversion of an optical image to an electrical image, with associated scanning. Farnsworth had received a patent (1 773 980) in 1930, and interference with the patent of Zworykin was declared in 1932.

'In 1935 the examiner of interference ruled that while Farnsworth had failed to establish that his invention was conceived prior to Zworykin's, Farnsworth was entitled to priority because Zworykin's device did not work in the manner claimed. The examiner further explained that Zworykin's camera tube failed to produce a scanned electrical image. The board of appeal upheld Farnsworth's contention in 1936. Shortly thereafter the patent examiner for the Zworykin application informed the Westinghouse (Zworykin) attorney that most of the patent claims were invalid as a result of the interference action in the Zworykin vs. Farnsworth and other prior art. Westinghouse requested a suspension of the action while a suit was instituted against RCA over H J Round's patent application[11], which had won favourable Patent Office action in other interferences. Zworykin, now working for RCA, was reluctant to testify in the case.'

Finally, in 1938, the District Court for Delaware ruled in favour of Westinghouse and ordered that a patent be granted.

Figure 16.1 illustrates Zworykin's system. Its most salient element was a very thin aluminium oxide film supported by a thin aluminium film on one side and a photosensitive (potassium hydride) coating with a large transverse resistance on the other. The image of the object was projected through a fine wire collector grid, in front of the aluminium oxide film, onto the photosensitised side of the film, while a high velocity electron beam scanned the opposite side. Illuminated portions of the photoelectric layer, which charged up negatively by photoemission to the collector between successive scans, were momentarily shorted to the aluminium coating or signal plate by the scanning beam penetrating to the insulating substrate. This resulted in a signal pulse, proportional to the illumination at the scanned element, in the signal plate and collector circuits. The process depended on bombardment-induced conductivity, a phenomenon investigated at a much later date by Pensak.

There is no evidence that any effort was made in 1923 to reduce the patent to practice, and until c. June 1924 it appears that Zworykin worked on other projects.

In April 1924 Campbell Swinton had three articles[12] published in *Wireless World and Radio Review* on 'The possibilities of television with wire and wireless'. These papers probably stimulated much interest and may have led General Electric in the US, Karolus in Germany and Takayanagi in Japan to commence, in 1924, their investigations of television. Significantly, electronic camera tube patents were applied for in 1924 by L and A Seguin of France, G J Blake and H J Spooner of the

Figure 16.3   First electronic camera tube made by Zworykin in c. 1924

Reproduced by permission of RCA

UK, A and W Zeitline of Germany and H J McCreary, and K C Randall of the US (see Chapter 15).

'Zworykin certainly learned of the Campbell Swinton articles from the Westinghouse patent library, and he was able to use it to persuade Westinghouse to reduce his patent application to practice as soon as possible.'[13] He began his practical work in June 1924. There were many difficulties to be overcome. Nevertheless Zworykin, using a modified Western Electric type 224A cathode-ray tube, worked with enthusiasm and in 1925 felt confident that his system could be favourably demonstrated to Kintner. He was 'very impressed by [its] performance'[4], but to further the work more effort, space and financial resources were needed. It was decided to show the system to H P Davis, the general manager of the company.

In the late summer/early autumn of 1925 Davis, Kintner and Schairer witnessed a demonstration of Zworykin's electronic television scheme. 'A small cross was held in front of the transmitter cathode-ray tube and its image appeared on a screen in the end of the receiver cathode-ray tube. The image of a pencil was also transmitted and received. Although the images were dim, of low contrast and of poor definition, Schairer was 'deeply impressed' but, unfortunately for Zworykin, Davis was not. Zworykin has described the consequence of the test: 'Davis asked me a few questions, mostly how much time I [had] spent building the installation, and departed saying something to Kintner which I did not hear. Later I found out that he had told him to put this "guy" to work on something useful.'

Still, the laboratory experience gained by Zworykin had a positive attribute, it enabled him to prepare another patent—one which was technically superior to his patent of 1923. The new application[14] differed from the 1923 submission in two important respects. First, the target comprised a mosaic of discrete photosensitive elements (globules) and, secondly, both the camera and display cathode ray tubes contained three-colour (Paget) screens. Of these changes the first was of fundamental importance since it would allow the camera tube to function according to the principle of charge storage.

In his original 1923 application Zworykin specified a thin layer of photoelectric material, but such a layer could not retain a charge distribution consonant with the image projected onto it. Essentially, the layer had to be subdivided into an array of many independent photocells so that each cell could retain an electric charge with a magnitude which was a function of the incident light flux. This was an imperative requirement of all camera tubes of the inconoscope type.

Following the submission of the 1923 patent application it seems that Zworykin and the Westinghouse patent department recognised the central worth of such a target and endeavoured to have the wording of the original application changed to accommodate globular targets. They were successful and when the patent was eventually published in December 1938 it mentioned: 'Preferably the photoelectric material is potassium hydride, deposited in such a manner that it is in the form of small globules each separated from its neighbour and insulated therefrom by the aluminium oxide.'

The immediate effect of the 1925 demonstration was the relocation of the television project to Dr F Conrad, who had participated in the engineering of Westinghouse's KDKA transmitter station, and who was highly regarded by Davis. Conrad's approach to his new task was to return to more conventional methods of scanning an object. He devised a system of television known as radiomovies in which 35 mm motion picture film was scanned to provide a source of video signals[15].

By August 1928 the system had been developed to a state where it could be demonstrated to senior members of the company.

Such was the interest and progress of television in the UK, Germany, France and the US at this time that the display of Westinghouse's television scheme attracted the scrutiny of senior executives from RCA, NBC and GE, as well as from Westinghouse. Among the spectators were D Sarnoff, Dr A N Goldsmith and E Bucher of RCA; M H Aylesworth and E B Taylor of NBC; Dr E F W Alexanderson and Dr W R G Baker of GE; and H P Davis, L W Chubb, O S Schairer, S Kintner and Dr F Conrad of Westinghouse. Curiously, Zworykin was not present even though two of his innovations—the caesium photocell and the modulated mercury arc lamp, for both of which he had applied for patents—were utilised.

During the laboratory demonstration on 8 August 60-line images, obtained from film, were transmitted, two miles by land line from the film scanning apparatus, to the KDKA transmitter where they were radiated back to the laboratory. Three frequencies were specified, *viz*: 2.00 MHz, 4.762 MHz and 3.33 MHz for the vision, sound, and synchronising signals, respectively. A 60 lines per picture, 16 pictures per second standard was adopted, and the vision signals were generated by means of a modified 35 mm standard cine film projector and a caesium photocell. At the receiver the mercury arc lamp and scanning disc enabled 'the radiopictures to be thrown upon a ground glass or screen, the first time this [had] been done with television apparatus'.

One year later, on 25 August 1929, KDKA began broadcasting radiomovies on a daily basis[16]. The advantage of Conrad's method stemmed from the use of transmitted light (through the film) rather than reflected light (from an object). As a consequence, the light flux on the cell could be many times larger that that for studio television. Several investigators, including C F Jenkins and D von Mihaly, devised scanning apparatus to take advantage of this fact.

Meanwhile, Zworykin had been working on the recording and reproduction of sound on ciné film. This work resulted in a new recording camera utilising a Kerr cell, which was manufactured, and the loss of two of Zworykin's associates who received handsome offers from a Hollywood film company. Zworykin was proffered a similar proposal but decided to remain in research: he obtained permission to transfer his efforts to the field of facsimile transmission. A notable feature of this project was his development, with Dr E D Wilson, of the gas-filled caesium photocell which was many times more sensitive that the commonly used potassium hydride cell. The cell aided Zworykin in the implementation of a new type of high-speed facsimile machine which enabled reproductions to be made rapidly on special paper without photographic development.

All these endeavours, and the publication of papers, enhanced Zworykin's standing in Westinghouse. He was given more independence in the choice of the problems on which his group worked and, of course, this choice concentrated more and more on television. The first problem to be considered was the design and construction of a hard vacuum cathode-ray picture tube. In this task he was to be much aided by an adventitious factor, and by work which was being undertaken elsewhere.

In 1928 all radio and television research carried out by General Electric, Westinghouse Electric Manufacturing Company and the Radio Corporation of America was on behalf of RCA, which was primarily a sales organisation; hence, Sarnoff's influence on developments was especially strong. Each firm had engineers working independently on the elucidation and engineering of the principles of television.

Of the senior executives in GE, WEM and RCA, Sarnoff was particularly forward looking. Just as he foresaw and expounded in 1915 the need for radio music boxes, so now (in 1928) he outlined his views on the prospects of television. His thoughts were expressed in an article, 'Forging an electric eye to scan the world', published on 18 November 1928 in the *New York Times*[17]. He wrote: 'Within three to five years I believe that we shall be well launched in the dawning age of sight transmission by radio.' He predicted that television broadcasting would be classified as radiomovies (following the policy adopted by WEM), and as radio television, in which vision signals would be generated directly (on the lines being pursued by, for example, Baird).

Sarnoff's use of the expression electric eye was intriguing. In the summer of 1928 he had visited France where it was known that television experimentation with cathode-ray tubes had been conducted by E Belin and Dr F Holweck, by Dr A Dauvillier and by G. Valensi. It appears that while in Paris Sarnoff, who closely followed television developments, saw something relating to television which provoked an interest in electronic television. He knew that Zworykin was a proponent of cathode-ray television and instructed him to visit Europe to determine 'the status of ideas applicable to cathode-ray television there'. The visit was to be, possibly, the most valuable ever to be undertaken by a television pioneer since RCA's television endeavours were to be greatly advanced by the knowledge and hardware which Zworykin acquired and by the appointment of an engineer of some considerable ability.

He sailed for Europe on 17 November 1928 with a schedule to visit England, France and Germany. Of these visits that which Zworykin made to the Laboratoire des Établissements Edouard Belin was of prime importance. Here he met E Belin, Dr F Holweck, G N Ogloblinsky and P E L Chevallier, and was shown one of Belin and Holweck's latest continuously pumped, all-metal cathode-ray display tubes[18].

In all cathode-ray tubes used either for electronic signal generation (camera tubes), or for display purposes (picture tubes) the fineness of the detail which can be determined and reproduced is related to the cross sectional areas of the electron beams incident on the photosensitive target and the fluorescent screen of the two forms of tubes. Clearly, the smaller the cross sectional areas of the scanning spots the greater the resolution, and hence the higher the definition, which is possible. Thus, the focusing of the electron beam in a c.r.t. is of central importance in a high-definition television system.

During the 1920s three methods of focusing were investigated: gas focusing, magnetic field focusing and electric field focussing. Some early work on television was carried out using the sealed-off, low potential, gas-filled Western Electric type 224 cathode-ray tube but, as noted previously, this produced a low intensity spot and it had an inherent limitation which made it fundamentally unsuitable for high-definition television reception.

Historically, the use of magnetic lenses long preceded that of electrostatic lenses for focusing. At the end of the 19th century a concentrating coil was utilised by Wiechert (in his work on the velocity of cathode rays) to form a narrow electron beam. Furthermore, it was Busch's investigations of the action of magnetic fields on electron beams that led him to formulate the general theorem that all axially symmetric magnetic and electric fields possess the property of electron lenses[19].

Since magnetic fields are unimpeded by the glass walls of c.r.t.s, experimentation on the action of magnetic lenses on electron beams is greatly simplified. Coil assemblies may be slipped over the cathode-ray tube with no disturbance of its internal structure, but with electrostatic lenses changes in their configuration can only be undertaken by opening up the tube. In addition, a uniform magnetic field can readily form a real image of an electron-emitting cathode whereas a uniform electric accelerating field does not have the same property. For these reasons magnetic lenses played an important role in the early development of cathode-ray oscilloscopes.

The focusing of an electron beam by an electrostatic field resembles that of a light beam by a lens system, Figure 16.4. There is a mathematical correspondence between the trajectory of a particle in a field of force and the path of a light ray in a refractive medium. (Technically the correspondence is between Fermat's principle of least time in light, and Hamilton's principle of least action in mechanics.) This analogy leads to the subject of electron optics, which differs from the optics of light principally because the deflection of an electron in a field is gradual whereas the deviation of a light ray at a refracting interface is abrupt[20], Figure 16.5.

Among the early investigators of electrostatic focusing there were W Rogowski and W Grosser who applied, in December 1923, for a patent for a c.r.t. based on this type of focusing; A Dauvillier[21] who studied the potential of such a method for

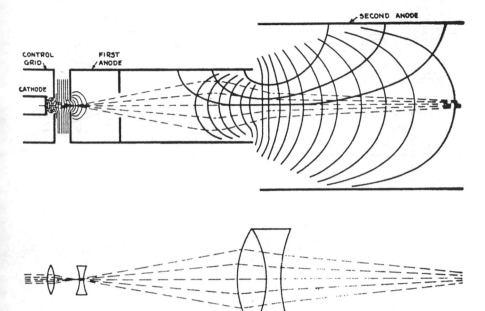

Figure 16.4   *Diagram showing the analogy between the focusing of a beam of electrons by the use of electron lenses, and the focusing of a beam of light by optical lenses*

Source: *Journal of the IEE*, **73**, 1933, p. 443

television purposes and filed a patent in February 1925; Holweck and Chevallier whose experimentation led to a patent application dated 4 March 1927; and R H George of Purdue University.

Holweck was an engineer of some distinction, and was renowned for his work on vacuum pump technology and demountable electron tubes: he was the chief engineer of the Madame Curie Radium Institute. In their work Holweck and Chevallier used a hard vacuum metallic c.r.t. and found that the electron beam could be focused if appropriate potentials (having a definite ratio) were applied to two apertured diaphragms.

Zworykin immediately recognised the superiority of Holweck and Chevallier's c.r.t. (compared to the Western Electric type 224) and the likelihood that their approach to the design of the c.r.t. would have a great impact on the evolution of an all-electronic television system. Some modifications to the construction of the tube would be necessary before a practical version suitable for the domestic market could be manufactured but, nevertheless, so convinced was Zworykin that the Holweck and Chevallier tube held the promise of providing the essential foundation for his cherished ideal that he sought to purchase a tube from the Laboratoire des Établissements Edouard Belin. He was successful and on 24 December 1928 arrived back in the US with the latest Holweck and Chevallier metallic, demountable cathode-ray tube, and a Holweck rotary vacuum pump. The c.r.t. was one of the most

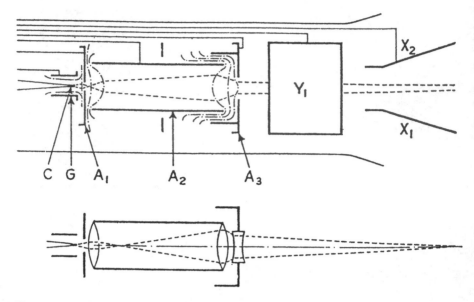

Figure 16.5   The elements of a cathode-ray tube: cathode, C, control grid G, anodes, A_1, A_2 and
A_3, the Y deflection plates, Y_1 and the X deflection plates, X_1 and X_2

Reproduced by permission of Chapman and Hall

valuable cargoes in the television field ever brought to the shores of the US. As one
historian has written[22]: 'Zworykin's trip to Paris changed the course of television
history.' The consequence of his journey would be the engineering of an electron
gun, with electrostatic focusing, which would be at the heart of all camera and
display tubes built and sold by the RCA and some of its licensees. However, there
was a difficulty: it would be essential for Westinghouse to arrive at suitable licensing
and/or patent rights with the Laboratoire des Établissements Edouard Belin.

On his return to New York, Zworykin reported to Kintner, the vice president of
Westinghouse. He suggested that since he (Zworykin) had gone to Europe on behalf
of RCA rather than the Westinghouse Electric Manufacturing Company he should
go to New York and discuss his work with David Sarnoff, vice president and general
manager of RCA[2].

The meeting took place in January 1929. 'Sarnoff quickly grasped the potentialities of my proposals,' Zworykin later wrote[23], 'and gave me every encouragement
from then on to realise my ideas.'

Recalling Sarnoff, 'a brilliant man without much education'[3], Zworykin has
recounted the following anecdote[4]:

> 'Sarnoff saw television as a logical extension of radio broadcasting,
> which was already commercially prosperous. His parting question, after
> my presentation, was to estimate how much such a development would
> cost. I said I hoped, with a few additional engineers and facilities, to be

able to complete the development in about two years and estimated that this additional help would cost about a hundred thousand dollars. This of course was too optimistic a guess; as Sarnoff has stated that RCA had to spend many millions before television became a commercial success.'

Fortunately, Sarnoff was inveighed by Zworykin's enthusiasm and arranged for Westinghouse to give Zworykin additional finances, staff and equipment. Work on the new television system commenced in February 1929 and, as Zworykin required a source of picture signals, a standard simplex ciné projector was adapted for the purpose.

Zworykin's appointment permitted him to concentrate entirely on research on basic electronic processes and devices essential to electronic picture signal generation and picture reproduction. He had an adequate staff of engineers and scientists to assist him and enjoyed also the close cooperation of other research teams in Camden, Harrison and New York which specialised in investigations of television system principles, circuitry, high frequency tube design, signal propagation and studio techniques.

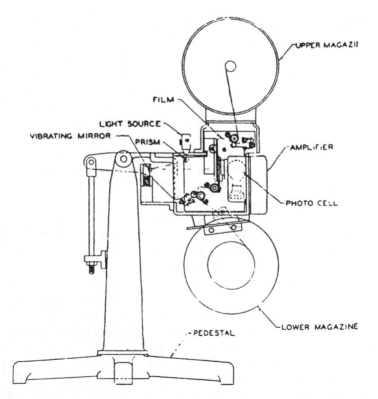

Figure 16.6   Details of the modified standard ciné film projector showing the location of the photocell, light source and vibrating mirror

Source: *Radio Engineering*, **9**, December 1929, pp. 38–41

Among the engineers[24] were H Iams, who had worked with Zworykin on facsimile transmission while at Westinghouse and had been engaged on the design of the television scanning deflection circuits sometime from November 1928, J Batchelor who joined in April 1929, A Vance (May 1929), G N Ogloblinsky (July 1929) and R Ballard (September 1929). W D Wright, who was later employed by the Gramophone Company (UK) and then EMI Ltd, was the optical engineer for the group, and Ballard was the radio receiver designer. Other tasks which were undertaken by the members of the group were the design and construction of the video amplifiers, deflection circuits and high voltage supplies by Vance, the fabrication of the picture tubes by Batchelor and A J Harcher and the general layout of the receivers by Iams. Other members of the team were S Sykpzak and a student named Pepper.

Zworykin lectured on, but did not demonstrate, his kinescope at the Rochester meeting of the Institute of Radio Engineers held on 18 November 1929. The production of moving images, obtained not from an iconoscope but from a film scanner-photocell unit, was described[25]. Westinghouse had been working on film scanners using vibrating mirrors from 1925 and as the principles of mechanical scanning were well understood it was natural that, lacking a suitable camera tube, this system should have been chosen for the tests with the kinescope.

The apparatus delineated by Zworykin employed a modified standard moving picture projector in which the intermittent motion device, the optical system and the light source had been removed. The film to be scanned was caused to move with a constant speed downwards, thereby providing the vertical component of scanning. In the apparatus, Figure 16.7, light from an ordinary six volt car lamp was focused by a condenser lens, L, upon a diaphragm, D, having a small orifice. From there the emergent beam was reflected from a vibrating mirror, M, and focused into a sharply defined spot on the moving film, F. The spot moved across the film horizontally so that, with the vertical motion of the film, the whole surface was explored. After passing through the film the light beam entered a photocell, C, which transformed the variations of optical density in the film into an electric current of variable amplitude.

The mirror vibrated at 480 Hz and consisted of a small steel rod with a vane placed between the U-shaped poles of an electromagnet, each leg of which was provided with a coil excited by a sinusoidal current having a frequency equal to the natural frequency of vibration of the rod. Necessarily, the velocity of the light spot across the film was nonuniform, being about 57 % higher at the centre than that for a spot scanning at a uniform speed, but practical tests showed that the nonlinear distribution of light across the picture was not readily apparent to the eye.

In the receiver the kinescope used an oxide coated cathode mounted within a controlling electrode, having a small hole through which the electron beam passed. The first anode accelerated the electrons to a velocity corresponding to 300–400 V. A further electrode, the second anode, consisting of a metallic coating on the inside of the glass bulb, accelerated the electrons to about one tenth the speed of light. The additional important function of this second anode was to focus, electrostatically, the beam into a sharp spot on the screen, Figure 16.8. The target wall of the bulb was approximately seven inches in diameter and was covered with a fluorescent

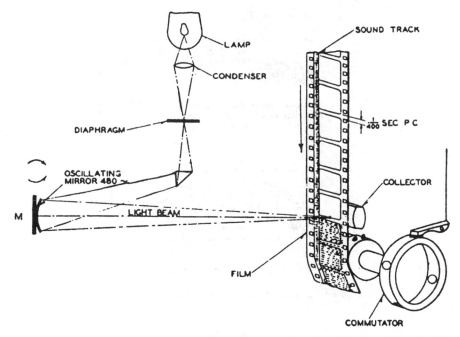

Figure 16.7 *Diagram showing the manner in which the whole surface of the picture is explored by the light reflected from the vibrating mirror*

Source: *Radio Engineering*, **9**, December 1929, pp. 38–41

material, such as Willemite, prepared by a special process to make it slightly con-
ductive.

Zworykin used deflecting plates and coils for frame and line scanning, respectively. These were situated between the first and second anodes so that the deflecting fields acted on the comparatively slow moving electrons: therefore the fields
could be smaller in magnitude than those which would have been needed to deflect
the beam under terminal velocity conditions. Similarly, the picture signals were
applied to the control electrode to achieve the highest sensitivity. The received
image was green.

In the 1933 version of the kinescope the tube diameter was nine inches, the
second anode operated at 4500 V, the vertical and horizontal scanning motions
were achieved by means of coils and the fluorescent screen material was a synthetic
zinc orthosilicate phosphor. This phosphor was chosen because of its luminous
efficiency (between 1.8 and 2.7%), its short time lag, its comparative stability and its
resistance to burning by the electron beam. Again, the image colour was green.

By November 1929 Zworykin's team had constructed six television receivers. One
of these was installed in his home at 7333 Whipple Street in Swissvale, Pennsylvania,
four miles from Westinghouse's KDKA station. Zworykin was permitted to use the
KDKA transmitter three nights a week from 02.00 a.m. to 03.00 a.m., to broadcast

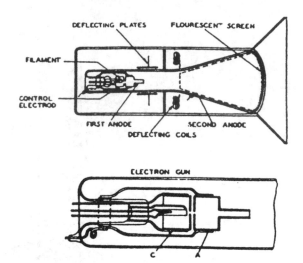

Figure 16.8   Details of Zworykin's 1929 cathode-ray display tube

Source: *Radio Engineering*, **9**, December 1929, pp. 38–41

carrier waves, modulated by vision signals and synchronising pulses derived from his ciné film projector equipment and sent to KDKA by land line.

At other times the station radiated, on a daily basis, television signals generated by Conrad's film projector. The received signals could be displayed using Zworykin's kinescope. With this the images had a size of 3 × 4 inches and were 'surprisingly sharp and distinct' notwithstanding the 60 lines per picture standard utilised[26].

Zworykin's endeavours, in 1929, showed that the nonmechanical method of image synthesis had the following desirable attributes[25]:

1   the image was visible to a large number of people at once, rather than to only one or two as with some mechanical scanners, and no enlarging lenses were required;
2   there were no moving parts and hence no noise;
3   the picture was brilliant enough to be seen in a moderately lit room;
4   the framing of the picture was automatic;
5   with the use of a fluorescent screen, the persistence of the eye's vision was aided;
6   the motor previously used, together with its power amplifier, was redundant, and the power consumed by the kinescope was no greater than that used in ordinary vacuum tubes;
7   'the inertialess electron beam was easily deflected and could be synchronised at speeds far greater than those required for television'.

Most of the advantages had been appreciated by earlier inventors but, lacking the resources and experimental facilities essential for the execution of their plans, they were perforce unable to implement them. Zworykin's main contribution at this time was the practical embodiment of the principles of cathode-ray oscillography to television reception. It was a major step in the march towards the target of high-definition television.

There can be no doubt that the short period (from January 1929 to September 1929) it took Zworykin[27] to engineer his kinescope resulted from his acquisition of the Chevallier c.r.t., and the work of R H George. This work was presented, in March 1929, at a meeting of the American Institution of Electrical Engineers, and was published in July 1929. George's hard vacuum cathode-ray oscillograph featured a special hot cathode electron gun, and made use of a new electrostatic method of focusing the electron beam. His solution to the problem of devising a satisfactory means of producing and focusing a high intensity beam over a wide range of accelerating voltages constituted the chief contribution of his investigation[28].

Chevallier applied, in France, for a patent[29] to protect his c.r.t invention on 25 October 1929: Zworykin's c.r.t. US patent[30] application was dated 16 November 1929. Both patents were for c.r.t.s of the hard vacuum, hot cathode, electrostatically focused type, with grid control of the fluorescent spot intensity and post deflection acceleration.

The prior disclosures by George and Chevallier did not enable Zworykin, and RCA, to establish an undisputed primary claim to the use of grid control, electrostatic focusing and post deflection acceleration. Since these characteristics were essential to the proper functioning of the kinescope it was necessary for RCA to purchase the rights of George's and Chevallier's patents.

In his paper[25] in *Radio Engineering* Zworykin stated that existing cathode-ray tubes could not be employed for picture synthesis since 'although they had scanning arrangements in two dimensions, they did not have means for varying the intensity of the picture'. His statement gives some perspective to the schemes of the Seguin brothers, Schoultz, Blake and Spooner, *et al.*, which incorporated cathode-ray tubes, but which did not use grid control.

Zworykin's lecture in November 1929 did not impress H E Ives of Bell Telephone Laboratories. The account of Zworykin's work was 'chiefly talk' said Ives. 'This method of reception is old in the art and of very little promise. The images are quite small and faint and all the talk about this development promising the display to large audiences is quite wild.'[31]

Ives's opinion was based on some practical experience. During its programme of work on television, Ives's group had undertaken an investigation (in 1926) on the appropriateness of utilising a cathode-ray tube in a television receiver[32]. Images of simple objects, such as a letter A, a bent wire, etc., had been reproduced on a modified cathode-ray tube. The received picture signals had been impressed on an extra grid in the tube and used to control the intensity of the electron beam incident on the fluorescent screen. In one type of tube, the grid was close to the hot filament and the picture signals acted on the beam before it reached the accelerating field. In the second type of tube, the cathode beam passed through two parallel wire gauzes just before striking the screen and the image signals were applied across these two gauzes.

Both forms of tube displayed images of the simple objects but they did not reproduce images of more complex objects, nor did they show half tones in a satisfactory
way. The difficulty was that a change of intensity of the electron beam from its
normal value altered both the focus of the beam and its point of incidence on the
screen. These two factors resulted in a serious distortion of the synthesised image.
The problem associated with gas-filled cathode-ray tubes were to become known to
other experimenters and, as noted in Chapter 15, several workers in the UK and
Germany sought to ameliorate the effects of gas-filled tubes by adopting velocity
modulation[33].

With Ives, the leader of the Bell Telephone Laboratories' television group,
holding such an adverse opinion on cathode-ray tube displays it was perhaps
inevitable that research and development effort on the evolution of an all-electronic
television scheme would not be a major part of the group's activities. Instead, in a
May 1931 memorandum[34], Ives confirmed his support for mechanical scanning by
suggesting three special projects, all based on this mode of reconstituting an image.
These were:

1   the demonstration of reception from an aeroplane;
2   the demonstration of direct scanning of some major outdoor event;
3   the demonstration of reception on film and thence projection in the theatre.

All of these recommendations, if approved, would have been based on existing principles and technology. Ives did not propose an investigation of electronic camera
tubes, and yet when his group eventually embarked on such a programme an important advance was made.

Essentially, by 1931, the analysis and synthesis of objects and images respectively
by mechanical scanning was unsuitable for the reproduction of high-definition
images. This inappropriateness was to be highlighted in 1936 when the BBC's
London television station was established. At first two alternate services were provided by the BBC, utilising equipment manufactured by Marconi-EMI Television
Company Ltd and by Baird Television Ltd. The M-EMI's apparatus was wholly
electronic whereas that of the Baird company employed mechanical scanners for its
spotlight camera, intermediate film studio camera and teleciné units. Within only
two months of the opening of the London station the BBC's studio staff and engineers had decided that the Baird systems, with the exception of the teleciné scanner,
could not compete on equal terms with the emitron cameras of M-EMI. From
February 1937 just that company's installation formed part of the UK's television
service[35].

The disappointment suffered by Baird resulted from his ignoring for too long the
inevitable move towards high-definition television and the use of cathode-ray tubes
in receivers and cameras. In October 1931 he was quoted as saying that he saw 'no
hope for television by means of cathode-ray bulbs', that 'the neon tube [would]
remain as the lamp of the home receiver', and that he was sceptical about the success
of the use of short waves in television 'because they covered a very limited area'[36].

The early successes of Ives's group were similar to those of Baird and his collaborators, although they were achieved quite independently. Both BTL and Baird at

the outset of their pursuits adopted the Nipkow disc (and its variants) as the main scanning element in their schemes. They both adapted their basic system to demonstrate spotlight scanning, large-screen television, daylight television, colour television, two-way television and multichannel television, *inter alia*. Moreover, the procedures which were perforce necessary to accomplish some successes were similar, see Table 16.1.

However, by 1931 it was evident that new approaches to the problem of seeing by electricity were needed if television broadcasting was to become widespread in popular appeal. This appeal would be encouraged when the Derby, tennis at Wimbledon, sporting and athletic events, news reports and so on could be televised to give adequate image detail.

Zworykin, Shoenberg and Farnsworth, among others, were very cognisant of this point and worked strenuously to perfect their companies' electronic camera tubes, the iconoscope, the emitron, and the image dissector respectively.

Tubes of the iconoscope and emitron types utilised the principle of charge storage. A charge storage tube is a hard vacuum cathode-ray tube in which a focused electron beam scans in two dimensions a special photosensitive mosaic target plate onto which an optical image is projected. The electron gun of such a tube is essentially identical with the electron gun of a picture display c.r.t. of the kinescope type. Either two-sided or single-sided target plates may be used but, in the early days of practical camera tube development, two-sided plates were favoured because of the simpler optical and electron gun requirements.

Figure 18.4 illustrates one form of camera tube[37]. Another version consists of an electron gun, a mosaic target plate, and two collecting electrodes, enclosed in a glass vessel from which all gases have been thoroughly removed. The plate comprises a conducting mesh screen coated with a thin dielectric layer to leave the interstices of the mesh open. These interstices are filled with plugs of silver onto which, on one side of the plate, a photoelectric material is deposited. The plate is mounted so that the photosensitised side of the mosaic of plugs faces a collector and the other side faces the electron gun.

In operation the scanning beam, when it strikes an unilluminated element of the mosaic, leads to the production of secondary electrons which, if the emission is saturated, exceed the number of primary electrons causing the secondary emission. Since the element is insulated it acquires an equilibrium positive charge, with respect to the anode, when the rate at which the secondary electrons leave the surface is equal to the rate at which primary electrons strike the surface.

If, now, the mosaic is nonuniformly illuminated (by an optical image of a scene or object) the mosaic elements emit photoelectrons which are collected by the positively charged electrode. The mosaic elements consequently accumulate positive charges. These are returned to the equilibrium condition by the scanning primary electron beam which restores the electrons lost by photoemission. Since each element is capacitatively coupled to the metallic mesh, any variation of charge on the element induces a corresponding variation of charge on the mesh. The rate of change of this charge gives rise to the signal current. Thus, during each scanning period (typically 1/25th of a second), all the mosaic elements follow a cycle of

Table 16.1   Some contributions by Bell Telephone Laboratories and by J. L. Baird to the development of television in the 1920s

Use of/demonstration of	Baird	Bell Laboratories
1 Nipkow disc	from 1923	from 1925
2 Glow discharge lamps	various experiments 1924	various experiments 1925
3 Means to reduce time lag of photocells	from 1925; used derivative of photocell current; patent 270 222, 21 October 1925	from 1925–6; used C–R coupling circuit to enhance high-frequency gain, internal memorandum 27 February 1926
4 Coloured filters on lamps to reduce discomfiture of persons televised	various experiments in 1925; demonstration of infrared radiation 23 November 1926	various experiments in 1925; mentioned in internal memorandum, 26 August 1925
5 Large Nipkow discs	utilised discs up to 8 feet (2.44 m) in diameter at some time during the period 1923–5	advantage of using discs up to ten feet (3.05 m) in diameter mentioned in an internal memorandum, 27 July 1925
6 Spotlight scanning	employed from 1926 to 1936; patent 269 658, 20 January 1926	employed from 1925–6; US patent applied for on 6 April 1927; UK patent 288 238, 18 January 1928
7 Two-way television	patent 309 965, 19 October 1927	UK patent 297 152, 17 June 1927; demonstrated from 9 April 1930 to 31 December 1932
8 Transatlantic television	demonstrated 9 February 1928	suggested as a publicity event in an internal memorandum 4 May 1927
9 Intercalated images to improve resolution	patent 253 957, 1 January 1926; various experiments 1924	mentioned in internal memorandum 9 September 1927
10 Colour television	demonstrated 3 July 1928	demonstrated July 1928
11 Large-screen television	demonstrated 28 July 1930	demonstrated 7 April 1927
12 Daylight television	demonstrated June 1928	need to work on natural-light scanning mentioned in an internal memorandum 4 May 1927; demonstrated 16 July 1928
13 Commutated lamp bank/display	patent 222 604, 26 July 1923. demonstrated 28 July 1930	demonstrated 7 April 1927
14 Use of arcs	demonstrated January 1931	mentioned in internal memorandum 11 February 1929; experiments 1929
15 Zone television	demonstrated 2 January 1931; patent 360 942, 6 August 1930	described in a paper by H E Ives, 'A multichannel television apparatus', J. Opt. Soc. Am., 1931, 21; demonstrated 1930–1

positive charge acquisition by photoemission and equilibrium charge restoration by primary electron beam scanning.

In electronic camera and display tubes for high-definition television it is essential to exclude gases to prevent contamination of the photoelectric and fluorescent screen materials; i.e. only hard vacuum tubes can be employed. Moreover, since in both camera and display c.r.t.s the image resolutions, by the analysing and synthesising electron beams, respectively, are dependent on their cross-sectional areas, it is essential that the electron beams be finely focused.

Zworykin's hard vacuum c.r.t (the kinescope), Figure 16.9, in which the electrons were electrostatically focused was well suited to adaptation as a camera tube. Basically, the fluorescent screen had to be replaced by a mosaic target plate and an additional collecting electrode incorporated into the tube. Initially, however, when Zworykin and Ogloblinsky (who had joined the inventor in July 1929) began their work on the iconoscope (the name given by Zworykin to the camera tube) they used the Holweck/Chevallier demountable, metallic tube which Zworykin had obtained in Paris. This had the advantage, compared to a glass tube, that it could be dismantled, the electrode configuration/mosaic modified and the tube reassembled for testing without the need to construct new tubes.

In his 1923 iconoscope patent Zworykin specified potassium hydride as the photoelectric substance: now, in 1929, following the 1928 discovery by Koller[38], the new, highly sensitive caesium–silver oxide photoelectric surface could be deposited onto his target plate.

According to Zworykin's biographer[39], a crude electronic camera tube was fabricated, and in a report Zworykin stated: 'A rough picture was actually transmitted across the room using cathode-ray tubes for both transmitter and receiver . . .

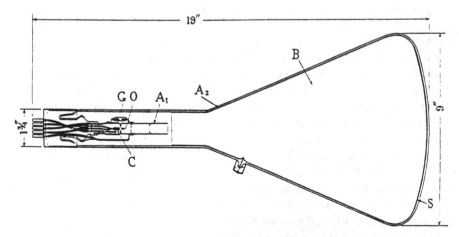

Figure 16.9   *Zworykin's kinescope (1933 version). The cathode, C, control grid, G, aperture, O, first anode, A_1, second anode, A_2 (deposited on the inside of the conical section of the tube) and the screen S are shown*

Source: *Journal of the IEE*, **73**, 1933, p. 445

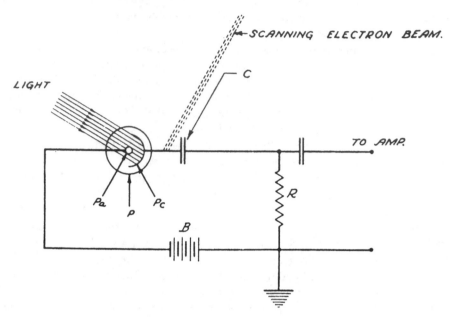

Figure 16.10 In a charge storage camera tube each photosensitive element of the mosaic has
           associated with it a capacitor which accumulates a charge during a time equal to the
           picture periodic time. At the end of each scan the photocell-capacitor combination is
           discharged by the scanning electron beam: the charging process then recommences

Source: *Journal of the IEE*, **73**, 1933, p. 440

The solution of the direct vision problem is thus considerably advanced and may be
the next point of attack in the practical development of television.' The report also
mentioned: ' . . . an interesting principle was found and verified experimentally.' It
seems highly likely that the principle referred to related to charge storage.

    With a single scanning cell, as used in the Nipkow disc-photocell arrangement
and also Farnsworth's image dissector tube, the cell must respond to light changes
in $1/(Nn)$ of a second where N is the frame rate and n is the number of scanned
elements. For N equal to ten frames per second and n equal to 10 000 elements (corresponding to a square image having a 100-line definition), the cell must react to a
change in light flux in less than ten millionths of a second. But if a scanned mosaic
of cells is used each cell has 100 000 millionth of a second in which to react, provided that each cell is associated with a charge storage element. (The principle of
charge storage is considered in more detail in Appendix 1.) Theoretically, the
maximum increase in sensitivity with charge storage is n, although in practice the
1930s iconoscopes only had an efficiency of about five per cent of n.

    This most important principle is implicit in a patent[11], dated 21 May 1926, of
H J Round, of the Marconi Wireless Telegraph Company. The main feature of the
patent was the provision of means whereby the picture elements (now called pixels)
of an image were effectively distributed, for scanning purposes, along the circumference of a circle, Figure 16.11.

Round suggested using a bundle of quartz or glass rods, arrayed so that at one end
of the bundle the faces of the individual ends of the rods formed a plane on which

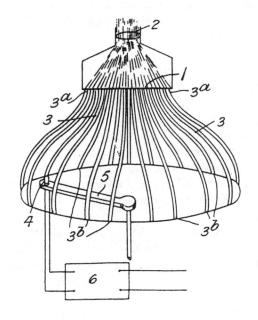

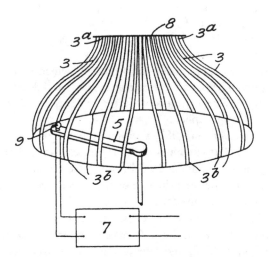

*Figure 16.11 Round's scheme: transmitter (top) and receiver (bottom). 1 is a picture through which
light is projected from a lens, 2, upon the ends, 3a, of a bundle of light tubes, 3.
These are arranged so that their other ends, 3b, lie along the circumference of a circle
around which a photocell, 4, carried by the rotating arm, 5, scans the transmitted
light flux. The receiver is a complement of the transmitter: 9 is a variable light source*

Source: British patent 276 084, 21 May 1926

the image of the object could be presented. The other ends of the rods were to be distributed as shown in Figure 16.11, and scanned by a rotating photoelectric cell. Round realised that this was not a practical arrangement and stated: 'In a modification a photoelectric cell is arranged opposite the "cipher" end of each of the rods in the transmitter, the energy from the cells being utilised to charge condensers, which are discharged at close intervals by means of a brush rotating 8 to 10 times per second.' Claim 5 of the patent refers to this arrangement. Necessarily, each photocell's output is integrated by the capacitor and the principle of charge storage is operative.

Curiously, charge storage is not described in the body of Zworykin's 1923 patent[8] although claims 1, 13, 14, 15, 16, 17 and 18 mention storing elements/devices. Claim 1, for example, refers to 'a cathode-ray scanning device including a plurality of elemental storage devices at said transmitting station . . . ' If Zworykin had appreciated the principle of charge storage in 1923 it is quite extraordinary that he did not state in the patent either its advantages or the means by which it could be implemented.

A possible explanation is that he became aware of the principle following the publication of Round's patent on 22 August 1927, realised that the suggested construction of his (Zworykin's) camera tube embodied the principle and attempted to have his patent application—the claims of which at that time were the subject of several patent interference actions—amended.

Abramson, who has carefully studied many television patent interference actions, has written[40]: 'Zworykin's patent lawyers included "a bank of condenser elements" on September 30, 1931, as claim 32, which ultimately became claim 13 (VKZ patent file, 1923 application). Unless one knows the background to this patent it appears that Zworykin did apply for a patent with "charge storage" in December 1923, when in fact this was added in 1932.'

On 1 May 1930 Zworykin applied for a patent[41] which related to the construction of mosaic targets. The patent was comprehensive and described many methods of fabricating photosensitive charge storage plates. This was Zworykin's first patent to mention charge storage explicitly.

Subsequently, in November 1931, an interference action was initiated between Zworykin and Round. The examiner of interferences awarded priority of invention to Round on 28 June 1935, and on 26 February 1936 the district court affirmed the examiner's verdict. Because the principle of charge storage was of such fundamental importance in the operation of the iconoscope, RCA purchased Round's patent.

When Zworykin's 1923 patent was eventually published on 20 December 1938 the principle was well known. Indeed, Zworykin had stated the principle in a paper presented in London in 1933, and several engineers/inventors—Mathes[42] (1926), Round[11] (1926), Gannett[43] (1927), Tihanyi[44] (1928), Jenkins[45] (1928), Gray[46] (1930) Takayanaki[47] (1930) and Konstantinov[48] (1930)—had mentioned the benefit of charge storage in their reports/patents before the end of the 1920–1930 decade.

Of these charge storage inventions that of Jenkins became well known since it was described in his 1929 book[49] 'Radiomovies radiovision, television': it was also the subject of a patent[45] filed on 16 July 1928. Interestingly, Jenkins actually constructed

a 'sensitive plate transmitter' which embraced the principle. It consisted 'of a sensitive area 2 feet square, studded with $\frac{1}{2}$ inch square lightsensitive cells, arranged in 48
horizontal rows, with 48 cells in each row', (i.e. a total of 2304 cells). 'Across each
cell circuit a small condenser [was] connected, the function of which [was] to
accumulate a charge from the constantly light-excited cell . . . ' The charge on each
cell was sampled every 1/15th of a second by rotary switches.

Jenkins recognised the limitations of his invention: ' . . . it is crude, as now constructed, but it performs, and I believe that because of the latitude possible therein,
this method—i.e. persistence of light activity for transient activity of elementary
areas—will ultimately survive.' He was correct in his forecast. Cameras of the icono-
scope/emitron charge storage type were highly successful in early high-definition
television studio productions; those of the image dissector form which did not
embody charge storage were never widely employed for direct studio broadcasts.

Significantly, Campbell Swinton, who died in 1930 and who was quite a prolific
writer of learned society papers, semipopular articles and letters to newspaper
editors, never claimed that his 1911 distant vision scheme embraced the principle of
charge storage. From this fact and Zworykin's apparent ignorance of the principle
until the late 1920s it may be concluded that the principle was not known explicitly
to Campbell Swinton at the time of his 1911 presidential address.

In January 1930 Zworykin and his group moved to RCA Victor at Camden, New
Jersey. The background history of RCA and its contributions to the evolution of
high-definition television will be considered in the next chapter.

References

1 BINNS, J.J.: 'Vladimir Kosma Zworykin', in 'Those inventive Americans' (The National
   Geographic Society, 1971) pp. 88–195
2 'Interview with Dr Zworykin on 3rd May 1965', transcription of tape, pp. 1–25, Science
   Museum, UK
3 ANON,: 'Vladmir Zworykin. The man who was sure TV would work', *Electronic Design*,
   1 September 1977, **25**, (18), pp. 112–115
4 ZWORYKIN, V.K.: 'Electronic television at Westinghouse and RCA', unpublished
   paper, 13pp., RCA Archives, Princeton
5 LANGER, N.: 'A development in the problem of television', *The Wireless World and Radio
   Review*, 11 November 1922, pp. 197–210
6 MARTIN, M.J.: 'Wireless transmission of photographs' (Pitman, London, 1921)
7 ABRAMSON, A.: 'Zworykin, pioneer of television' (University of Illinois Press, Chicago,
   1995)
8 ZWORYKIN, V.K.: 'Television system'. US patent 2 141 059, 29 December 1923
9 Patent interference 64 026, US Patent Office, Crystal City, Washington, DC
10 ANON.: 'Patent controversies in the history of radio', Smithsonian Institution,
   Washington, DC, US, 5 October 1968
11 ROUND, H.J.: 'Improvements in or relating to picture and the like telegraphy', British
   patent 276 084, 21 May 1926
12 CAMBELL SWINTON, A.A.: 'The possibilities of television with wire and wireless', *The
   Wireless World and Radio Review*, 9 April 1924, pp. 51–56; 16 April 1924, pp. 82–84;
   23 April 1924, pp.114–118
13 Ref.7, p. 49
14 ZWORYKIN, V.K.: 'Improvements in or relating to television systems'. British patent
   255 057, application date 3 July 1926 (UK); US patent 1 691 324, filed 13 July 1925

15 Ref.7, p. 67
16 ANON.: 'Television from film goes on the air', *The New York Times*, 25 August 1929, p. 15:7. Also: 'Cathode-ray television receiver developed', *Sci. Am*, February 1930, p. 147; and ' "Crystal globe" reception', *Wireless World*, 27 November, **25**, p. 595
17 SARNOFF, D.: 'Forging an electric eye to scan the world', *The New York Times*, 18 November 1928, X, 3:1
18 Ref.7, p. 71
19 BUSCH, H.: 'On the operation of the concentration coil in a Braun tube', *Arch. Electrotech.*, 1927, **18**, p. 583
20 MYERS, L.M.: 'Electron optics' (Chapman and Hall, London, 1939)
21 DAUVILLIER, A.: 'Procédé et dispositifs permettant de réaliser la télévision'. French patent 592 162, filed 29 November 1923; and first addition 29 653, filed 14 February 1924
22 Ref. 7, Chapter 6
23 ZWORYKIN, V.K.: 'The early days: some recollections', *Television Quarterly*, November 1962, **1**, (4), pp. 69–72
24 Ref.7, p. 79
25 ZWORYKIN, V.K.: 'Television with cathode-ray tube for receiver', *Radio Engineering*, December 1929, **9**, pp. 38–41
26 Ref.7, p. 83
27 ZWORYKIN, V.K.: 'Television with cathode ray tube for receiver'. 9 September 1929, a report, Westinghouse Electric and Manufacturing Company, archives
28 GEORGE, R.H.: 'A new type of hot cathode oscillograph and its application to the automatic recording of lighting and switching surges', Trans. AIEE, July 1929, pp. 884
29 CHEVALLIER, P.E.L.: 'Kinescope'. US patent 2 021 252, filed 20 October 1930
30 ZWORYKIN, V.K.: 'Vacuum tube'. US patent 2 109 245, filed 16 November 1929
31 IVES, H.E.: Memorandum to H.P. Charlesworth, 16 December 1929, case file 33089, p. 1, AT&T Archives, Warren, New Jersey
32 GRAY, F.: 'The cathode-ray tube as a television receiver'. Memorandum to H.E. Ives, 16 November 1926, case file 33089, 1–2, AT&T Archives, Warren, New Jersey
33 BEDFORD, L.H., and PUCKLE, O.S.: 'A velocity modulation television system', *JIEE*, July–December 1935, **75**, pp. 63–82, discussion pp. 83–92
34 IVES, H.E.: 'Future program for television research and development'. Memorandum for file, 18 May 1931, case file 33089, 1–11, AT&T Archives, Warren, New Jersey
35 BURNS, R.W.: 'British television, the formative years' (Peter Peregrinus Ltd, London, 1986) Chapter 17, pp. 394–422
36 DUNLAP, O.: 'Baird discusses his magic', *The New York Times*, 25 October 1931, section IX, p. 10:1
37 ZWORYKIN, V.K., and MORTON, G.A.: 'Television, the electronics of image transmission' (Wiley, New York, 1940)
38 KOLLER, L.R.: 'S.1 photocathode', *Phys. Rev.*, 1930, **36**, p. 1639
39 Ref.7, p. 82
40 Ref.7, p. 248
41 ZWORYKIN, V.K.: 'Photoelectric mosaic'. US patent 2 246 283, 1 May 1930
42 MATHES, R.C.: Internal memorandum, 8 May 1926, quoted by Abramson (1987)
43 GANNETT, D.K.: Internal memorandum, 19th July 1927, AT&T Archives, Warren, New Jersey
44 TIHÁNYI, K.: 'Improvements in televison apparatus'. British patent 315 362, 10 July 1929
45 JENKINS, C.F.: 'Cell persistence transmitter'. US patent 1 756 291, 16 July 1928
46 GRAY, F.: 'Proposed television transmitters'. Internal report, 20 May 1930, AT&T Archives, Warren, New Jersey
47 TAKAYANAGI, K.: Japanese patent 93 465, 27 December 1930
48 KONSTANTINOV, A.: Russian patent 39 380, 28 December 1930
49 JENKINS, C.F.: 'Radiomovies, radiovision, television (Jenkins Laboratories Inc, Washington, DC, 1929) pp. 85–87

Chapter 17

RCA, Sarnoff and television, (1919–1932)

The Radio Corporation of America, prior to the commencement of the Second World War, played a vital and central role in the advancement of television.

It was formed on 17 October 1919 to combat the growing influence of the Marconi Wireless Telegraph Company in international communications[1]. After the First World War, the United States had become increasingly aware of the importance of communications for military, commercial and public uses, but at that time the only nonmilitary source of such services available to the country was an organisation owned and controlled by foreign (British) interests, namely, the MWT Company of America. The possible complete dependence upon such a business was a matter of much concern, particularly to the US Navy Department, as it was probably the main user and advocate of long distance wireless communications. This apprehension was compounded when the General Electric company (US), which held some fundamental and extremely vital patents covering the Alexanderson alternator[2], made it known that it was negotiating exclusive licensing rights to the alternator with the British company. The company wished to buy 24 Alexanderson alternators, 14 for the American Marconi company and the remainder for the British firm.

When the US Government learned in March 1919 of these negotiations and the consequential prospect of an even tighter control on international wireless communication by the Marconi company, Admiral Bullard, the director of naval communications, informed General Electric's O D Young of the Government's desire for an American dominated organisation. The Government proposed that such a body be set up to provide the essential international communication facilities needed by the United States, and that it should be given the necessary GE licences to use the Alexanderson alternator. GE readily agreed and on 17 October 1919 the Radio Corporation of America was incorporated to operate the appropriate stations and to market equipment, while GE concentrated on manufacturing. A few days later, on 20 November 1919, RCA acquired a controlling interest in the American Marconi company and the Government's objective was realised.

Young, who had been head of GE's legal department, was made chairman of the board of the new company, E J Nally, who had been vice president and general manager of the American Marconi Company, was installed as its president, D Sarnoff was appointed the first commercial manager and Dr E F W Alexanderson became RCA's chief engineer[1].

The immediate task facing the new corporation was the establishment of a wireless point-to-point communication service. This service was to be based on the stations which the Government had taken over from the Marconi company during the war—from April 1917—and which, in February 1920, had been handed over to RCA for commercial use. But, additionally, the expansion of the service demanded the construction of new stations. There were difficulties, however. Many of the most important patents in this field were held by GE, Westinghouse and the AT&T Company. Consequently, the RCA had to take steps to obtain either the ownership or the licence rights to many different patents in order to make its commercial operations a success.

The acquisition of these was not easy for no person or firm controlled even a substantial percentage of them and the Corporation had perforce to institute extensive cross licensing agreements with all the above mentioned companies. RCA and GE arrived at congenial arrangements in 1919, and over the next two years similar contracts were drafted between RCA and Westinghouse, AT&T, United Fruit and the Wireless Speciality Company. By June 1921 RCA had rights to more than 2000 patents in the radio field[1]. Some of these had been suggested by the Government in the best interests of the country—a fact which had a bearing on later antitrust actions.

As a result of these settlements RCA could operate point-to-point radio communications, though not exclusively, and market receivers, and GE and Westinghouse had the exclusive right to manufacture these receivers—60 % for GE and 40 % for Westinghouse: AT&T retained the exclusive right to manufacture, lease and sell transmitters. The ownership of RCA at this time was as follows: GE, 30 %; Westinghouse, 20%; AT&T, 10 %; United Fruit Company, four per cent others, 36 %[3].

The possession of all these patents enabled the Corporation to engage not only in wireless communications but in many other radio activities[4]. Sarnoff, the aggressive, knowledgeable and able commercial manager, had as early as 1916 considered the desirability of a public broadcasting service being set up by the Marconi company. He was at that time its assistant traffic manager and, in an internal memorandum to Nally, he outlined his suggestion[5]:

'I have in mind a plan of development which would make radio a household utility in the same sense as a piano or phonograph. The idea is to bring music into the home by wireless . . . Should the plan materialise, it would seem reasonable to expect sales of 1 000 000 "radio music boxes" within a period of three years. Roughly estimating the selling price at $75 per set, $75 000 000 can be expected.' The Marconi company did not heed this prophetic forecast. Now, in 1920, the position was different. The corporation, spurred on by Sarnoff, seized the opportunity which neither GE nor Westinghouse had grasped and commenced the marketing of a small radio receiving set. 'It was produced and sold amazingly well at a rather high unit price. It [was] considered to be one of the factors that started the broadcasting ball rolling.'

At the same time RCA began strengthening the network of radio stations. By the end of 1923 two stations had been built and opened—one in New York and the other in Washington[6]—and another station followed in 1925: 'Broadcasting flourished and competition was soon intense'. At this juncture RCA, GE, Westinghouse and others ran into a stone wall in the shape of AT&T for, despite the earlier cross-licensing agreements, the telephone company chose not to provide a wire-line distribution (network) service to radio stations which were competing with AT&T stations[4]. However, eventually, on 1 July 1926, the telephone firm withdrew from broadcasting and gave up its rights to manufacture radio sets, RCA acquired the key AT&T station WEAF for $1 000 000 and, together with its other stations as a base, formed the National Broadcasting Company. In return RCA agreed to use the AT&T wire-line network and not to compete with the company for telephone business. RCA, GE and Westinghouse owned 50 %, 30 % and 20 % respectively of the NBC. The first full year of operation, 1927, saw a total of 48 stations, nationwide, achieve a business of nearly $4 000 000. This figure increased to almost $10 000 000 in 1928[4].

The formation of RCA occurred at a most propitious time—the birth of domestic radio broadcasting. With sales booming the corporation could afford to establish a research and development section. Indeed, such a section was vital for the well-being of the corporation. Its products had to be tested and evaluated and new ones had to be developed. Much work had to be carried out, for radio broadcasting was in its infancy and television was just becoming a reality. A technical and test department was therefore set up in 1924 with Dr A N Goldsmith as its chief engineer. It was located at Van Cortland Park and was based on a staff of about 70 engineers, technicians, carpenters, administrative and service personnel.

RCA was blessed in having a far-seeing vice president in the person of David Sarnoff. In a report to the RCA board of directors, dated 5 April 1923[7], he wrote: 'I believe that television, which is the technical name for seeing as well as hearing by radio, will come to pass in due course . . . It may well be that every broadcast receiver for home use in the future will be equipped with a television adjunct . . . which . . . will make it possible for those at home to see as well as hear what is going on at the broadcasting station.' His optimistic outlook for the eventual creation and growth of a television industry never wavered, and his encouragement and financial support of Zworykin's ideas was vital in ensuring the pre-eminence which RCA achieved in the 1930s in the US.

Meanwhile, problems arose between RCA and its manufacturing associates, GE and Westinghouse, concerning product standardisation, production scheduling and control and competitive pressures on profitability. RCA had no effective management control over the costs of the products of its two autonomous manufacturers and the additional mark up which it imposed for the distribution process created 'significantly increasing profit problems on merchandising only'.

Consequently, in February 1929, RCA acquired for $154 million the assets—including a large manufacturing facility in Camden, NJ—of the Victor Talking

Machine Co., and by the autumn of the same year agreement[8] in principle had been reached among GE, Westinghouse and RCA that it would become a 'unified, highly self-sufficient organisation'. As part of this understanding GE and Westinghouse transferred some of their radio facilities and staff to RCA.

The corporation described its activities to its stockholders in its 1930 annual report[9] as follows: 'For convenience in administration of a business of such diversified character, Radio Corporation of America has become largely a holding company . . . In the radio sales and manufacturing fields [the] corporation is represented by RCA Victor Company Inc, and RCA Radiotron Company Inc; in the field of wireless telegraphic communication by RCA Communications Inc, and Radiomarine Corporation of America; in broadcasting by the National Broadcasting Company Inc; and in sound recording and reproduction for talking motion pictures by RCA Photophone Inc; RCA Institute Inc. trains students for radio work; E T Cunningham Inc distributes radio tubes; and the Radio Real Estate Corporation of America has charge of real estate holdings.'

The corporation was staffed, in 1930, by 22 000 employees; it had a gross income of $137 000 million and earned $5.5 million. At the end of 1930 it possessed total assets worth $169 million and total liabilities of $40 000 000[4].

On 14 April 1929, Dr Goldsmith reported on the progress which the Van Cortland Park group was making on mechanically scanned television systems and intimated that a television receiver would be available for purchase at an early date. The newspaper report[10] of his statement so disturbed General J G Harbord, president of RCA, that he admonished Dr Goldsmith and ordered[11] him to clear his publicity releases with RCA before issuing them to the press. It would seem that this course of action stemmed from RCA's policy of not interfering with the profitable radio broadcasting operation of NBC.

Nonetheless, the future prospects for a television industry were appealing in their potential scope. The radio trade, prior to the depression, had experienced a tremendous growth, and more was planned. In 1929 it was anticipated that five million wireless sets and eight million valves would be sold. There were approximately 18 million telephones in US homes and eight million receiving sets, and it was the hope of the radio trade that where there was a telephone there would be a radio set—and eventually a television set. At the New York Radio Show (1929) the attendance had been a record 306 000 visitors, many of whom were primarily interested in television and had attended in the expectation of seeing television.

Then the great depression hit the United States. Gross income for RCA in 1930 fell by $45 million relative to 1929's figure and earnings were down by $10 million. The 1929 gross income figure was not to be equalled until 1942[1], and during the years 1932 and 1933 the corporation suffered a loss. Its work force was cut by 4000, from 22 000, in the period 1930 to 1934. Even so RCA, fared better than many US businesses during this time, and its directors maintained their belief in the inevitability of a public television service. It is significant that both RCA[12] and EMI continued to finance expensive television research teams over this difficult period.

Changes were introduced in the research organisation after Sarnoff became president of RCA on 3 January 1930[13]. Dr W R G Baker of GE became the chief engineer, and by mid-1930 the Van Cortland Park activity had largely been dispersed and all the basic television investigations had been concentrated at Camden, New Jersey. The official transfer date of GE and WEM staff to RCA was 1 April 1930.

Of the research groups which were established at the Camden site, one group under the direction of a former GE engineer, Dr E W Engstrom, was called the general research group, and the other, directed by A F Murray, was named the research division. Murray's division comprised sections which were concerned with radio receivers, acoustics and television, under the supervision of Dr G Beers, Dr I Wolff, and Dr V K Zworykin, respectively. The two groups totalled initially 45 technical specialists and both teams proceeded, almost immediately, to tackle television research and development problems.

Zworykin's main task was to engineer a reliable, sensitive, high-definition electronic camera tube. At this time the only other activity, anywhere, on such camera tubes was that being conducted by Farnsworth at his Green Street laboratories in San Francisco. Since Zworykin had been admirably rewarded when he had visited, at the end of 1929, the Laboratoire des Établissements Édouard Belin, Paris, it was reasonable to anticipate that a visit to Farnsworth's laboratory would likewise be beneficial. Zworykin was not disappointed, although the knowledge which he gained during his April visit was not so vital to his objective as the knowledge he acquired of Holweck and Chevallier's work on hard vacuum cathode-ray tubes (see Chapter 16).

Zworykin and Ogloblinsky recommenced their work on two-sided iconoscopes in May 1930. Contemporaneously, Dr Wilson, of Zworykin's section, investigated the performance of image dissector tubes fabricated with photocathodes sensitised with caesium-silver-oxide coatings. These tubes soon produced good 80-line images. Indeed, the tubes were much more sensitive (by a factor of c. 100, because of the use of Koller's process) than the image dissectors which Farnsworth was making. Wilson's success contrasted with the difficulties which Zworykin and Ogloblinsky were experiencing in preparing blemish-free, two-sided mosaic target plates. Zworykin referred to this matter in his 1940 book, with G A Morton, titled 'Television'. 'The screen wires [were] coated with an insulating layer of vitreous enamel lacquer, or ceramic. Great care [had to] be taken to avoid cracks or pin-holes as such imperfections [generated] strong spurious signals . . . The technical difficulties associated with the making of these targets with the high degree of perfection required [were] very great.' Because of this Zworykin's group, in May 1931, decided to experiment with single-sided targets.

Meantime, it appears that some rivalry, perhaps antipathy, had arisen between the GE television group, led by Alexanderson, and the former Westinghouse television group, led by Zworykin. A trial of their television systems was arranged on 15 July 1930 in an attempt to resolve the conflict and evaluate the disparate methods. The trial was held at the Test House, Collingwood, New Jersey. GE demonstrated its 48 lines per picture, 20 pictures per second, mechanical flying spot scanner and standard mechanically scanned home receiver; and WEM exhibited its new 80 lines per picture, 12 pictures per second, modified film projector and kinescope receiver. The trial completely vindicated Zworykin's approach to the television problem. A few days later a decision was agreed to appropriate 90 % of the budgeted research funds to Zworykin's television system and just ten per cent to that of GE.

The new funding allocation markedly strengthened and enhanced Zworykin's team. An additional five engineers were appointed and the glass blowing section was augmented. Furthermore, a special development television test department was set up at RCA Victor, Camden, under the direction of W A Tolson who had been the manager of GE's television project. The department included three most capable engineers from GE, namely R D Kell, A V Bedford and M Trainer, and its function was to engineer and integrate the various designs and prototypes issuing from Zworykin's laboratory into a practical television system.

RCA's annual report for 1930 commented on the importance of this television activity:

> 'While television during the past two years has been repeatedly demon-strated by wire and by wireless on a laboratory basis, it has remained the conviction of your corporation that further research and development must precede the manufacture and sale of television sets on a commer-cial basis. In order that the American public might not be misled by purely experimental equipment and that a service comparable to sound broadcasting should be available in support of the new art, your corpo-ration has devoted its efforts to intensive research into these problems, to the preparation of plant facilities and to the planning of studio arrangements whereby sight transmission could be installed as a separate service of nationwide broadcasting.'

This was also the view of Electric and Musical Industries. RCA and EMI did not wish to put forward a low-definition service as others had done; rather they wanted to perfect a television system having a real entertainment value for the general public before starting television broadcasting.

Both RCA and EMI were manufacturing organisations marketing a range of products, with the resultant realisation of profits. They were not dependent only on the sale of television transmitters and receivers for their whole wellbeing, and could afford to engage in expensive research and development work. The two companies, although wishing to establish a medium/high-definition television service as quickly as possible, were not in the invidious position of some companies, which operated experimental low-definition television systems, as regards the necessity for a partic-ularly early beginning for the transmissions. And so the two large industrial com-plexes were able to adopt different policies compared with those of the smaller firms.

RCA's 1930 policy on television was certainly influenced by Zworykin who had convinced Sarnoff of the desirability of creating an all-electronic system. Sarnoff had been favourably impressed by Zworykin's work at Westinghouse and, at a meeting of the two men in early 1929, had been completely converted to the advantages of such a system. He had offered his full support to Zworykin. This commitment was stated explicitly in the 1930 annual report:

'Television must develop to the stage where broadcasting stations will be able to broadcast regularly visual objects in the studio, or scenes occurring at other places through remote control; where reception devices shall be developed that will make these objects and scenes clearly discernible in millions of homes; where such devices can be built upon a principle that will eliminate rotary scanning discs, delicate hand controls and other movable parts; and where research has made possible the utilisation of wavelengths for sight transmission that would not interfere with the use of the already overcrowded channels in space.'

The terms of reference of RCA and its stated policy on television had an important influence on the post-1930 advancement of television at the Bell Telephone Laboratories: its work on all-electronic television was delayed until 1933.

Following a period of six years (1925–1930) as a centre of excellence the activities of its television laboratories declined in importance after c. 1931. This decline stemmed from certain constraints which had been imposed on the business of the AT&T Co. and also from a certain lack of direction of the research and development programme.

In May 1931 Ives wrote a memorandum[14] on 'Future program for television research and development'. At that time BTL's television projects were not aimed at any specific advances. Its work was mainly exploratory in the realm of fundamentals. Ive's conviction was that the only real and lasting field for television, if it could be developed, was that of home entertainment. Consequently, Ives felt that the one worthwhile problem in television research and development was the improvement of broadcasting methods and the engineering of terminal apparatus which would produce the high grade of image that would be satisfactory. He realised the magnitude of the task required and wrote: 'The technical difficulties are enormous and the possibility of overcoming them is veiled in obscurity.' Nevertheless, the problems could be faced with interest and enthusiasm by his staff but a 'damper [was] put on thought and planning along these lines by the knowledge that television for home entertainment [lay] outside the chosen sphere of activities of the Bell System. It was debarred therefrom by its present relations'.

Ives was thus faced with a dilemma. Either a drastic decision to drop the whole subject of television as one foreign to the Bell System could be taken, or a programme of research, on both sending and receiving equipment, could be initiated as though the whole field was open to the Laboratories, in 'the hope or with the definite intention of so changing [the] policy and contract relations that [the company would] ultimately go into [this] field'.

The position relating to television enterprises and publicity was a sensitive one vis-à-vis the AT&T-RCA contract. Caution in publicity matters was exercised and in 1931 no announcement of television progress was given in the annual report of the AT&T company. There was a reference to cables which would permit an extremely wide band of frequencies to be transmitted but without any statement of its application in a television system. No mention of either wideband cables or television was given in the 1932 and 1933 reports: these noted the depressed business conditions and the Depression respectively. Thereafter nothing was said, in the pre-war annual reports, of the company's television ventures[15]. Again, after 1932 the editor, R T Barrett, of the *Telephone Almanac* was required to delete from future issues of the journal all references to Bell System's television activities[16]. Publicity of these was not renewed until 1946.

The lack of public disclosure did not connote a cessation of work on television. On the contrary: from 1931 to 1935 (inclusive) the expenditure of the company on television advancement amounted to $592 400[17].

The Radio Corporation of America was one of the first to recognise the necessity for conducting investigations in the (43–80) MHz bands and during the 1930 to 1940 decade it carried out a number of field tests, in 1931–1932, 1933, 1934 and 1936–1939. These tests were conducted with the same thoroughness and engineering excellence as the Bell Telephone Laboratories' 1927 trials. An important feature of RCA's enquiries was the use of a systems approach, an approach which is concerned with the formulation and understanding of the individual units which make up the whole system and with the interconnection of the units to form an integrated system.

The 1931–1932 test was based on the cathode-ray tube television system then being developed by the research branch of the company and the results were published in a series of papers in the Proceedings of the Institute of Radio Engineers for December 1933[18,19,20,21].

There were many points to be investigated:

1   the number of lines per picture for good and interesting television;
2   the number of pictures per second for flicker free reproduction;
3   the propagation characteristics of very high frequency television signals and their susceptibility to interference signals;
4   the use of cathode-ray tube reception;
5   the use of studio and ciné film scanners for signal generation;
6   the design of circuits having bandwidths appropriate to medium definition television;
7   the synchronising of the receiving apparatus with the transmitter.

RCA chose as its location for the tests the Metropolitan Area of New York and installed its studio and transmitting equipment in the Empire State Building (on the 85th floor, 1000' above street level). The short wave transmitting antenna was fixed to the top of the airship mooring mast (1250' above the street level). By the end of 1931 the system had been put into position and testing started during the first half of 1932.

Figure 17.1 shows a block diagram of the equipment layout. It was essentially unambitious in concept for the cathode-ray tube receiver had been demonstrated in 1929, the mechanical scanner of the Nipkow type had been in use for several years, as had the spotlight scanning of both studio and film scenes, and some work had been carried out on the transmission and reception of short waves.

However, no other organisation by 1931–32 had engaged in such a large scale test of this type and made its findings known, and hence RCA, with its firm judgment

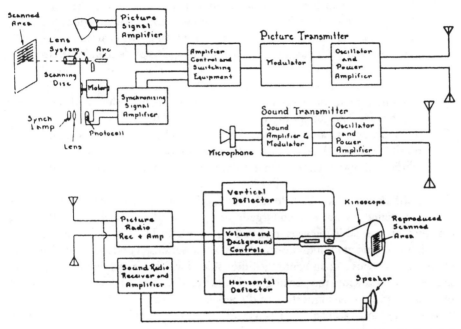

Figure 17.1   Schematic diagram of RCA's experimental television system (1931–1932)

Source: *Proc. of the IRE,* **21**, (12), December 1933, pp. 1652–1654

that television was inevitable, was compelled to proceed with its own research and development programme. As Engstrom, the group leader, observed[18]: 'The equipment used . . . was in keeping with the status of television development at that time.'

The basic parameters of the system were selected to be 120 lines per picture, sequentially scanned, and 24 pictures per second.

A number of receivers were placed in and around New York, many of them in the homes of the technical personnel, in order to collect subjective viewer reaction data as well as objective engineering measurements. Separate sound and vision transmitters, widely spaced in frequency, were used to simplify the apparatus, and both line synchronisation and frame synchronisation were employed.

Engstrom summarised some of the major observations and conclusions of the 1931–32 field test as follows[18]:

'The frequency range of (40–80) MHz was found [to be] well suited for the transmission of television programmes. The greatest source of interference was from ignition systems of automobiles and airplanes, electrical commutators and contactors, etc. It was sometimes necessary to locate the receiving antenna in a favourable location as regards signal and source of interference. For an image of 120 lines, the motion picture scanner gave satisfactory performance. The studio scanner was adequate for only small areas of coverage. In general, the studio scanner was the item which most seriously limited the programme material. Study indicated that an image of 120 lines was not adequate unless the subject material from film and certainly from studio was carefully prepared and limited in accordance with the image resolution and pick up performance of the system. To be satisfactory, a television system should provide an image of more than 120 lines . . . The operating tests indicated that the fundamentals of the method of synchronising used were satisfactory. The superiority of the cathode-ray tube for image reproduction was definitely indicated. With the levels of useful illumination possible through the use of the cathode ray tube, the image flicker was considered objectionable with a repetition frequency of 24 per second.'

During the period of RCA's field tests Zworykin and his group had been endeavouring to engineer a single-sided target plate camera tube. In such a tube the plate consisted of a very thin (between one and three mils thick) mica sheet, onto one side of which a mosaic of minute silver globules, photosensitised and insulated from one another, was deposited; and onto the other side of which a metal film (known as the signal plate) was sputtered. Each globule was capacitatively coupled to the signal plate and hence to the input stage of the video amplifier.

When an optical image was projected onto the mosaic each photosensitive element accumulated charge by emitting photoelectrons. The information contained in the optical image was stored on the mosaic in the form of a charge image. As the electron beam uniformly scanned the mosaic in a series of parallel lines the charge associated with each element was brought to its equilibrium state ready to start charging again. The change in charge in each element induced a similar change in charge in the signal plate and, consequently, a current pulse in the signal lead. The train of electrical pulses so generated constituted the picture signal.

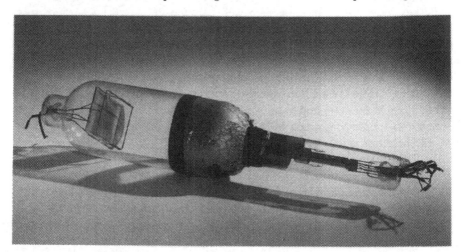

Figure 17.2   Early iconoscope camera tube

Reproduced by permission of RCA

Although promising results were obtained just one month after Zworykin's group transferred its efforts to the fabrication of single-sided plates much experimentation was required before the preparation of the mosaic was perfected. As with many successful undertakings, an adventitious factor aided the group. On one occasion Essig accidentally left one of the silvered mica sheets in an oven too long and found, after examining the sheet, that the silvered surface had broken up into a myriad of minute, insulated silver globules. This, of course, was what was needed: by October 1931 good results were being obtained with the new processing method. According to Iams, the first satisfactory tube (no. 16) was fabricated on 9 November 1931. One of the early experimental tubes was employed by the National Broadcasting Company for three years but general adoption of the iconoscope had to wait until 1933 when it could be manufactured to have uniform and consistent properties.

References

1 MACLAURIN, W.R.: 'Invention and innovation in the radio industry' (Macmillan, New York, 1949) p. 100
2 BAKER, W.J.: 'A history of the Marconi Company' (Methuen, London, 1970) pp. 180–181
3 ARCHER, G.L.: 'Big business and radio' (Americal Histroical, New York, 1939) p. 8
4 BITTING, R.C.: 'Creating an industry', *J. SMPTE*, **74**, pp. 1015–1023
5 Quotation given in Ref.4, p. 1016
6 WARNER, J.C.: 'Radio Corporation of America. Part 1—The years to 1938', 1938, pp. 2–8, RCA Archives, Princeton
7 SARNOFF, D.: Report to the RCA Board of Directors, 5 April 1923, RCA Archives, Princeton
8 RCA annual report for 1929, RCA Archives, Princeton
9 RCA report for 1930, RCA Archives. See also: Refs.4 and 6
10 ANON.: 'Images dance in space, heralding new radio era', *The New York Times*, 14 April 1929, sect.XI, p. 1:17
11 HARBORD, J.G.: Letter to Dr A N Goldsmith, 17 April 1929, GE Archives, Schenectady
12 Ref.1, pp. 206–220
13 LYONS, E.: 'David Sarnoff' (Harper and Row, New York, 1966) p. 188
14 IVES, H.E.: 'Future program for television research and development'. Memorandum for file, 18 May 1931, case file 33089, 1–11, AT&T Archives, Warren, New Jersey
15 ESPENSCHIED, L.: 'Announcement of television developments in the AT&T Co. annual reports'. Memorandum, 19 February 1954, 1–3, AT&T Archives, Warren, New Jersey
16 ESPENSCHIED, L.: 'When television was in eclipse in the Bell system—case 37014'. Memorandum for file, 18 February 1954, 1–3, AT&T Archives, Warren, New Jersey
17 ANON.: 'System of television, case 33089, work authorization estimated costs'. From file on 'Television and radio history picture transmission', AT&T Archives, Warren, New Jersey
18 ENGSTROM, E.W.: 'An experimental television system', *Proc. IRE*, December 1933, **21**, (12), pp. 1652–1654
19 ZWORYKIN, V.K.: 'Description of an experimental television system and the kinescope', *Proc. IRE*, December 1933, **21**, (12), pp. 1655–1673
20 KELL, R.D.: 'Description of experimental television transmitting apparatus', *Proc. IRE*, December 1933, **21**, (12), pp. 1674–1691
21 BEERS, G.L.: 'Description of experimental television receivers', *Proc. IRE*, December 1933, **21**, (12), pp. 1692–1706

Chapter 18

RCA and all-electronic television, (1933–1935)

With progress being made in the development of electronic scanners, it was to be expected that RCA's second field test would embrace an appreciation of the device. In the New York tests the major limitation to adequate television performance had been the studio scanning apparatus, since the lighting in the studio had been of too low an intensity to give a satisfactory signal-to-noise ratio; only when motion picture film was being scanned could a reasonable ratio be achieved. Fortunately, the sensitivity of the iconoscope was sufficient to allow of a further increase in the number of lines scanned per picture, in addition to permitting outdoor, as well as studio, scenes to be on television.

RCA's second experimental television system, incorporating the iconoscope, was built at the Camden works and operated there during the first few months of 1933. The picture characteristics were based on 240 lines per picture, sequentially scanned at 24 pictures per second, with an aspect ratio of 4:3 (width to height)[1]. Again, separate sound and vision transmitters were used[2] instead of a method which would have transmitted both the sound and picture signals on a single carrier by means of a double modulation process. The corporation had considered this latter alternative and had concluded that all the schemes available were subject to the possibility of serious crosstalk in some portion of the circuit, either in the transmitter or in the receiver, and were inefficient in transmitter power utilisation. A further disadvantage was that the transmission channel would be very wide, necessitating wideband radio and intermediate frequency systems in the receiver, resulting in costly construction and difficulty in tuning.

In the Camden tests the picture and sound carrier frequencies were 49 MHz and 50 MHz respectively. These values enabled the use, in the receiver[3], of a single radio frequency tuning system consisting of two coupled radio frequency circuits, having sufficient bandwidth to accept both carriers and their side bands simultaneously, and a heterodyne oscillator, the output of which when mixed with the carriers produced two intermediate frequencies spaced 1 MHz apart. A further factor which led to the adoption of this solution with adjacent carriers was the realisation that, in any national television network employing a considerable number of transmitters (to ensure adequate population coverage), the sound and vision carrier frequencies would have to be chosen so as not to interfere with each other and with television stations operating on similar frequencies.

The probable future need for a national television network dictated the nature of the propagation path which the corporation adopted in one of its tests, for it was aware that groups of stations would have to be tied together to form a network service, using as the interconnecting link either a special land wire or radio. RCA was certainly looking ahead.

Thus, two separate field tests took place during 1933: one to acquire experience with the system fundamentals and with the terminal apparatus, using a 240-line standard, and another to determine the characteristics of a relay system employing a 120-line standard[4].

In the first case, the sound and picture transmitters, which had nominal power outputs, were located in one of the RCA Victor buildings, and the studio and control equipments were housed in another building about one thousand feet away. Most of the receiving tests were conducted at a site, four miles from the transmitter, called Collingswood. One of the problems which had to be solved was the provision of facilities for outside broadcasts and so, in the Camden system, a pick-up point was selected approximately one mile from the studio. Here, an outside programme was televised and relayed to the main studio and transmitter by radio.

In the second of the tests indicated above, RCA used the 120-line standard system which had been installed in the Empire State Building for the 1931–32 trial programme[5]. The object was to relay the 44 MHz television signals radiated from this building's antenna to Camden, a distance of 86 miles, where they were rebroadcast. A preliminary study of some survey maps had shown that one intermediate relay station would be needed and this was subsequently located at Arney's Mount, an isolated hill rising to 230′ above sea level. The summit was high enough to give line of sight working to Camden, 23 miles away, and was only about 200′ below the line of sight from the New York antenna, 63 miles distant. This unequal division of the total relaying distance was justified for several reasons: first, the great height of the 44 MHz antenna in New York (1250′ above sea level); secondly, the fact that the second lap of the relay was to be carried on 79 MHz, which would probably require a direct optical path; and, finally, because a strong signal would be needed in Camden to override severe induction interference from the factories there.

RCA's testing programme was thus comprehensive in concept: line and radio links; film, studio and outside broadcasts; relay working; and electronic working— all were included. The publication of the test data and conclusions did not take place until November 1934, and hence did not aid Marconi-EMI, or Baird Television in their deliberations with Lord Selsdon's Television Committee. This committee had obtained its evidence from the two British companies in June 1934 and, by November 1934, its report was almost ready for reading by the Postmaster General. Nonetheless, Marconi-EMI had put forward some proposals which were vindicated by RCA's field tests. In addition, the Hayes group had suggested the use of d.c. working and interlacing which did not form part of the Corporation's investigations.

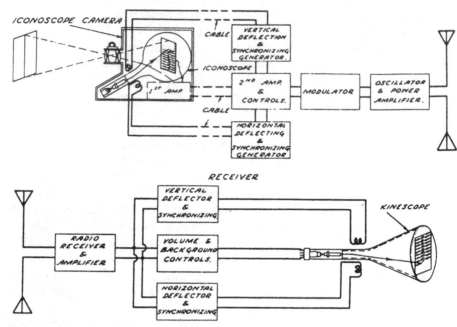

Figure 18.1   Block diagram showing a television system which incorporates an iconoscope and a kinescope

Source: *Proc. of the IRE*, **22**, (11), November 1934, pp. 1241–1245

Engstrom stated some of his group's findings as follows[1]:

1   'The use of the iconoscope permitted transmission of greater detail, outdoor pick-up, and wider areas of coverage in the studio. Experience indicated that it provided a new degree of flexibility in pick-up performance, thereby removing one of the major technical obstacles to television . . . The iconoscope type pick-up permitted a freedom in subject material and conditions roughly equivalent to motion picture camera requirements [Figure 18.2].

2   'The choice of 240 lines was not considered optimum, but all that could be satisfactorily handled in view of the status of development . . . The increase of image detail widened very considerably the scope of the material that could be used satisfactorily for programmes. Experience with this system indicated that even with 240 lines, for critical observers and for much of the programme material, more image detail was desired. The desire was for both a greater number of lines and for a better utilisation of the detail capabilities of the system and lines chosen for the tests.

3   'As in the New York tests, much valuable experience was obtained in constructing and placing in operation a complete television system having standards of performance abreast of research status. Estimates of useful field

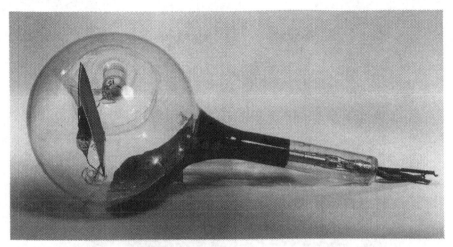

Figure 18.2   First production type of iconoscope of the type later used commercially

Reproduced by permission of RCA

strengths were formulated. The need for a high power television transmitter was indicated.'

From the end of June 1933, news of Zworykin's iconoscope was printed in several newspapers and journals. Zworykin, Figure 18.3, described, but did not demonstrate, the electronic camera tube at a meeting of the Institute of Radio Engineers held on 26 June in Chicago. The meeting was attended by Dr F Gray, of Dr H E Ives's group at Bell Telephone Laboratories. Gray wrote a short memorandum[6] dated 6 July 1933, on what he had learned from the lecture, and on the same day Ives sent a note[7] to Dr H D Arnold, the director of research at BTL. 'The device,' stated Ives, 'involves some principles which are new in the television art and represent a considerable advance in the direction of attaining a television transmitting device suitable for general use in such places as moving picture camera might be used.'

He felt that Zworykin had taken an important step in the right direction and opined: 'It is not at all improbable that there will emerge from his development a television transmitter which will make high grade transmission of television material suitable for entertainment a practicable possibility.'

Several days later (20 July) Ives[8], noted: 'The consensus of opinion at our conference yesterday [on camera tubes] appears to be the feeling that we would do well to start some work along these lines.' On 31 July Gray[9] outlined a programme of work and soon a beginning was made on an enquiry into the design of an electronic camera tube. Unfortunately for the prestige of BTL[10], this effort began three years too late. On 20 May 1930 Gray[11] had written a comprehensive memorandum (21 pages long with 11 figures) on a 'proposed television transmitter' based on charge storage, in which he had put forward 11 different methods whereby signals might be generated by various photoelectric means. One of his objects in listing the proposals

Figure 18.3   Dr V K Zworykin with one of his electronic camera tubes

Reproduced by permission of RCA

was 'to aid in deciding the direction of certain future researches on television'. Had Gray's inventiveness and resourcefulness been acted upon, it is possible that the Laboratories, with its excellent experimental facilities and staff, would have demonstrated an electronic camera contemporaneously with Zworykin's demonstration.

Ives's team seems, however, to have been reinvigorated by its new task. Within a year: Gray[12] had proposed a cathode-ray transmitter that utilised the stored charge flowing through a photoconductive film during a complete image cycle, and had put forward various ideas[13] for projecting images from the screen of a cathode-ray tube, with an optical reflecting power and transparency which could be affected by the electron beam (cf. the eidophor); Teal[14] had reported on experiments undertaken with new photoelectric emitters ($Cs–Cs_2S–Ag$ and $K–K_2O–Ag$ surfaces) suitable for use in an iconoscope; a conference[15] had been held at which seven novel methods for displaying large images in theatres had been discussed; Ives[16] had suggested a very simple form of iconoscope for generating picture signals from continuously moving motion picture film; and Hefele[17] had demonstrated a basic form of electronic camera.

By the end of July 1934 photosensitive targets, of the charge storage type, had been prepared by two methods. In one form, one side of a very thin sheet of mica had been coated with metal to form one plate of a capacitor and the other plate (the reverse side of the mica sheet) had consisted of tiny globules of pure sodium deposited by carefully controlled distillation. The second method had consisted of ruling a platinum film, sputtered on the face of a mica sheet, into small elements 0.004 inches square and then depositing a very thin layer of sodium onto the platinum squares to enable each one to be photosensitive.

A specially constructed cathode-ray gun was used to scan these targets. Hefele reported: 'Shadowgraphs of various objects focused onto the sensitive plate of the transmitting tube were clearly recognised . . . These encouraging results mark a forward step in our study of nonmechanical television transmission. Attention is being paid to the development of higher sensitivity photoelectric mosaics and to the enlargement of the picture detail by the transmission of more picture elements which involves research in cathode-ray structures as well as amplification and trans-mission of a wide frequency range.' Twelve months later, on 26 July 1935, Hefele demonstrated to several senior members of Bell Telephone Laboratories a 72-line photoconducting television camera tube. It contained a film of red mercuric iodide, cooled by solid carbon dioxide to render the film stable and so permit easier working conditions over long periods of time. The electron gun of the tube was designed by Gray and the received images were displayed on a cathode-ray tube of C J Davisson's design. In view of the 'excellent images' obtained, it was decided to attempt 'immediately' to construct a 240-line camera tube[18].

By 6 August 1935 Davisson had submitted a design for a 240-line electron gun (of 5 μA beam current) to give a spot size of 0.008 inches diameter so that an existing 2 × 2 inch photoconducting surface could be employed: and Jensen and Strieby had agreed to provide two 240-line sweep circuits by 1 September.

Hefele's demonstration appears to have been the first ever given of a photocon-ductive camera tube. The demonstration showed that Bell Telephone Laboratories had the expertise, the experimental facilities and the essential ideas to engineer an all-electronic television system. Nevertheless, their accomplishments were in some respects approximately two years behind those which had been obtained by RCA and EMI. The period of comparative inactivity of Ive's group (from January 1931 to July 1933) had put it at a disadvantage relative to Zworykin's group at RCA, and Shoenberg's group at EMI, where work on high-definition television systems was being pursued with some zeal, notwithstanding the depressed business conditions.

In the United Kingdom the 1933 news of Zworykin's inconscope seems to have been received with some eagerness by McGee and Tedham of EMI. They had con-structed, unofficially at EMI, a single-sided camera tube which had worked briefly in 1932, but had not succeeded in persuading their management of the worthwhile nature of their work[19]. Now, with Zworykin's 'momentous step forward', McGee and Tedham were allowed to take up this aspect of television research.

Zworykin's first published account of a television signal generator tube was the paper received by the Institution of Electrical Engineers (UK) on 17 July 1933.

Figure 18.4, taken from that paper, illustrates the iconoscope. The electron gun used was similar to that adopted for the kinescope and the signal plate was formed by a metallic coating on one side of a mica sheet. On the other side the mosaic was composed of a very large number of minute silver globules, each of which had been photosensitised by caesium by means of a special process. The sensitivity and colour response of the mosaic were of the same order as that of a corresponding high vacuum caesium oxide photoelectric cell. From the spectral response it was obvious that the iconoscope could be used with either ultraviolet or infrared radiation in addition to visible light. Zworykin noted that the sensitivity of the tube was about equal to that of photographic film operating at the speed of a motion picture camera with the same optical system. Some of the tubes constructed were satisfactory up to 500 lines 'with a wide margin for future development'.

Apart from the iconoscope, the paper also dealt with Zworykin's kinescope. However, the paper was deficient on practical constructional details and did not reproduce any photographs of images taken from the screen of the kinescope. No mention was given of the special processes used in making the mosaic; the mechanism of signal generation described was simplistic and 'was soon found to be very far from accurate'; keystone distortion and its correction was not referred to; and the presence of shading signals, which Shoenberg[20] was to call 'this evil', was not touched upon. Thus, a research laboratory seeking to construct an inconoscope had the advantage of knowing that such a tube could work, as well as the disadvantage of being compelled to devise the necessary fabrication procedures *ab initio*. This was the position of McGee's group in the summer of 1933.

During the 1933 RCA field tests it was found that the resolution of the iconoscope was considerably better than that which the rest of the system was capable of handling: further field tests would be necessary to explore the performance and

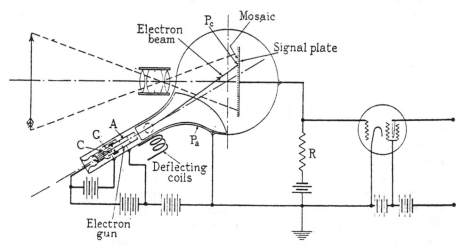

Figure 18.4   Sketch of the iconoscope camera tube. The cathode, C, control grid, G, first anode, A, second anode, P_a and the mosaic elements, P_c, are indicated

Source: *Journal of the IEE*, **73**, 1933, p. 441

limitations of the camera tube. For its 1934 tests RCA made three changes to the specification of the 1933 system, *viz.*: the number of lines was increased to 343; an all-electronic synchronising generator was employed; and interlaced scanning was introduced, giving a frame rate of 60 Hz[21,22].

The first and third of these points were based not only on the 1933 findings but also on Engstrom's study of television image characteristics. The researchers of Weinberger, Smith and Rodwin of RCA, of Gannet of AT&T and of Wenstrom of the US Signal Corps, on television image analysis have already been considered. These researches had used still pictures. There was a need to consider the problem of acceptable line and frame standards based on the use of moving images. Engstrom undertook this investigation and made his conclusions known in an important paper[23] published in December 1933. His conclusions were based on:

1   the resolving properties of the eye;
2   practical viewing tests of motion film specially prepared to give pictures having 60, 120, 180, 240, lines per picture, and a normal projection print quality.

For his elementary studies of some of the properties of vision, Engstrom used the chart shown in Figure 18.5. This was placed at a viewing distance which just allowed, for example, the two vertical lines of the pattern in the bottom left-hand part of the chart (corresponding to a 120-line picture) to be resolved. By repeating this test for each of the various patterns, Engstrom was able to relate the equivalent number of scanning lines per inch of picture height to the distance at which they

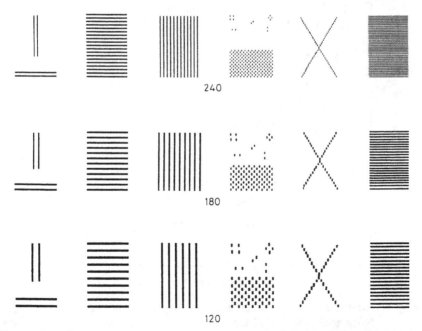

Figure 18.5   Form of chart used by Engstrom in his study of television image characteristics

Source: *Proc. of the IRE*, **21**, (12), December 1933, pp. 1631–1651

could just be resolved for each of the different types of pattern. His results, which were based on the performance of three observers, are shown in Figure 18.6. The theoretical curves corresponding to visual acuities (for that part of the field of view which falls on the fovea of the retina) of 0.5, 1.0 and 2.0 minutes of arc are shown superimposed on the observational data of Figure 18.6 in Figure 18.7. Thus, for example, an observer sitting at a distance of six feet from a television picture and having an acuity of vision of one minute of arc would be able to resolve a maximum of 48 lines per inch of screen. Hence, a picture having a height of nine inches (a 15 inch screen) would require 432 lines to define it, Figure 18.8.

In the second part of his investigation Engstrom used an ingenious technique to make ciné films having a detail structure equivalent to that of television images. His films included:

1   head and shoulders of girls modelling hats;
2   close-up, medium and distant shots of a baseball game;
3   medium and semiclose-up shots of a scene in a zoo;
4   medium and distant shots of a football game;
5   animated cartoons;
6   titles.

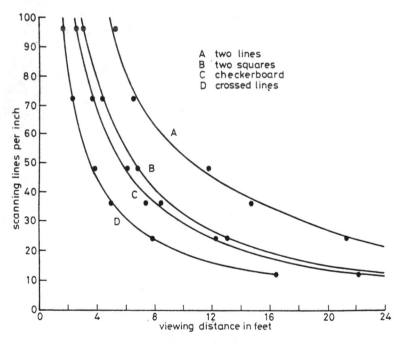

Figure 18.6   *Experimental results of resolution tests (averages of three observers). For each type of 'scanning line detail' of Figure 18.5, for example two squares, a viewing distance was chosen at which the details could just be resolved; graph B for the example given*

Source: *Proc of the IRE*, **21**, (12), December 1933 pp. 1631–1651

Each of these films was made to correspond to line standards progressing from 60 to 240 lines per frame. Viewing tests were made with projected pictures of various heights and, for pictures of a given height and line structure, observations were made for each type of subject matter on the film. In taking the observations, viewing distances were chosen at which the lines and detail structure just became noticeable. At closer viewing distances the picture structure became increasingly objectionable although at the viewing distance chosen the picture detail was satisfactory. It was noted by Engstrom that the type of picture subject did not influence the viewing distance selected by more than ten per cent.

In this work Engstrom set as his standard the ability of the eye to see the elements of detail and picture structure. Another less exacting standard was the ability of the images, having various degrees of detail, to tell the desired story. Taking as a standard the information and entertainment capabilities of 16 mm home movie film and equipment, Engstrom estimated the quality of the television images in comparison as:

60 scanning lines	entirely inadequate
120 scanning lines	hardly passable
180 scanning lines	minimum acceptable
240 scanning lines	satisfactory
360 scanning lines	excellent
480 scanning lines	equivalent for practical conditions

Engstrom's findings are summarised in Figure 18.8, which includes all the necessary information to determine the number of scanning lines required if the viewing dis-

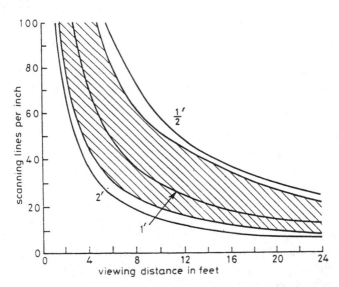

Figure 18.7   Theoretical curves for visual acuities of ½, 1 and 2 minutes of arc superimposed on the observational data shown in Figure 18.6

Source: *Proc. of the IRE*, **21**, (12), December 1933, pp. 1631–1651

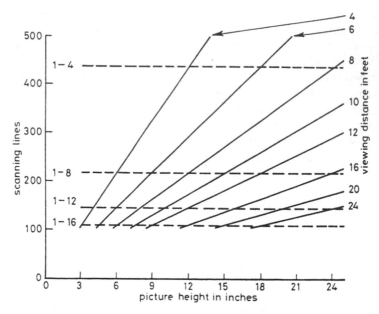

Figure 18.8   Relationship between scanning lines and picture size for several viewing distances. The broken lines indicate picture height to viewing distance ratios. Tests showed that viewing conditions were satisfactory for ratios between 1:4 and 1:8

Source: *Proc. of the IRE,* **21**, (12), December 1933, pp. 1631–1651

tance and picture height have been decided upon. His conclusions agreed effectively with those of Wenstrom[24], particularly with regard to the number of lines per picture needed to give excellent reproduction. Both investigators found that 360–400 lines/picture were required for this purpose.

RCA's choice of 343 lines per picture for their 1934 field trials indicated their desire to advance to high-definition television by stages instead of by a bold leap from 240 lines per picture to c. 400 lines per picture which was the strategy adopted by Shoenberg of EMI.

Engstrom and his colleagues did not publish any papers on the 1934 field test and so their first-hand observations are not immediately available for study. However, on 24 October 1934 some members of the British Television Committee (Lord Selsdon, Colonel A S Angwin, Mr N Ashbridge and Mr F W Phillips) left Britain for the United States on a fact finding visit. They arrived in the host country on 30 October and, during their stay, were given a demonstration of the RCA system. The delegation subsequently reported[25]: 'Demonstrations were given in the RCA Victor Laboratories of studio subjects such as a pianist, illustrated talks, topical films of plays and current sporting events, and also an outdoor scene. Reproduction was on a cathode-ray tube giving an image 6 by 8 inches with 343 scanning lines and 30 frames per second. Very good reproductions were obtained. All the studio and film subjects were comparable and perhaps slightly superior as regards absence of flicker to corresponding demonstrations in [the UK]. For the reproduction of an outside scene, the pick-up camera was directed towards a bridge 100 yd or so from the roof of the building. At the time of the demonstration, the sky was heavily overcast, the light conditions being such that photography would only be possible with a long exposure of highly sensitive film. Despite the poor light, a reproduction of the scene was obtained on the television receiver, and whilst the definition was faint, it was quite possible to distinguish motor cars moving across the bridge. Under daylighting conditions it was considered the reproduction would have been clear and distinct.'

The reference to flicker is significant since in its 1934 tests RCA used interlaced scanning for the first time. Until 1934 all television demonstrations had been based on sequential scanning: but as the brightness of cathode-ray tube screens increased the phenomenon of flicker became objectionable.

The various toys and optical models put forward during the Victorian era to show motion[26], and the importance of persistence of vision (see Chapter 4), also produced a sensory effect called flicker. When the retina of the eye (adapted to darkness) is suddenly exposed to a steady bright field of view, the sensation produced rises rapidly to a maximum and then falls to a lower constant value. With the removal of the stimulus, the sensation does not disappear immediately but takes a finite time to decay below the limit of perception. The extent of the overshoot increases, and the time taken to settle down to the final value decreases with increasing intensity of the retinal illumination. Consequently, if the eye is exposed to a source of rapidly varying intensity, the effects of the definite rates of growth and decay of the sensory response (or the persistence of vision) may prevent flicker from being noticeable. Flicker is absent when the luminous variations are regular in nature and have a frequency above the critical frequency.

William Henry Fox Talbot discovered the flicker effect and the law which characterises it, namely, if the frequency of a varying light source is sufficiently high for flicker to be imperceptible, the eye is able to integrate the brightness over the cycle of variations. The effect is as if the light for each cycle were uniformly distributed over the period of the cycle.

The highest frequency at which flicker can just be detected, the critical flicker frequency, is nearly a linear function of the field's brightness (over the range appropriate for television), and the sensitivity of the eye to flicker is noticeably increased when the field of view is enlarged from a few degrees to an image of the size encountered in cinemas. The sensitivity to flicker is also greater for averted vision, when viewing large fields of varying brightness. When the illumination is raised, an increased flash rate is required to give the subjective impression of steady illumination.

Flicker is of importance not only in television reproduction but also in motion picture projection and, as the implementation of the latter art considerably predated the former, some knowledge was available on picture frame rates for flicker free viewing when television advanced to the stage of a public broadcast service.

Edison in 1894 used a picture rate of 48 per second[27]. Since the film lengths utilised were just 50 feet long, the duration of his peep show images was a mere 13s. Later, the picture rate was reduced to 16 frames per second, presumably to conserve film and reduce costs, and this rate remained roughly the norm throughout the era of silent films (with sound films the speed is 24 frames per second).

The use of the above picture frequency and a single bladed shutter gave rise to an appreciable flicker effect and means had to be found to maintain the picture rate with a higher viewing frequency. The solution was to adopt a three-bladed shutter which gave three light-dark changes during the projection of each film frame, Figure 18.9, and thus a viewing picture frequency of 48 flashes per second. A reasonably flicker-free image was obtained under normal circumstances but, for many years after, the colloquial term 'the flicks' remained in popular usage as a name for motion pictures. With the introduction of the talkies and the adoption of 24 pictures per second, only a two-bladed shutter was necessary to give the same flicker-free frequency.

In television the early workers were limited to the utilisation of a channel band-width of the order of 10 kHz and, as the bandwidth for television reproduction is proportional to the square of the number of lines employed and directly propor-tional to the picture scanning rate, a compromise had to be decided upon between the need to use a high picture rate for flicker-free viewing and a large number of lines for good definition. Baird adopted a 30 lines per picture, 12.5 pictures per second standard, but naturally this gave rise to some image flicker. The first German television standard also used the same picture rate.

But when cathode-ray tubes began to be made and used for television reproduc-tion in the early 1930s, capable of giving bright pictures, the low picture rate was found to be unsatisfactory for prolonged viewing. Television engineers in several countries increased the rate to 25/30 pictures per second, but with interlaced scan-ning the frame rate could be increased to 50/60 frames per second, while the picture

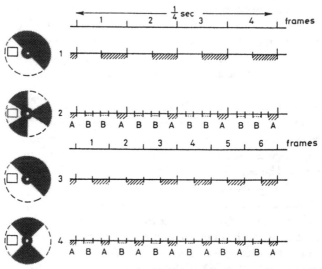

Figure 18.9   In early 16 frames per second cinematography the single-bladed shutter, 1, used in the camera gave rise to a flickering effect on the projector screen. This effect was reduced by employing a three-bladed shutter, 2, and by moving the film in the pro-jector during the dark intervals (marked A). With 24 frames per second cinematog-raphy a two-bladed shutter, 4, and an intermittent film motion (during the periods marked A) gave the same flicker frequency as in 2

rate was maintained at 25/30 pictures per second. This approach paralleled the practice which had been adopted in the motion film industry. However, there is an important difference between the projection of a ciné film and the reproduction of a television image. In the former case the whole of the picture is shown at any instant of film projection, whereas in the latter the image is built up line by line. The employment of the technique used in the film industry to achieve a high frame rate clearly was not possible, and so the principle of interlacing was adopted.

The wisdom of this choice was confirmed in April 1935 when Engstrom published his second paper[28] on 'A study of television image characteristics' which dealt particularly with the determination of the picture frequency for television in terms of flicker characteristics. Engstrom's investigations were comprehensive: he made measurements using a sector flicker disc and also a kinescope; he carried out tests on progressively scanned rasters and interlaced rasters and he considered the effect of room illumination, the persistence characteristic of the screen material and the spectral characteristic of the kinescope screen on his observations. These indicated that, when a cathode-ray tube screen was progressively scanned and the screen illumination was only one foot candle, a picture rate of '38 pictures per second was required for just noticeable flicker, and 35 pictures per second for noticeable but satisfactory flicker; 28 pictures per second resulted in disagreeably objectionable flicker. Thus a standard of 24 pictures per second could not be justified. These data also indicated that 48 pictures per second would be satisfactory from the standpoint of flicker for values of illumination likely to be encountered in television'.

For a system of television using an interlaced scanning pattern Engstrom concluded that 'satisfactory flicker conditions exist if each picture consists of two groups of alternate lines (equivalent to 48 pictures per second)' when the actual rate is 24 pictures per second.

In another RCA paper[29], published in June 1936, the authors (Kell, Bedford and Trainer) found that an integer ratio between the alternating current mains frequency and the picture frequency was very desirable for progressive scanning and was almost imperative for interlaced scanning.

The scanning system used by RCA, and also by EMI, was that devised by R C Ballard[30] of the Radio Corporation of America and patented by him in 1933. In the EMI version of the Ballard system the picture was completely scanned in 1/25 second by means of two downward traversals or frames, each 1/50 second duration, and the lines constituting one such frame were arranged to lie between the lines of the other frame so as to give good picture detail. In Figure 18.10 the lines show the track of the scanning spot, which moved under the influences of a regular downward motion (frame scan), with a rapid return, and a regular left to right motion (line scan), and with a very rapid flyback.

Assuming that the commencement of one frame scan was at A, the combination of the above motions caused the scanning spot to trace lines AB, CD, EF, . . . JK, when at the point K the return stroke of the frame motion began and returned the spot to L at the top of the raster. $202\frac{1}{2}$ lines were scanned during this frame scan. Then the downward motion started again causing the spot to trace lines LM, NB, PQ . . . TV until at U the frame scan returned the spot to A. Because point L was

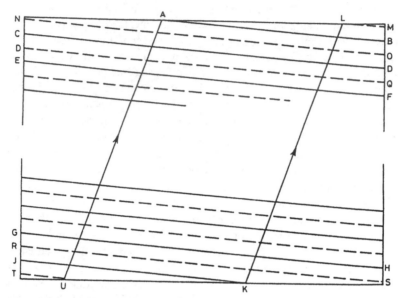

Figure 18.10 Interlaced frames of the 405-line picture

arranged to be half a line ahead of point A, the lines traced out on the second scan were half way between the lines scanned during the first frame. After two frame scans 405 lines had been traced out and the cycle recommenced. The essential feature of the Ballard system was that the complete picture was scanned in two frames and, as each frame contained an integral number of lines plus a half, the two frames were bound to interlace. The process is analogous to the device used in film projection, whereby it is arranged that although only 24 pictures are shown per second, each picture is thrown onto the screen two or three times so as to raise the light flicker frequency to 48 or 72 flashes per second.

Ballard's patent on interlacing was not the first on the subject. The terms interlacing and intercalation were introduced by Latour[31] and Baird[32], respectively, simultaneously in 1926, and a form of this type of nonsequential scanning dates from 1914. Several inventors, including Hart[33], Stephenson and Walton[34], Latour[31], Baird[32], von Ardenne[35] and the Telefunken Company[36] had proposed systems of nonsequential scanning for various purposes prior to the publication of Ballard's patent.

References

1 ENGSTROM, E.W.: 'An experimental television system', *Proc. IRE*, November 1934, **22**, (11) pp. 1241–1245
2 KELL, R.D., BEDFORD, A.V., and TRAINER, M.A.: 'An experimental television system. Part II, The transmitter', *Proc. IRE*, November 1934, **22**, (11) pp. 1246–1265
3 HOLMES, R.S., CARLSON, W.L., and TOLSON, W.A.: 'An experimental television system. Part III, The receiver', *Proc. IRE*, November 1934, **22**, (11), pp. 1266–1285

4 YOUNG, C.J.: 'An experimental television system. Part IV, The radio relay link for television signals', *Proc. IRE*, November 1934, **22**, (11) pp. 1286–1294
5 ENGSTROM, E.W.: 'An experimental system', *Proc. IRE*, December 1933, **21**, (12), pp. 1652–1654
6 GRAY, F.: 'Note on Zworykin's iconoscope'. 6 July 1933, case file 33089, pp. 1–3, AT&T Archives, Warren, New Jersey
7 BUCKLEY, O.E., and IVES, H.E.: Memorandum to H D Arnold, 6 July 1933, case file 33089, pp. 1–2, AT&T Archives, Warren, New Jersey
8 IVES, H.E.: Memorandum to O E Buckley, 20 July 1933, p. 1, case file 33089, AT&T Archives, Warren, New Jersey
9 GRAY, F.: 'A suggested outline for development work on a cathode ray transmitter'. Memorandum for file, 31 July 1933, case file 33089, pp. 1–3, AT&T Archives, Warren, New Jersey
10 BURNS, R.W.: 'The contributions of the Bell Telephone Laboratories to the early development of television' (Mansell, London, 1991) pp. 181–213
11 GRAY, F.: 'Proposed television transmitters'. Internal report, 20 May 1930, AT&T Archives, Warren, New Jersey
12 GRAY, F.: 'A proposed cathode ray transmitter'. Memorandum for file, 3 January 1934, case file 33089, p. 1, AT&T Archives, Warren, New Jersey
13 GRAY, F.: 'Projection of images from a cathode-ray tube'. Memorandum for file, 15 January 1934, case file 33089, pp. 1–5, AT&T Archives, Warren, New Jersey
14 TEAL, G.K.: 'A new photo-electric emitter suitable for use in the iconoscope'. Memorandum for file, 25 April 1934, pp. 1–2; and 'The potassium-potassium oxide-silver matrix as the photo-sensitive element of the iconoscope'. Memorandum for file, 14 May 1934, case file 33089, pp. 1–2, AT&T Archives, Warren, New Jersey
15 'Notes of a conference held in Dr Buckley's office on large image schemes for theatre showing'. 20 June 1934, case file 33089, pp. 1–8, AT&T Archives, Warren, New Jersey
16 IVES, H.E.: 'Iconoscope for transmission from motion picture film'. Memorandum for file, 6 July 1934, case file 33089, AT&T Archives, Warren, New Jersey
17 HEFELE, J.R.: 'Television transmission system using cathode-ray tubes'. Memorandum for file, 31 July 1934, case file 33089, pp. 1–4, AT&T Archives, Warren, New Jersey
18 Nix, F.C.: 'Photo-conducting television transmitter'. Memorandum to H E Ives *et al.*, 6 August 1935, case file 33089, p. 1, AT&T Archivves, Warren, New Jersey
19 McGEE, J.D.: 'The early development of the television camera'. Unpublished paper, pp. 1–72, Archives of Imperial College of Science and Technology, London
20 GARRATT, G.R.M., and MUMFORD, A.H.: 'The history of television', *Proc. IEE*, 1952, **99**, part IIIA, pp. 25–42
21 BEAL, R.R.: 'Equipment used in the current RCA television field tests', *RCA Rev.*
22 CLEMENT, L.M., and Engstrom, E.W.: 'RCA television field tests', *RCA Rev.* July 1936, pp. 32–40
23 ENGSTROM, E.W.: 'A study of television image characteristics', *Proc. IRE*, December 1933, **21**, (12), pp. 1631–1651
24 WENSTROM, W.H.: 'Notes on television definition', *Proc. IRE*, September 1933, **21**, (9), pp. 1317–1327
25 'System of television demonstrated in [the] USA'. Report of the Television Committee, Appendix IIIA (unpublished), Memorandum A, p. 2, 1936, Post Office Records Office, UK
26 CORK, O.: 'Movement in two dimensions' (Hutchinsons, London, 1963) Chapter 8, pp. 121–136
27 ANON.: Article on motion pictures, Encyclopaedia Brittanica, 1963, **15**, p. 851
28 ENGSTROM, E.W.: 'Determination of frame frequency for television in terms of flicker characteristics', *Proc. IRE*, April 1935, **24**, (4), pp. 295–310
29 KELL, R.D., BEDFORD, A.V., and TRAINER, M.A.: 'Scanning sequence and repetition rate of television images', *Proc. IRE*, 1936, **24**, (4), pp. 559–575
30 BALLARD, R.C.: 'Improvements in or relating to television systems'. British patent 420 391, 19 July 1933
31 LATOUR, M.: 'Improvements in the transmission of photographs or other images to a distance'. British patent 267 513, application date (UK) 8 March 1927
32 Television Ltd., and BAIRD, J.L.: 'Improvements in or relating to television and like systems'. British patent 289 307, dated 15 October 1926

33 LAVINGTON HART, S.: 'Improvements in apparatus for transmitting pictures of moving objects and the like to a distance electrically'. British patent 15 270, application date 25 June 1914
34 STEPHENSON, W.S., and WALTOM, G.W.: 'Improvements relating to apparatus for transmitting electrically scenes/representations to a distance'. British patent 281 766, application date 18 April 1923
35 VON ARDENNE, M.: 'Improvements in television'. British patent 387 087, application date (UK) 21 December 1931
36 Telefunken: 'Improvements in or relating to television and picture receiving systems'. British patent 380 602, application date (UK) 22 September 1931

Chapter 19

EMI, Shoenberg and television, (1931–1934)

In May 1934 the UK's Postmaster General set up a Television Committee[1] 'to consider the development of television and to advise on the relative merits of the several systems and on the conditions—technical, financial and general—under which any public service of television should be provided'.

The report of the committee was published in January 1935 and recommended that Baird Television Ltd and Marconi-EMI Television Company Ltd should be given an opportunity to supply, subject to conditions, the necessary apparatus for the operation of their respective systems at a television station to be established in London. One month later, on 4 February 1935, Mr I Shoenberg, the director of research of the Marconi-EMI company put forward to the Television Committee his company's proposals for an all-electronic television system. It was based on a specification of 405 lines per picture and 50 frames per second, interlaced to give 25 pictures per second, and incorporated electronic camera tubes called emitrons.

Subsequently, from February 1937, following the adoption of the Marconi-EMI system, the 405-line format became the television standard for the UK and was used until 1985. By any criterion the company's achievement was a quite outstanding one.

EMI was formed in April 1931 to acquire the ownership of the Gramophone Company Ltd and the Columbia Graphophone Company Ltd[2]. The shareholders in these two companies exchanged their shares for an equal number of shares in EMI. At that time EMI owned all of the shares of the two companies and its issued capital consisted of 5 805 749 ordinary shares of 50p each, and 460 000 cumulative redeemable preference shares of £1.00 each.

The Columbia company, from about 1898 until 1922, was managed in the UK as a branch of an American company called the Columbia Graphophone Manufacturing Company. Then, in 1917, the Columbia Graphophone Company Ltd was formed as an English company, its shares being owned by the American company. In 1922 the whole of the share holding of the English company was acquired by British nationals and, from that date, it was carried on in the UK and

Figure 19.1   Sir Isaac Shoenberg (1880–1963)

Reproduced by permission of Thorn EMI Central Research Laboratories

in many other countries as a British company with British capital. Three years later, in 1925, the British company, but a curious turn of the wheel of fortune, secured the whole of the issued capital of the American company.

The early history of the Gramophone Company has been given in Chapter 14. There it was noted that, when the Radio Corporation of America procured in 1929 all of the shares of the Victor Talking Machine Company, the corporation became the beneficial owner of the shares held by the Victor company in the Gramophone Company Ltd. The number of EMI shares in the name of the Victor Talking Machine company in 1934 was 1 700 000. These were ordinary shares of 50p each and represented 29.28 % of EMI's total issued ordinary share capital, and 27.13 % of the total issued shares of EMI, including the preference shares.

In 1934 the board of directors of EMI comprised ten members who were all British with the exception of the Marchese Marconi (Italian) and Mr D Sarnoff, the president of the Radio Corporation of America. All of the directors (Mr A Clark, chairman, Mr J Broad, Mr H L H Hill, the Marchese Marconi, the Rt Hon Lord Marks, Sir Arthur Roberts, Mr D Sarnoff, Mr E de Stein, Mr L Sterling, managing

Figure 19.2   Dr J D McGee (centre) with Mr I Shoenberg (left) and a visitor to EMI's research laboratories

Reproduced by permission of Thorn EMI Central Research Laboratories

director and Mr E R Ll Williams) had been members either of the board of the Gramophone Company or of the board of the Columbia Graphophone Company. Sarnoff replaced the two Victor company directors in 1929. He was not an active member of the board and attended only two board meetings from 1931 to 1934. There was never any suggestion of interference with or direction of the policy of EMI from Sarnoff.

Apart from the shares held by RCA, there were on the register of EMI in 1934 a few other shareholders, with addresses outside the United Kingdom, whose holdings represented a total which was just one per cent of the company's issued share capital.

The shares of EMI were dealt in on the stock exchanges of London and of several provincial cities and also on the New York Stock Exchange. There, the method of dealing with shares differed somewhat from that in London. In America, share ownership passed by a simple endorsement of the share certificate, but a condition imposed by the New York Stock Exchange when shares of an English company were being handled was that a separate depository should issue so-called American shares. These shares were quoted and dealt in daily on the exchange.

Before the formation of EMI the Columbia company's shares were quoted on the New York Stock Exchange, and hence the EMI directors decided that its shares also should be listed on that exchange. In accordance with the usual practice, the depository system applied and the depository chosen was the Guaranty Trust Company of New York. This company held a number of shares in EMI (in the name of an English nominee company) and issued American share certificates for a similar

Figure 19.3   Mr A D Blumlein (1903–1942)

Reproduced by permission of Thorn EMI Central Research Laboratories

number. The number of shares held in 1934 by Guaranty Nominees Limited of London was 1 241 605 ordinary 50p shares, and the number of individual holders of American shares (in 1934) was 8279. This figure, together with the 13 000 registered shareholders in England, gave a total of 21 279 shareholders. The American shares were gradually returned to the UK, and in 1933 and the first half of 1934 more than 600 000 shares had been returned.

Experience showed that all of the 8279 American shareholders did not vote in the same way and it was thought highly improbable that they would ever do so. However, assuming the improbable, their unanimous vote in 1934 would have represented 21.39 % of EMI's issued ordinary shares and 19.28 % of the total issued shares, including the preference shares. The shareholding of RCA and of Guaranty Nominees Ltd of London totalled 2 941 605 ordinary 50p shares and equalled 50.67 % of the issued ordinary shares and 46.95 % of all the shares.

The immediate effect of the merger was the combining of the research groups of the former companies which now constituted EMI. Shoenberg's Columbia team

Figure 19.4   The Central Research Laboratories of Electric and Musical Industries Ltd at Hayes

Reproduced by permission of Thorn EMI Central Research Laboratories

which included A D Blumlein, P W Willaims, E C Cork, H E Holman, H A M Clark and others, joined the HMV group at Hayes, England. This group comprised the research manager, G E Condliffe, W F Tedham, C O Browne and W D Wright.

After the formation of EMI, the directors agreed that HMV's television development effort[3,4,5] should continue to be an item of the new company's research and development programme. One of the first questions which had to be tackled was whether this work should proceed on mechanical or electronic lines. Mechanical scanners had the advantage that they had been made successfully, whereas the electronic scanner did not exist as a practical device. On the other hand, electronic scanning had many potential advantages, as Campbell Swinton had indicated, particularly for high-definition television. The HMV's exploratory work on television had been based on the strategy that they should aim to develop an effective cathode-ray tube picture receiver, but should leave any endeavour to progress a television camera to others.

EMI felt that the most promising attack on the problem of home television would be along the lines initiated by HMV. There were thought to be divers factors which made it desirable for EMI to proceed independently of RCA and these were listed by A Whitaker (who had visited RCA in April 1930) in a memorandum to W M Brown *et al.*, dated 22 April 1931[6]. The reasons are given here *verbatim*, because in

Figure 19.5   One of the vacuum pumping stations in the research laboratories of EMI. An emitron camera tube can be seen left of centre

some writings on television there are innuendos that EMI's link with the RCA gave EMI much knowledge about the television activities of the US company and that the emitron camera tube which EMI subsequently developed was really a copy of RCA's iconoscope camera tube.

1   'We are not satisfied that they [RCA] are covering all the salient difficulties with satisfactory lines of investigation.

2   'It is only possible to ensure at intervals when one of our staff visits Camden, that our information is at all up to date, reports on this line of work being of a very scanty nature: and

3   'that it is disputed, within the RCA organisation, to whom the rights of man-ufacturing the cathode-ray elements belong. Radiotron claim that it is within their province: if Radiotron take this work over, as seems likely, we shall, of course, get very little, if any, direct information.'

With these points in mind it was agreed EMI should carry on with television development on the following lines[6]:

1   'getting 150-line scanning equipment, and 150-line cathode-ray receiving equipment into operation by the end of this year [1931];

2   'that we should commence a small amount of experimental work on the production of cathode-ray oscillographs suitable for television;

3   'that we should keep in mind for investigation and development as soon as possible, the short wave transmitting and receiving gear which will be essential for the commercial utilisation of television services.'

As an indication of the time scale of this programme Whitaker suggested: first, the experimental system should be completed by December 1931; secondly, the cathode-ray tube work should be well advanced by June 1932; and thirdly, the prob-lems of a short wave radio link should be solved by June 1933. Thus, EMI would have to make provision for the commercial utilisation of its television work by the middle of 1933.

Actually, this programme was soon modified. Although the view had been taken that the company's business was in the field of receiving apparatus and not trans-mitting apparatus, it was essential to have, for the testing of the receivers, a source of video signals. Browne's 150-line equipment was unnecessarily complicated, with its multichannel requirements, and was not really suitable for this aspect of the work. Consequently, four months after the merger, work started on the design of a single picture channel, 120 lines per picture, 24 pictures per second mirror drum film scanner[7].

By the end of August 1931 W D Wright had built a cathode-ray tube receiver, using saw tooth scanning, and in a report dated 18 August he had recorded: 'Some good receptions of the Baird television broadcasts have been obtained, but not consistently from day to day, nor during any one broadcast.'

RCA and EMI at this time were following independent programmes of television development. In a 'Report on RCA Victor television' dated 15 April 1932[9] the differences between the lines of attack on the television problem were highlighted.

These differences are given below in some detail to show that EMI's approach to the problem was not based on copying that of RCA.

Transmitters

'The information available on the RCA Victor transmitters indicates that a straightforward attack has been made on the problem. . . . . Beyond the fact that choke modulation is being used, we have no actual information as to design and performance. We are thus unable to compare it with our present system of grid modulation . . . RCA Victor do not use any special aerial in order to ensure efficient radiation in a horizontal direction . . . [We] have decided to investigate the possibility of erecting a vertical aerial array, which will concentrate the radiation horizontally . . . '

Scanning equipment

'There is nothing novel under this heading, except the use of the cathode-ray scanning tube. The film scanner used by RCA Victor is the customary Nipkow disc with 60 holes . . . The studio equipment is the usual spotlight scanner constructed on ordinary lines . . . At present we have no actual details of the components . . . [We] consider the mirror drum to be superior to the Nipkow disc since much greater accuracy can be obtained . . . '

Scanning circuits

'Our nonoscillating circuits appear to have advantages from the point of view of cost, in that the RCA Victor receivers employ eight valves for the separation etc of the synchronising signals, while our circuit employs only five.'

Cathode-ray tubes

'Our work on the cathode-ray tubes has followed very closely the work of Dr Zworykin at Camden. Victor are proposing to standardise on two types, [*viz.* a 9" tube and a high intensity tube giving a c. 2" picture]. In our opinion, Victor appear to have standardised on these two types, and on their electron gun, a little too early. There are considerable possibilities of development in the small tube . . . Our experience so far has been that it is possible to produce a very sharp focus with a small tube, and if these can be obtained bright enough at low voltages, it should be possible to project an image equivalent to the 9" tube, and thus give a better picture than the 9" tube itself.

'The Victor Engineering reports state that cathode-ray tubes in which the ratio second anode current/first anode current equals unity are considered to be efficient, while recent experiments carried out by us on the design of an electron gun have indicated that it is quite a simple matter to obtain values 8 to 10 for this ratio by suitable design of aperture shapes and electrode lengths.

'It is quite probable, therefore, that we shall be able to design tubes working at lower voltages which are very much brighter than those produced up to the present.

'Victor have already standardised on their artificial Willemite mixture as a commercial screen material. In our opinion this action has been a little premature. The white screens which we have produced by mixing calcium tungstate with a little Spodumene and Willemite, have been satisfactory for colour and for brightness.'

Scanning tubes

'The reports from Victor make it clear that the results of research under this heading are the most outstanding of any.

'Our preliminary examination of the problems involved in direct studio and outdoor scanning, definitely confirms the view expressed by Dr Zworykin that development of this type of scanning device will be essential for successful studio etc. television. We have already made a start on this work, but it will be some little time before any results can be obtained in view of the complexity in the design and the necessity for developing the manufacturing technique.'

Thus, in 1931, though the progression of television by RCA and EMI was proceeding independently along similar, but not identical lines, it would seem that the UK company was not receiving detailed practical know-how from the RCA. Many years later, several senior members of EMI's research and development staff— including Mr I Schoenberg, Dr J D McGee, Dr H G Lubszynski, Dr L F Broadway and Mr S J Preston—confirmed the independence (from RCA) of EMI's research and development on television.

In September 1931 A G D West, of the research and design department, visited Zworykin's RCA laboratory in Camden, New Jersey. West was shown the current results of RCA's television work and reported[10]: ' . . . television is on the verge of being a commercial proposition. They [RCA] intend to erect a transmitter on top of a New York Skyscraper in the autumn of 1932.' West had seen a televised image about 6 by 6 inches in size and had observed that its quality was comparable to that of an ordinary cinema picture viewed from a position near the extreme back of a large theatre. The planned selling price of the receiver was to be c. \$470 (c. £100) and all the receiving apparatus for sound and sight was to be contained in a single cabinet of the size of an ordinary radiogram. The vision and sound wavelengths were to be six metres and four metres respectively.

West's report had an immediate effect: EMI reviewed its position concerning television. As Whitaker noted: 'Television is apparently coming rather more quickly than even the most optimistic of us have considered likely during the last couple of years.' On the one hand, in the UK, Baird Television had achieved a great deal of favourable publicity for its 30-line system; the BBC had reluctantly given the Baird company some degree of official recognition by allowing it to use the BBC's transmitters for experimental broadcasts during nonprogramme hours; there seemed to be the prospect that Baird Television would be granted permission for its equipment to be housed in the BBC on a permanent basis, and additionally it had been stated that short Baird transmissions would take place during normal BBC broadcasts. On the other hand, the depressed business conditions of 1931 had led to a curtailing of

EMI's expenditure on television and because of this EMI could not give the demonstration at the end of the year which it had originally planned; there was a possibility that, if Baird's system became firmly established in the BBC, future television standards—and 'there [were] at least a dozen points which [needed] to be considered in great detail with a view to securing standardisation'—could be influenced by the standards of the Baird system; and there was a 'most urgent necessity' for EMI to establish 'a real demonstrated competition to the Baird system, which at the same time [would] give [EMI] prestige in television'.

For Whitaker there appeared to be only one short-term solution to the problem of gaining a position of authority in television matters and that effectively was to buy in expertise. RCA was further advanced than EMI in its work on seeing by electricity and so Whitaker suggested that approval should be given for the expenditure of roughly $50 000 (c. £13 000) for the purchase of transmitting equipment (including 4 m and 6 m, 2 kW transmitters for sound and vision, studio equipment and demonstration receiving sets), and for the expenditure of not more than £2000 for the installation of this equipment in England. Whitaker's figure of $50 000 was based on an approximate quotation which he had received from RCA. A later quotation received on 2 December 1931 gave a total figure of $83 732[11].

EMI's executive committee was persuaded by Whitaker's arguments and decided to recommend[12] to the board of directors the proposal outlined by him. However, this was adamantly opposed by Shoenberg, then head of the patent department.

By October 1931 it had become clear that technical cooperation with RCA was going to be 'very difficult and unreliable'[13]. The RCA rights on the transmitting side belonged to the Marconi Company and a favourable offer had been received by EMI from the company for the hire of radio transmitting apparatus. Hence, the idea of attempting to work in parallel with RCA faded out and Whitaker's proposition was not implemented. Instead, a programme of work[14] on television began under I Shoenberg, independent of RCA except for the patents to which EMI was entitled[15].

EMI's great success was undoubtedly due to the very powerful team which Shoenberg, its director of research, assembled. Professor J D McGee (who joined EMI in 1932), in his 1971 memorial lecture on 'The life and work of Sir Isaac Shoenberg, 1880–1963' described how Shoenberg came to Great Britain[16]:

> 'Isaac Shoenberg was born of a Jewish family in Pinsk, a small city in north west Russia. His great ambition in life was to be a mathematician, and it was this ambition that brought him later to England. But because of the difficulties facing a young Jew in Russia at that time, he had to settle for a degree in engineering: reading mathematics, mechanical engineering and electricity at the Polytechnical Institute of Kiev University. However, he was to retain his interest in mathematics throughout his life and, about 1911, he was awarded a Gold Medal by his old university for mathematical work.
>
> 'After leaving university, he worked for a year or two in a chemical engineering company, but very soon joined the Russian Marconi Company and so—in the first decade of this century—he became

involved in wireless telegraphy. From 1905 to 1914 he was chief Engineer of the Russian Wireless and Telegraph Company of St Petersberg. He was responsible for the research, design and installation of the earliest radio stations in Russia.

'In 1914 he resigned his good job and emigrated to England where, in the autumn of the same year, he was admitted to the Royal College of Science, Imperial College, to work under either Whitehead or Forsythe for a higher degree in mathematics.'

Unfortunately, the outbreak of the First world War brought this plan to a premature end and Shoenberg had to seek employment. He found a post with the Marconi Wireless and Telegraph Company at the princely salary of £2 per week. His abilities and potential were soon noted and he subsequently became joint general manager and head of the patents department. His resignation from this position and appointment with the Columbia Graphophone Company has been outlined by McGee[16].

'Shoenberg had no formal training as a musician, but he had an intense natural love of music; this made him a keen connoisseur of recorded music; and the technique of recording. This in turn resulted in a friendship with Sir Louis Sterling, and an invitation, in December 1929, to join the Columbia Graphophone Company and to put into practice his ideas on recording. He was clearly successful in this and soon the Columbia Company was competing effectively with the formerly unchallenged, prestigious firm The Gramophone Company or HMV. So began the association of Sterling with Shoenberg which was, to my mind, crucial in the field of television in the following decade. The shrewd business man and financier, Sterling, completely trusted Shoenberg, an engineer, scientist or applied physicist with a large dash of the visionary.'

When Shoenberg took over the direction of EMI's television project in October/November 1931 the current activities[17] included work on a film scanner, photocell amplifiers and cathode-ray tube reception under the supervision of C O Browne (assisted by three scientific assistants and two laboratory boys), and the construction of a cathode-ray tube (according to the RCA Victor design) by W F Tedham and an unknown highly skilled glassblower. This staffing was quite inadequate for the programme of research and development which Shoenberg had in mind. He has written[15]:

'When we started our work in 1931, the mechanically scanned receiver was the only type available and was under intensive development. Believing that this development could never lead to a standard of definition which would be accepted for a satisfactory public service, we decided to turn our backs on the mechanical receiver and to put our effort into electronic scanning.'

This could possibly be attained by the use of either soft or hard cathode-ray tubes but Shoenberg, remembering the vagaries and instability of the soft valves of the 1912–1916 period, decided against the soft cathode-ray tube and determined to direct EMI's research towards the evolution of hard vacuum camera and receiver cathode-ray tubes.

Shoenberg's first task was to strengthen the research team and to obtain a licence[14] from the General Post Office which would allow EMI to transmit experimental broadcasts from its research building at Hayes. In this task Shoenberg was considerably aided by the depressed economic conditions which prevailed in the United Kingdom in the early 1930s. Because of this situation EMI never needed to advertise its research and development posts, and all the staff were hand picked.

One of the earliest recruits was Dr (later Professor) J D McGee[18]. He had been for three years a research student at the Cavendish Laboratory, University of Cambridge, engaged on nuclear physics research under Lord Rutherford and Sir James Chadwick. Sometime during 1931 a group from the laboratory had been invited to see a demonstration of the mechanical scanning system being developed by J L Baird. At about this time McGees' tenure at the Cavendish came to an end and he had to seek employment.

He has stated that even with the backing of Rutherford and Chadwick it was not easy to find a post. Eventually Chadwick advised him to accept a position which EMI had offered to him and he joined the research department on 1 January 1932.

Further appointments were made and by June 1934 the research department comprised 32 university graduates, 32 laboratory assistants, four glass blowers, four girl pump operators, one coil winder, three mechanics, 25 instrument and tool makers, five girl assistants, seven draughtsmen designers and one designer draughtsman, a total of 114 persons[19]. Of the 32 graduates, nine had PhDs—despite the fact that PhDs were not particularly common in the early 1930s—and ten had been recruited direct from Oxford and Cambridge universities.

Many of these engineers/scientists were highly individualistic and exceptionally capable. A D Blumlein[20], who was killed in 1942, was possibly the greatest British electronics engineer of the 20th century; E C Cork, in charge of aerials research, had at one time to choose between a career either as a concert pianist or as an engineer; C O Browne, in charge of the studio transmitting section, excelled at tennis and played at Wimbledon; and M B Manifold of Cork's section, had a unique clause in his contract to have Wednesday afternoons off to permit him to engage in fox hunting. He also raced at Brooklands. Several senior engineers/scientists, namely Drs J D McGee, H Miller and O Klemperer became university professors and two, Dr McGee and J L Pawsey, were later elected Fellows of the Royal Society.

The two most senior members of Shoenberg's staff were G E Condliffe, formerly of HMV, and A D Blumlein who had been with the Columbia Graphophone Company. Condliffe was an efficient and able research manager, and Blumlein had a roving commission. Such was his genius that he had Shoenberg's authority to enter any of the laboratories and engage in discussion with any of the research staff. He had a seminal effect on all work undertaken in the research department and contributed greatly himself to all aspects of EMI's research projects. Being so full of ideas and of a modest disposition he went out of his way to give credit to others. This generosity engendered complete trust so that staff would discuss with him even the most 'half baked ideas'[21].

Funding was generous. Any equipment considered necessary to advance the research projects was agreed at once. Initiative was encouraged.

Moreover, working conditions for the staff were very agreeable. Shoenberg believed in giving his engineers time to relax and to reflect: the annual vacations varied up to eight weeks in duration depending on seniority and length of service. For most firms in the 1930s a holiday allowance of two to three weeks was the norm.

Such munificence had its reward for EMI. The staff were highly dedicated and, for many, standard working hours meant nothing. Blumlein, for example, would work until late in an evening if he was pursuing a particular idea, and if a thought occurred to him on a Saturday or Sunday he would return to the laboratory and conduct practical work to test the idea's validity. Later, during the war, he and others would on occasions, sleep in their laboratories.

To support his powerful research team Shoenberg succeeded in persuading the EMI Board to invest about £100 000 per year in EMI's research and development work on television. This growth in the Company's TV endeavours was the harbinger of an ominous situation for Baird Television Ltd which could not match EMI's staff and financial resources.

The challenge which faced Shoenberg—and which was being faced by Farnsworth and by Zworykin—was immense. Some indication of the difficulties which had to be overcome before a reliable all-electronic television could be commercially marketed have been given by McGee[16]. After his appointment in 1932 he became responsible for the design and fabrication of EMI's camera tube known as the emitron. Hence, his comments are based on first hand experience.

1   'Photoelectricity: in spite of the fact that Einstein had made this an honest physical phenomenon some twenty five years previously, we had still not begun to understand the physics of photosensitive materials. The physics of the solid state was still in its infancy. The technology of photoemissive and photoconductive materials was 'rule-of-thumb'—almost black magic. The first reasonably efficient photoemissive surface, the silver-caesium (Ag-AgO-Cs or S1) photocathode, had only just been discovered by Koller in 1930; it was to be four years before Gorlich discovered the antimonide photocathodes.

2   'Vacuum technique was pretty primitive, but was just beginning to be adequate. Glass-to-metal seals were very unreliable, and Pyrexglass was just becoming available. For example, when I mentioned to a former Cavendish colleague—recognised as a leading vacuum expert—that Tedham and I were trying to make a hard c.r.t. with a thermionic-oxide cathode, he exploded in rather rude laughter.

3   'Electron optics had not been formulated as a subject. Although Busch had given the first clue in 1927, even he referred to the magnetic lens for a c.r.t. as the 'concentrating coil'. Only with the publication in 1932 of the classic paper by Knoll and Ruska did we realise the complete analogy between light-optics

and electron-optics. This was a fundamental requirement for both the transmitting and receiving cathode-ray techniques for television.

4   'The physics of the solid state, on which the physics of phosphors depended, was still unformulated and hence the physics of phosphors was in much the same state as that of photoelectric materials, yet the former were as important for the effective display of a TV picture on a c.r.t. as the latter were for the generation of picture signals.

5   'Secondary electron emission was known as a phenomenom, but very little was known about the important details of the effect. However, it was of fundamental importance in all cathode-ray tubes in which a stream of electrons bombard highly insulating targets or fluorescent screens. How does the target avoid being driven more and more negative until the electrons can no longer reach it? Only, of course, because of secondary electron emission.

6   'Radio communication techniques were unbelievably primitive compared with today. Amplifiers of reasonable bandwidth, scanning circuits, pulse circuits, even smoothed h.t. power supplies, radio transmitters with bandwidths of more than a few kHz with a carrier frequency of 50 MHz and so on, all had to be invented as we went along.'

In view of the state of the art in 1931 the endeavours of Farnsworth, Zworykin and Shoenberg to implement an all-electronic television system, deserve much praise.

A report, written by Condliffe and dated 6 November 1931, on 'Programme of work for the advanced development division' lists the c.r.t. (cathode-ray tube) developments which were being progressed[17]: construction of c.r.t. according to the present RCA Victor design; improved tube design for lower modulation voltages; investigation of fluorescent screens for rate of decay of fluorescence and colour; development of commercial methods of manufacture; development of high voltage tubes for image projection; development of a cathode-ray scanning device—this was considered to be an important line of development for the production of scanning devices for outdoor use; construction of large vacuum photocells for studio work. The list illustrates the intentions and commitment of EMI regarding all-electronic television at the end of 1931.

With some of this work EMI was not starting *ab initio* for the activities of Tedham, at HMV, had been fruitful. His investigations[22] of photoelectric cells for talking film and television usage had led to the fabrication of a vacuum cell which had a higher sensitivity than that of any cell which could be purchased in 1931. This project had given Tedham first hand experience of Ag-AgO-Cs photosensitive surfaces which were to be used in EMI's camera tubes. And Tedham when employed by HMV had made a start on the construction of c.r.t.s. Associated with these ventures were the techniques of glass blowing, chemical deposition and vacuum technology, techniques which were to be greatly expanded as Shoenberg's research programme enlarged.

Sixty years on it is perhaps hard to appreciate the state of this technology as it existed in the early 1930s. Fortunately, Dr G Lubszynski, who joined EMI in 1933, has left a first-hand account. Since very few such accounts are extant, some of Lubszynski's recollections are reproduced below[21].

Figure 19.6   Final stage of glass blowing after assembly of an emitron, in the EMI research laboratories

Reproduced by permission of Thorn EMI Central Research Laboratories

'The standard vacuum system was all Pyrex, backed by an oil pump of the rotating vane type. All the glass work, including diffusion pumps and even vacuum taps, was made in-house by the glass blowing section of about 10 glass blowers. Emitron bulbs and optically flat windows were bought in, but gun necks, pinches etc, were made and sealed together by our glass blowers. Pumping and processing an emitron took at least a day. Pressure was measured in a McLeod gauge also made in-house. Ionisation gauges came very much later. The amount of asbestos flying around makes me wonder how so many of us managed to survive. It was probably because we did not know the dangers. Ovens and pump stands were made in our workshop. Tubes were baked for hours at 400°C and we would have been horrified by present techniques of pumping vidicons as low as 125°C to avoid melting the indium seals. Gun cathodes were a law to themselves. Some would emit after hours of torturing. Others would be cracked out, a new cathode mounted and hopefully sealed in again. Precision drawn glass tubing for gun necks did not exist and it was left to the glass blower's skill to point the gun to the centre of the target. The proximity of the flat window to the gun neck created problems of stress in the glass work. The mica assemblies were rolled up, inserted through the side appendix, left to spring open and sealed to the bulb at the four corners. We had a "dust free" room approximately 6' × 8' inhabited by two girls in white overalls. The components were handed in and the complete assemblies handed out through a hatch, carried along a corridor to the glass blowers for sealing in. How we could ever produce tubes without spots I do not know to this day [1986] . . .'

When Lubszynski left the Telefunken Research Laboratories in April 1933 to join later EMI, 'hard vacuum cathode-ray tubes were still a glint in the eye', but when he first visited the EMI Research Laboratories in July 1933 'there were literally dozens of batches of hard tubes of all sizes on life test'.

McGee found on his arrival at EMI, in January 1932, that Tedham had advanced remarkably rapidly considering the resources at his disposal, and already had electrostatically-focused cathode-ray tubes operating well. Moreover, McGee soon learnt that Tedham was very familiar with the literature of the whole field, and especially with Campbell Swinton's ideas.

By May 1932 they had formulated some notions on 'cathode-ray tubes for scanning'. The formation of the mosaic for a camera tube was described by Tedham in a patent opening memorandum dated 12 May 1932.

'The little photoelectric elements are best made up of caesium on oxidised silver, and it is with the formation of the "mosaic" of silver backing that this invention is chiefly concerned.

'Various methods have been proposed to form this backing. A thin silver layer may be formed chemically or otherwise, on the plate, and this may be subdivided by ruling in two directions. This method is rather slow and elaborate. Methods have been proposed in which silver salts are dusted on to the plate and reduced to silver or in which a thin silver layer is caused to agglomerate into spherules by heat treatment.

'The use of reduced photographic images on a Schumann plate has also been suggested.'

With these methods, mounting of the mosaic plate was commonly effected by sealing, by heat, the support wires into the glass, but as Tedham noted: 'This process is damaging to the silver mosaic, which [becomes] unevenly "burnt"—the silver taking on a whitish appearance.'

Clearly, Tedham was familiar with the divers methods of making the mosaic and had undertaken some work on it fabrication. He suggested a new method using a mesh screen, and in a memorandum[23] to Shoenberg, dated 12 May 1932, Condliffe mentioned: 'This has been tried out experimentally, and shows very considerable promise. The other methods described, particularly the one covered by the [RCA] Victor Docket, have also been tried out and proved inferior.' He also noted that the method followed closely that employed in the deposition of the silver surface in photoelectric calls, 'but the use of a fine mesh screen for a scanning tube is, I think, sufficiently novel to form the basis of a good and valuable patent.' The patent[24] was applied for on 25 August 1932.

At this stage in the early history of EMI's development of an electronic camera tube the exact sequence of events is not known. According to McGee, work on such a tube, at that time, was not part of the official policy of the company and permission was not given to them to construct a tube[25]. Nevertheless, Tedham and McGee could not resist the temptation to ignore official policy and as they had available most of the required components and techniques they constructed during the autumn of 1932 an experimental camera tube, Figure 19.7. This was based on their patent[24] of 25 August 1932.

The target of the tube was an aluminium disc (3) with an insulating anodised layer on the surface facing the electron gun (6). A metal mesh of about 50 meshes per inch was mounted near this surface and held as close to the surface as possible. In the neck of the tube a heating filament carrying a silver bead enabled silver to be evaporated through the mesh to form a pattern of square patches of silver on the aluminium oxide. Surrounding this a continuous ring of silver was evaporated, Figure 19.7.

Following evaporation the silver bead and stencil mesh were removed and an electron gun, which used electrostatic focusing of the type which Tedham was using

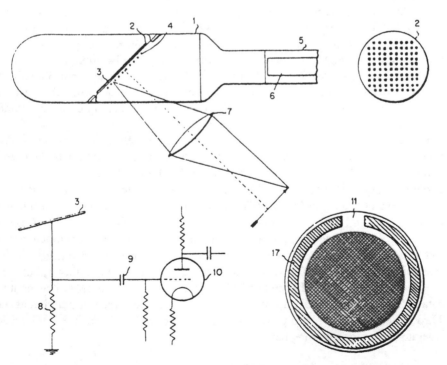

Figure 19.7   Single-sided mosaic target plate camera tube (from Tedham and McGee's patent of 25 August 1932)

in his cathode-ray display tubes, was sealed into the same narrow glass neck. The plane of the target was at an angle of approximately 45° to the axis of the tube so that the optical image could be projected normally onto it, even though the light rays had to pass through the curved glass wall of the tube.

After further baking and pumping, the silver mosaic and peripheral ring were oxidised in a glow discharge in oxygen. Caesium vapour was then admitted to the tube while it was being baked at c. 180°C and the increase in photosensitivity of the mosaic surface measured by means of the conducting ring. When 'moderate sensitivity' had been obtained, the tube was sealed off from the pump.

McGee has written[25]: 'We borrowed a signal amplifier from C O Browne and drove the scanning coils on the neck of this tube in parallel with those on a display tube from the same scan generator. We projected a simple draughtboard picture onto the mosaic and after some minor adjustments a picture appeared on the receiver—it appeared as if by magic . . . The definition was as good as one could expect from such a coarse mosaic, there was little image lag and it seemed reasonably sensitive.' There were also some spurious signals, that were a feature of the later emitron tubes, which came to be known as shading signals.

Regrettably for Tedham and McGee, the picture soon began to fade due to gas evolution from the electron gun cathode. No photograph of the image or of the apparatus was taken but the tube has survived and is in the Museum of Photography, Film and Television, Bradford.

On 22 August 1932 Shoenberg became director of research. It seems that having spent £20 000—a very large sum in 1931–32—on an apparatus which would generate television signals from film, EMI was loth to spend more money on something which would, or might, render this apparatus obsolete. Lubszynski has written[26]: 'Legend has it (both McGee and Condliffe telling me years later) that Shoenberg forbade work on tubes and that Condliffe always rang Tedham when there was a danger of Shoenberg coming over [to the laboratories] whereupon all tubes and other signs were quickly put into cupboards.'

Certainly, EMI by the end of 1932 was keen to initiate a television service using film images rather than studio generated images. On 21 October 1931 Condliffe and Whitaker had talks[27] with Colonel Angwin and Mr F Gill of the engineer in chief's department, General Post Office. They sought permission to set up dual short wave transmitters, operating on 4 m and 6 m, at a maximum power output in each antenna of 2 kW, the bandwidths of the radiated signals being 500 kHz and 25 kHz. Angwin did not foresee any objections provided that EMI did not deliberately attempt to transmit material of high entertainment value and did not cause interference with the transmitter stations of the War Office and Admiralty situated in Westminster, London.

A few weeks later on 8 December 1931 a licence[28] was sent to EMI allowing it to establish, for experimental purposes, a wireless sending and receiving station at the Hayes premises of the company[29]. In compliance with this licence EMI could operate within the following bands:

c.w. and telephony	62.01–61.99 MHz
television	44.25–43.75 MHz

and at a power of up to 2 kW into the antenna. (The licence details show that some-time between 21 October and 8 December 1931 EMI had abandoned its plan to use RCA transmitting equipment which worked[10] on 4 m and 6 m.)

The practical association between EMI and the Marconi Wireless Telegraph Company Ltd, which was to prove so beneficial to both companies, began late in 1931 when Shoenberg invited MWT to supply a low power v.h.f. transmitter, complete with a modulator, for use with EMI's film scanner. N E Davis, MWT's transmitter expert, was instructed to prepare a transmitter and modulator for this purpose and subsequently the units were sent to EMI in January 1932. The transmitter had an output of 400 W at a frequency of 44 MHz and employed grid modulation of the final stage.

Davis was also responsible for the installation[30] of the transmitter and antenna system, which consisted of a self supporting half-wave dipole, Franklin feeder and terminating arrangement[31]. According to Davis, since the EMI research group was 'completely without knowledge' of ultra short wave transmitters, methods of modulation, and so on, he remained at Hayes until April 1932 to advise and generally assist in overcoming any initial problems[32]. Soon afterwards EMI's first 120-line, 25 frames per second film transmissions were successfully transmitted.

On the 11 November 1932, Shoenberg invited[33] the BBC's chief engineer, Mr N Ashbridge to a private demonstration of both the transmission and reception of television. 'In my humble opinion,' wrote Shoenberg, 'they would be of quite considerable interest to you.'

Ashbridge visited the Hayes factory on 30 November and was shown apparatus for the transmission of films using four times as many lines per picture as Baird's equipment and twice as many pictures per second. He was impressed and thought the demonstrations represented by far the best wireless television he had ever seen and that they were probably as good as or better than anything that had been produced any where else in the world. He wrote[34]:

'. . . there is not the slightest doubt that a great deal of development, thought and expenditure had been expended on these developments. Whatever defects there may be they represent a really remarkable achievement. In order to give some idea of the cost of such work, I might mention that the number of people employed is only slightly less than that in the whole of our research department.'

The actual demonstration consisted of the transmission of a number of silent films, over a distance of approximately two miles, by means of an ultra short wave transmitter using a wavelength of 6 m and a power of about 250 W.

On the quality of the images Ashbridge reported:

'The quality of reproduction was good, that is to say one could easily distinguish what was happening in the street scenes and get a very fair impression of such incidents as the changing of the guard, the Prince of Wales laying a foundation stone and so on. A film showing excerpts from a play was in my opinion not so good. Also, it was possible to follow what was going on all over the stage. On the other hand, excerpts from a cartoon film were definitely good. I think they could have given a better demonstration had they been in possession of better films. The ones they showed had been in use for several years. The size of the screen is about 5 inch × 5 inch but they have a second machine which magnifies this by about four times in area. The quality of reproduction can be compared with the home cinematograph but the screen is smaller.'

EMI had asked the BBC to take up this system—on an experimental basis—for about seven or eight months and then later for regular use on their programmes. On this issue the chief engineer had the following observations to make[34]:

1   'There is no doubt that the film is the only way in which we can develop the television of actualities. I cannot see any method developing in the immediate future so as to allow the direct televising on a satisfactory basis of, say, the finish of the Derby or the Wimbledon tennis matches at any rate within the next few years.

2   'Transmission by film would be entirely satisfactory for plays and sound could be added on another shortwave channel if it were not suitable for broadcasting.

3   'The above remark (2) would mean that our technique would need to change considerably and that, if the system were established on a programme basis, it would mean practically the establishment of a further alternative programme and this would be extremely costly, particularly since it would be necessary to transfer everything onto a film at any rate at the present stage of development.

4   'Provided that sufficient apparatus were available, a film could be made and re-transmitted within a matter of only a few hours.

5   'It seems to me highly desirable to develop the system so that direct television could be carried out of a studio performance. There seems some doubt as to the immediate possibility of this.

6   'No incident could be televised on this system which occurred at a distance from the transmitter on account of the wide frequency band involved, that is to say that ordinary music line cannot be used.

7   'The above implies that the studio and transmitter should always be fairly close together, that is to say a matter of yards, not miles. The range of these ultra short waves cannot approach the ranges we are in the habit of reaching with our ordinary programmes, but it would be possible technically to erect a number of transmitters in a number of important towns and cities.

8   'The cost of a receiver to take short wave channels, one for television and one for sound with ordinary broadcasting facilities in addition is estimated by EMI at about £100 so that the service if established would have to be looked upon as a luxury service, possibly entailing a special licence.

9   'If we took up the transmitting side we ought not to have difficulties in

connection with development on the receiving side, having regard to the fact that EMI are prepared to make all the receiving apparatus and even manufacture their own cathode-ray tubes.'

EMI was very keen that some form of television service should be started on ultra short waves, and following Ashbridge's visit to Hayes, Mr Alfred Clark, the chairman of the firm, paid a visit to the BBC to have discussion with the director general[35]. Clark was anxious to know what standards would be adopted for television. He hoped Reith would say that the number of pictures per second and the number of lines per picture would be 25–30 and 120–180, respectively. EMI would then have been in a position to have started an experimental service on ultra short waves early in 1933, and probably before Baird Television was in a position to do so.

The emergence of a competitor in the form of EMI caused J L Baird and his associates much unease. They could not acknowledge for some considerable time that EMI's television system was being engineered by British workers in a British factory using British resources. For them the Radio Trust of America, through its associated companies in London, was the mainspring of EMI's progress. A very noticeable bitterness is evident in both Baird's and Moseley's letters during the early 1930s on the progress, and support from the BBC, of the Hayes company. Baird was always ready to point out that his firm could match the advancements towards high-definition television which were being made by EMI and that therefore the pioneer company, i.e. Bairds, should be supported.

In a letter dated 6 December 1932 to the Director General of the BBC, Baird mentioned[36]:

'We have spent considerable time developing apparatus for use with ultra short waves, both with cathode-ray and mechanical scanning, and have gone as far as 240-line scanning. This apparatus, however, is entirely unsuitable for the wave band owing to the immensely high

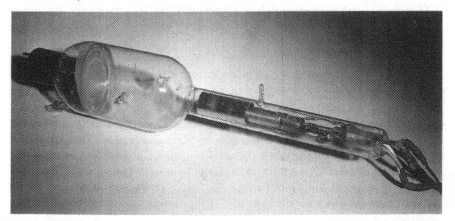

Figure 19.8   An early experimental emitron camera tube

Reproduced by permission of the Science Museum

frequencies involved, but the pictures produced are of course immensely superior.'

Baird raised this matter because he understood that EMI had been giving demonstrations of apparatus under laboratory conditions and he did not want these demonstrations compared with those which the BBC was sending out on medium waves.

'Our own results in the laboratory are far superior to those which we are sending out through the BBC, but such results are only of academic interest until the ultra short wave channel is sufficiently developed to pass the very high frequencies involved.'

Baird's concern to stress this point was evidently his belief 'that the American Radio Trust through its subsidiaries [was] endeavouring to create the impression that it [had] something which [was] superior to ours'. The inventor said that his company could supply transmitters with from 90 to 240 lines as soon as the ultra short wave service had been evolved by the BBC's engineers.

Obviously it was not true to say that the EMI system was American based. As Shoenberg[37] later told Lord Selsdon and his colleagues at a meeting of the Television Committee: 'We are making all our equipment at Hayes, right down to the last screw, including the cathode-ray tubes and photocells.' Shoenberg was eager to clarify the issue of the relationship of EMI and RCA because of the rumours which were being spread about that EMI was a foreign company or was controlled by foreign interests.

When RCA was formed in 1919 the Marconi Company made an agreement with RCA and by virtue of this the Marconi Company gave RCA all the patent rights which it owned in America. Similarly, RCA and GE gave the rights in their patents to the Marconi Company[38]. The field of the agreement covered wireless communications. Then in 1929 the Marconi Company sold its interests in radio receiver patents and other patents which related to home entertainment to the Gramophone Company, the Marconi Company reserving to itself the rights for the transmission of traffic or broadcasts.

Of course, when EMI was created the rights of the Gramophone Company went to EMI, but the former company had no right from RCA 'to any manufacturing or laboratory information'[39]. Indeed, during 1933–34, RCA had tried very hard to induce EMI to pay a certain percentage on its turnover, for the rights to RCA's manufacturing and laboratory information. Shoenberg had urged the company not to agree to this: he did not think it was worthwhile. All that EMI obtained apart from the patent rights was the advantage of seeing the patent specifications which RCA was filing in the UK, so that EMI's researchers were able to read and study them before they were published. 'I wish to assure you,' Shoenberg informed[37] the Television Committee in June 1934, 'that all the development of television has been done here, even in connection with fundamental points where we very often disagree with American practice. They are trying now to obtain our patents for the USA but we are not under any obligation to give them the patents.'

Figure 19.9   Sir Noel Ashbridge (1889–1975)

Reproduced by permission of the BBC

As part of its television engineering programme, EMI wanted its apparatus to be operated in Broadcasting House, London, by the BBC, while EMI continued to experiment and incorporate any improvements into the equipment.

Although Ashbridge had been surprised by the advances of EMI and was entirely in favour of further research being carried out in television, he disliked the rush tactics of the company. 'They are too much like the Baird Company's tactics in the early days,' he said, 'and the object I think may be the following: to try and rush us into establishing a service as quickly as possible so that they can get well ahead in the receiver market for television sets, although no doubt they would be prepared to licence other people when it paid them to do so.'[34]

Ashbridge thought that the BBC should make it clear that it was not in favour of a regular service, mass production and all the rest of it being started next summer (i.e. 1933), and felt that at least six months' further experimental work was required. The position[34] was essentially this:

'Let us assume that they persuade us to establish a regular programme next September: it would cost us shall we say £20 000 a year to carry this on even in a limited way and they would probably try and make us pay for the television transmitter and no doubt we should pay for the two ultra short wave transmitters, further accommodation and so on: they on their side might sell a thousand receivers in the first year to a few people who made a fetish of buying anything new and have the money to spend. This might go on for several years, EMI selling a thousand or so receivers each year and ourselves paying £20 000 a year to keep the service going. They would be making a small profit and we should be making a large loss, even if an extra licence fee were paid for television reception. This is very unattractive from our point of view. On the other hand, if a very large number of receivers were sold, that is to say if the public took the scheme up with enthusiasm, then I think we should be more or less bound to go ahead, partly because we are a monopoly concern and partly because in any case television may eventually help us in our main objective. What, however, I dislike is the nature of our paying out a very large sum of money in order to let EMI make a profit of a few thousands a year for a year or two with the possibility of making a much larger profit later on but with practically no risk whatever, except the danger of losing the sum they have spent on development. In any case, they would gain considerable publicity. However, you may say that we are entirely justified in making what may amount to a loss of £100 000 in order to establish television at the end of a period of three or five years.'

Ashbridge's conclusion was clear: 'I feel strongly against trying to dump a rather doubtful service on the public so soon as next autumn.'

Against this background of the prospect of higher-definition television, the BBC had to consider the position of the Baird 30-line programmes on ordinary broadcast wavelengths in the medium band. These programmes had been started, first, to determine whether a good producer could make anything of them from the programme point of view in relation to public interest, secondly, to gain experience of the technical and programme problems presented by such a service and, thirdly, to give Baird the opportunity of progressing further in his research.

It had been recognised for many years that 30-line television would never be a satisfactory long-term solution to the broadcasting of television, and with the advent of medium definition (120–180 line) television reproduction, the BBC had to give careful examination to the future of the Baird process and in particular the mass production of low-definition receivers. Ashbridge did not think that the programmes which had been broadcast were such as would hold public interest on a permanent basis and he therefore felt that the BBC's agreement with Baird Television should be terminated in March 1934.

EMI's desire to install equipment in Broadcasting House was entirely reasonable. The company's experiments had been carried out under country conditions and it was necessary for a further period of experimental tests to be undertaken in a large city in the pursuit both of transmission and reception. No tests on u.s.w. transmission with 120-line television modulating signals had been effected in the UK from a built-up area. Thus, the influence of other large buildings and objects on the characteristics of the propagation of the waves was a necessary investigation before a regular service could be started. Also, of course, a high building was required in order to give an adequate range. Neither Reith nor Ashbridge had any objection to such tests[40].

Baird Television Ltd was formally informed of the proposed cooperation between the BBC and EMI at a meeting held on 27 January 1933[41]. Naturally, Baird and Moseley expressed their disquietude at such a prospect. The tenor of concern was given in Moseley's letter[42], dated 28 January 1933, to the Postmaster General, Sir Kingsley Wood. 'I wonder whether, in the welter of cynicism of modern politics, there is any sincerity in the pleas for "British First"?.' He was strongly of the opinion that British pioneers should be encouraged, not by finance but by every other legitimate means.

'I am not satisfied that the BBC realises its duty to the country in this respect,' he wrote, and observed: 'The BBC which holds a monopoly by virtue of a charter granted by HM Government, seems to me to be extraordinarily cynical where the rights of a British sister science are concerned.' On the question of the 'tentacles of the Radio Trust of America' extending throughout the world, Moseley feared that, if the PMG did not take steps immediately, one of them might 'force a means of "muscling-in" through the back doors of the BBC'. If this were attempted, it would be a public scandal and Moseley would not hesitate to call a public meeting.

The controversy, which is described in abundant detail in the author's book 'British television, the formative years', was resolved when[43]:

' . . . the PMG [believed] that it would be right to postpone a decision in regard to the institution of tests of the EMI apparatus at Broadcasting House until the demonstration of Baird's apparatus [had] taken place. The arrangement was that both EMI and Bairds would give demonstrations to be witnessed by the BBC and the Post Office and that a decision on the installation of EMI apparatus at Broadcasting House should be postponed until the results of these demonstrations have been considered[43]'.

The demonstrations were arranged for the 18 and 19 of April 1933 at Long Acre and Hayes, respectively[44], and were seen by Admiral Sir Charles Carpendale, Mr Noel Ashbridge, Mr H Bishop, all of the BBC, and Colonel A G Lee, Colonel A S Angwin, Mr A J Gill, all of the engineering side of the Post Office, and Mr L Simon, Mr F W Phillips, Mr W E Weston and Mr J W Wissenden of the administrative side.

The Baird apparatus was demonstrated over wires between neighbouring rooms in the Long Acre premises. It was described by Simon in the following terms[45]:

'The transmitting apparatus was of a makeshift type and, at the receiving end, pictures about three inches by three inches were produced in black and white on the broad end of a funnel-shaped cathode-ray tube in two cases, and by Nipkow disc in a third. Films were fed into the transmitters: but the received pictures were in all cases indistinct, jerky and erratic. It was stated that arrangements were being made for the presentation of a picture nine inches by five inches. The best that could be said for the demonstration was that it was an interesting experiment in picture transmission with rather crude apparatus.'

Baird Television still did not seem to appreciate the need to give professional demonstrations. In the company's favour it might be fairly urged that the equipment had been assembled in less than two months; but on the other hand it had fixed the time limit for the demonstration, and in a letter dated 17 February Baird had stated that his affiliated company in Germany had already supplied a similar transmitter to the German Post Office[46].

At Hayes the EMI apparatus was demonstrated by wireless transmission, the transmitting apparatus being at the works and the cathode-ray tube receiving set in a cottage two miles or so away. Simon observed that the complete receiving apparatus for sight and sound was complicated and involved the use of 25 valves; but it was claimed that the number of valves could be reduced and the apparatus otherwise simplified so as to reduce the cost of a television set to about £80 or £100[45]. (Baird claimed that his company's receiving set could be manufactured in bulk for about £30 or £40.) The demonstration consisted of the reproduction of films on a screen giving an image size of 6½inch each way. In one case the receiving set gave a black picture on a white background and in the other on a green background.

'The action on both pictures could be followed clearly throughout, without the guidance afforded by the accompanying speech; but the detail on the green background was superior to that on the white. A very high degree of stability was achieved. The company are experimenting with various substances on the cathode-ray tubes with the object of securing a black and white picture without loss of detail. They are also experimenting in the further magnification of the received pictures and a demonstration screen picture nine inches square was shown. The Post Office engineers, who were present at the previous demonstration in February last, considered that marked improvements had been achieved.'

Obviously, the EMI engineers were advancing rapidly in the art of television. They had excellent facilities at their disposal and a highly trained, very competent and inventive staff. In these respects they had a great advantage over the Baird company.

How long Shoenberg's putative embargo on the development of camera tubes persisted is not known. It appears to have been very short lived. There is documentary evidence that in November 1932 and subsequently McGee was undertaking some fundamental laboratory investigations on the secondary electron emission characteristics of 'experimental mosaic tubes'. His laboratory notes show that he was studying the phenomena in relation to silver oxide and caesium surfaces. The interaction of photoelectric and secondary electron emission was little understood in 1932 and so some understanding of the mechanism was necessary before an electron camera tube could be designed and engineered. McGee's early (November 1932) conclusion[47] on such a development was not too sanguine:

'The photoelectric emission is superimposed on this [secondary emission] with considerable confusion as a result. It is difficult to see how reasonably accurate reproduction of light intensities can be expected.'

Observations on the secondary emission properties of aluminium and of silver targets were made in January 1933. His laboratory notebook records, for 2 February 1933[48]:

'A tube was constructed as shown in the diagram. The target was an Al [aluminium] plate quite clean on the side on which the electron beam impinged and on the other side covered with a Ag [silver] layer on which a Cs-CsO [caesium-caesium oxide] photoelectric cathode was formed. A grid was placed opposite the photoelectric cathode. The tube was quite hard and gave a fairly good p.e. emission. The usual measurements were made of the secondary emission from the Al plate under the [influence of the] incident beam.'

It seems that while McGee was endeavouring to understand the physics of camera tube mosiac plates, Tedham was concerned with the chemistry of their preparation. A memorandum[49] from Condliffe to Shoenberg, dated 7 February 1933, refers to a method of forming the silver photocathode for cathode-ray transmitting tubes. It was an adaptation of the method of depositing silver disclosed in the BTH photocell patent and had 'very considerable advantages over the method of forming the mosaic surface disclosed by Zworykin in [a] Victor Docket, since the tube is not opened up, and there is accurate control over the size of the particles of silver deposited'.

Condliffe's final paragraph gives an indication of EMI's policy at this time. He, and presumably the research staff, could foresee that direct television would eventually partly replace television from films and opined: 'We shall ultimately be interested in cathode-ray scanning tubes, and completion of this patent is recommended.'

Both single-sided and double-sided mosaic plates were being examined by EMI early in 1933. A note[50] written by Tedham outlines the use of an aluminium disc, having an oxide coating (c. 0.002 inches thick or less), which acts as a capacity electrode to the silver spots on the outside of the oxide layer. 'This method gives a strong light plate with a higher capacity between each spot and the aluminium than can be obtained between electrodes separated by a thin glass disc (which is very weak if thinner than about 0.010 inches).' It was essential for the capacitance to be as large as possible, since the successive potential changes accompanying the discharging of the spots were shared between this capacitance and the grid cathode capacitance of the signal amplifier.

The first of EMI's experimental camera tubes with a double-sided mosaic target was specified in a memorandum[51], written by McGee, early in 1933. Such tubes have two important advantages: first, the scanning beam and the associated secondary electron emission are separated from the photoelectric emission. The target potential can be stabilised at, say, near the anode potential, by the scanning beam, and the photoelectrons can be collected from the photemissive surface on the other side of the target. Secondly, such an arrangement has the practical advantage that the electron beam can fall symmetrically onto the scanned surface and the optical image can be projected normally on the other, photosensitive, side.

The main features of McGee's invention are shown in Figure 19.10: his description follows[51,52]. A glass envelope, 1, has a plane glass window, 9, at one end and a neck at the other in which is fitted a standard electrostatic focusing electron gun. The thermionic cathode of this gun is preferably raised to a negative voltage and the second anode is kept at earth potential so that a beam of electrons, 21, is projected into the chamber with a velocity corresponding to the voltage applied between the anode, 7, and the cathode. Inside and parallel to the glass window, 9, is fitted a grid of metal wires which is also earthed and close to this grid is the mosaic plate, 11, on which the electron beam falls.

The plate, 11, is made up of a large number of metallic conductors of small diameter and a length equal to the thickness of the plate placed side by side and insulated from one another by a thin insulating layer. The ends of those conductors which face towards the grid are made photosensitive. A metallic ring, 12, is placed on the other side of the plate, 11, to collect the secondary electrons and is connected to the grid of the first valve of the amplifier.

The simplistic theory of operation of this device was that the removal of photoelectrons liberated by light from the mosaic elements would cause them to become positively charged, and that these charges could be discharged by electrons of the beam as it scanned each frame, and the residue of the beam thus modulated would be collected on the electrode, 12. This current passed through the signal resistor to produce the picture signal.

McGee's invention was patented[53] on the 5 May 1933. His assistant G S P Freeman constructed tubes of this design but they were never made to operate suc-

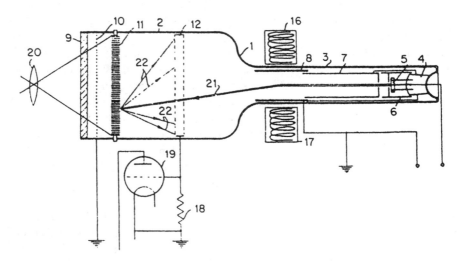

Figure 19.10 Early McGee (EMI) double-sided mosaic signal generating tube

Reproduced by permission of Thorn EMI Central Research Laboratories

cessfully, although McGee, in 1979, believed that the design was physically sound[54]. It was the difficulty of making the mosaic target of sufficiently fine structure that made it impracticable and led subsequently to more attention being given to a single-sided target.

These brief examples of the work in progress at EMI in 1932 and early 1933 are given to show that EMI's research and development work was not dependent on copying RCA's television activities. Dr Broadway, who was appointed in August 1933 has written (in 1958)[55]:

> 'My recollection is that some information on pick-up tubes had been received from RCA in the form of dockets, but that such information was rather scanty and its significance was not necessarily apparent; indeed, it was felt at the time that some of the explanations given by Dr Zworykin were incorrect and there was no other means of knowing what was going on in Dr Zworykin's laboratories.'

Following the Baird and EMI demonstrations of April 1933, a conference was held at the Post Office, between the BBC and the Post Office representatives, on 21 April 1933. It was agreed that[56]:

1   the EMI results were vastly superior to those achieved by the Baird company;
2   the results were incomplete because of the different transmission methods (line and wireless) used in the two cases (also the effect of electrical interference and absorption could not be tested);
3   further tests by wireless in a town area were essential to determine the range of reception and the effect of interference;
4   whatever system of synchronisation was adopted in the first instance for a public service might be liable to standardise the type of receiving equipment;
5   a test of one system could not be a reliable judgement on the results achieved by the other.

During the discussion the BBC said it was anxious to start trials of the EMI system, but considered the inability of the system at the moment to produce direct television a disadvantage, as the cost of film—about £30—might be prohibitive. When asked by the Post Office whether it would be prepared to give equal trials to both systems the BBC replied that it would prefer to test the EMI system first and only give a trial to the Baird system 'when—if ever—Baird succeeded in producing results equal to, or better than to those of EMI'.

No definite recommendations were made at the conference as to the future course of action to be taken. Television of high definition was now inevitable, although the BBC considered that progress towards a public service would necessarily be slow.

Unfortunately, the Baird company failed to realise that there was no future for low-definition television and persisted in its efforts to advance the sale of 30-line receivers.

J L Baird met the PMG, Sir Kingsley Wood, on 23 May 1933 and told him that his company were putting 100 £70 television sets on the market. He assured the

PMG that a warning would be given to the public about the possible discontinuance of the 30-line transmissions in March 1934[57].

On the question of short wave television broadcasting, Sir Kingsley Wood said that he would probably arrange for an experimental transmission by the two competing systems, but that Baird must face up to the possibility that 'the other system might prove superior, as there was some reason to believe . . . '.

Here was a clear warning to the company that its previously held position as the leader of British television developments could only be recovered by accepting a challenge: a challenge to produce a high-definition ultra short wave television system preferably using cathode-ray tube receivers and cameras.

Following several representations, particularly from Reith, the BBC's director general, the PMG reiterated his view that to give EMI only across to Broadcasting House might bestow the coup de grâce to the Baird concern[58]. Subsequently, both firms were allowed to install experimental equipment: Baird Television was permitted first use of the accommodation and transferred some of its apparatus to Broadcasting House in September 1933. However, its stay was not entirely successful and the company removed its apparatus in December 1933[59].

An indication of the progress of the company's efforts is given in a report written by Colonel Angwin, of the Post Office. After a visit to the Long Acre laboratories on 20 November 1933, he observed[60]:

> 'A cathode-ray tube was used at the receiver, this being of a type specially developed by the General Electric Company, having a diameter of 12 inch and giving a picture of 10 inch × 8 inch in area. A very marked improvement in definition and stability was observed compared with that previously demonstrated by Bairds and it is now considered to be quite equal in quality to that shown at Hayes by Electric and Musical Industries Ltd.

The demonstration had consisted of the transmission, over a short line, of three topical films at a definition of 180 lines per picture, 25 pictures per second.

Some of this progress probably stemmed from the appointment of Captain A G D West[61] as the technical director, in July 1933. He had been placed in this position by the Ostrer Brothers (of the Gaumont British Company) who now virtually owned Baird Television Ltd.

Born in 1897, West served as a wireless experimental officer in the forces during the 1914–18 war. After obtaining his MA (with distinction in the mathematical tripos) while at Cambridge University, he gained a BSc degree from London University and then returned to Cambridge as a research student in the Cavendish Laboratory. From 1923 to 1929 he was head of research at the BBC and was responsible for much original research in the fields of radio transmission and acoustic measurements. He left the BBC to take a similar position with the Gramophone Company, and in 1932 entered the film industry as chief recording engineer of APT Studios, Ealing. He was one of the founder members who formed, in 1931, the British Kinematograph Society, and subsequently became its second president.

In the same month that West was designated technical director, the Baird company had taken a lease at the Crystal Palace, Norwood, and rented part of the ground floor and the south tower from the manager, Sir Harry Buckland. The site was ideal for ultra short wave transmission experiments as the Palace was sited on one of the highest points in London. When the company acquired the accommodation it transferred over its work from The Studio, Kingsbury and the Long Acre laboratories. The Postmaster General granted the company a research licence to transmit on the following frequencies: 6040 kHz at 500 W, 1930 kHz at 250 W, 48.00–50.00 MHz at 500 W, and this enabled it to send experimental transmissions to Film House in Wardour Street, London[62].

July 1933 was also the month during which Zworykin delivered his paper in London on the iconoscope. One effect of this was that McGee was given approval in September to expand his small group, to advance EMI's camera research and development, but it did not include Tedham who now had to devote his efforts to cathode-ray display tubes[18]. Soon afterwards he suffered ill health and had to resign his position with the company.

In reviewing the situation at this time McGee has said, ' . . . it was quite clear that we could learn nothing from Zworykin's paper'. No practical information had been given and the preparation of the mosaic by a 'special process' had not been described. Apart from the knowledge that successful development was feasible, McGee's group had to develop all the necessary mosaic-making techniques itself.

At that time the only efficient photocathode surface was the S1 (Ag-AgO-Cs) which had been discovered by Koller in December 1930. McGee and Tedham knew from their 1932–33 experiments that such a surface could be formed on a silver surface by first oxidising the silver in a discharge in oxygen and then exposing it to caesium vapour at about 160°C. The difficulty in forming the mosaic lay in the preparation of the array of many thousands of individually insulated silver islands on the substrate material. There were three possible ways of achieving this:

1 by cross ruling a silver layer which had been deposited by evaporation *in vacuo*;
2 by evaporating silver onto a substrate through a stencil mesh;
3 by aggregating a thin layer of silver by heating to c. 600°C in air.

Of these methods McGee had had experience of the second since this was the method which Tedham and he had adopted when fabricating their 1932 camera tube. The technique had worked well but in 1932/33 the meshes available did not have a small enough pitch to give a fine-grained mosaic for a high-definition picture. (Indeed, such metal meshes were not obtainable until after the Second World War, when H E Holman, of EMI, developed the means to make meshes having more than 10^6 apertures per inch square, and a low shadow ratio. These were used as stencils in forming the transparent mosaic antimonide photocathodes for the c.p.s. emitron tubes.)

McGee has recorded[25] that the first ten experimental tubes produced at EMI used ruled silver mosaics on mica. While at the Cavendish Laboratory he had acquired a good deal of practice in splitting thin sheets of mica, for nuclear physics work, and so it was not long before he had sheets of mica of size $3'' \times 3'' \times 20\,\mu m$. These were metallised on one surface by painting with platinum and baking, followed by evaporation of silver, *in vacuo*, on the other surface. The surface was then ruled in the workshop using a fine sapphire point, in a machine adapted from EMI's record ruling equipment, at about 100 lines per inch in two orthogonal directions. This had the disadvantage that the silver surface was exposed in a not very clean atmosphere for many hours.

There were many problems in the processing of the tube which had to be overcome as a result of the rather incompatible processes which were necessary: degassing of the tube and electron gun; activation of the thermionic cathode; oxidation of the silver mosaic; and 'last and most difficult the introduction of [the] caesium in just the right amount to activate the silver mosaic to give good photosensitivity without impairing the mosaic insulation by bridging over with caesium the gaps between [the] adjacent elements'. By operating the tube while it was still connected to the vacuum pump the sensitivity and mosaic insulation could be monitored and adjusted either by adding caesium, or removing it by baking.

Of the first ten experimental tubes, number six gave 'a very presentable picture' on 24 January 1934. McGee's group demonstrated, for the first time, its all-electronic television system to the company chairman, Mr A Clark, and the director of research, Mr I Shoenberg, on 29 January 1934.

Meanwhile, Dr L Klatzow had been investigating the process of aggregating the silver films on mica to form an array of silver globules. These films soon appeared to be so much superior to those formed by ruling, from the point of view of uniformity, cleanliness and fineness of structure of the mosaic that from tube number 12 onward all mosaics were prepared by this process.

EMI applied for patents for both the stencil[24] and aggregate methods, but was not successful in obtaining a patent for the latter method. S Essig, of RCA, had also discovered the aggregation process and his patent application[63] preceded that of McGee by only a short time.

Progress was now rapid. 'Reasonable' tubes were being made by February 1934. When tube number 14 was made it was so much better in picture quality and sensitivity that a very experimental camera directed through the window of the laboratory, on 5 April 1934, enabled a daylight outside broadcast picture to be obtained[18].

The iconoscope, or emitron as EMI called its version of this type of camera tube, had several defects which had to be overcome before a practical camera could be marketed. Given a single-sided mosaic onto which both the scanning electron beam and an image of the scene to be televised were incident, it is apparent from Figure 19.11 that either the optical image must be projected normally onto the mosaic target and the scanning beam projected obliquely, thereby giving rise to keystone distortion of the raster, or the converse. In either case, a considerable depth of field is required, either of the electron lens or of the camera lens, respectively.

In 1934/35 the target area of the mosaic was 5 × 4 inches. This called for a long focal length lens, about six inches, to cover it. At full aperture, f/3, the optical depth of field was insufficient for the second of the above alternatives. However, by lengthening the electron gun and by stopping down, the electron lens could be designed to

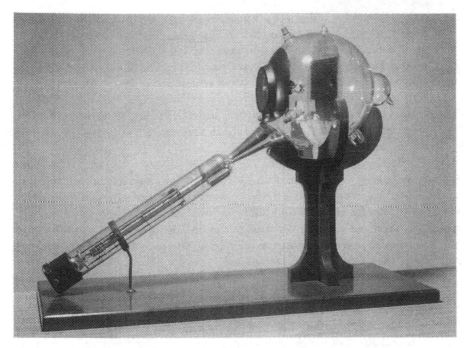

Figure 19.11 A completed emitron camera tube

Reproduced by permission of the Science Museum

give a uniform beam focus even when the beam was incident obliquely onto the mosaic. For this reason the arrangement given in Figure 19.11 was chosen. Necessarily, the line scanning signals had to be modulated to give a constant width scanning raster on the mosaic.

The adoption by RCA and EMI of the same constraints for the configurations of their camera tubes naturally led to similar shapes and the innuendo that EMI had copied Zworykin's iconoscope. This imputation probably gained further credence because of the known RCA association with EMI. But, McGee[18] has stated 'categorically that there was no exchange of knowhow between the two companies in this field during the crucial period 1931 to 1936'. Lubszynski, too, has confirmed[21] the independence of McGee's team from any RCA influence. He has written:

'There is a large difference between a patent description and the technology to make it work. If I may give two examples: Essig of RCA patented a method of making a mosaic target by evaporating a silver layer on mica and coagulating it into tiny islands by baking in a furnace. I remember the many weeks McGee and I spent in making the method work by varying the essential parameters of layer thickness, baking times and temperature without causing the mica to blister.

'At the other extreme is over-description such as the famous Plumbicon patent where, I remember, about 36 methods of making such a layer were claimed. Each one would have kept a team of engineers busy for months to establish its feasibility. The old hands at the game will remember how long it took rival laboratories to succeed in making Plumbicons. We could, in the end, make good tubes but had not solved the problem of life time when I retired and a few months later EMI gave up the development.

'In view of these examples, it can justly be claimed that the early development of camera tubes at EMI took place independently of RCA.' (See also Appendix 4)

Actually, of course, there was simply no need for EMI to copy any other company's work. Shoenberg had created a brilliant research team of physicists, engineers, mathematicians and chemists who were the equal, at least, of any other research group. As noted earlier many members of the EMI team possessed a PhD or had had an Oxbridge university education. The group also included Blumlein, who was nothing less than a genius and who 'showered idea upon idea on the team'[21].

After McGee's group fabricated its first batch of emitrons it was found that the signals obtained tended to become submerged in 'great waves' of spurious signals associated with some secondary emission effects. Shoenberg[15] has said that there was a temptation to continue the work using mechanical scanners. 'Instead we decided that the potentialities of the electron scanning tube justified a great effort to over-

Figure 19.12 A Marconi 10 kW transmitter sited on the 5th floor of the research laboratories of EMI (1934)

Reproduced by permission of Thorn EMI Central Research Laboratories

come the problems it [secondary emission] presented at that time.' Improvements to the tube made these spurious signals more manageable and Blumlein and McGee invented[64] at about this time the ultimate solution to the problem—cathode potential stabilisation. However, time did not permit it to be worked out in practice and Shoenberg had to prevail upon the circuit engineers to deal with the unwanted signal components. 'Great credit,' recalled Shoenberg[15], 'is due to Blumlein, Browne and White for the resourcefulness with which they devised tilt, bend and suppression circuits to combat this evil.'

Curiously, Zworykin did not refer to these signals, or to keystone distortion, in his 1933 IEE paper[65].

In addition to these limitations the early emitrons had a low photosensitivity, estimated at, perhaps, 25 % of the ideal. Moreover, it was obvious to McGee that the efficiency with which even this low photosensitivity was being utilised to produce picture signals was also very small—approximately only a few per cent.

Nonetheless, with all its faults the emitron did have two major virtues: first, there was practically no visible lag in the picture and hence moving objects were reproduced clearly and, secondly, it had a very acceptable gamma which resulted in quite pleasing images even when the scene illumination was excessively contrasty. The reason for both these observed facts did not become clear for some time.

Many visits to EMI were made by important personages during the formative period of EMI's system. These included: Sarnoff, of RCA; the Prime Minister; Ashbridge and Kirke of the BBC; Vandeville of MWT; Reith, the Director General of the BBC; representatives of the Post Office, *et al.* Of these private demonstrations, that given to Ashbridge in January 1934 is of some importance since his account[66] of his visit is extant. He was very impressed:

> 'The important point about this demonstration is, however, that it was far and away a greater achievement than anything I have seen in connection with television. There is no getting away from the fact that EMI have made enormous strides.'

The demonstration consisted of the transmission of films, with 150 lines per picture, from the EMI factory at Hayes to the recording studios in Abbey Road, a distance of approximately 12 miles.

> 'This was by means of an ultra short wave transmitter with a power of 2 kW on a wavelength of approximately 6.5 m. The results were extremely good and there was no question in my mind that programme value was considerable. The receivers used appeared to be in a practicable form and looked very much like large radio gramophones. On the other hand, it has to be said that the aerial arrangements were very elaborate, being directional in order to cut out interference.'

Shoenberg told Ashbridge that the policy of EMI was to develop television energetically and that the company believed that there was a great commercial future for the firm which was first in the field with something practicable.

Ashbridge mentioned to Reith that there could be absolutely no comparison between the way in which EMI was handling the problem as compared with Baird Television. 'Supposing therefore the BBC wished immediately to establish a television system, it would be almost unnecessary to consider the rival merits of the two firms from the point of view of who supplied the transmitter.' This, of course, excluded political deliberations.

During Shoenberg's and Ashbridge's discussion the ensuing facts emerged:

1   'EMI no longer wished to install apparatus in Broadcasting House but to go directly to the next step beyond that and to experiment with transmitters of much higher power and not less than 4 kW as compared with the 500 W we have here.

2   'They [were] still very anxious for [the BBC] to take the whole operation of the transmitting side and of course the provision of programmes when it [came] to regular transmission.

3   'They [were] prepared to loan to us not only the televising gear, but the wireless transmitter of 4 kW as well, if the BBC [could] find accommodation for it.

Figure 19.13 Marconi low-definition television apparatus of the early 1930s. This is a lens drum scanner for transmitting a 100-line picture

Reproduced by permission of the Marconi Company Ltd

4   'They [emphasised] that they [were] not interested in the transmitting side and [were] prepared to allow us complete access to every detail of information in connection with their transmitting apparatus.'

EMI's view that there would be sufficient business on the receiving side of television for it not to be particularly concerned with the transmitting side was a point of view which Clark had mentioned to Reith early in 1933. This view seemed to change two months later for H A White, the managing director of Marconi Wireless Telegraph. Co Ltd told Reith that his company had concluded an agreement with EMI to collaborate in the field of television transmissions.

In his letter[67] of 23 March 1934, White said: 'A separate company (Marconi-EMI Television Co Ltd) will be formed to supply apparatus and transmitting stations. This company will have the benefit of extensive research work done by both organisations. The shares of the new company will be subscribed and held in equal parts by the Marconi and EMI.' Reith did not quite understand this position[68] but the most likely explanation is that EMI's high-definition studio and receiving equipment was useless without a suitable ultra short wave transmitter capable of transmitting without distortion the wide sideband signals. Since EMI itself had almost no experience in this field it was perhaps natural that it should consider some form of cooperation with the leading manufacturer of transmitters in the United Kingdom.

The new company was established in May 1934 and its board of director comprised representatives of the two companies, *viz*[69]:

> The Right Honourable Lord Inverforth (Chairman)
> Alfred Clark
> The Marchese Marconi
> I. Shoenberg
> L. Sterling
> H.R.C. Van der Velde
> H.A. White

EMI agreed that the secretary and chief accountant of MWT should act for the joint company and also that MWT should be engaged in the selling and contractual work.

The most surprising aspect of this merger was that MWT initially did not wish to link up with EMI[70]: it seems that the Chelmsford firm preferred an association with Baird Television Ltd. According to some autobiographical notes[71] left by Baird: 'The Marconi company got in touch with us in 1932 and were anxious to join forces. We had numerous meetings, I went up to Chelmsford and was shown round their television research department. Many meetings and luncheons followed and the whole stage was set for a merger.'

Baird's too brief account is not only imprecise in its chronology but also gives no reason why the Wireless Telegraph Company should have wished to consider an association with the comparatively recently formed television company. Perhaps MWT felt it lacked ideas in the new field of technology; the company certainly had not shown anything which could be described as highly original at the 1932 British

Association meeting in York[72]—although the company's exhibits were excellently designed and engineered: perhaps the wireless firm found Baird's patents a stumbling block to its development of mechanical television systems. It may be that Marconi Wireless Telegraph felt that a combination of its engineering skill and Baird's inventive ability would produce a powerful organisation for the furtherance of an industry which, in 1932, appeared to have an assured and prosperous future.

There seems little doubt that had a merger taken place, the fortunes of the Baird company would have been drastically changed. The television company would have had access not only to the patents and expertise which the MWT company had established in the fields of antenna, transmission line and transmitter design, but also to some highly important television patents, including British patent 369 832 which covered a television transmitting system using an iconoscope type of electron camera (Zworykin's invention). These patents would have been denied to EMI, and together with a lack of experience in designing transmitters might have retarded that company's television development programme.

However, Shoenberg was a shrewd, capable research director. He had been joint general manager of MWT and knew well the wealth of talent, expertise and experience which existed in the company. And so with his own group's undoubted abilities in research and development it seemed logical that an amalgamation should occur. Consequently he put the proposition to MWT. 'At first they did not take very kindly to my suggestion, but at last we hired from them a transmitter, about 200 W, and they made us pay £1000 a year for it. After that we worked for about 18 months and, when we started getting results, we went again to the Marconi Company and suggested to them that, since we were each of us concerned with half the complete transmitter, obviously closer collaboration should prove beneficial to both parties. As a result of those conversations, the two companies decided to form between them a private company . . . ' (Details of the Marconi Company's contributions are given in Burns *op. cit.*)

There was a further advantage of the merger for EMI for the company was given a licence to use MWT's patents. With its other agreements the Marconi-EMI Television Company had exclusive rights to all the patents relating to television which had originated with the General Electric Company of America, with the Radio Corporation of America, with Telefunken of Germany, with Marconi Wireless Telegraph Company and with Electric and Musical Industries.

In one respect the task of MWT during its period of cooperation with EMI was very much easier than for the latter company. MWT had acquired its expertise and supremacy over a period of many years in the fields of wireless telegraphy, wireless telephony, radio broadcasting, facsimile transmission and low-definition television. But EMI was effectively starting *ab initio* in many of its investigations. It had to devise and engineer many techniques associated with the fields of vacuum physics, electron optics, wideband electronic circuits, pulse forming and shaping circuits, thin film deposition and so on. Hence, EMI's research and engineering effort during the 1931–1939 period far surpassed ('probably of the order of 50:1') the corresponding labours of MWT[73].

Returning now to Ashbridge's thoughts on television, there were several points which for him were fairly clear:

Figure 19.14 Marconi low-definition television receiving apparatus of 1931/32. Horizontal movement of the spot was produced by the lens drum scanner and vertical movement of the spot was achieved by the mirror drum

Reproduced by permission of the Marconi Company Ltd

1   Broadcasting House was not suitable for short wave television experiments on high powers, (e.g. up to 20 kW);
2   the BBC would have to take other premises, suitable for future development;
3   the first television service would consist mainly of film transmissions and would be established first of all in London and afterwards in other places;
4   the problem of finance on both programme and technical would require consideration.

On the technical side, the chief engineer of the BBC thought that the organisation should aim to install transmitting stations of about 20 kW power output initially in London and then in Birmingham, Manchester, Newcastle, etc. Twin transmitters would be required to provide both a sound and a vision channel. In addition, he visualised the need for a system to rapidly distribute the films to each centre, since long distance lines could not be used. But all of this would involve not only the BBC but also the public in very considerable expense, as the receivers could not cost much less than £80 to begin with. The service would have to begin as a luxury service and an extra licence fee of as much as £2.10p might not go very far in the early stages towards meeting expenses. Ashbridge suggested the financial difficulties might become so serious that the only way of getting enough revenue would be having sponsored programmes, or possibly a separate service, not financed out of the 10/- licence fee but controlled by the BBC which gave sponsored programmes with concessions to one or more firms like EMI on a territory basis.

On the programme side, Ashbridge considered that the most attractive kind of programme which he could visualise would be a sort of newsreel such as one sees in certain film theatres. 'If we could somehow obtain such films and give a performance each night for about an hour or an hour and a half, this might constitute the main programme feature; film plays would also of course be transmitted, but it would be difficult to obtain say three hundred different plays and probably the same play might have to be repeated several times accompanied by news and informative matter generally.' This type of programme would presumably be resisted by the regular film-producing companies, both of the news collecting type and others, but Ashbridge imagined that this resistance 'could be overcome by making (the films) ourselves or paying a large sum to a film company'.

Ashbridge stressed that his report was intended to be merely an introduction to a discussion of the whole matter, but he had raised it because he thought that, if the BBC could not see its way to going on to make television an established service on a considerable scale, it would be better not to be involved in serious preliminary steps.

Undoubtedly, the chief engineer had been most impressed by what he had seen at Hayes and felt it necessary to consider the television situation in a new light, even although the development of a possible service might take some years.

Two months after EMI's demonstration, Baird Television showed[74] its cathode-ray tube receiver to a party comprising the Prime Minister, BBC representatives, Colonel Lee, Colonel Angwin and, because the cathode-ray tube was of GEC manufacture, GEC personnel. Transmissions were given from the Crystal Palace using a wavelength of about 8.5 m for the vision signals and a longer wavelength for the sound signals. The pictures transmitted consisted of an introductory talk given by a speaker (showing the head and shoulders), a violin solo by a lady violinist (also limited to head and shoulders), a talk on architecture illustrated by large scale photographs and short extracts from two films.

Colonel Angwin thought[75] that the standard of reception was approximately the same as that attained in the demonstration given on 20 november 1933, with some limitations due to the radio link. 'Some interference was obvious from electrical sources, but as far as the radio link was concerned, the conditions were fairly good. The receiving aerial was at the top of the four-storey building and fairly remote from motor car interference.'

The detail of both the head and shoulder subjects televised and the close-ups of the films was reasonably good. For larger scenes the detail was much improved. It was not a demonstration to inspire Ashbridge, who wrote of his disappointment. The Prime Minister, however, congratulated Mr Baird on the success which he had obtained and the very great advance on his earlier attempts.

'The film transmission given by EMI is appreciably better than that shown by the Baird Company,' Ashbridge observed. 'On the other hand, however, no opportunity has been available so far to compare a demonstration under absolutely strictly

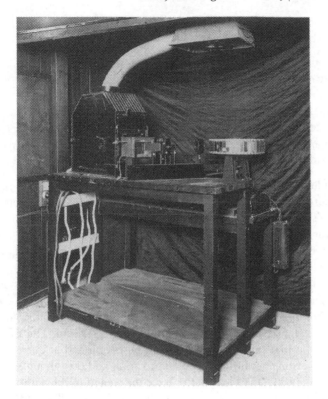

Figure 19.15 Marconi low-definition television receiver for large screen projection. It used a Kerr cell light control valve (1931/32)

Reproduced by permission of the Marconi Company Ltd

comparable conditions. Moreover, the EMI Company have not so far attempted a demonstration with living objects.[76']

In an attempt to settle the rival claims of the two companies, Reith on 15 March 1934 wrote to Kingsley Wood[77], the Postmaster General, and proposed a conference 'between some of your people and some of ours to discuss the future arrangements for the handling of television'. Reith thought that there were three aspects to discuss: the political, 'using the term in a policy sense and for want of a better one', the financial and the technical. He nominated Admiral Sir Charles Carpendale and Mr N Ashbridge. Kingsley Wood[78] agreed and put forward the names of Mr F W Phillips and Colonel A S Angwin. The decision to ask for a conference had not been precipitated by the Baird demonstration alone, for Phillips had noted[79] four days before this that the BBC would probably be seeking an interview shortly to discuss the whole question of the future arrangements in regard to television.

The informal meeting, which was chaired by Phillips, and to which Mr J W Wissenden of the GPO had also been invited, was held at the General Post Office on 5 April 1934[80].

A number of general questions were examined by the BBC and GPO representatives, including:

1   the method of financing a public television service;
2   the use of such a service for news items and plays;
3   the relative merits of some of the systems available including those of the EMI, Baird, Cossor and Scophony companies;
4   the arrangements necessary to prevent one group of manufacters obtaining a monopoly of the supply of receiving sets;
5   the possible use of film television to serve a chain of cinemas.

The use of film for television purposes was of some importance at this time because only the Baird company had shown direct television. EMI had refrained from exhibiting this form since it regarded the development in the early part of 1934 as unsuitable for commercial exploitation. The Company also felt that a film had a more lasting commercial value, its view of direct television was that by its very nature it was essentially transient.

With two rival companies campaigning for the creation of a television service— EMI for a new BBC station and Baird Television for a station of its own—it was agreed by the conference that a committee should be appointed to advise the Postmaster General (PMG) on questions concerning television. The BBC representatives were keen that this committee should be established at once since difficult questions were arising, and would continue to arise, and they thought that it would be helpful for the BBC and the GPO to have the weight of the authority of a committee behind them in any decision which they might take. The PMG agreed and the Television Committee was constituted.

Elsewhere, in the US, Germany and France, the move towards the goal of high-definition television images continued to make steady progress. At the 1933 Berlin Radio Exhibition the receivers on view displayed images based on 90 and 180 lines per picture, 25 pictures per second. A report[81] noted that the image detail and intensity were generally good to excellent, and bright to very bright respectively, and that image flicker was not in evidence. Three manufacturers exhibited receivers showing white and black images, and four manufacturers—Fernseh, Telefunken, Loewe and von Ardenne, together with the RPZ—had cathode-ray tube receivers on their stands.

As in the US and the UK, there was in Germany a convergence of ideas and the implementation of those ideas in the progressive trend towards a practical system of television which would have real entertainment value. But in one respect television research and development effort in Germany differed from that being followed by EMI in the UK, and by RCA and Farnsworth in the US: there was no activity even in 1934 in the field of all-electronic television systems. The preferred method of generating images was by means of film scanning using either motion film industry stock or film which had been exposed and subsequently processed by the intermediate film (i.f.) method, see Table 19.1. Moreover, all the film image analysing scanners employed by the exhibitors at the 1934 Funkausstellung were of the mechanical disc or mirror drum types. Direct scanning of an extended scene (i.e. other than

Table 19.1   Transmitters and receivers on view at the 1934 Berlin Funkausstellung

Firms	Fernseh AG			Tekade			Telefunken		RPZ		Loewe	von Ardenne	RRG
Exhibit no.	a	b	c	a	b	c	a	Karolus	a	b	a	a	a
Transmitters													
Type	film	film	spotlight	7 metre transmission	film	film	film	flood light	7 metre transmission	film	film	7 metre transmission	intermediate film
Scanner	disc	disc	disc		disc	disc	disc	mirror drum		disc	disc		disc
Lines	180	180	180		90	120	180	96		180	180		180
Elements	40 000	40 000	40 000		5400	9600	40 000	12 000		40 000	40 000		40 000
Pictures per sec.	25	25	25		25	25	25	25		25	25		25
Ratio	5:6	5:6	5:6		3:2	3:2	5:6	3:4		5:6	5:6		5:6
F_{max} kHz	500	500	500		67	120	500	150		500	500		500
Amplifier	LF	HF:DC	LF		LF	LF	HF	LF		DC	?		Loewe
Trans. channel	wire	wire	wire		wire	wire	wire	4 wires		wire	wire		wire
Receivers													
Scanner	disc	cathode-ray tube	cathode-ray tube	mirror screw	mirror screw	mirror screw	cathode-ray tube	mirror drum	cathode-ray tube		cathode-ray tube	cathode-ray tube	Loewe cathode-ray receiver
Light source	Kerr cell			neon lamp	Kerr cell	Kerr cell		Karolus cell					
Lines	180	180	180	180	90	120	180		180		180	180	180
Elements	40 000	40 000	40 000	40 000	5400	9600	40 000	1200	40 000		40 000	40 000	40 000
Pictures per sec.	25	25	25	25	25	25	25	25	25		25	25	25
Ratio	5:6	5:6	5:6	5:6	3:2	3:2	5:6	3:4	5:6		5:6	5:6	5:6
Image size	9×12ft	25×30cm	12×15cm	22×18cm 9×11cm	18×15cm	18×15cm	10×12cm 18×22cm	3×4ft	18×21cm		13×17cm	13×17cm	13×17cm
Colour	white	sepia	pale yellow	pink	white	white	white	white	sepia		sepia	pink	pink
Detail	fair	excellent	v. good	good	v. good	excellent	excellent	good	good		excellent	good	excellent
Intensity	v. bright	v. bright	v. bright	fair	v. bright	v. bright	ex. bright	good	bright		v. bright	fair	v. bright
Flicker	none	none	none	none	none	none	considerable	none	slight		slight	none	none
Synchronisation	mains	impulse	impulse	mains	mains	mains	mains	mains	mains		impulse	impulse	impulse

head and shoulders) was not part of the research and development efforts of the German television companies. Rather, the major company, Fernseh, persevered with the cumbrous, complex and costly i.f. system. However, notwithstanding its acknowledged expertise based on the experience of its constituent members—Zeiss Ikon, Bosch, Loewe and Baird Television—Fernseh's i.f. process demonstrations left much to be desired and did not compare well with other 180 line demonstrations. 'The chemical difficulties seem enormous and it will be some time before this system is perfected,' wrote one commentator on the 1934 Berlin Radio Exhibition[82].

The progress of German television engineering in the 1930s could easily be followed, anywhere, because of the excellent and profusely illustrated annual reports on the exhibitions written by the foreign secretary, E H Traub, of the (UK) Television Society. These were presented at lecture meetings of the Society and were subsequently published in the Society's journal. Since Traub visited and commented upon the Funkaustellungen from 1932 to the onset of hostilities in 1939 his reports permit trends in the technology of German television to be readily identified. On the 1934 exhibition, see Table 19.1, Traub noted:

1   'the excellence of the Fernseh and Telefunken pictures;
2   'the impracticability of all cathode-ray receivers with the exception of Loewe;

Figure 19.16 A television picture was broadcast across the world in 1932. The head and shoulders of Mr A Longstaff (the seated figure) were clearly recognisable at Sydney, Australia. The definition was 30 lines per picture

Reproduced by permission of the Marconi Company Ltd

3   'the fine results achieved on the mirror screw in spite of economy of lines;
4   'the disadvantage of ultra short wave transmissions;
5   'the importance of the art of presentation;
6   'the absence this year of a small projection receiver for home use;
7   'the objection to flicker by the public;
8   'the enormous amount of money that seems to be available in Germany for television research;
9   'the importance of transmitting the d.c. component of the image;
10 'the lack of original ideas.'

His views on specific models of receivers were:

1   'The detail [of the Fernseh 180-line, $10'' \times 12''$ cathode-ray tube receiver] was extraordinarily fine. Substantiating this, I would mention that a film was shown in which a musical score appeared on the screen, and it was possible to distinguish the sets of 5 lines and the various notes inscribed on them . . . The colour was pleasant although lacking somewhat in true blacks . . . [Only] line transmission and not radio reception, which was not attempted by this firm, [was demonstrated].

2   'The 180-line [Tekade] mirror screws were demonstrated with neon lamps as light source on radio reception. Some difficulty was experienced in picking out the signal in the hall due to screening but at times the pictures were very good.

3   'As regards detail, colour and contrast, [the Telefunken] pictures were probably the best in the show. The colour was absolutely black and white, and it is believed that this is the first time that this has been achieved in a cathode-ray receiver. The general pictorial impression was extraordinarily realistic . . . the only difficulty with the image, as such, was very intense flicker, so that it was rather tiring to watch it for any great length of time.

4   'The picture quality [of the Loewe receiver] has increased considerably, due to the use of high vacuum cathode ray tubes in the place of gas-filled tubes. The picture size was slightly smaller than last year.
    'This receiver . . . can now be regarded as having reached a certain finality in development. It is the only receiver at the moment that approaches to be a commercial proposition . . . The sale price to the public, it is stated, will be DM 700 complete, but it is not yet for sale.

5   '[The new von Ardenne high vacuum cathode ray tube] undoubtedly represented a big advance on his older tubes [but] it did not come up to the standard of the commercial tubes. The colour was reddish and the well known "handkerchief" effect [a flapping of one edge of the picture] was noticeable. The detail, however, was good.'

At the end of the Funkaustellung, on 30 September 1934, the German Post Office issued a communiqué setting out the prospects for general television broadcasting[83]. Television broadcasting throughout the country would entail the installation of a considerable number of transmitters. For economic reasons the number would have to be kept as small as possible and so experiments would be initiated to determine the usable range of ultra short wave sight and sound transmissions. A transportable television transmitter had been ordered and from the summer of 1935 reception tests would be carried out from the Brocken, the highest peak in the Harz mountains. If good results were obtained and a range of 100 to 150 km achieved, then towns such as Hanover, Braunschweig, Magdeburg, Kassel and Erfurt could be provided with television.

In addition, the German Post Office in collaboration with industry had succeeded in constructing a new television cable which would be laid and tested in the near future in Berlin. If the tests proved favourable, television might be introduced by the aid of cables over long distances in the same manner as radio. There was also the prospect of two-way television being realised by their use. A 1934 news report[84] mentioned that such a service was being considered and that the first transmission link would be between Berlin and Munich, where instruments would be installed in the central telegraph offices[85]. But: 'The process at present, is so expensive that the service would be mainly of experimental value.'

Since expenditure was an imperative factor in all these plans, the German Post Office felt it desirable to state that the proposals advanced would involve a programme of several years before regular, public television broadcasting could be introduced.

Meantime, in the United Kingdom, good progress had been made, by the end of September 1934, by the Television Committee, towards the creation of a public television service. The work of the committee will now be considered.

References

1 BURNS, R.W.: 'British television, the formative years' (Peter Peregrinus Ltd, London, 1986) Chapters 13, 14 and 15
2 Notes of a meeting of the Television Committee held on 27th June 1934. Evidence of Messrs Clark and Shoenberg on behalf of EMI and The Marconi-EMI Television Co. Ltd, minute 33/4682, Post Office Records Office
3 WHITAKER, A.: Report on a visit to America, c. April 1930, pp. 17–18, 23–24, EMI Archives
4 WHITAKER, A.: Letter to Dr A N Goldsmith of RCA, 22 August 1930, pp. 1–3, EMI Archives
5 GOLDSMITH, A.N.: Letter to Mr A Whitaker, 4 September 1930, pp. 1–2, EMI Archives
6 WHITAKER, A.: 'Note on the development of television'. 22 April 1931, pp. 1–2, EMI Archives
7 WRIGHT, W.D.: 'Report of television progress to date'. 31 August 1931, pp. 1–10, EMI Archives
8 See the handwritten version of Ref.7 which includes a circuit diagram of the sweep generator
9 CONDLIFFE, G.E.: 'Report on RCA Victor television'. 15 April 1932, EMI Archives
10 WHITAKER, A.: 'Note on television'. 1 October 1931, pp. 1–3, EMI Archives
11 Cable from Victor, Camden, received 2 December 1931, EMI Archives
12 Executive Committee minute no. 20851, 1 October 1931, EMI Archives
13 MITTEL, B.: 'Television publicity'. 16 February 1936, p. 3, EMI Archives
14 WHITAKER, A.: 'Note on a visit to Engineer-in-Chief's department, General Post Office [on] 21 October 1931'. 23 October 1931, pp. 1–3, EMI Archives
15 SHOENBERG, I.: Discussion on 'The history of television', *J. IEE*, 1952, **99**, Part IIIA, pp. 41–42
16 McGEE, J.D.: '1971 Shoenberg Memorial lecture', *R. Telev. Soc. J.* May/June 1971 **13**, (9)

17 CONDLIFFE, G.: Summary of 'Programme of work for advanced development division'. 6 November 1931, pp. 1–3, EMI Archives
18 McGEE, J.D.: 'The early development of the television camera'. Unpublished manuscript, p. 16, IEE Library
19 ANON.: Laboratory staff, research department, 12 June 1934, EMI Archives
20 BURNS, R.W.: 'A.D.Blumlein—engineer extraordinary', *Eng. Sci. Educ. J.*, February 1992, **1**, (1), pp. 19–33
21 LUBSZYNSKI, G.: 'Some early developments of television camera tubes at EMI Research Laboratories', *IEE Conf. Publ.*, 1986, (271) pp. 60–63
22 TEDHAM, W.F.: 'Report on an investigation of the factors affecting emission in photo-electric cells'. Report GD.2, 17 September 1930, pp. 1–12
23 CONDLIFFE, G.: 'Re: PO382'. 12 May 1932, memorandum to I. Shoenberg, EMI Archives
24 TEDHAM, W.F., and McGEE, J.D.: 'Stencil mesh mosaic'. British patent 406 353, 25 August 1932
25 Ref.18, pp. 27–30
26 LUBSZYNSKI, G.: Private letter to A E Jennings, 24 March 1981
27 Ref.1, pp. 252–253
28 WISSENDEN, J.W.: Letter to EMI Ltd, 8 October 1931, BBC file T16/65
29 'Statement re-television'. 10 October 1931, p. 1, EMI Archives
30 CONDLIFFE, G.: 'Re: television—present programme'. 14 December 1931, pp. 1–2, EMI Archives
31 CONDLIFFE, G.: 'Television transmitter'. 4 January 1932, p. 1, EMI Archives
32 DAVIES, N.E.: 'Marconi-EMI television 1931 to 1937'. Appendix 2, 24 April 1950, Marconi Historical Archives
33 SHOENBERG, I.: Letter to N Ashbridge, 11 November 1932, BBC file T16/65
34 ASHBRIDGE, N.: Report on television demonstration at EMI, 6 December 1932, BBC file T16/65
35 Director General (BBC): Memorandum to the chief engineer, 1 January 1933, BBC file T16/65
36 BAIRD, J.L.: Letter to Sir John Reith, 6 December 1932, BBC file T16/42
37 Notes of a meeting of the Television Committee held on 8 June 1934. Evidence of The Marconi-EMI Television Company, minute 33/4682
38 See Chapter 14
39 Report, to Mr A Clark, on a visit to the United States 12 June 1931, pp. 1–8. Also see 'Instructions for trip to the United States'. 12 May 1931, pp. 1–3; and 'Extracts from cables from Hayes'. p. 1 EMI Archives
40 Director General (BBC): Memorandum to the chief engineer, 9 January 1933, BBC file T16/42
41 Minutes of a meeting held with representatives of Television Ltd, 27 January 1933, BBC file T16/42
42 MOSELEY, S.A.: Letter to Sir Kingsley Wood, 28 January 1933, minute 4004/33, Post Office Records Office
43 PHILLIPS, F.W.: Letter to Sir J F W Reith, 13 March 1933, BBC file T16/42
44 PHILLIPS, F.W.: Letter to Sir J F W Reith, 10 April 1933, BBC file T16/42
45 SIMON, L.: Memorandum on Baird and EMI demonstrations, 27 April 1933, minute 4004/33, Post Office Records Office
46 BAIRD, J.L.: Letter to Sir J F W Reith, 18 February 1933, BBC file T16/42
47 McGEE, J.D.: Laboratory notebook, entry for 14 November 1932, EMI Archives
48 McGEE, J.D.: Laboratory notebook, entry for 2 February 1933, EMI Archives
49 CONDLIFFE, G.: Memorandum to I Shoenberg re: EMI case 782, 7 February 1933, EMI Archive
50 TEDHAM, W.F.: 'Method of deposition of a mosaic electrode', EMI case 782 for completion, 24 February 1933, EMI Archives
51 McGEE, J.D.: 'A transmitting tube for television', EMI Archives
52 Ref.18, pp.20, 69
53 McGEE, J.D.: 'Double-sided mosaic tube'. British patent 419 452, 5 May 1933
54 McGEE, J.D.: 'Electronic generation of television signals', in LOVELL, A C B (Ed.): 'Electronics' (The Pilot Press, London, 1947) Chapter IV, pp. 135–211
55 BROADWAY, L.F.: 'Early developments at EMI on camera tubes'. 25 March 1958,

478   *Television: an international history of the formative years*

2pp., EMI Archives
56 Notes on a meeting held at the GPO, 21 April 1933, BBC file T16/42
57 Notes of a telephone conversation between the BBC and the Postmaster General, 23 May 1933, BBC file T16/42
58 KINGSLEY WOOD, H.: Letter to Sir J F W Reith, 22 May 1933, BBC file T16/42
59 Ref.1, chapter 12, pp. 274–301
60 ANGWIN, A.S.: Memorandum on a visit to the Baird laboratories, 20 November 1933, minute 4004/33, Post Office Records Office
61 Obituary, 'Albert Gilbert Dixon West', *British Kinematograph*, September 1949, **15**, (3), pp. 73–74
62 GPO: a letter to the BBC, 27 July 1933, BBC file T16/42
63 ESSIG, S.: 'Aggregated silver mosaic'. British patent 407 521, 24 February 1932
64 BLUMLEIN, A.D., and McGEE, J.D.: 'Improvements in or relating to television transmitting systems'. British patent 446 661, 3 August 1934
65 ZWORYKIN, V.K.: 'Television with cathode-ray tubes', *J. IEE*, 1933, **73**, pp. 437–451
66 ASHBRIDGE, N.: Report on television, 17 January 1934, BBC file T16/65
67 WHITE, H.A.: Letter to Sir J F W Reith, 23 March 1934, BBC file T16/65
68 REITH, J.F.W.: Memorandum to N Ashbridge, 26 March 1934, BBC file T16/42
69 Ref.2, p. 3
70 Ref.1, p. 319
71 BAIRD, J.L.: 'Sermons, soap and television' (Royal Television Society, London, 1988)
72 Press release (MWT Co.): 'Demonstration of Marconi television'. 5 September 1932, Marconi Historical Archives
73 SMITH, G.B.: 'Historical relationship between Marconi, RCA and EMI'. 1 September 1950, Marconi Historical Archives
74 GREER, H.: Letter to Sir J F W Reith, 7 March 1934, BBC file T16/42
75 ANGWIN, A.S.: Memorandum, 12 March 1934, minute 4004/33, Post Office Records
76 ASHBRIDGE, N.: Report on demonstration to the Postmaster General and C(A), 12 March 1934, BBC file T16/42
77 REITH, J.F.W.: Letter to Sir Kingsley Wood, 15 March 1934, BBC file T16/42
78 KINGSLEY WOOD, H.: Letter to Sir J F W Reith, 20 March 1934, BBC file T16/42
79 PHILLIPS, F.W.: Memorandum to H Napier, 8 March 1934, minute 4004/33, Post Office Records Office
80 Notes on 'Conference at General Post Office, 5 April 1934', minute Post 33/4682, Post Office Records Office
81 TRAUB, E.H.: 'Television at the 1932 Berlin Radio Exhibition', *J. Television Society*, 1932, pp. 273–285
82 TRAUB, E.H.: 'Television at the 1933 Berlin Radio Exhibition', *J. Television Society*, 1933, pp. 341–351
83 EDWARDS, R.P.F.: Memorandum to the Department of Overseas Trade, 16 October 1934, Post 33/5534, file 6, Post Office Records Office
84 Press Service of the Reichpost Ministry: 'The position of television in Germany', Post 33/5534, file 13, Post Office Records Office
85 BURNS, R.W.: 'Prophecy into practice: the early rise of videotelephony', *Eng. Sci. Educ. J.*, 1995, supplement to **4**, (6), December 1995, pp. S33–S40

Chapter 20
Progress in the UK and abroad (1934–1935)

The first meeting[1] of the Television Committee was held on 29 May 1934. It comprised important persons from the BBC, the GPO, the Department of Scientific and Industrial Research and industry, namely[2]:

> The Right Honourable Lord Selsdon (chairman), Figure 20.1
> Sir John Cadman (vice chairman)
> Colonel A S Angwin, assistant Engineer-in-chief, GPO
> Mr O F Brown, Department of Scientific and Industrial Research
> Vice-Admiral Sir Charles Carpendale, controller, BBC
> Mr F W Phillips, assistant secretary, GPO
> Secretary: Mr J Varley Roberts

To ensure impartiality no representatives of either Marconi-EMI or Baird Television were invited to join the committee. It was decided that all the committee's meetings should be held in private and that reports of future meetings should not be issued to the press[3]. The terms of reference were as given in Chapter 19.

As part of the committee's brief was to advised the PMG on the relative merits of the several systems of television, it was necessary for the committee to consider not only British television developments but also those of other countries. Hence, on 18 May 1934, letters were sent to the Telegraph Administrations of France, Germany, Italy and to the Federal Radio Commission, US, requesting information on a number of questions regarding their country's television schemes. Since the responses to these questions were made by the heads of the appropriate government departments it may be assumed that their replies were factually correct and not subject to the exaggerations and distortions which sometimes coloured the statements and views of inventors and newspaper reporters. For this reason, and to provide an overview of the position of television, in the summer of 1934, in the US, Germany, France and Italy, details from the responses follow[4].

Figure 20.1   Members of the Television Advisory Committee (which followed the Television Committee) at Hyde Park Corner in 1937. They are examining one of the camera installations for the Coronation Day Procession. Sir Noel Ashbridge is on the plinth, and he is talking to Lord Selsdon

Reproduced by permission of the BBC

Position in the US

(Dr C B Jolliffe, chief engineer, FRC, June 1934)
The Federal Radio Commission had only licensed television stations (see Table 20.1) on an experimental basis according to certain regulations of the FRC and no fee was charged. Licencees were not permitted to commercialise their programmes in any way whatsoever and no advertising time could be sold. All the stations were

Table 20.1 Development of television standards in Germany from 1929 to 1951

Date of introduction	Number of lines	Frames per second	Aspect ratio	Broadband (kHz)	Frequency band	Sender
March 1929	30	10.0	1:1	4.5	MW	Berlin-Witzleben
September 1929	30	12.5	4:3	7.5	MW	Berlin-Witzleben
May 1930	30	12.5	4:3	7.5	LW	Königs Wusterhausen
January 1931	48	25	4:3	38.4	KW	Döberitz
August 1932	90	25	4:3	94.5	UKW	Berlin-Witzleben
End 1933	180	25.	4:3	378	UKW	Berlin-Witzleben
Late autumn 1938	441	25*	4:3	2200	UKW	Berlin (Amerikahaus)
September 1951	625	25*	4:3	5000	VHF	Hamburg (Heiligengeistfeld), Berlin-Tempelhof (FTZ)

*With line jumping process.

run on 'the basis of the developments which they were making in television' and as a consequence 'very few' of them had regular programmes, or a regular listening audience. There were 'one or two of these experimenters' who maintained regular programmes of one or two hours per day.

Apart from the principal experimenters (The RCA Victor Company of Camden, New Jersey, and Philco Radio and Television Company of Philadelphia) there were several other research laboratories but these were 'far behind' either of the two leading companies. Some of the experimenters were working with Nipkow discs or mirror drums of various kinds with 'rather mediocre results'. Almost universally they used 60 lines per picture, 24 pictures per second, scanned from left to right and from top to bottom, with an aspect ratio of width to height of 3:4 or 5:6.

The largest number of receiving sets of these types was in Chicago, but the total number was probably not greater than 200. Some of the sets had been made commercially by the licencees, and others had been constructed by enthusiastic home constructors. 'Results have been relatively poor and it [was] difficult for any one to maintain sustained interest in the pictures which it [was] capable of broadcasting by this method.' Only the head and shoulders of one or two persons could be televised and the amount of light required was 'quite excessive'. The details were 'very poor' and the method was incapable of using outside pick up.

Of the principal experimenters, the RCA Victor Company had described its systems in a series of papers and had developed receiving sets which could be used in the home. However, no attempt had been made to sell the sets and the only use which had been made of them was by the engineers of RCA who observed their own transmissions. There was no advertised schedule of the transmissions and no endeavour had been made to obtain public reaction. The costs of producing a sustaining programme of public interest seemed to be the stumbling block to commercialisation.

Philco Radio and Television had not published any papers on its developments and had made no plans for public commercialisation.

The position in Germany

(The Reichspost Minister, July 1934)

The development of television had advanced so far that it had been possible to erect a properly functioning experimental installation in Berlin from which regular test transmissions on ultra short waves were being carried out. In addition, picture signals were still being broadcast on long waves by the 'Deutschland' transmitter at Königs Wusterhausen. The characteristics of the two transmissions were:

	No. of lines per picture	No. of picture/s	Aspect ratio	Scanning	Maximum modulation	Carrier	Line and frame sync. pulses
Berlin	180	25	5:6	horizontal	500 kHz	42.95 MHz	after each line and frame
Königs Wusterhausen	30	12.5	3:4	horizontal	7.5 kHz	191.0 kHz	after each line and frame

Up to July 1934 the programme material had comprised films. Regular television broadcasts were radiated from the experimental installation at Berlin-Witzleben as follows.

	Monday	Tuesday	Wednesday	Thursday	Friday
9.00–11.00	Telekino	Telekino	Telekino	Telekino	Telekino
11.00–12.00	–	Music	–	Music	Music
15.00–16.00	Telekino	–	Telekino	–	–

(An extension of the transmission period to the period 20.30–22.00 hours was envisaged.)

From Königs Wusterhausen televised films were sent out on Mondays from 9.05 to 10.00 hours and on Thursday from 15.00 to 16.00 hours.

For reception, receivers with Braun tubes were principally used although mirror screw receivers were also produced. However, the general public did not yet participate in the reception of television transmissions, and, with the exception of a few amateurs, private persons were not in possession of receivers.

Since the construction by amateurs of 180-line television receivers was difficult, only a few had the necessary skill to assemble a receiver. Four large firms were engaged in the development and manufacture of sets and it was anticipated that several more firms would be added. When sufficiently good results were obtained from Berlin-Witzleben and the industry could construct sufficient receivers, television installations would be erected at other places. No decision had been made regarding a receiving licence fee.

Position in France

(*Le Directeur de l'Exploitation Télégraphique, PTT, September 1934*)
Television in France was in an experimental state and the results that had been obtained so far were not of a quality suitable for a regular public service. Spotlight scanning, by means of Nipkow discs or mirror wheels, was used and the image definition corresponded to either 30 lines per picture or 60 lines per picture, scanned horizontally at 25 pictures per second, with an aspect ratio similar to 'talkie' films, *viz.* 18 × 21 (sic). Line synchronising pulses continuous with the vision signals were transmitted; framing was automatic. The vision signals were invariably radiated at 695 kHz and the associated sound signals (from the Eiffel tower) were sent out at a frequency of 216 kHz.

In the Paris area the number of amateurs who could receive, without paying a licence fee, the 30-line signals was c. 500. Only a very few could receive the 60-line transmissions. Generally, the amateurs built their own Nipkow disc or mirror drum apparatus. From the station of the École Superieure des Postes Télégraphes et Téléphones vision broadcasts were made at the following hours:

Monday	17.00 to 18.00 hours
Tuesday	15.00 to 16.00 hours
Friday	15.00 to 18.00 hours

Apart from these broadcasts, laboratory investigations of 120 and 180-line images were in progress. It was envisaged that these would be radiated on a wavelength of c. 7 m. Summarising, the carrier frequencies and bandwidths were:

No. of lines	Carrier frequency	Bandwidth
30	695 kHz	12 kHz
60	695 kHz	48 kHz
120	of the order of 30 MHz	of the order of 250 to
180	(transmitter in construction)	500 kHz

Position in Italy

(April 1935 (sic))

No regular service of television was in operation in Italy although the Ente Italiano Audizioni Radiofoniche (EIAR), the exclusive concessionaire for the broadcast service, was particularly interested in the development of television and had carried out continuous experiments for several years.

Two systems had been adopted: a 90-line direct transmission system utilising indirect scanning and a 180-line film system. In both cases Nipkow discs and horizontal scanning were used. The aspect ratio was 4:5 and the picture rate was 25 pictures per second. Both line and frame synchronising pulses were transmitted. The vision and sound carrier wavelengths, of the irregular broadcasts, were 6.90 m and 6.15 m, respectively, and the maximum modulation frequency was 500 kHz.

Mirror screw scanners were usually employed in the 90-line receivers and cathode-ray tubes in the 180-line receivers.

An estimate of the probable number of persons possessing receiving sets could not be given but the number was negligible.

Regulations which would in due course govern the transmission and reception of television were being considered.

Apart from these statements the Television Committee may have seen the paper 'Television—a survey of present day [US] systems' which was published in the October 1934 issue of the widely read magazine *Electronics*. *Electronics* had undertaken its comprehensive survey of the field in an attempt to dispel some of the 'confusion which surrounds television development' and had enlisted the cooperation of the companies known to be actively engaged in television research. Each company's laboratory had been visited and its engineers interviewed. They included W R G Baker of RCA, A F Murray of Philco, P T Farnsworth of Television Laboratories, A Zillger of National Television, W H Priess, W H Peck and J V L Hogan. From these, and other sources, *Electronics* concluded: 'The radio industry, demoralised more than once by ill-timed ventures into new fields, is determined that the new art shall be introduced when the time is ripe, and not before. But there is no definite agreement on just what constitutes the "ripe" time; in fact, there is a wide divergence of opinions.'

Table 20.2 *Power ratings of the television transmitters of the German Post Office in Berlin, 1932 to 1943*

Sender	Start date	Performance	No. of lines	Manufacturer	Remarks
Witzleben, Berlin Radio tower	10.8.1932	16 kW	90	Telefunken	Only transmitted pictures; from 18.4.1934 used as audio transmitter
Witzleben Berlin Radio tower	18.4.1934	20 kW	180	Telefunken	Picture transmitter; transmitters 1 and 2 destroyed by fire 19.8.1935
Witzleben Berlin Radio tower	23.12.1935	14 kW	180	Telefunken	In operation until late autumn 1938
Charlottenburg, Berlin Amerikahaus	Late Autumn 1938	14 kW, later 40 kW	441*	Telefunken	Later to be C. Loreriz's audio transmitter; both transmitters destroyed 26.11.1943
Neuköln, Berlin Karstadt Building	1938	1 kW	441*	RPF**	

*With line jumping process.
**Research of the German Post Office.

The Television Committee worked with commendable speed and examined 38 witnesses, some of them on more than one occasion, and in addition sent delegations to the United States (headed by Lord Selsdon), and to Germany (headed by O F Brown) to investigate and report upon progress in television research in those countries. By September 1934 the committee had so advanced its deliberations that it was able to commence the preparation of its report. The detailed work of the committee in interviewing British witnesses has been dealt with in the author's book, *op.cit.*, and is not repeated here.

The delegation which visited the United States comprised Lord Selsdon, Colonel Angwin, Mr Ashbridge and Mr Phillips. The members were given demonstrations of the all-electronic television systems which were being developed by the Radio Corporation of America and by the Philco Radio Company, and of the mechanical scanning system of the Peck Television Corporation[5]. In addition, the members of the committee had talks with representatives of RCA, the Federal Communications Board, AT&T, NBS, CBS, the International Telephone and Telegraph Corporation and Western Union Telegraph Company. These private, confidential discussions were especially important to the visiting party since they enabled the Television Committee to contrast the case for an early UK television service as presented by British workers with the frankness of others.

At the RCA Victor Laboratories the demonstration centred on the televising of a number of subjects, including a 'pianist, illustrated talks, topical films of plays and current sporting events' and an outdoor scene. The cathode-ray tube images were 6 × 8 inches in size and were based on a 343 lines per frame, 30 frames per second standard. As stated earlier: 'Very good reproductions were obtained. All the studio and film subjects were comparable and perhaps slightly superior as regards absence of flicker to corresponding demonstrations in [the UK].'

At the Philco Radio Company the demonstrations given to the committee were 'comparable with those of the RCA-Victor Company'. Philco had recently abandoned the Farnsworth system in favour of the Zworykin iconoscope and had been licensed to use RCA's patents while developing the system on its own initiative. In one outside broadcast television test the camera was located on the roof of the Philco building and directed to a tower and other objects within the range of 100 to 200 yards. 'The light intensity during this demonstration was that of a normal autumn or bright winter day and was sufficient to give a clear and well defined image on the television receiver.'

From their talks Lord Selsdon and his three colleagues felt that the television situation in the US in October-November 1934 could be summarised as follows[5]:

1   Television was now 'almost ripe' for a public broadcasting service; the difficulties of creating such as service were financial rather than technical.
2   The cost of providing attractive programmes would be very considerable; and since the distances to be covered would be very great—because of the limited range of high-definition television transmissions—many stations would be needed even if the service were restricted to regions of dense populations.
3   In the absence of licence revenue, the service would presumably have to be supported by advertising revenue. But, initially and for some time to come, receiving sets would be costly and the number of persons who could, or would, purchase them would be relatively small. This state of affairs would deter prospective advertisers who would not be willing to pay high rates for television 'time', and so it would be impossible to finance the service from advertising income.
4   New York would be the best prospect for a TV station, but if a service were started in that city by one of the principal broadcasting organisations, its competitors might feel obliged to provide similar facilities. Moreover, there would be strong public pressure for services to be provided in other important centres: disregard of such pressures might involve loss of good will to which the broadcasting companies attached much importance.
5   The provision of a service at just a few centres might 'induce many persons to refrain from buying new radio sets in the belief that, before too long, combined sound and television sets would be available at reasonable prices. Serious injury might thus be done to the radio trade in all parts of the country'.

6   Because of these difficulties, the early establishment of a television service in the US did not seem to be imminent, and no authority for any television service on a commercial basis had been given by the FCB.

7   There was no evidence to suggest that the US Government was seriously considering such a service, and the Federal Communications Board—which was responsible for granting licences—did not seem to contemplate 'taking the initiative in any way'.

All of these points, and especially the pessimistic views of the radio companies and AT&T, did not augur well for an immediate start to a widespread television service. Furthermore, there was apprehension among the manufacturers that the commencement of a service in the United Kingdom or in Germany would lead to public pressure for a service in the United States.

A delegation of four members visited Germany in November 1934. These were O F Brown and J Varley Roberts of the Television Committee, and A J Gill (staff engineer of the Post Office) and H L Kirke (head of the research department, BBC) representing Col A S Angwin and N Ashbridge respectively.

In Germany (in 1934) two departments of state were concerned with television, namely, the Propaganda Ministry and the Reichspostministerium. The latter was concerned only with the technical development and wireless transmission of television and with the task of carrying out, as far as possible, the policy set by the Propaganda Ministry. The Reichs Rundfunk Gesellschaft (the broadcasting company) would be responsible for providing the programmes and operating the studio equipment when a public television service was inaugurated. However, the policy in 1934 was not to begin such a service until it had been established that an efficient service could be given, and that a receiver could be marketed at a price which the public generally could afford to pay.

At this time the Reichspost vision and sound transmitters—located at Witzleben, a suburb of Berlin—radiated daily experimental transmissions on wavelengths of 6.7 m (sound) and 6.985 m (vision). The crystal controlled transmitters were of Telefunken manufacture and were rated at a peak power output of 16kW, Tables 20.1 and 20.2. Quarterwave vertical antennas with horizontal semicircular counterpoise networks placed at the top of the 430' high self-supporting steel towers were used with each transmitter.

The television laboratories of the Reichpostministerium were about 400 m from the antenna tower. A Fernseh film scanner and a Fernseh spotlight scanner were employed to generate the picture signals, the bandwidth of which extended down to zero frequency. The spotlight scanner was provided with a 150 A arc lamp, giving an output of 25 000 lumens; the disc ran at 6000 r.p.m. in a partially evacuated enclosure.

Comparative tests were demonstrated to the Television Committee representatives of a person's head and shoulders using definitions of 30, 60, 90, 120 and 180 lines per picture. Remarkably, the Reichspost were of the opinion that 90 lines per picture gave adequate definition.

Visits were also made by the representatives to the three main companies concerned with television, namely, Telefunken, Fernseh and Loewe. At Telefunken, demonstrations of television transmission and reception, over both line and radio links, with the company's apparatus were witnessed. The black and white c.r.t. images viewed were 'definitely brighter than any other seen in Germany', but suffered from 'very noticeable' flicker. No iconoscope-produced images were shown although Telefunken was now experimenting with the Zworykin camera tube. Fernseh showed its intermediate film pick-up van and i.f. large screen projection receiver, and Loewe demonstrated its television receiver.

Discussions were held with senior members of the Reichspostministerium, the Telefunken, Fernseh and Loewe Companies, and the Reichs Rundfunk and from these the conclusions reached were[6]:

1   'Television appears to have reached approximately the same stage of technical development as in [the UK].

2   'No public service seems likely to be started in Germany for about a year.

3   'There would seem to be no intention on the part of the Government—in the near future at any rate—of subsidising public television in Germany to any appreciable extent.

4   'When a public television service is started, it will probably be on a small scale and increase gradually.

5   'It is considered very desirable that the transmitting points for television should be elevated.

6   The use of ultra short waves is not regarded as presenting any specially difficult problems with regard to fading and interference.

7   No decision has yet been reached regarding the financing of a public television service. The surplus from sound broadcasting licence revenue will probably be drawn upon for the purpose—in the first instance at any rate.

8   'Cable capable of carrying high-definition television over long distances can be made, but the cost of manufacture is uncertain.

9   'At least one serviceable combined sound and television receiver of excellent design (Loewe) can be produced in quantities forthwith at a price to the public of £30/set.

10   'The intermediate pick up process has proved satisfactory in practice, while the alternative iconoscope method has yet to be proved.

11   'Interlaced scanning has been satisfactorily demonstrated and appears to be completely successful in removing flicker.'

No visits were made by the Television Committee to either the USSR or to France. Regular, experimental television transmissions had commenced in 1931 in Leningrad, Moscow, Tomsk and Odessa but these were of low definition and based on the Baird standard of 30 lines per picture, 12.5 pictures per second. There is no evidence that any innovative television research of a high quality was being undertaken in the USSR in the early years of the 1930s.

In France, too, little innovative television work was in progress during the first half of the 1930–1939 decade[7]. The position here was described in a review article published in the December 1934 issue of the journal *L'Industrie Française Radio-Électrique*, which was forwarded to the committee by the commercial counsellor to HM Embassy, Paris.

In the article, the author, M Chauvièrre, complained that all the representations, regarding the need for state encouragement and support for television, he had made had proved unrewarding:

> 'A petition, signed by 500 persons, which was forwarded by M Chauvièrre to the Minister for Posts late in 1933, and a letter to M Belin, met with no greater success. The latter replied that he was personally in favour of television emissions provided that they took place after midnight, because such emissions, like those of telephotography, were of a nature considerably to disturb listening in general. As regards the petition to Monsieur Laurent Eynac, he received Monsieur Chauvièrre personally and after he had visited the radio-television laboratory established in the offices of his department, a note was issued to the Press to the effect that he had decided that the question of the organisation of regular radio-television emissions should be investigated by the PTT and Eiffel Tower Station in order to enable amateurs to follow the progress of the new technique. Unfortunately a week later the government of which Monsieur Laurent Eynac was a member was defeated and nothing more has been done,'

An experimental low-definition (60 lines per picture, 25 pictures per second) service began on 26 April 1935 but by this date the Television Committee had already submitted its report to the Postmaster General and, of course, such a service was of no interest to the committee.

A world survey of television, conducted by the BBC's chief engineer in 1936, showed[8] that only five countries—France, Germany, the UK, the US and the USSR—had transmitted, by the beginning of that year, some form of television broadcasting. Thus, Lord Selsdon's committee had gained first-hand experience of all the important research and development television activities (with the exception of those of Farnsworth) being progressed anywhere.

The Television Committee submitted its report to the Right Honourable Sir Kingsley Wood on 14 January 1935. It had cost roughly £965 to prepare and had taken seven months to produce. The report itself was quite short, a mere 26 pages, but this did not include notes of the formal evidence, which were presented in Appendix 2 (volumes 1 to 4) because much of it contained secrets of commercial value which had been given in confidence and under promises of secrecy. For similar reasons, Appendix 3 which included reports on developments in the US and Germany, and Appendix 4 which gave descriptions of each television system that the committee had examined, were not published. Appendix 5 dealt with certain financial details that were considered to be of a confidential nature.

The report was well received by the national press. 'The British Post Office which has shown so much enterprise, in all directions, deserves to be congratulated on the adoption of the first scheme in any part of the world to bring television within the reach of the public,' noted the *North Western Daily Mail*[9]. For the *New Statesman and Nation*[10] the report was 'well set out, concise, surprisingly brief and matter of fact. Indeed reading the short sensible recommendations it is hard to realise that here is another scientific miracle about to be let loose on the world . . . The Television Committee has carried out an extremely difficult task with most commendable thoroughness'. 'The report [was] only an unbiased estimate of the present stage of development of television, but it has the merit, in addition, of being frank in its recognition of some of the difficulties,' observed *Wireless World*[11].

In the US the news of the publication of the television report 'fell like a bombshell on the ears of American radio engineers'. 'Either the Britons are masters of the art of dissimulation or they are satisfied that they have something really good for Lord Selsdon and his staff did not indicate during their American tour that they thought England was ahead of this country in television development. On the contrary, those who met them socially and in the laboratories aver that they were effusive in their praise of comparative American advances.'[12]

German reaction took the form of emphasising that Germany had not lost the lead. The Secretary of State responsible for the German Post Office television tests published an article pointing out that the right of being the premier country in television belonged to Germany: 'For the past year and a half,' he wrote, 'high-definition television on ultra short waves [has] been broadcast in Berlin.'[12] He omitted to tell his readers, however, that reception of these experimental transmissions was limited to the laboratories of interested firms and that suitable receivers were not available on the German market.

Surprisingly, the Radio Manufactures Association (RMA), which might have been expected to welcome the birth of a new industry, viewed the recommendations of the report with some reservations; it declared that 'the time [was] not yet ripe for television to take the place of radio broadcasting'[13]. Possibly this caution reflected the boom which the radio industry was experiencing at the time. Philco Corporation, for example, had during the period 5 August to 31 December 1934 increased its company's sales by 78 % compared with the similar period in 1933. 'We are living and doing business in the most prosperous country in the world,' the general sales manager of the Philco Radio and Television Corporation Ltd noted[14] at a dealers' convention in Glasgow on 30 January 1935. 'A buying wave is sweeping over the whole of Britain. More people are buying than ever before, and still half the homes in Britain have not yet heard radio music,' he continued.

As a consequence of this boom, radio manufacturers were expanding their factory and plant facilities for the production of medium and long-wave receivers and the BBC was building three new transmitting stations for sound broadcasting. Clearly, the industry did not want the bubble of prosperity pricked by the sudden invasion of a new entertainment medium, and the view that the buying public might cease to purchase wireless sets in the belief that they would become rapidly redundant was thus a very real one—and one that had to be discouraged. 'We shall not see television for at least two or three years [and] when it does come, it will be local. It will necessitate the building of booster stations all over the country,' Philco's general sales manager observed. 'It is only fair to the general public that this should be pointed out.'

But whereas the trade was apprehensive about the possible outcome of the report, many newspaper columnists romanced on the prospects television offered. 'How splendid it would have been if last night's relay of Mascagni's new opera "Nero" from La Scala, Milan, could have been received in Everyman's drawing room simultaneously with a television motion picture of the stage action.'[15] The public 'want the running commentary on, say, the Derby, or a centre court match at Wimbledon, or a magnificent state ceremony, or ski-ing in the Tyrol or Christmas day on an Australian farm illustrated on their television sets with a picture of the event'.

This view of the joys to come was somewhat encouraged by the opinion of the Television Committee: 'The time may come when a sound broadcasting service entirely unaccompanied by television will be almost as rare as the silent film is today,' it stated in its report[2]—adding, probably as a sop to the industry, 'but in general sound will always be the more important factor in broadcasting'.

Not only was the radio trade uneasy about the possible effects on its sales, the cinema and theatre industry too was anxious. *Variety News*[16], in an editorial dated 7 February 1935, asked: 'What is going to happen to the music hall when we can see as well as hear the performers in our homes by means of television allied with radio? Experience suggests that such a transmission would keep millions of people at home, and consequently away from music halls and theatres, and from many other places of amusements. A televising of the English Cup Final might have a similar effect on the "gates" of professional football matches. A televising of a Garbo film premiere might affect the takings of every cinema in the kingdom. So every form of public entertainment will be up against it when television comes!'

The editor of *Variety News* quickly made up his mind about television: 'As likely as not it may prove to be an enemy. In that event, there is nothing like taking time by the forelock. No general goes into battle—or he should not—without ascertaining as quickly as possible the enemy's dispositions.'

These views seemed to portend an ominous beginning for the new industry, but fortunately the writer, after discussing the wisdom of the cinema/theatre industry embarking on a programme of hostility, counselled 'cooperation with the menace' as the best plan of action. For, he pointed out, cooperation has made broadcasting helpful to living entertainment; it was cooperation that had brought music hall artists to the microphone and so enhanced their popularity that when they appeared in person in the music halls their value as a box office asset had been enormously increased. 'And what the radio has done in that respect, television may do still more.'

The principal conclusions and recommendations of the report are summarised below:

Type of service

1 'No low-definition system of television should be adopted for a regular public service.

2 'High-definition television has reached such a standard of development as to justify the first steps being taken towards the early establishment of a public television service of this type.

Provision of service
Operating authority
3   'In view of the close relationship between sound and television broadcasting the Authority which is responsible for the former—at present the British Broadcasting Corporation—should also be entrusted with the latter.

Advisory committee
4   'The Postmaster General should forthwith appoint an advisory committee to plan and guide the initiation and early development of the television service.

Ultra short wave transmitting stations
5   'Technically, it is desirable that the ultra short wave transmitting stations should be situated at elevated points and that the masts should be as high as practicable.
6   'It is probable that at least 50 % of the population could be served by ten ultra short wave transmitting stations.

Patent protection
7   'It is desirable in the general interest that a comprehensive television patent pool should eventually be formed.

Initial service
8   'A start should be made by the establishment of a service in London with two television systems operating alternately from one transmitting station.
9   'Baird Television Limited and Marconi-EMI Television Limited, should be given an opportunity to supply, subject to conditions, the necessary apparatus for the operation of their respective systems at the London station.

Subsequent stations
10   'In the light of the experience obtained with the first station, the Advisory Committee should proceed with the planning of additional stations—incorporating any improvements which come to light in the meantime until a network of stations is gradually built up.
11   'The aim should be to take advantage, as far as possible, of all improvements in the art of television, and at the same time to work towards the ultimate attainment of a national standardised system of transmission.

Finance of service
12   'The cost of providing and maintaining the London station up to the end of 1936 will, it is estimated, be £180 000.
13   'Revenue should not be raised by the sale of transmitter time for direct advertisements, but the permission given in the British Broadcasting Corporation's existing licence to accept certain types of sponsored programmes should be applied also to the television service.
14   'Revenue should not be raised by an increase in the 10s [50p] fee for the general broadcast listeners licence.
15   'There should not be any separate licence for television reception at the start

of the service, but the question should be reviewed later in the light of experience.

16   'No retailer's licence should be imposed on the sale of each television set, but arrangements should be made with the trade for the furnishing of periodical returns of the total number of such sets sold in each town or district.

17   'The cost of the television service—during the first experimental period at least—should be borne by the revenue from the existing 10s licence fee.'

Following the presentation of the report to Parliament, the Postmaster General rapidly constituted the Television Advisory Committee.

News of the early establishment of a public television service in the UK seems to have had an invigorating effect elsewhere. Whereas in 1934 when the Television Committee received reports[17] on developments in the US and France there were no imminent prospects, or even plans, for public broadcasting in these countries, the position now, in 1935, was much more optimistic. Georges Mandel, the new minister of the French Postes, Télégraphes et Téléphones, and David Sarnoff, the president of the Radio Corporation of America, outlined their proposals in speeches made in April and May 1935 respectively.

On 17 April Mandel convened a press meeting and announced[18] that from a station of the PTT a new television service would begin which would operate in parallel with the radio diffusion transmissions. The service would use a wavelength of c. 175 m and the line standard would be 60 lines per picture (scanned horizontally) at a frame rate of 25 pictures per second. Mandel envisaged that, after a short time, the standard would increase to 90, then to 180 and finally to 240 lines per picture. To effect this plan a seven metre transmitter would to be installed at the PTT broadcasting station at 103, Rue de Grenelle, Paris. Barthélemy was the engineer chosen to install the equipment.

Mandel's announcement was timely, for earlier in the month the French magazine *Les Echos* had commented[19]:

> 'It is useless to deny that France, so far as television is concerned, is behind [the UK, the US and Germany]. But this is not the fault of our engineers. Up to quite a short time ago, the heads of nearly all the wireless telephone manufacturing firms declined to take any interest in television. They preferred, no doubt, to let foreign engineers develop their processes, and then to acquire manufacturing licences. This explains the smallness of the capital (some three millions [French francs]) invested in television enterprises in France. By contrast the British television enterprises had spent at least 200 million francs, and the Baird firms alone [had] a technical personnel that [included] 60 engineers.'

Within 10 days of Mandel's statement the first official television broadcast[20] in France was inaugurated from the PTT station. Barthélemy's system was employed and the 500 W transmitter was one constructed by the Compagnie des Compteurs. The broadcast, which lasted barely 20 minutes, was witnessed by a large crowd, and featured Mlle Beatrice Bretty, an actress of La Comedie Francaise, who 'raconte les souvenirs d'une tournée theatrale qu'elle vient d'effectuer en Italie'.

180-line television broadcasting[21,22] began on 17 November 1935, from the Paris PTT station. The old radio studio had been modified for television use and had been fitted with six 5 kW projector lamps and twelve 1 kW lamps. Since the studio was only 36′ in length special means had had to be installed to air condition it and so cool the actors and equipment. Ships' ventilators mounted on the studio floor, together with a chamber where artificial rain was produced at a low temperature (by a refrigeration plant) and fans ensured a constant circulation of air. A temperature of 27°C rather than 55°C without cooling could be maintained despite the 42 kW of floodlighting.

Direct scanning, instead of spotlight scanning, was accomplished with a double spiral Nipkow disc rotating at 3000 r.p.m., and for synchronisation a separate disc, directly coupled to the scanner, was utilised. Synchronisation pulses were transmitted at the end of each line and at the end of each picture. The camera room was adjacent to the studio, being separated from it by a sheet of glass. This room also housed the various amplifiers and control equipment.

From 103 Rue de Grenelle the picture signals were sent via an underground coaxial cable to the transmitter which was situated 2.5 km away in the north column of the Eiffel tower. A 320 m long coaxial cable carried the modulated transmitter output to the antenna mounted on the top of the tower. The vision and sound signals were radiated at 6.52 m and 7.14 m respectively and from the summer of 1936 the transmitter output power was 2.5 kW. The antenna consisted of four equidistant vertical conductors, each one wavelength in length, spaced on an imaginary 8′ diameter horizontal circle.

On the 18 November[23], from 17.30 to 19.30 hours, a special demonstration was given to the press. Actors from the Theatre Français, dancers of l'Opéra, music hall artistes and personalities including Elvine Popesco and Sacha Guitry were televised. The general public was able to view television in six public television rooms situated at the Office National du Tourisme, the Salon de la France d'Outre-Mer au Grand Palais, the Maison du Vᵉ arrondisement and the Conservatoire National des Arts et Métiers.

Criticism of the televised programmes was soon forthcoming. According to one commentator they were monotonous: during the formative period no innovations either in the presentation of the artists or in the setting of the scenes had been introduced; the announcer always used the same introduction, reading from the same small piece of paper, when presenting artists and from the initial image of the performer it was known what his/her performance would be. If the image comprised the bust of the artist then the artist would recite or sing and if shown full length another standard offering would be given[24].

In Germany, on 22 March 1935, the Reichs Post Zentralamt (RPZ) officially opened a regular public television service from Berlin[25]. The picture standard was 180 lines per picture, 25 pictures per second. According to the researches of Engstrom of RCA, such a standard was the minimum acceptable for television. Another, German, method of classification described an image based on the standard as '84% recognisable'.

The total equipment for the broadcasts comprised two 4 kW ultra short wave transmitters, for vision and sound, operating at frequencies of 44.30 MHz and 42.50 MHz respectively, the antennas mounted at the top (453′ above ground) of the Funkturm—a tower opposite the Reichrundfunk Gesellschaft building—together with one room in the RRG building which contained (in July 1935) just one set of film scanning apparatus, with the necessary monitoring equipment, and a demonstration room.

Programmes were broadcast, according to the schedule shown below[26], solely with the object of providing demonstrations in large rooms situated at six or eight different locations in Berlin. The public was admitted to these demonstrations for a 20 minute period free, on production of a ticket for which they had to wait several weeks.

The programmes consisted of ordinary short films, obtained without charge from the film trade. No studio programmes were broadcast and no films were made specially for the service. One report stated that the films would be supplied free of charge until the number of viewers reached 5000, after which the question would be re-examined. Although an outside broadcast van, containing a Fernseh intermediate film installation, was available it was not used for programme purposes (in July 1935).

Programme schedule (July 1935)

	Monday	Tuesday	Wednesday	Thursday	Friday	Saturday	Sunday
10.00–12.00	T&M	T&M	–	T&M	T&M	T&M	T&M
12.30–13.30	M	M	–	M	M	M	M
15.00–16.30	T&M	–	–	–	T&M	–	–
17.00–20.25	M	M	M	M	M	M	M
20.30–22.00	T&M	T&M	T&M	T&M	T&M	T&M	–
22.00–24.00	M	M	M	M	M	M	M

T&M = Television and music; M = Music

On the quality of reproduction, the BBC's chief engineer opined[27]: 'Having regard [to the line standard] the quality of reproduction is fairly good, but of doubtful programme value.' He witnessed one of the public demonstrations which consisted of, in the first period, a news reel having 'probably just about programme value', and, in the second period, a short film play 'which was definitely not of programme value'. 'There did not seem to be any comments on the part of the audience, or any applause. The sound channel was very poor, and the quality almost untelligible [sic].'

No television receivers were on sale in July 1935.

For the future, German television broadcasting planning[28] was based on the need to provide throughout Germany a minimum electric field strength of 1 mV/m from transmitters operating in the 5.5 m to 7.5 m band. Theoretical considerations and practical tests suggested that 21 to 30 transmitting stations, with broadcast ranges varying from c. 60 to c. 120 miles, would permit most of the country to be

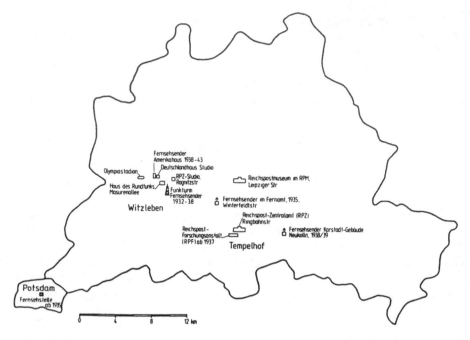

Figure 20.2   Location of the important television broadcasting stations in Berlin in the 1930s

Source: *Archiv fur das Post-und Fernmeldewesen*, Bonn, August 1985, p. 196

provided with satisfactory to good television reception. The number of noninterfering carrier frequencies necessary would be five or six and these would be distributed geographically among the transmitting stations so no mutual interference would arise. Figure 20.2 depicts the sites of the stations which would include locations on the Brocken (1128 m) in the Harz mountains and the Zugspitze (2962 m) in the Bavarian alps.

An intriguing German television policy decree was published on 12 July 1935. According to the Reichsgesetzblatt[29]:

> 'The further development of television makes [it] urgently necessary to unite in one [body] the responsibilities of the various government authorities concerned.
>
> 'In view of the special importance of television as regards the provision of safety in flying and as regards national air defence, I order: competence in the sphere of television is transferred to the Reichs Minister for Air, who will exercise control in consultation with the Reichs Minister of Posts.'

The decree was signed by the Chancellor, Herr A Hitler, the Reichs minister of posts, Dr J Goebbels and the Reichs minister for air, General H Goering.

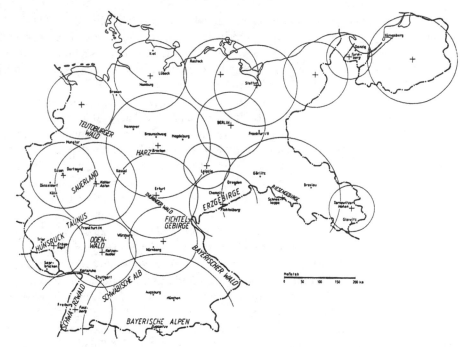

Figure 20.3   *Proposed locations of the television transmitting stations according to the 1935 plan of the German Post Office*

Source: *Archiv fur das Post-und Fernmeldewesen*, Bonn, August 1985, p. 202

Neither the air attaché nor the assistant military attaché at the British Embassy, Berlin, could offer any suggestion as to the meaning of this association of television with air defence. 'The fact that in Germany antiaircraft defence is in the hands of the Air Ministry, however, suggests the possibility of some technical development for utilising television apparatus in connection with AA guns or other defences,' wrote a member of the Embassy.

Actually, the interest, by the Air Force, in the military applications of television was not unique. In the United Kingdom the Admiralty[30], in 1923, had initiated at the Admiralty Research Laboratory, Teddington, a research and development programme which had as its objective the use of television in aircraft as an aid to improving the accuracy of naval gunfire. A television camera and transmitter in an aircraft would—it was hoped—send images of the 'fall of shot' and the target to the naval vessel firing its guns. Nothing came from this venture: the very crude definition available in the 1920s was quite unsuitable for the Admiralty's purpose. The Air Ministry and the War Office took a slight interest in this work.

Later, from c. 1937, EMI and Baird Television engineered lightweight airborne television equipment which could be employed for aerial surveillance[31,32]. Both the French and the Russian governments were attracted to this type of television and endeavoured to place contracts with the British companies but the onset of hostilities in 1939 led to an abandonment of the developments.

In August 1935, at the Berlin Radio Exhibition, Fernseh showed a 320-line Nipkow disc scanner in operation[33]. This was the first occasion in Germany when such a line standard was demonstrated. The scanning disc, having 80 holes, ran *in vacuo* at 6000 r.p.m.. One observer thought the received image—on a 15" diameter cathode-ray tube—was 'the finest picture he had ever seen'. His only criticisms concerned 'the very pronounced flicker and the distortion due to screen curvature'. Other exhibitors used 90-line and 180-line standards, see Table 20.3.

Direct scanning of extended studio or outdoor scenes, other than by the intermediate film process was still not part of the German research and development effort. Much reliance was placed on the generation of images from film stock. The Reichs Rundfunk Gessellschaft exhibited a Fernseh i.f. projector incorporating a high voltage cathode-ray tube but Traub reported that 'the results were . . . rather a little disappointing, and there was very little improvement [from last year]'. He was inclined 'to suspect the system as being unproductive of good results' since no doubts could be expressed concerning the competence of the system's designers and engineers[33].

During the Funkaustellung a fire started in a hall adjacent to the television exhibition hall. The exhibition hall was undisturbed but the sound and vision transmitters were completely destroyed. Resumption of working of the Berlin-Witzleben station was announced by the German News Bureau on 24 December 1935[34].

Prior to this date television signals had been radiated from a small station[35] in the centre of Berlin; they had a range of about 4 km. Sir Frederic Williamson, director of postal services in the UK, witnessed a demonstration of 180-line television, in December 1935, in one of the public viewing rooms which had been set up at various post offices in Berlin, Tables 20.4 and 20.5. He was not impressed[36].

'I was first shown a reproduction of a film. This was not at all satisfactory. The horizontal lines were very visible, the figures were very indistinct, the flicker was extremely pronounced, there was a very noticeable distortion of vertical lines (doorways, window frames, picture frames) and the sound was extremely indistinct.

'I was next shown direct television from the central station. In this case the only image was the head and shoulders of the operator, a telephonist. This was much more successful. The lines were not so noticeable; the image, no doubt owing largely to the greater scale, was much clearer, and the sound was very distinct. I spoke to the demonstrator by telephone and saw the reply being given in the image.

'If what I saw was representative, and there is no reason to think the contrary, it would appear that Germany generally is a good deal behind the present stage of development in this country.'

When the Witzleben station was rebuilt and reopened the line standard was still 180 lines per picture, 25 pictures per second, and the wavelengths were 7.053 m and 6.772 m for the sound a vision transmissions respectively. The opening 90 minute programme featured Willi Schaeffen, compère and humourist, Else Elster, who sang some popular songs and Carl de Vogt, who played the flute.

Table 20.3   Details of the exhibits of the 1935 Berlin Funkausstellung

Facts	Fernseh AG		Telefunken			Tekade			Loewe		Lorenz	Müller	RPZ		HHI	RRG
Exhibit no.	a	b	a	b	c	a	b	c	a	b	a	a	a	b	a	a
Transmitters																
Type	film		film	floodlight		film	film		film				spotlight			film
Scanner	disc	Berlin 180-line radio transmissions	disc	slow mirror drying	Berlin transmissions	disc	disc	Berlin transmissions	disc	Berlin transmissions	Berlin transmissions	Berlin transmissions	disc	Berlin transmissions	Berlin transmissions	disc
Lines	320		180	100		90	90		180				90			180
Elements	123 000		40 000	10 000		10 000	10 000		40 000				8000			40 000
Pictures per sec	25		50 interlaced	50		25	50 interlaced		50 interlaced				25			25
Aspect ratio	5:6		5:6	1:1		5:6	5:6		5:6				1:1			5:6
Fmax Kc	1560		500	2½ kc per line		130	130		500				100			500
Amplifier	LF		carrier frequency	?		LF	LF		DC				DC			LF
Transmission wire channel	wire		wire	wire		wire	wire		radio and or wire				wire			wire
Receivers																
Scanner	Cathode-ray tube	Cathode-ray tube	Cathode-ray tube	lamp screen	Cathode-ray tube	mirror screw	mirror screw	mirror screw	Cathode-ray tube	Cathode-ray tube	Cathode-ray tube	Cathode-ray tube	Cathode-ray tube	Cathode-ray tube	Cathode-ray tube	CR tube & interm'd. film
Light source or valve	1					Kerr cell	Kerr cell	Kerr cell								
Image size	25 × 30 cm	18 × 22 cm	5 × 18 cm	2 × 2 m	18 × 22 cm	25 × 30 cm	18 × 22 cm	15 × 18 cm	19 × 21 cm	19 × 21 cm	18 × 22 cm	15 × 18 m	15 × 15 cm	18 × 22 cm	18 × 22 cm	3 × 4 m
Colour	sepia	sepia	yellow	pale pink	white	white	white	white	pale blue	pale blue	white	green	white	sepia	white	white
Definition	excellent	v. good	good	good	v. good	v. good	good	v. good	good	v. good	v. good	fair	good	good	good	good
Intensity	v. bright	bright	v. bright	ext. bright	v. bright	ext. bright	ext. bright	bright	ext. bright	ext. bright	v. bright	bright	fair	fair	bright	v. bright
Overall flicker	considerable slight	slight	none	none	considerable slight	slight	none	none	none	considerable slight	slight	slight	none	none	slight	none
Synchronism	impulse	impulse	impulse	mains	impulse	mains	mains	mains	impulse	impulse	impulse	impulse	impulse	impulse	impulse	impulse

Table 20.4   Locations of television viewing rooms in Berlin in July 1936

Building and town	Operator	Opening date
Reichspostmuseum, Leipziger Str., Ecke Mauerstr.	DRP	9.4.1935
Postamt Potsdam, Am Kanal 16	DRP	13.5.1935
Hause des Rundfunks, Masurenallee	RRG	15.5.1935
Reichsverband der deutschen Rundfunkhändler, Potsdamer Str. 123b	RRG	15.5.1935
Reinickendorfer Str. 113	RRG	15.5.1935
Lichtenberg, Parkaue 6-7	RRG	15.5.1935
Postamt W 30, Geisbergstr. 2	DRP	Sept. 1935
Postamt Schöneberg, Hauptstr. 27	DRP	Sept. 1935
Postamt Charlottenburg, Berliner Str. 62	DRP	18.7.1936
Postfuhramt, Artilleriestr. 10	DRP	18.7.1936
Postamt Steglitz, Bergstr. 1	DRP	18.7.1936
Postamt Neukölln, Richardstr. 119	DRP	July 1936
Postamt Lichtenberg, Dottistr. 12	DRP	July 1936
Postamt Pankow, Wollandstr. 134	DRP	July 1936
Postamt Reinickendorf-West, Berliner Str. 99	DRP	July 1936

DRP = German Post Office, RRG = Reichs-Rundfunk-Gesellschaft.

This line standard, inadequate though it was, and the equipment needed to implement it, remained operational for at least a further two years. When the new Berlin television studios were completed, in April 1938, in the Deutschlandhaus on Hitler's Square, the official description stated: 'Near the actual stage is the stage manager's room and a waiting room with only a dim blue light, where artists assemble before going on the stage, to accustom themselves to the semidarkness in which they must act.' This account[37] implies spotlight scanning which the BBC rejected as unsatisfactory a few weeks after the opening of the London television station in November 1936.

Table 20.5   Locations of theatres showing large-screen television in Berlin

Building	Location	Opening date	Number of seats	Screen size
Postamt NW 27	Turmstr./Ecke Lübecker Str.	Autumn 1935	294	3 m × 4 m
Reichspostmuseum	Leipziger Str./ Ecke Mauerstr.	1936	120	1 m × 1,2 m (1,8 m × 2,0 m ab 1938)
Bechstein-Saal	Linkstr., Nähe Potsdamer Platz	January 10 1942	300	No result probably cinema size

By December 1935 the UK Television Committee had made excellent progress towards the establishment of a high-definition television service. A decision had been agreed that both 240-line (Baird Television) and 405-line (Marconi-EMI) systems would be tested; tenders for complete transmitting and studio television equipments had been invited and quotations had been received from the rival companies; and contracts had been accepted by the commencement of 1936.

References

1 Notes on 'Conference at General Post Office, 5 April 1934', minute post 33/4682
2 Report of the Television Committee, Cmd. 4793, HMSO, January 1935
3 Minutes of the first meeting of the Television Committee, 29 May 1934, minute 33/4682
4 Paper 3, the Television Committee, minute 33/4682
5 Ref.2, Appendix IIIA (unpublished), 'Visit to the United States', pp. 1–8
6 Ref.2, Appendix IIIB (unpublished), 'Visit to Germany, 4–11 November 1934', pp. 1–13
7 ABRAMSON, A.: 'The history of television, 1880 to 1941' (McFarland, Jefferson, 1987)
8 BURNS, R.W.: 'British television, the formative years' (Peter Peregrinus Ltd, London, 1986) pp. 406–408
9 ANON.: A report on 'Television', *North Western Daily Mail*, 5 February 1935
10 ANON.: Report, *New Statesman and Nation*, 9 February 1935
11 Editorial, *Wirel. World*, 1935, **36**, (6), p. 129
12 Current topics, *Wirel. World*, 22 February 1935, p. 192
13 ANON.: 'The coming of television. Radio manufacturers warning', *The Times*, 25 January 1935
14 ANON.: 'Television two years away. Radio chief says it cannot come before', *Daily Express*, 31 January 1935
15 ANON.: 'Television prospects', *Northern Whig*, 1 February 1935
16 ANON.: 'Is television a menace to the "halls"?', *Variety News*, 7 February 1935
17 Commercial Counsellor to HM Embassy, Paris: 'Television in France'. 6 December 1934, post 33/5534, file 9
18 PAUCHON, B.: '50 ans de television en France', *Revue de l'UER*, December 1986, (220), pp. 3–12
19 ANON.: 'World progress in télévision'. Les Echos, 1 April 1935, Post 33/5143, file 1
20 ANON.: 'The French 60-line transmission', *Television and Short-wave World*, June 1935, p. 359
21 ANON.: '180-line television from the Eiffel tower', *Television and Short-wave World*, January 1936, pp. 4–5
22 Le DUC, J., and Barthélemy, R.: 'Technical details of the Eiffel tower television', *Television and Short-wave World*, September 1936, pp. 494–495, 512
23 Reports in *The Times* for 18 November 1935, p. 12a, 20 September 1935, p. 11g and 14 December 1935, p. 16
24 Ref.23, p. 15
25 ANON.: 'High-definition television in Germany', *Wirel. World*, 22 March 1935, pp. 288–289
26 ANON.: 'Reception of the Berlin ultra-short wave transmitter'. 11 July 1935, post 33/5143, file 8
27 'Report by Chief Engineer [BBC] on visit to Berlin 26th/27th June 1935'. BBC file E1/797/1
28 SCHOLZ, W.: 'National television plans for Germany', *Wirel. World*, 24 May 1935, pp. 525–526
29 NEWTON, B.C.: Letter to the Right Honourable Sir Samuel Hoare, from the British Embassy, Berlin, 15 August 1935, post 33/5143
30 BURNS, R.W.: 'Early Admiralty and Air Ministry interest in television'. Conference papers of the 11th IEE Weekend Meeting on the 'History of electrical engineering', July 1983, pp. B/1–B/17

31 HERBERT, R.: 'Airborne television', *Radio Electron. World,* February 1985, pp. 16–17
32 'Television transmission from aircraft'. AVIA 13/1263, PRO, Kew
33 TRAUB, E.H.: 'Television at the 1935 Berlin Radio Exhibition', *J. Television Society,* **2,** Part III, pp. 53–61
34 'Resumption of working of the television transmitter Berlin-Witzleben'. German News Bureau, 24 December 1935, post 33/5143, file 6
35 ANON.: 'The ultra-short wave television transmitter at Berlin', *Television and Short-wave World,* September 1935, pp. 532–533
36 Television Advisory Committee, paper 22, 12 December 1935, post 33/5536, file 24
37 Quoted in HUBBLE, R.W.: 'Four thousand years of television' (Harrap, London, 1946), p. 153

Chapter 21

The London station and foreign developments, (1935–1938)

The first meeting of the Television Advisory Committee (TAC) was held on 5 February 1935 with Lord Selsdon in the chair. Colonel A S Angwin, Mr Noel Ashbridge, Mr O F Brown and Mr F W Phillips, who were members of the original Television Committee, now served on the new committee in addition to the new member, Sir Frank Smith. Mr J Varley Roberts retained his position as secretary[1]. A Technical Subcommittee (TSC) was set up with Sir Frank Smith as Chairman, Roberts as secretary and Angwin, Brown and Ashbridge as members.

Among the immediate matters to be resolved were the specification of the television apparatus and the location of the London station. The first of these was referred by the TAC, at its first meeting, to the TSC[2]. At its second meeting, on 15 February 1935, the representatives of Marconi-EMI (Messrs Shoenberg, Condliffe and Blumlein), who had been invited to attend and give their views, advanced the following proposals for the specification[3]:

1   Scanning: 405 lines interlaced to give 50 frames per second, the frames to be scanned from top to bottom.
2   Modulation direction: high carrier to represent white, low carrier to represent black.
3   Background brightness components: d.c. working at the transmitter; 70 % of the transmitter characteristic to be used for vision signals, the peak output to represent white and 30 % of the peak output to represent black. The suggested degree of perfection of d.c. working at the transmitter was to be such that during any one transmission the absolute carrier output representing black was to vary by no more than ± 3 % of the peak transmitter output.
4   Synchronisation: synchronising signals to be pulses in the blacker than black direction extending downwards from 30 % of peak carrier to zero carrier; all pulses to be rectangular, the duration and shape being as shown in Figure 21.1; the synchronising pulses to extend substantially to zero carrier, residual carrier to be no more than two per cent of peak transmitter output.

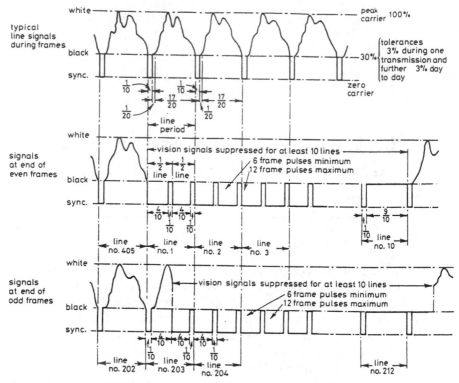

Figure 21.1   Marconi-EMI Television Ltd video waveform

This meeting of the TSC was the first occasion on which Shoenberg put forward his company's intention to provide a system of television using 405 lines. It was a momentous decision to make.

Professor J D McGee, a former member of Shoenberg's team, has related some of the problems which Shoenberg had to resolve before he arrived at this heroic conclusion[4]:

> 'There was, for example, a long debate as to whether d.c. or a.c. amplification should be used—and then, if a.c., how could the d.c. level (black level), be re-established? And as the picture definition was increased the bandwidth increased, so that the radio transmission frequency necessary also increased, limiting the range to approximately line of sight. Would this be acceptable? Then, should positive or negative modulation be used? And, as the brightness of the picture increased, flicker became a serious problem so that the alternative of sequential or interlaced scanning had to be decided. These problems landed squarely on Shoenberg's desk.
>
> 'As the number of lines crept up from 120 to 180 and then 240, the required picture signal bandwidth increased; this increased progressively the problems of amplifiers, of transmitters, of tubes and of

achieving adequate service area. It was clear that the quality of the picture increased very noticeably as the number of lines increased. Since it was possible to scan an electron beam at these much higher speeds, it was natural, indeed inevitable, that the advantages of the electronic system should be exploited to the maximum. But what was the practical maximum? It would clearly be difficult, if not impossible, to push the number of lines much further than 240. No one knew what disastrous snags we might meet if we attempted to reach still higher definition. Higher definitions were *terra incognito*, and not just the rather obvious line of development that it may now appear, 35 years later.

'To us, then young men, it was a challenge and an adventure, but to Shoenberg it must have been a very worrying problem. On him fell responsibility to the company and to his staff to make the right decision ... At this point Shoenberg made what was probably the biggest—and I consider the most courageous decision—in the whole of his career: to offer the authorities concerned a 50 frame/second, 405 line/picture television system. Remember that this meant a 65 % increase in scanning rate and a corresponding decrease in scanning beam diameter in the c.r.t., a nearly three-fold increase in picture-signal bandwidth and—worst of all—a five fold increase in the signal/noise ratio of the signal amplifiers, and this lists only a few of the resulting problems.

'The cynic may say that this was a piece of gamesmanship planned to overwhelm our competitors. But no one who knew Shoenberg or who was aware of the real state of technical development at that time would give this idea a moment's credence. No, it was the decision of a man who, having taken the best advice he could find, and thinking not merely in terms of immediate success, but rather of lasting, long-term service, decides to take a calculated risk to provide a service that would last ...

'To us this decision was a stimulating challenge. To Shoenberg it must have been a heavy and worrying burden. In later years, he often recalled how colleagues in the higher management of the company had seriously questioned his decision, and had warned him that should he fail to fulfil his contract it would be disastrous for the company.

'Yet I cannot remember that he ever showed his worries to us at all obviously. The nearest perhaps was one day when things were particularly sticky. That day he finished up a rather depressing review of our progress with the comment "Well gentlemen, we are afloat on an uncharted ocean and God alone knows if we will ever reach port".'

Shoenberg's choice of 405 lines per picture was probably based on the following points:

1 Wenstrom had shown that the definition of home motion pictures corresponded to approximately 400 lines, and also that 400-line television was

'perfect for a single face, excellent for small group or detached objects, and good for full theatre stage or outdoor spectacles'.

2   Engstrom had concluded that 360 lines/picture gave an excellent image and 400 lines/picture was equivalent 'for practical conditions' to a home movie film.

3   343 lines/picture television had been demonstrated in 1934 by the Radio Corporation of America.

4   An odd number of lines per picture was required for the interlacing method adopted.

5   The number of lines per picture had to be compounded from integers preferably less than ten for ease of signal generation.

6   Shoenberg was keen that a line standard should be chosen which would allow its full potentialities to be appreciated as research and development work made it possible for the system to be improved.

Taking the standard mentioned in point 3 above as a base, the only odd integers from 343 to 405 (point 4) which can be factorised are:

$$343 = 7 \times 7 \times 7 \qquad 365 = 5 \times 73 \qquad 387 = 3 \times 3 \times 43$$
$$345 = 5 \times 3 \times 23 \qquad 369 = 3 \times 3 \times 41 \qquad 393 = 3 \times 131$$
$$351 = 3 \times 3 \times 13 \qquad 371 = 7 \times 53 \qquad 395 = 5 \times 79$$
$$355 = 5 \times 71 \qquad 375 = 5 \times 5 \times 3 \times 3 \qquad 399 = 3 \times 7 \times 19$$
$$357 = 3 \times 7 \times 17 \qquad 381 = 3 \times 127 \qquad 405 = 5 \times 3 \times 3 \times 3 \times 3$$
$$363 = 3 \times 11 \times 11 \qquad 385 = 5 \times 7 \times 11$$

Using point 5, the only numbers which can conveniently be used are 343, 375 and 405. Presumably, Shoenberg did not wish to use 343 as he could have been accused of copying the RCA system. Consequently, he was left with a choice between 375 and 405, and points 1, 2 and 6 favoured the larger number.

Mr S J Preston, formerly patents manager of EMI, once observed[5] that Shoenberg made his resolution to adopt 405 line definition 'knowing that receivers available at that time could not be expected to deal with the full bandwidth which 405 line scanning would require but his view was that it was better that the early pictures should be somewhat lacking in definition along the line so that later developments in receiver design which he was sure would take place could be usefully employed without any change of standards'.

Baird Television Ltd (BTL) advocated a 240 lines per frame (progressively scanned), 25 frames per second picture with an aspect ratio of 4 to 3, and with the scanning taking place from right to left and from top to bottom[6].

BTL recommended progressive scanning at the present stage of the art for the reason that it was not considered proved that interlaced scanning eliminated or reduced flicker, 'and even if interlaced scanning had advantages it [did] not follow that interlacing by scanning alternate lines [was] the most suitable mode of intercalation, of the many which [had] been and [were] being experimented with'. The company stated that the process of interlaced scanning did not present serious technical difficulties, but felt that the improvement in flicker, gained by this method, was not sufficient to warrant the extra complication involved at the receiver and suggested that the question should be deferred until further experimental work had been carried out.

On the issue of the number of lines per frame to be used, BTL considered that the 240-line standard represented the economical and practical limit which could be recommended for standardisation for at least three years. 'It [was] definitely true that the optimum definition for a picture transmitted and received by radio on a 240-line basis [had] by no means been demonstrated. When this [had] been done it [would] show a picture a long way ahead of what [had] been demonstrated by any system and [would] give ample definition, clearness and quality of all types of scenes as well as for close-ups' noted BTL[7].

BTL was very anxious that the number of lines should not be raised beyond 240, and when questioned as to its views on the adoption of 400 lines the company stated a number of grave difficulties which would be encounted both at the transmitter and receiver and indicated that all forms of mechanical scanning would have to be abandoned. The company felt that at the 'present state of the art, as known to them, a more pleasing result would be obtained by the faithful transmission of a 240-line picture rather than by transmitting a 400-line picture which would, of necessity, suffer during the process'.

BTL backed up this statement by giving a number of facts as a result of testing the reactions of a large number of people.

With a bright 240-line cathode-ray tube picture 12 × 9 inches with a picture frequency of 25 frames per second, the lines were not seen at four feet or more; and at six feet or more the definition seen by the eye was equal to that of an actual photographic enlargement of the picture being televised—the dimensions, brightness and contrast being the same.

In addition, as the number of lines increased beyond 240:

1   the mechanical and optical problems associated with the design and operation of picture scanners for transmission became increasingly difficult;
2   the increased bandwidth required lowered the gain per stage, which meant more stages of amplification, 'with a consequential increase of interference noises';
3   the increased frequency range involved the use of a higher carrier frequency as it was 'generally accepted that the modulation frequency should not be higher than $2\frac{1}{2}\%$ of the carrier frequency';
4   the cost of receivers increased as more stages were required with a corresponding increase 'not only in the cost of components but in the cost of manufacture and inspection, particularly with regard to the "ganging" of the circuits'.

Also, as regards the cathode-ray tube, BTL felt that it was already difficult to obtain in production a spot size small enough for the reproduction of a 240-line picture and, furthermore, the illumination of the fluorescent screen of a cathode-ray tube decreased with increase of scanning frequency.

A final point in BTL's evidence concerned the lack of experience of television problems, of most receiver manufacturers, even from an experimental point of view.

In spite of this 'it was believed that standardisation on a 240-line basis would enable manufacturers generally to be in a position to produce receivers within the next nine months but there [was] no doubt that the receiving problems which [arose] with a greater number of lines than 240 would hold up the quantity production of television receivers for at least eighteen months'.

BTL's list of the hurdles to be overcome tended to highlight the tremendous technical advances which had been made by Marconi-EMI and especially by Shoenberg's research and development team. It was unlikely that the Baird company knew precisely the degree of success that the Hayes group had achieved in high-definition television since it was EMI's policy not to encourage publication of the results of its research work. However, BTL was aware that EMI was associated with RCA, and Dr Zworykin of that company had recently published a paper on 'Television with cathode-ray tubes' in the *Journal of the Institution of Electrical Engineers*[8]. In this paper he had stated that some of the iconoscope tubes actually constructed had been satisfactory up to 500 lines 'with a wide margin for future improvement'.

This information, together with the ideas contained in EMI's patents and the Postmaster General's utterance to Baird that he must face up to the possibility of there being a system superior to his own, should have led BTL to form some conclusion, albeit a rather ominous one, concerning EMI's progress.

EMI was led to consider interlacing when its cathode-ray tube receivers began giving brighter pictures and it was obvious that 25 Hz flicker would be unacceptable in practice[9]. Prior to the beginning of 1934 the company was working on a 180-line picture with sequential scanning at 25 frames per second, using cathode-ray tubes which were not at all bright. With a poorly illuminated picture at the receiver, flicker was not objectionable, but as the brightness of the screen improved the flicker could not be tolerated.

The scanning method which EMI adopted was that devised by R C Ballard of the Radio Corporation of America and patented by him in 1933. His procedure of producing interlaced scanning has the great advantage that both frame scans (which generate an image) are absolutely identical 'so that the method of scanning at the receiving end remains exactly the same as for noninterlaced scanning and the receiver is blissfully ignorant as to whether interlaced or ordinary scanning is taking place'[10]. EMI considered this point to be very important, from the position of the universality of a television system, because it was most desirable not to tie down the receiver to any peculiar characteristics of the transmitter. The Ballard patent covered the only method of interlaced scanning which possessed the above feature. Shoenberg knew of no other: 'And if I knew of any other I should have patented it,' he told Lord Seldon[11]. 'I might also tell you that we have a pretty large patent department that watches all patents which have any relevance to our business and that no piece of apparatus can be built by the company in any way whatsoever without the approval of that patent department.'

The Ballard patent was certainly a master patent and EMI was fortunate in being able to use it. The utilisation of interlacing to diminish flicker was one of the three most important aspects of the M-EMI system: the others were the transmission of the synchronising signals in the same channel as the picture signal and d.c. working.

In any scene that is being televised by a television camera/scanner the camera/photocell tube generates signals having a value corresponding to the brightness of the various elemental areas which comprise the scene. The output from the camera/photocell tube is therefore unidirectional; albeit a fluctuating current above a certain datum line (corresponding to a completely black scene). This fluctuating, unidirectional current may be analysed into an alternating component which depends on the relative brightness values of the different parts of the picture and a direct component which conveys information on the brightness of the whole of the picture area. Thus, the signals resulting from the televising of a completely black sheet of paper and a brightly illuminated white sheet of paper are the same except for the magnitude of the direct component in the two cases (assuming that the sheets of paper completely fill the fields of view).

The EMI team thought[11] that it was essential to transmit the direct component so that the receiver would accurately reproduce the brightness values of the original picture although in the United States the practice was to use the alternating component only[12]. This practice possibly arose because of the difficulty in designing satisfactory stable high gain d.c. amplifiers. The consequence was that the resultant signal varied about a mean value rather than about a datum value which represented black.

Now with an a.c. television signal the amplitude of the signal corresponding to black relative to the mean signal is a variable quantity depending on the nature of the picture. Thus, both the signal due to a white dot on a black background and that due to a black dot on a white background are as shown in Figure 21.2, and from these it is apparent that in the first case the signal corresponding to black is close to the mean line whereas in the latter case this signal is remote from the mean line. A change in the nature of the picture thus causes a wander in the amplitude of the black signal so that, as the d.c. component has been removed, the valves of an amplifier require to be biased in such a way that they can handle not only the apparent maximum amplitude but absolute values in excess of the difference between black and white. Indeed, a television waveform devoid of its direct component can drift about its datum line by approximately 60% of its amplitude. Such a signal requires

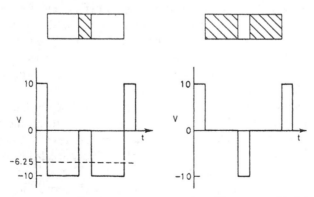

Figure 21.2   A black spot on a white background and a white spot on a black background lead to the production of line signals which have differing mean levels

larger and more powerful valves to handle it and the apparatus becomes unnecessarily costly, or, for a given available peak transmitter output, much smaller useful signals are obtained at the receiving point.

Shoenberg has written[13]: 'A great controversy raged for some months among my staff as to whether we should use d.c. amplification with the d.c. component. In the early days P W Willans was the protagonist of the restoration method and later on Blumlein advocated it.' Willans had been with the Marconi Wireless Telegraph Company and was probably familiar with its work on the transmission of synthetic half-tone pictures by d.c. keying the radio transmitter. It was Willans who proposed the solution[14], in 1933, to the problem of d.c. reinsertion following Shoenberg's decision to adopt d.c. restoration. In 1952 Shoenberg noted: 'It seems strange now that this matter should ever have been one for controversy, but at the time there were cogent arguments on both sides and in the end I made the decision in favour of restoration.' This ruling meant that definite carrier values represented black, white and the synchronising signal. There was no mean carrier value as this depended on the picture brightness and so the system of transmission was analogous to telegraphy rather than telephony.

In addition to reproducing a more realistic image, the transmission of the d.c. component enabled the effective radiated power from the transmitter to be increased by a factor of between eight and 12[11].

Marconi-EMI's and Baird Television Ltd's image standards were made public on 7 June 1935. The announcement was not met with universal acclaim. Indeed, some of the comments expressed by experts were quite critical of the TAC[15].

1   'The adoption of two standards of transmission having no mathematical relationship to each other is such a grotesque step in the wrong direction that I feel that the introduction of a television service should rather be postponed *ad infinitum* until a common standard can be arrived at.'

2   'Such an enormous increase in frequency band appears to the writer to be an impractical proposition . . . '

3   'It is in the matter of service that I see the greatest difficulty . . . The television receiver is likely to contain more valve stages, higher voltages and still more unfamiliar equipment than a modern semi-standardised broadcast receiver, and may therefore be expected to present a still more acute service problem.'

4   'Even with the types of screen material most commonly used, I feel that the interline shimmer of an interlaced picture is more trying to the eyes than the 25 frame flicker of a sequential scan . . . '

5   ' . . . a standard employing an odd number of lines should not be allowed . . . and presents a serious obstacle to those developing the mechanical systems.'

6   ' . . . an easy calculation shows that from the necessary viewing distance [for three people watching a television screen] the eye will hardly be capable of resolving a fully detailed 405 line picture.'

7   ' . . . a maximum modulation frequency of 1 MHz represents the practical limit of transmission on a wavelength of not less than 6 m. It is impractical to go below that wavelength for a variety of sound reasons.'

Figure 21.3   Alexandra Palace, north London, with the television mast

Reproduced by permission of the BBC

Figure 21.4   A close-up view of the Franklin antenna array

Reproduced by permission of the Marconi Company Ltd

All of this was, of course, based on a total ignorance of what M-EMI had achieved and shown to the members of the TAC. As a matter of policy, no technical papers on the 405-line system had been published by Shoenberg's staff and no demonstrations had been given to the general public. The experts had given their opinions on the knowledge which they had gleaned from their own experiences of 180-line television. They were to be pleasantly surprised.

On 3 June 1935 it was announced that the location of the London television station would be Alexandra Palace. The ground level at the Palace was 306' above sea level and it was thought possible to remove the cupola of the south-east tower and build on to the brick work a 225' lattice mast, giving an antenna height of 615'[16].

Alexandra Palace possessed excellent accommodation which comprised suitable rooms for studios, *inter alia*, on both ground and first floors:

Ground floor	First floor
67′ × 50′	30′ × 30′
70′ × 50′	24′ × 30′
57′ × 40′	65′ × 30′
35′ × 30′	
40′ × 30′	
30′ × 30′	

The BBC considered the accommodation provided by the first floor rooms to be sufficient for the transmitter rooms, two studios and two spare rooms, and the ground floor rooms provided space for the main studios. Access to Alexandra Palace was good and a car could make the journey from Broadcasting House (in the centre of London) to the Palace in 20 minutes (without the negotiation of heavy traffic—in

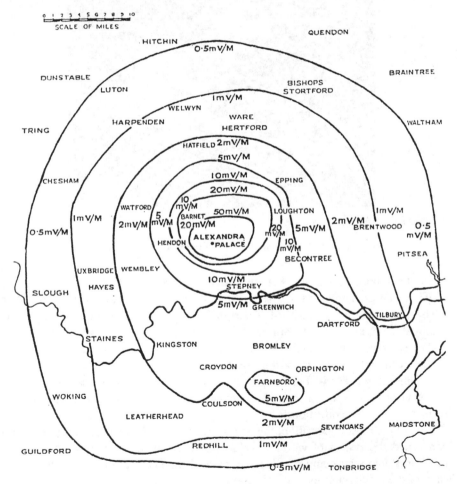

Figure 21.5   Map showing the approximate field strength of the Alexandra Palace transmitter at a height of ten to 15 metres above ground level

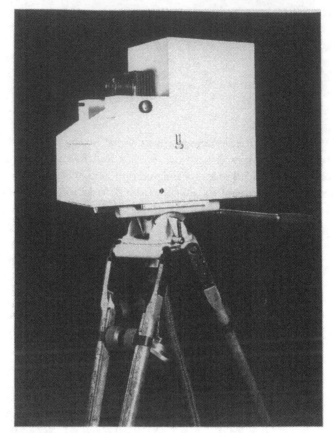

Figure 21.6   EMI's emitron camera

Reproduced by permission of the Marconi Company Ltd

1935). In addition, the governors of the Palace were ready to assist the BBC in every way within reason and the site was so situated that the service area for a transmitter at this point covered primarily 'that part of London in which the residents [might] reasonably be expected to be among the first to acquire television receivers'.

The site recommendation of the TAC was made only after a careful study of a model of the relief contours of the London area, which had been prepared in the Post Office Engineer-in-Chief's department. This was made in wet sand and pieces of twine were stretched across to determine the obstructions which existed for reception in various directions[17].

One of the most important of the issues which had to be considered by the TAC concerned the system of vision waveform generation to be used for film transmission, studio scenes and outdoor scenes, and the lighting intensity required for each of these different modes of television.

Marconi-EMI's tender covered the supply of six electronic cameras (known as emitrons), four for studio work and two for scanning film, for these activities. The

Figure 21.7 *The photograph shows the internal arrangement of the camera tube and the electronic circuits*

Reproduced by per mission of Thorn EMI Central Research Laboratories

company considered that adequate illumination for studio operation would be produced by 50 W per square foot of floor area using incandescent lamps from either an a.c. or d.c. supply[18]. (M-EMI's system is shown in Figure 21.8).

Marconi-EMI's preference for the emitron camera showed that the company had advanced the development of this apparatus since its meeting with the Television

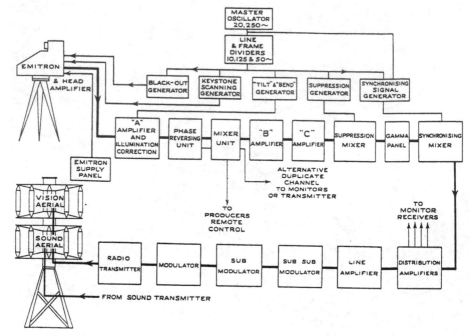

Figure 21.8   Schematic diagram showing the layout of the Marconi-EMI television transmitter at Alexandra Palace

Source: *Journal of the IEE*, **83**, 1938, p. 768

Committee in June 1934, for then Shoenberg had told the committee that direct television as practiced at that time did not satisfy him scientifically. 'We think that at the moment we have a cathode-ray scanner which can be used for direct vision with the same limitations as the disc is now used, that is to say, within a prescribed area determined by the arrangement of the photocells, I am rather anxious not to over state it.' Later, at the same meeting, when referring to the M-EMI system, he mentioned that he had recommended interlaced scanning 'because I think finally, in five years time, say, the iconoscope or something of that kind is coming along which will enable you to do away with the intermediate film'[19].

BTL's tender[20] comprised equipment for studio television by the following methods:

1.   Spotlight scanner

'This apparatus is suitable for the transmission of close-ups of speakers and announcers. It is proposed that about half the area of the room adjacent to the studio should be subdivided to provide a small studio, spotlight scanner room and monitor room. The spotlight scanner cannot be used actually in the studio because it involves the use of a disc running at 3000 rev/min, and it is noisy. It therefore works through a glass window. The monitor room also has a glass window so that the control engineer can see the performer and also the latter can see his own image on a monitor tube, and so avoid moving out of the spotlight beam.'

2 Electron image camera

'This would be used in the main studio for the televising of larger scenes, and must be situated within 20′ of its scanning and amplifying apparatus. It is proposed to form a small room, partly in the corner of the studio and partly on the colonnade, to house the associated apparatus. The camera is mounted on a movable truck, which can be moved within a radius of 20′ of the apparatus.'

3 Intermediate film apparatus (Figure 21.9)

'This is also suitable for the televising of large scenes, but it is of bulky nature and requires connection with the water mains and drainage. It is, moreover, noisy and must be enclosed in a sound proof booth. It is suggested that an intermediate film camera room be built on the colonnade, with a sort of "bow window" looking into the studio, so that the camera may be "panned" to take in any part of the studio. By this means three sets could be effectively covered.'

Thus, two of the three camera scanners were noisy and static and required a soundproof room. On the other hand the emitrons were small, noiseless, easily

Figure 21.9   The Baird intermediate film scanner installed at Alexandra Palace

Reproduced by permission of the BBC

portable and, very important, fairly sensitive. This was shown by the studio lighting requirements in the two tenders[21]:

Marconi-EMI

1	Roof lighting	18 kW
2	Directional lighting	6 kW total = 24 kW

Baird Television

1	Supply arc for two teleciné transmitters and intermediate film transmitters	28.5 kW
2	Supply arc of spotlight transmitter	31.5 kW
3	Supply studio lighting for electron camera and intermediate film transmitter	94.4 kW

Figure 21.10 Marconi-EMI film scanner, August 1936. The photograph shows an emitron camera coupled to a modified film projector

Reproduced by permission of the BBC

Figure 21.11 The Baird telecine scanners on test at Crystal Palace prior to being installed at Alexandra Palace

Reproduced by permission of the BBC

An additional disadvantage of the intermediate film process was the cost of the 35 mm film stock and processing chemicals which amounted to £48 per hour or £12 per hour using split 35 mm film stock. Against this the cost of servicing the emitron cameras with tubes was £2.10.0 d per transmission tube hour.

BTL's tender highlighted the cumbrous nature of the company's equipment vis à vis the mobile emitron camera and gave a realistic perspective of the efforts of others—in Germany, France, the USSR and the US—who were advocating or progressing mechanical scanning television systems. Even the Farnsworth image dissector camera which had been under continuous development from 1927 did not compare favourably with the emitron.

Following scrutiny by the TAC both companies suffered some suggested cuts in the estimated running costs of their apparatus. Shoenberg agreed to consider reducing the charge for servicing the emitron tubes to £2 per transmission hour, with a minimum annual charge of £1 000 per annum and to reconsider this charge at the end of six months. He estimated the life of such tubes to be 50 hours and thought that it might be possible to evolve a lower grade of emitron for rehearsal purposes which would have a longer life[22]. However, this decrease of ten shillings per hour in serving costs was relatively minor compared with those which Ashbridge would have wished for in the operation of the intermediate film process. He viewed with grave concern the very high running cost of the system even if split 35 mm film stock were used and suggested that if the electron camera held out any early promise of success it might perhaps be expedient to consider its adoption for the public service in place of the intermediate film[23]. The difficulty was that this system was preferred to the camera from the point of view of both performance and reliability. Nevertheless, Baird Television agreed to supply and install the camera free of charge in place of the film system when once they were satisfied with its development. In the meantime BTL said that it would approach Ilfords and Kodak to bargain for a reduction in the price of film; there seemed some hope of obtaining a reduction which would bring down the cost of intermediate film working to £9 per hour.

The plan to establish a high-definition television station was certainly a bold one and put Great Britain in the forefront of television progress anywhere in the world. A particularly important factor for the growth of world television during the formative years was that the forthcoming trial between M-EMI and BTL would effectively involve comparative tests of the major schemes of picture signal generation, namely, emitron and image dissector cameras, studio spotlight scanning, intermediate film, and film scanning by two different methods—Nipkow disc and emitron camera. It was likely that other television administrations would gain much useful knowledge, for the establishment of television in their countries, from the experience which was about to be gained by the BBC.

In a survey, published in 1936, in *Radiodiffusion*, the six-monthly review of the International Broadcasting Union, Sir Noel Ashbridge gave details of the various systems of television which were being used experimentally at the beginning of that year. He had written to those responsible for broadcasting services in a number of countries all over the world and had summarised the data received from 27 administrations in his paper. The countries comprised Australia, Canada, Japan, New Zealand, South Africa, the USSR, the US and 20 European states. Of these, television programmes were being transmitted, either regularly or irregularly, in only six countries, namely, France, Germany, Holland, the USSR, the UK and the US. Extracts from Ashbridge's survey now follow.

France

In Paris, two experimental television transmitters [were] working; one, a low-definition system, on a wavelength of 180 m, with a power of 700 W, [transmitted] 60 lines, 25 frames per second; the other, a high-definition system, on a wavelength of 7 m, with a power of 1 kW, [transmitted] 180-lines, 25 pictures per second (sequential scanning). A new transmitter having a power of 10 kW in the aerial [was] being built in the Eiffel Tower, and [would] be put into service in the spring. The accompanying sound [was] broadcast from one of the Paris medium-wave transmitters.

'The Nipkow disc method for scanning [was] used, and the picture was scanned horizontally (line scanning). Transmissions [were] not limited to films. The format of the picture [was] square.

'A special cable [connected] the studios, in Rue de Grenelle, to the Eiffel Tower transmitter. A carrier current circuit on 1800 kHz was used, the bandwidth transmitted being approximately 500 kHz. The service [was] carried out by the State Broadcasting Service. It [was] stated that the price of a receiver [was] approximately three times that of a normal broadcast receiver. Public viewing rooms [were] available on Sundays.'

Germany

'A large amount of experimental work has been done on television in Germany. Demonstrations of television have been given at the annual Radio Exhibition since 1928. In March 1935, a public service of television transmissions was inaugurated, and although no receivers were available to the public at the time, a number of public televiewing rooms were established at various points in Berlin, and considerable public interest was aroused.

'These took place three times a week, one and a half hours programme being given at each transmission. Two ultra short wave channels were used, with a power of 7 kW.

'The transmitters were established at the base of the Funkturm, Witzleben, the ultra short wave transmitting aerials being supported at the top of this tower. Programmes consisted both of film transmission and direct television by the indirect film method. 180-line pictures were transmitted, 25 per second, sequential scanned. Wavelengths of 6.7 m and 7.0 m [were] used.

'The experimental television transmitters were destroyed in a fire which took place during the Radio Exhibition in 1935, but new transmitters [had] replaced them, and a regular service [was] again being given.

'Receivers [were] now available to the public at the price of approximately 600 to 1800 RM.'

Holland

'Although no public television service [had] been established, experimental sound and vision transmitters [were] installed at the Philips Laboratories at Eindhoven.

'The power of both the vision and the sound channel [was] approximately 500 W, the waves used being 41.208 MHz and 43.200 MHz. In the past, experiments [had] been made with 180 and 360 lines, sequential scanning, 25 pictures per second, but in the future experiments [would] be made with 375 and 405 lines, interlaced scanning, 50 frames per second, 25 complete pictures. An iconoscope camera, developed and improved in the laboratory, [was] used for these experiments. The size of the reproduced picture [could] be 5 × 6 or 7½ × 9 inches.

'No receivers [were] at present available to the public, nor [were] there any public demonstration rooms. No special cables capable of transmitting a wide band of frequencies [were] at present available, apart from the coaxial cable used between the studio and the transmitter, a distance of 300 m.'

The USSR

'At the present time regular low-definition television programmes [were] being transmitted from Moscow (30 lines, 25 pictures per second). The picture [was] divided into 1200 elements. Two long-wave broadcasting transmitters [had] been used, one for sound and the other for vision.

'Transmission [were] made during the night hours and specially selected cinema films, concerts, short scenes etc., [were] transmitted. The format of the picture [was] 3 × 4, and the direction of line scanning [was] horizontal.'

The US

'At the present time there [were] no stations regularly transmitting television programmes in the United States of America, although low definition transmissions [had] taken place irregularly during the past few years.

'A very large amount of research work [had] been carried out in various research laboratories, and economic considerations alone [were] responsible for there being no high-definition service in operation.

'The Radio Corporation of America [proposed] during 1936 to carry out experimental work in which the National Broadcasting Company [would] operate a television system for demonstration purposes, without the sale of equipment to the public. The RCA system [would] be used, the peak power of the vision channel at maximum modulation being 32 kW, the power of the sound channel 8 kW, Copenhagen rating. Sound [would] be transmitted on 52 MHz, and vision on 49 MHz. The number of lines [would] be 343, 30 complete pictures per second, scanned twice interlaced (60 frames per second). The iconoscope camera [would] be used, both for film transmission and direct pick up. The format of the pictures [would] be 3 × 4.

'No special television cables [were] yet available, but the American Telephone and Telegraph Company [proposed] to construct about 90 miles of concentric cable between New York and Philadelphia, capable of transmitting a band 1 MHz wide. Concentric cable for the interconnection of plant can be purchased.

'It [was] contemplated that ultra short waves and micro waves [would] be used for outside broadcast pick ups when the system [was] operated for public reception.

'The Columbia Broadcasting System, which operated a low-definition television service three years ago, [had] announced that it [was] keeping a close watch on developments and [would] await the outcome of the RCA experiment.'

Ashbridge's definitive survey showed that the approaching trial at Alexandra Palace of M-EMI and BTL equipment would be unique; not only in determining the ease of use, from a television producer's viewpoint, of the divers equipments, but also in testing under actual studio conditions the experimental findings of Wenstrom and of Engstrom that the definition of 400-line image was equivalent for practical purposes to that of home motion films.

Installation of the apparatuses at Alexandra Palace proceeded more slowly than the TAC had anticipated[24,25,26,27] but by August 1936 sufficient work had been undertaken to allow television broadcasts from the Palace to be received at the Radio Exhibition at Olympia[28]. The period of the exhibition was 26 August to 5 September.

Sir Noel Ashbridge's account[29] of the events on the opening day is especially engrossing since it is an objective account and did not need to be written to have a wide popular appeal.

'The demonstration at Olympia started at 12 o'clock on Wednesday, August 26th, with a programme on the Baird system using the spotlight apparatus and film projector. The first transmission was technically quite good but, of necessity, it had to consist of a lot of film, and I think the public would have liked more direct material. However, it was a very smooth demonstration. In the afternoon there were frequent break-downs lasting five or ten minutes at a time but when the transmission was going it was good . . . The first EMI demonstration took place this morning (27 August), and was, on the whole very good. I think, however, that I have seen better EMI transmissions, but there was no definite defect. The programme was of course more interesting as it included outside scenes and a good deal of variety in the studio . . . We are of course gathering a great deal of information both from the crowd and our own observations and we have senior staff posted at every important point where information can be collected . . . On the whole the Press is favourable and helpful and is inclined to gloss over defects. There is of course a tremendous amount of notice being taken of tele-vision but I should not say at the moment that opinion is entirely unanimous that the transmissions are quite fit for regular programmes. What is more serious, is the high price of receivers. The lowest price seems to be eighty guineas, at the Exhibition at any rate. So far as the transmissions and programmes are concerned, I think that great strides will be made during the month of September because most of the difficulties are due to minor defects.' [scc Figure 21.12]

Programmes on the BTL system consisted mainly of film with televised announce-ments by the flying spot method. These included: a new Paul Rother documentary film, 'Cover to cover' about books, which brought to the screen A P Herbert, W Somerset Maugham, Julian Huxley, T S Eliot, Rebecca West and other well known authors; 'Here's looking at you' a variety half hour produced by Cecil Madden and G More O'Ferrall—performed in the Alexandra Palace studios; a scene from 'As you like it' in which Elisabeth Bergner as Rosalind and Laurence Oliver as Orlando were seen in the Forest of Arden (this was the first British Shakespearean film and cost £150 000 to produce); an excerpt from the Alexanda Korda film 'Rembrandt' featuring Charles Laughton in the title role and Gertrude Lawrence; a number of other film extracts and the Gaumont British news[30].

The Marconi-EMI programme was much more interesting than BTL's owing to its wider scope and included outside scenes of the Alexandra Palace grounds, variety acts in the studio and, naturally, a considerable amount of film. All this was a foretaste of what television would be like after the inauguration of the public service in November. From overheard comments it soon became obvious that the public

'A pleasant and informal manner'
(the BBC's requirement for television announcers, 1936)

Figure 21.12 'A pleasant and informal manner' (the BBC's requirement for television announcers, 1936

Figure 21.13 General view of one of the studios at Alexandra Palace showing the television orchestra being televised by an emitron camera (seen on the right behind the grand piano)

Reproduced by permission of the BBC

generally were not very interested in film excerpts, the comment being 'Oh; that's not television, that's pictures'.

Ashbridge wrote, apropos this point:

'Perhaps the most interesting thing which emerged from the demonstration was the lack of interest in films, and this I consider due almost entirely to the fact that insufficient detail is visible. Films produced especially for television could avoid this defect to some extent. One could observe that interest increased immediately the picture became a close-up, whether it was film or direct. Some people considered that films were better reproduced by the Baird system than the Marconi-EMI and that there was more clarity, and this is remarkable having regard to the fact that the Baird system has much more flicker, and theoretically less definition. There is no doubt that general opinion is that a good variety turn is better than an interesting film from the point of view of television. People were not so much impressed by outside scenes simply because the detail was insufficient and the general effect artificial. That is to say, trees were recognisable as trees, but they did not look like the real thing, and the light values were obviously false.'

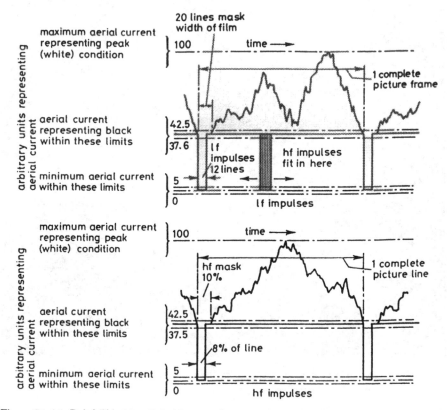

Figure 21.14 Baird Television Ltd video waveform

In addition to their own reports, the TAC had the benefit of a useful report[31] from A C Cossor Ltd. As a designer and manufacturer of radio receivers the company was particularly concerned with any departures from the published waveform standards of the two television companies, since the general public might attribute picture faults to the set manufacturers rather than the contracting authorities. In the main the M-EMI equipment was 'all that could be desired'. But: 'unfortunately,' said the receiver manufacturer, 'there [were] several rather important errors in the waveform transmitted by the Baird system'. (These have been enumerated in Burns, *op.cit.*)

Some of these deficiencies were to persist after the inauguration of the London Station in November 1936, and Cossor felt compelled to write[32] to the Television Advisory Committee in that month about the discrepancies between BTL's published waveform specification and the actual transmission.

The comparison which Cossor made between the two contracting companies' studio equipment was also ominous from Baird's point of view:

> 'The direct transmissions were rather a revelation. We were extremely surprised to find so much entertainment value obtainable in this way. It is, however, our opinion that the spotlight studio is extremely inflexible and that since it involves almost perfect stillness on the part of the artist, we think that its value for television transmission is almost nil. On the other hand, the use of the iconoscope for indoor and outdoor sets presents great flexibility and is capable of providing first class results. Particularly were we struck with the extremely neat and pleasing results obtained by fading from one iconoscope to another. The depth of focus, range of vision and ability to pan in any direction which the iconoscope possesses makes it an extremely valuable piece of apparatus, and in our opinion makes the spotlight studio a complete anachronism.'

All of this was well known to the BBC, but as it came from a radio and television set manufacture, the report lent added weight to Ashbridge's observations. It seemed, at the time, that the result of the forthcoming trial period between the two companies was a foregone conclusion.

Contemporaneously with the demonstrations of 240-line and 405-line television at Radiolympia, the 11th Olympic Games were being held in Berlin, Germany. Television coverage was provided by the Reichspost and by the Fernseh company. Both organisations used outside broadcast mobile units containing intermediate film apparatus, and electronic cameras. Those employed by the Reichspost were of the iconoscope type, and had been supplied by the Telefunken company, and the one utilised by Fernseh featured the Farnsworth image dissector tube. The latter camera 'delivered very sharp signals, free from interference components, but only in bright weather'.

All three television systems were based on the 180-line standard. However this was much too coarse a standard to provide enjoyable viewing of outside sporting events and some unfavourable comments were reported. An account[33] titled 'Television shows relay', in the *New York Times* said:

' . . . America's phenomenal relay victory in the Olympic Stadium today didn't look so good.

'It was evident that the boys were running fast, but Jesse Owens looked pale and Frank Wykoff's legs wobbled. The stadium's spectators in the background sort of melted together, and the whole scene shook like jelly . . . The eye strain was considerable and many didn't stay for the remainder of the afternoon's events.'

Since no television sets were on sale in Germany in 1936 the television transmissions could only be seen in the 25 television rooms and theatres in Berlin.

Taken together, both the Radiolympia Show and the Olympic Games highlighted the inadequacies of the 180-line standard. In effect, the 180-line transmissions vindicated the decisions of Shoenberg and of the Television Advisory Committee to adopt the 405-line standard, despite the protestations of some sections of the trade, as one of the two standards which would be tested following the inauguration of the London television station.

In one field of television activity[34], namely, two-way television (video telephony) Germany began to lead the world in 1936. On 1 March Herr von Eltz Rubenach, the minister of posts, opened[35] the telephone-television service connecting Berlin and Leipzig, a distance of about 100 miles. A wideband cable was utilised to transmit the vision and sound signals. Two public call offices in each city allowed members of the public to communicate with others at a cost of DM 3.50 for a three-minute session. This sum included the notification of the desired person at the other end of the link. (In 1936 DM 20 = £1.00 at par; and in the UK the average weekly wage was £2.70.) The call offices in Berlin were situated in the Kolumbus Haus on the Potsdamer Platz, and at the busy corner of the Kantstrasse and the Hardenberg in the West End; and in Leipzig they were located in the central post office on the Augustplatz, and on the fair grounds. The service remained open from 08.00 to 20.00 hours during the Leipzig Spring Fair: this closed on 7 March 1936.

Great interest was aroused by the service and, because the images were based on a 180 lines per picture, 25 pictures per second standard, reports on the quality of the reproduction were good. 'The head and shoulder image of a person is clearly reproduced. The effect is comparable to a small size projection of a substandard cinema film . . . Details like the hands of wristwatch or ring on the hand holding the telephone are said to be clearly visible.'

Mechanical scanning was employed throughout. The equipment used in Berlin was constructed by the German Post Office Laboratory and that used in Leipzig by the Fernseh A G, in which Baird Television Ltd had held a quarter of the shares until 1936.

Two months after the end of the Leipzig Fair, Dr Ohnesonge, the State Secretary at the Ministry of Posts, announced[36] further extensions of the Berlin-Leipzig television-telephone cable from Berlin to Hamburg and from Leipzig to Munich, giving a total length of 620 miles. By September 1937 the Berlin-Leipzig cable service had been extended to Nuremberg (a total distance of c.300 miles)[37]. The intention was to use it not only for two-way television but also for the direct relaying of the events of the National Socialist German Workers' Party's rally at Nuremburg for

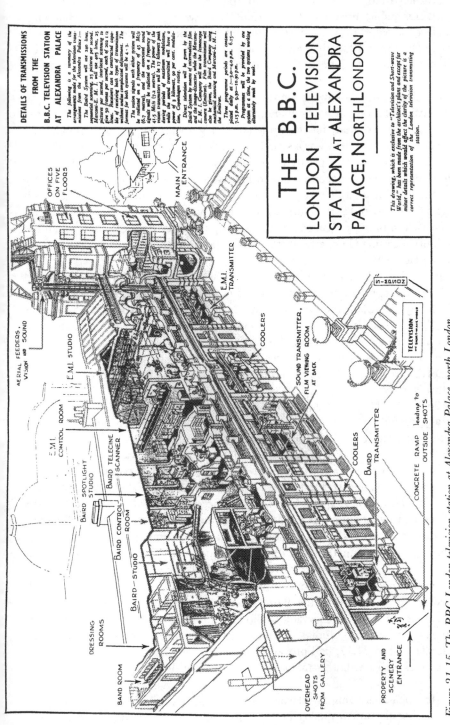

Figure 21.15 The BBC London television station at Alexandra Palace, north London

Source: *Television and Short-wave World*, **9**, October 1936, pp. 576–577

broadcasting from the Berlin television station. The extension to Munich was opened, for public use, on 13 July 193838. News reports said that Hamburg, Frankfurt, Cologne and Vienna were also to be incorporated into the system. Later, from 15 December 1938, the Deutsche Reichpost operated a local two-way service in Hamburg which by June 1940 was handling c. 20 calls per day. The broadband cable between Hamburg and Berlin was used occasionally for a two-way link between these cities. Following the outbreak of hostilities in 1939 further work on the extension of the service was abandoned.

The opening ceremony for the world's first, high-definition, regular, public television broadcasting system was a singularly modest affair and represented rather an anticlimax to the months of strenuous activity and preparation which had been undertaken by the two contracting organisations and the Television Advisory Committee. A low cost, low key approach was adopted for the proceedings which would be remembered in the future as an historic occasion, 'not less momentous and not less rich in promise than the day, almost exactly fourteen years ago, when the British Broadcasting Company, as it then was, transmitted its first programme from Marconi House'[39].

The opening programme, arranged by the BBC and approved by the TAC, lasted hardly one quarter of an hour and consisted of a few short speeches, of four minutes each, by Mr Norman, the chairman of the BBC, Major Tryon, the Postmaster General, and Lord Selsdon[40], the chairman of the Television Advisory Committee. The Chairmen of both the Marconi-EMI and Baird Television companies were allowed half a minute each to say, literally, a few words, and after an interval of five minutes the programmes continued with the showing of the British Movietone News and then variety.

Although Mr Norman's and Major Tryon's speeches tended to be, somewhat naturally, platitudinous in character, Lord Selsdon in his speech endeavoured to give some assurance to the future purchasers of television sets regarding the stability[41] (for at least two years) of the transmission standards. He was keen to point out that the service would cover the Greater London area, with a population of about ten million people, and said that he was unwilling to lay heavy odds against a resident in Hindhead viewing the Coronation procession. For all three speakers the future of the new medium was bright and held 'the promise of unique if still largely uncharted opportunities of benefit and delight to the community'.

The experiences gained by the BBC of the operation of both television systems under service conditions from 2 November to 9 December 1936 were described in an important report[42] written by Gerald Cock, the BBC's director of television. It proved highly damaging to the Baird interests; indeed it meant the end of the company as a supplier of television studio and transmitting equipment for the corporation's stations and studios. Cock stated that the Marconi-EMI apparatus, Figures 21.16 and 21.17, had proved capable of transmitting both direct and film programmes with steadiness, and a high degreee of fidelity:

'Its apparatus being standardised throughout, reproduces a picture of consistently similar quality and requires only one standard of lighting,

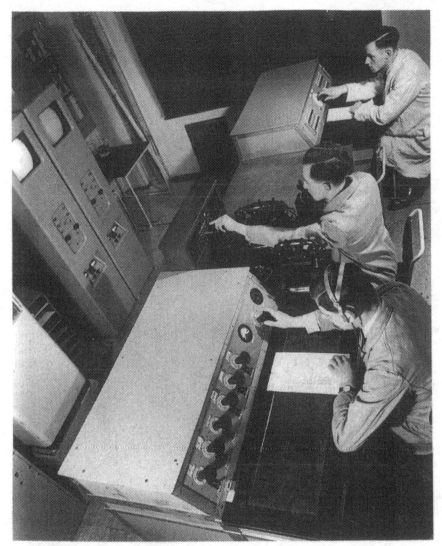

Figure 21.16 EMI vision control apparatus at Alexandra Palace

Reproduced by permission of Thorn EMI Central Research Laboratories

make-up and tone contrast in decor. Its studio control facilities are convenient and comparatively simple. It has proved reliable, and has already established a large measure of confidence in producers, artists and technicians. Outside broadcasts and multicamera work have added considerably to the attractions of programmes. With improved lighting, additional staff and studio accomodation, single system working by Marconi-EMI would make a service of general entertainment interest immediately possible.'

Figure 21.17 EMI camera and sound control equipment at Alexandra Palace

Reproduced by permission of Thorn EMI Central Research Laboratories

On the BTL system the director of television found it difficult to say anything complimentary. The programmes were being transmitted under practically experimental conditions and the prospects of anything approaching finality in the studio stages of transmission seemed remote. He noted:

'Alterations in apparatus were constantly taking place. Breakdowns, with little or no warning, and, even more serious, sudden, unexpected, and abnormal distortions are a frequent experience. In such cases, it is difficult and embarrassing to make a decision to close down, since there is always the possibility that faults may be corrected within a short time. This inevitably leads to criticism of television by those who may only have observed it in adverse conditions.'

In studio operations the use of the spotlight scanner, intermediate film process and the electron camera (the image dissector camera) required a different lighting technique and make-up for each picture generating system and added to the problems of the producers. Cock's views on these methods were as follows:

Spotlight scanner

'This apparatus is limited to double portrait reproduction. Distortion of picture tone and shape still appears to be intrinsic and unavoidable. No reading is possible in a spotlight studio, nor could any artist depending upon looks and personality be expected to televise by this method. The result is a caricature of the image televised.'

Intermediate film method (see Figure 21.18 and 21.19)

'This is extremely intricate, and depends upon so many processes that it causes continual anxiety. It is inflexible and rigid in operation, being confined to "panning" in two planes. Changes of view can only be effected at the cost of lens changes and "blackouts"; otherwise the picture is static. Its quality is variable; the delay action is extremely inconvenient for timing and other production purposes. The maximum continuous running time is at present limited to approximately sixteen minutes, which adds to the difficulty of arranging programmes. The cost for film alone is £12 per hour (rehearsal or performance). Sound, recorded on 17.5@mm film and subject to development at high speed, is invariably of bad quality, whole sentences having occasionally been inaudible. It is consequently an unsuitable method of presenting any programme item in which quality of music or speech is important. Mechanical scanning produces line bending and twisting in a variable degree. Black and white contrast is however generally good.'

Electron camera

These cameras have quite recently been improved, but future progress is likely to be seriously handicapped by destruction of the Baird research plant and technical research records in the Crystal Palace fire. Bending and twisting of lines are pronounced. The cameras are in a somewhat primitive stage of development and are still without facilities for remote (outside) or "dissolved" work. Breakdowns have been frequent. Electron cameras do not appear likely seriously to compete with emitrons at any rate for a considerable time. They have advantages over other Baird apparatus in being instantaneous in action; in permitting good sound transmission;

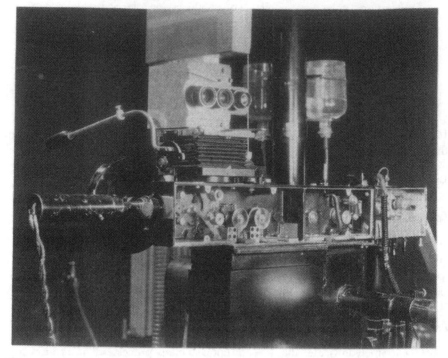

Figure 21.18 Experimental version of the intermediate film apparatus of Baird Television Ltd

Reproduced by permission of BFI Stills, Posters and Designs

in mobility; and in the elimination of mechanical scanning. At present their operation seems somewhat precarious.'

On the BBC's experience with BTL's teleciné scanner for 35 mm commercial film, Cock commented:

> 'Originally, this apparatus gave the best picture obtainable by the Baird system and possibly (flicker apart) was better than the Marconi-EMI for reproduction from standard film. Line bending and twisting have since been noticeable due perhaps to mechanical scanning and the difficulty of maintenance in first class condition. Its contribution to programmes is limited by the present restricted use of 35 mm film.'

This very damaging report, from BTL's point of view, was discussed by members of the TAC at its 34th meeting on 16 December 1936. They had to decide whether the time had now arrived for them to make a definite decision on the question of transmission standards. The contracts made with both companies contained a reference to the London experimental period which was defined as terminating on the date on which a judgement was reached concerning the system to be employed at the London station. In the light of Cock's adverse report on the working of the BTL

Figure 21.19 The intermediate film equipment of Baird Television Ltd. The Nipkow disc scanner is housed in the chamber at lower centre left

Reproduced by permission of BFI Stills, Posters and Designs

equipment the committee had to decide whether further expenditure on the advancement of a system which appeared unlikely to survive was prudent. Failing this it could suggest that the system should be temporarily suspended until the apparatus could be handed over in a reliable and efficient state—and within a reasonable time limit. Cock foresaw that artists and celebrities might refuse to be televised in BTL programmes, thus making the problem of producing programmes still more difficult. There was no doubt that the uncertainties and limitations of the equipment were having a deplorable effect on the production staff whereas with the Marconi-EMI system the apparatus was sufficiently advanced and reliable to enable interesting and entertaining programmes to be devised and transmitted with a 'high degree of reality and with complete confidence on the part of the producers and studio organisation'.

With Cock's and Cossor's reports in mind, the TAC quickly came to a conclusion: they would recommend to the Postmaster General the adoption of Marconi-EMI's transmission standards as the standards for the London station, at any rate for the next two years[43]. Baird Television was to be given the opportunity of making representations, before the public announcement, to the committee on 23 December. The discontinuance was to date from 2 January 1937[44]

Baird Television's reply[45] to the TAC's letter of the 16 December was surprisingly moderate and low key in tone, far different from the letters which Moseley used to write on behalf of the Baird Companies when events were not in Baird's favour. Apart from a mild reference to the very short time which had elapsed since the start of the transmission and an expression of the company's belief that the committee could not have made its recommendations on the basis of any defects or inferiority of the Baird transmitter, the letter contained no objection to the committee's decision. Indeed, the Baird directors were, in principle, in accord with the view that one standard was preferable to two provided certain conditions were satisfied. So comparatively mild was their reply that it gives the impression that they were not dissatisfied with the recommendation. The directors, of course, would have been well aware of the broadcast quality of their system vis-à-vis that of Marconi-EMI, and although Cock's report was private and confidential to the BBC and the TAC, nevertheless Baird's engineers must have given some indication to their directors that all was not well with their Alexandra Palace equipment. Unhappily those engineers had their difficulties compounded by the disastrous fire which had occurred on 30 November at the Crystal Palace and which destroyed much apparatus and workshop/laboratory facilities.

Margaret Baird[46], in her biography of her husband has written: 'The board was relieved by the turn which events had taken. Clayton, Ostrer's accountant, observed that the transmissions through the BBC had done nothing but lose money, the company's only hope of making money being the sale of receivers.' This may indicate the reason for the mildness of the Baird Company's letter of 22 December.

Moseley in his book[47] on J L Baird also confirmed this opinion: 'The directors argued that transmissions were not of much consequence: it was the sale of Baird receivers that matters; that brought in the money.'

The BBC's last transmission using the Baird system was sent out on 30 January 1937. 'And so after all these years we were out of the BBC,' wrote Baird in his autobiography[48]. The Television Advisory Committee's judgement was a 'terrible blow' to Baird. 'It seemed that he had been forgotten by the world,' his wife noted in her book. 'He bore up with practically no mention of his troubles to anyone and presented himself to his assistants and the press with his usual calm exterior, but I knew that inwardly he was seething.'

Baird had been a sufferer in adversity for many years and did not intend giving up his life's work. If the BBC would not allow him to transmit from Alexandra Palace, it seemed to him that now 'being out of the BBC [he] should concentrate on television for the cinema with Gaumont British installing screens in their cinemas and working towards the establishment of a broadcasting company independent of the BBC for the supply of television programmes to cinemas . . .'[48].

The history of cinema television is beyond the scope of this book but it has been the subject of one of the author's papers[49].

Cock's frank assessments on the operation of the main systems of television picture signal generation are of especial interest in the history of television. As the BBC's director of television he had a major responsibility to the corporation and the viewing public to create a service which viewers would find attractive, compelling and enjoyable. His task could not be founded on any personal allegiance to a particular company or television scheme but had to be based solely on the quality of the images produced by the divers studio apparatuses and the operational ease of capturing the visual scenes by cameras. And, obviously, his thoughts on the results achieved and displayed in peoples' homes had to be consonant with those of numerous unbiased observers.

The BBC's decision probably caused no astonishment. Spot light scanning had been employed in the US, France, Germany, the UK and elsewhere from the late 1920s and its limitations were well known. Again, the intermediate film process had been much investigated and developed by Fernseh and others during the first half of the 1930–1940 decade and certain deficiencies of the process had been highlighted at the Berlin Radio Exhibitions, (see Chapters 11 and 13). Also, as noted in Chapters 15 and 16, Farnsworth's image dissector tube lacked charge storage and could not compete in sensitivity with tubes of the emitron/iconoscope types. All these points were known generally at the beginning of 1937.

Effectively, those countries which, in 1937 and subsequently, aspired to operating high-definition television broadcast stations had to use studio equipment based on the emitron and iconoscope cameras. Only EMI and RCA, and their licensees, world-wide, manufactured such equipment: and so, during the formative years of television growth, prospective television administrations, perforce, had to found their nascent television systems on the work of RCA and EMI. Pre-war, the influence of these companies was particularly evident in France, the USSR, Japan and Germany.

High-definition television, based on a line standard of 455 lines per picture, was first shown in France on 10 July 1937 at the International Exposition, Paris[50]. During the exposition the Postes, Télégraphes et Téléphone's new apparatus provided televised presentations, at the radio pavilion, of athletic events, actualities, dramatic art and dances, and interviews. At the end of the exposition the equipment was transferred to the PTT studio in the Rue de Grenelle[51].

From September 1937 high-definition television broadcasts were sent out from the new 30 kW (peak power) transmitter installed in an underground room at the base of the Eiffel Tower[51,52]. The transmitter was supplied by Le Materiel Téléphonique, an associate company of Standard Telephone and Cables, London. At first the station operated at a reduced power but in January 1938 the output was raised to 15 kW[53]: this enabled the reception distance to be extended from 15 km to 25 km. The carrier frequency of the radiated signals was in the band 40 MHz to 50 MHz, 46 MHz being used in May 1938.

A November 1938 report[54] on the Eiffel Tower station mentioned that just one room was available for studio work because of difficulties in allocating space in the main PTT building.

In 1937 EMI studio equipment, based on the emitron camera, had been ordered by the French Thomson-Houston Company[55] with whom EMI had previously completed arrangements for the handling of their system in France. The equipment was submitted to the French Government for comparative tests and from January 1938 was used for experimental broadcasting. It was then purchased by the PTT for its regular television transmissions.

It seems that EMI's system was well received, for, in January 1939, M Jules Julien, the French minister of the PTT, was quoted[56] as saying: ' . . . there [was] nothing comparable in either America or Germany to what [was] being done by the BBC in England.' The brief news report further stated the intention, in France, of erecting a chain of stations to cover the whole country: the next stations being at Lyons and Lille. Both of these would be identical with the Eiffel Tower station.

High-definition television, pre-war, in the USSR was based on the RCA system. The link between the US company and the communist state originated in 1929 when an agreement was signed for the exchange of technical information. This led, in 1930, to RCA sending a transmitter engineer, a Mr L Jones, to Moscow to assist the Russians in establishing a radio broadcasting service[57]. Further strengthening of the association came about when Zworykin, in 1933, 1934 and 1935, paid visits to the country of his birth.

During his 1934 visit Zworykin was asked, not unexpectedly, by the head of the communications industry if RCA Victor would be prepared to sell and install in Moscow a television transmitter and some receivers. The request seems initially to have been one of a general nature, but towards the end of 1935 Amtorg, the Soviet trading corporation, ordered $2 000 000 worth of radio apparatus and machinery from RCA.

The report[58] on this deal, in the December 1935 issue of the *New York Times*, mentioned the approval of the US army and navy, and the State Department, and the fact that the equipment would be manufactured in the RCA plants at Camden and Harrison, New Jersey. As part of the contract, a Soviet delegation stayed in the United States from 1936 to 1937 to oversee the production, and accept the various items, of the television system. A brief report in the 2 May 1937 edition[59] of the *New York Times* noted that the RCA apparatus was expected to reach Russia early in May for use in the construction of the television centre in Moscow.

The USSR television plan provided for the setting up of three stations in Leningrad and Kiev, as well as in the capital. A common television standard was not adopted for the three stations: the Leningrad station, the equipment for which was being manufactured in Soviet factories, was based on a line standard of 240 lines per picture, whereas the Moscow station used the RCA definition of 343 lines per picture. According to the above report, the television schedule for 1937 would include television transmissions from 'sport stadiums and squares in Moscow'.

Another news item[60] said that the first regular experimental television broadcasts would begin in Moscow in June 1938. Cinema films and prominent artists would be televised. About 100 television receiving sets existed in the clubs, political and cultural centres, and houses in Moscow; and plans for the mass production of cheap sets were in the course of execution. The Leningrad television factory was expected to manufacture 200 large receivers of the type which had been adopted for use on collective farms. However, even though 12 RCA engineers from Camden and Harrison, under the management of L Jones, had departed from the US to supervise the Moscow installation and train the Russian technicians and engineers in its operation and maintenance, the RCA installation was still incomplete in August 1938.

An impression[61] of Soviet television was given in 1939 by C K Freeman, the American stage director:

Programmes are a combination of film, concert and dramatic entertainment, with visiting scientists and Soviet bigwigs being propped on occasion in front of the iconoscope. I witnessed, among others, an impressive array of talent from the Vachtangov theatre in four vignettes of that theatre's repertoire . . . Since the talent field is state-controlled there is no limit to the extent of able people at the beck and call of the programme directors.

'Accustomed . . . to perfect studio acoustics and "dead" sound, one is amazed at the appearance of hardwood floors and wood panels in the Moscow studio . . . Sound interference is due also to the lack of sufficient incandescent lighting, though arc lamps now being used for studio programmes are fast being replaced. The producers are likewise hampered by the scarcity of cameras. A pretty tricky dramatic show I watched was covered with but one unit, dollying, panning, and moving in for close-ups with facility . . .

'The Moscow station has a tower 150 m high and diffusion around 30 km, with a record reception gained of 70 km. There is a splendid control room modelled after RCA specifications.'

Zworykin's electronic camera tube design was also at the heart of Japan's high-definition television system. During the summer of 1934 Professor Kenjiro Takayanagi, of the Hamamatsu Higher Technical College, visited RCA's Camden site and was shown its 343-line system[62]. He had had an interest in television from the mid-1920s and had endeavoured to fabricate a camera tube. His lack of success, which he ascribed to an inadequate vacuum technique, led him to concentrate on the implementation of a gas filled cathode-ray tube television receiver. Takayanagi claimed to have reproduced his first cathode-ray tube image—a Japanese character—on 25 December 1926[63]. He probably used a Nipkow disc to generate the image signals since a later (September 1928) report referred to a 40-hole, 14 revolutions per second, Nipkow disc operating in a spot light scanner configuration with two GE photocells. The 4 cm by 4 cm greenish c.r.t. image contained 1600 picture elements: recognisable faces were said to have been synthesised. Further development of his system led Takayanagi and a colleague, Professor Tomomasa Nakashima, to transmit, from 1 November 1931, televised images from the Tokyo radio station JOAK. The image standard was now 80 lines per picture, 20 pictures per second, and the carrier wavelength was 84.5 m.

Thus, when Takayanagi visited RCA in 1934, he had acquired, from first hand experience, quite a wealth of knowledge about television matters generally and cathode-ray tubes more particularly. He also visited the laboratories of Farnsworth, General Electric, Westinghouse and Bell Telephone; and, in England, Baird Television Ltd. Takayanagi travelled to Berlin but was restricted to viewing the exhibits at the 1934 Berlin Funkausstellung. His tour led to a paper, 'Recent development of television technic in Europe and America' (March 1935), and a determination to fabricate an iconoscope tube[62].

In 1934 the principal television research and development centres in Japan were:

1   the Engineering College of Waseda University, near Tokyo;
2   the Hamamatsu Higher Technical College, about 100 miles from Tokyo;
3   the Electrotechnical Laboratory of the University of the Ministry of Communications, Tokyo.

The three centres were subsidised by the Broadcasting Corporation of Japan (Nippon Hoso Kyokai), and their joint findings were pooled for the benefit of the NHK.

On 8 December 1936 the *Japan Times and Advertiser* newspaper reported that a 500 000 yen government appropriation for television research based on Zworykin's work had been approved: Takayanagi would be the engineer in charge. And, on 25 August 1937, the United States Department of Commerce reported the formation of a Japanese company—the Nissan Television Kaisha—to operate the patents of EMI held by the Japan Industry Company, a holding corporation[64,65].

An earlier, March 1937, report to the US Department of Commerce (from its assistant trade commissioner in Tokyo), referred to a laboratory for television research being constructed, near Tokyo, by the Japanese Broadcasting Corporation; it was expected to be completed by the end of the month. The NHK had allotted $1 500 000 for television research in 1937 so that when the Olympic Games were held in Tokyo in 1940 the Corporation would be able to transmit televised images of the games to all points within a 12 mile radius of the capital[66].

News of the NHK's impending tests, in preparation for its temporary broadcasts in July 1938 in Tokyo were given by the Tokyo press on 2 February 1938. Later on 17 May 1939 the first television programme for experimental purposes was broadcast from a transmitter in the research laboratory of the NHK, at Kamata, on the outskirts of Tokyo. According to an American writer, 'the equipment and standards [used] seemed to be just about the same as those in use by RCA at the highly publicised opening of the New York World's Fair' on 30 April 1939[64].

The plan to televise the Olympic Games was, of course, cancelled and all work relating to television was suspended during the hostilities.

The influence of RCA was also most marked in pre-war Germany. Until the beginning of 1936 the television efforts of the various German television pioneers and companies had been concentrated on the engineering of systems designed to analyse a film image, scene or object by mechanical methods based, in the main, on the utilisation of the Nipkow disc spot light scanner, the lens drum scanner, the Nipkow disc film scanner and the intermediate film process. In 1936 this position changed. At the 11th Olympic Games held in Berlin in July/August 1936 the scenes were televised, on a standard of 180 lines per picture, 25 pictures per second, by the German Post Office using several intermediate film processing trucks and three iconoscope cameras provided by Telefunken AG. This company was a licensee of RCA and received all its patents, as well as those of Westinghouse, and General Electric. Two of the electronic cameras covered the principal athletic tack events, and one was sited at the Olympic swimming pool. Fernseh, too, contributed electronic camera equipment for the games. Its Farnsworth image dissector camera generated good signals, but only in very bright light conditions. As mentioned in

Chapter 20, the television transmissions were not a great success because of the inadequate television standard in operation.

Both the Telefunken iconoscope and the Fernseh image dissector types of cameras were exhibited at the 1936 Berlin Funkausstellung; the former type now functioned at 375 lines per picture[67]. Interestingly, and significantly, the Fernseh company, which had signed an agreement with Farnsworth Television to employ and sponsor the development of the image dissector camera, did not demonstrate the camera, in its studio role, at the 1937 Berlin Radio Exhibition. Instead, Fernseh showed two forms of iconoscope cameras. The explanation given by one commentator noted the image dissector camera's lack of sensitivity, 'in spite of secondary electron multiplication', for general use. A noteworthy feature of this exhibition was the display of apparatus working on a definition of 441 lines per picture, 50 pictures per second. Presumably, the German companies had been influenced by the BBC's confirmed 405-line standard[68].

An indication that a high-definition television service was imminent was given in the 14 January 1938 issue of *World Radio* by its Berlin correspondent[69].

'Very shortly, Germany, with the introduction of the 441–line standard and the opening of three ultra short wave stations, will again become one of the leading television countries. New studios with full lighting equipment and with iconoscope cameras for use in interiors will at last enable the programme builders to tackle properly the problems of television entertainment. It [is] hoped that work with 441 lines in these studios will start in April [1938].'

Actually, when the Berlin television studios, in the Deutschlandhaus on Hitler's Square, were completed in April 1938, no 441-line studio apparatus was in operation. The 441-line television service was not officially inaugurated until 1 November 1938[70]. Telfunken manufactured the iconoscope cameras for studio productions, and Fernseh produced the image dissector cameras for film programmes. It was reported that receivers would be on sale to the public in 'about a fortnight's time' and that the price would be approximately Dm1000 (£50 at par). As on previous occasions the timescale proved to be too optimistic.

On a comparison of the facilities in the studios at the Deutschlandhaus and Alexandra Palace, two officials of the German service, who were given a conducted tour of the London station, 'were very much surprised at the complexity of [the BBC's] studio equipment at Alexandra Palace and said they thought it would be a long time before their equipment in Berlin reached anything like this complexity'[70].

Curiously, although several firms—Fernseh, Telefunken, Loewe, Lorenz, Tekade and von Ardenne—and the German Post Office had been active in the television field for quite some time, and although the television medium appeared to be eminently suited for propaganda purposes, the apparent cautious approach of the German Post Office did not permit rapid progress to be made towards a regular high-definition television service. No television receivers were on sale in 1938. This putative caution probably stemmed from the extensive rearmament of the Nazi

State which thereby precluded the allocation of industrial resources to television receiver manufacture.

On 31 January 1933 Adolf Hitler became Chancellor of the Third Reich. Soon afterwards, his loyal friend Hermann Goering, together with Erhard Milch (who in 1926 had been appointed chairman of Deutsche Lufthansa) embarked on a plan which would lead to the recreation of the German Luftwaffe. In 1933 no combat aircraft were constructed by the German aircraft industry, but in 1934 and 1935 the numbers of military aeroplanes produced were 840 and 1923 respectively. Official news of the reformation of the German Luftwaffe was announced on 1 March 1935. With its 20 000 officers and men, and 1888 machines the new airforce was a potential threat to peace. The head of the Luftwaffe was Goering[71]: he was also responsible for all television activities in Germany.

From the outset Goering's plan was to build-up a modern airforce comprising fast, single-engined and twin-engined monoplane fighters and bombers of the types which had been, and were being, designed by the aircraft companies of Heinkel, Dornier, Junkers, Messerschmidt and Focke Wolfe. Contemporaneously with the implementation of this policy, the German army and navy were being vigorously developed and increased in size. An early portent of the expansion of German military power was given on 7 March 1935 when 35 000 German troops marched without opposition into the demilitarised zone of the Rhineland. All these air, land and sea forces required modern radio and telecommunications apparatus. In addition, the Luftwaffe's aircraft had to be provided with navigation aids and means to enable the aircraft to operate at night. Similarly, it was necessary for the German navy's fleet to be installed with radio direction finders, gun fire control systems, and listening equipment. The burden on German radio/electronic firms, following Hitler's rise to power, must have been very great; and it seems reasonable to suppose that the provision of television receiving sets to the general public was delayed as a consequence of the pressure imposed on industry by the rearmament policy.

Thus, whereas high-definition television sets were on sale in Great Britain from 1936, the first mass produced set for purchase by a member of the German public was not available until August 1939[72], one month before the onset of hostilities. Essentially, the growth of military might was not conducive to an early growth of a media form which was nonurgent in its impact on society, even if it had some propaganda potential. Newspapers, films and radio transmissions could adequately disseminate Goebbel's political pronouncements to the masses when compelled to do so. As the Berlin correspondent[73] of *World Radio* wrote just before the war commenced:

'It seems sad that after 16 good years of development, lively invention and keen technical activity, German radio progress must halt . . . so as to save foreign exchange and to liberate workers for other and sterner duties. The radio industry is busy with contracts for communication apparatus required by the ever growing giant army, for Germany will have between one and a half and two million men under arms in August.'

References

1 Television Advisory Committee, minutes of the first meeting, 5 February 1935, Post Office bundle 5536
2 Technical Subcommittee, minutes of the first meeting, 11 February 1935, minute post 33/5533
3 Technical Subcommittee, minutes of the second meeting, 15 February 1935, minute post 33/5533
4 McGEE, J.D.: 'The life and work of Sir Isaac Shoenberg 1880–1963', *R. Telev. Soc. J,* 1971, **13**, (9), May/June
5 PRESTON, S.J.: 'The birth of a high definition television system', *Telev. Soc. J.*, 1953, **7**, July/September
6 Technical Subcommittee, minutes of the fourth, fifth and sixth meetings held on 26 February 1935, 1 March 1935 and 8 March 1935, minute post 33/5533
7 Baird Television Ltd: 'Additional information desired by the Advisory Committee'. Technical Subcommittee, 8 March 1935, minute post 33/5533
8 ZWORYKIN, V.K.: 'Television with cathode ray tubes', *J.IEE*, 1933, **73**, pp.437–451
9 BLUMLEIN, A.D.: 'The transmitted waveform', *J.IEE*, 1938, **83**, pp. 758–766
10 Television Committee, evidence of Marconi-EMI Television Co Ltd, 8 June 1934, minute post 33/4682
11 Television Committee, evidence of A Clark and I Shoenberg on behalf of EMI Ltd and the Marconi-EMI Television Co Ltd, 27 June 1934, minute post 33/4682
12 LEWIS, H.M., and LOUGHREN, A.V.: 'Television in Great Britain', *Electronics*, October 1937, p. 32
13 GARRATT, G.R.M., and MUMFORD, A.H.: 'The history of television', *Proc. IEE*, 1952, **99**, Part IIIA, pp. 25–42, see discussion
14 Electric and Musical Industries and WILLIANS, P.W.: 'Improvements in or relating to signalling systems, such for example as television systems'. British patent 422 906, application date 13 April 1933
15 Technical Sub-committee, minutes of the sixteenth meeting, 5 June 1935, minute post 33/5533
16 ASHBRIDGE, N.: 'Summary of information relative to suggested sites for London high definition television transmitter'. Television Advisory Committee, 5 March 1935, Post Office bundle 5536
17 ASHBRIDGE, N.: 'Address on the occasion of the 8th annual meeting', *J. Television Society*, 1936, **2**, part VI
18 Electric and Musical Industries, response to questionnaire of the technical subcommittee on 'Proposed vision transmitter', minute post 33/5533
19 Notes of a meeting of the Television Committee held on 8 June 1934, evidence of the Marconi-EMI Television Co Ltd; represented by Messrs. Shoenberg, Condliffe, Blumlein, Agate, Browne and Davis
20 Baird Televison Ltd, response to questionnaire of the Technical Subcommittee on 'Proposed vision transmitter', minute post 33/5533
21 Television Advisory Committee, minutes of the seventeenth meeting, 14 June 1935, Post Office bundle 5536
22 Television Advisory Committee, minutes of the nineteenth meeting held 23 July 1935, Post Office bundle 5536
23 Television Advisory Committee, minutes of the twentieth meeting held on 24 July 1935, Post Office bundle 5536
24 Television Advisory Committee, minutes of the twenty-third meeting held on 2 October 1935, Post Office bundle 5536
25 Television Advisory Committee, minutes of the twenty-fourth meeting held on 19 December 1935, Post Office bundle 5536
26 Television Advisory Committee, minutes of the twenty-fifth meeting held on 20 January 1936, Post Office bundle 5536
27 'London television station progress'. BBC notes, 26 May 1936, Post Office bundle 5536
28 Television Advisory Committee, minutes of the thirtieth meeting held on 23 July 1936, Post Office bundle 5536
29 ASHBRIDGE, N.: Letter to J Varley Roberts, 27 August 1936, minute post 33/5536

30   BURNS, R.W.: 'British television, the formative years' (Peter Peregrinus Ltd, London, 1986) p. 413
31   A C Cossor Ltd: 'Notes on the BBC television transmissions to Olympia'. c. September 1936, minute post 33/5536
32   A C Cossor Ltd: Letter to the Secretary (TAC), 13 November 1936, Post Office bundle 5536
33   ANON.: 'Television shows relay', *New York Times*, 10 August 1936, p. 12:6
34   BURNS, R.W.: 'Prophecy into practice: the early rise of videotelephony', *Eng. Sci. Educ. J*, December 1995, **4**, (6), pp. S33–S39
35   ANON.: 'Public television in Germany', *Nature*, 7 March 1936, **137**, p. 391
36. ANON.: 'Television-telephone in Germany', *The Times*, May 1936, 13c
37   ANON.: 'Television-telephone over 300 miles', *The Times*, 7 September 1937, 10d
38   ANON.: 'Seeing by telephone', *The Times*, 13 July 1938, 15d
39   BBC chairman: Speech at opening ceremony, 2 November 1936
40   Television Advisory Committee, minutes of the thirty-second meeting held on 15 October 1936, Post Office bundle 5536
41   SELSDON, Lord: Speech at television opening, 2 November 1936
42   Cock, G.: 'Report on Baird and Marconi-EMI systems at Alexandra Palace'. TAC paper 33, 9 December 1936
43   Television Advisory Committee, minutes of the thirty-fourth meeting, held on 16 December 1936, Post Office bundle 5536
44   Draft of public announcement, Post Office bundle 5536
45   Secretary (Baird Television Ltd): Letter to J Varley Roberts, 22 December 1936, TAC paper 34, Post Office bundle 5536
46   BAIRD, M.: 'Television Baird' (HAUM, South Africa, 1974)
47   MOSELY, S.A.: 'John Baird' (Odhams, London, 1952)
48   Baird, J.L.: 'Sermons, soap and television' (Royal Television Society, London, 1988)
49   BURNS, R.W.: 'The history of television for public showing in cinemas in the UK', *IEE. Proc. A, Sci. Meas. Technol.* 1985, **132**, (8), pp. 553–563
50   PAUCHON, B.: '50 ans de l'television en France', *Revue de l'UER*, December 1986, (220), pp. 3–12
51   See reports in *The Times* for 18 March 1937, p. 15f, 19 March 1937, p.8e and 27 May 1937, p.15d
52   ANON.: 'A super-power television transmitter for the Eiffel tower', *Television and Short-wave World*, May 1937, pp. 287–288
53   ANON.: 'The world's most powerful television station', *Television and Short-wave World*, May 1938, pp. 261–264
54   ANON.: 'A visit to the Eiffel tower television station', *Television and Short-wave World*, November 1936, p.699
55   CLARK, A.: Report and accounts for the year ended 30 September 1938, EMI Ltd Directors' Report. EMI Music Archives
56   ANON.: 'French praise for BBC television', *Television and Short-wave World*, 24 January 1939 **12**
57   ABRAMSON, A.: 'Zworykin, pioneer of television' (University of Illinois Press, Chicago, 1995) p. 128
58   Ref.57, p. 146
59   ANON.: 'Moscow television centre to use American devices', *New York Times*, 2 May 1937, XI, p. 12:2
60   ANON.: 'Television here and abroad', *New York Times*, 19 June 1938, **IX**, 8:1
61   Quoted *in* HUBBLE, R.W.: 'Four thousand years of television' (Harrap, London, 1946) pp. 147–148
62   Ref.57, pp. 134–135
63   Ref.57, p. 60
64   Ref.61, pp. 156–157
65   ANON.: 'Radio in Japan to be extended', *New York Times*, 12 December 1936, **XI**, 12:7
66   ANON.: 'Television research expanded in Japan', *New York Times*, 7 March 1937, **XI**, 12:6
67   TRAUB, E. H.: 'Television at the Berlin Radio Exhibition, 1936', *J. Television Society*, 1936, pp. 181–191
68   TRAUB, E.H.: 'Television at the Berlin Radio Exhibition, 1937', *J. Television Society*, 1937, pp. 289–297

69 Quoted in ref.61, p. 153
70 HOIED: 'German television service'. Memorandum, November 1938, BBC file T8/32
71 WOOD, D., and DEMPSTER, D.: 'The narrow margin' (Hutchinson, London, 1961) pp. 35–51
72 ANON.: 'German television receivers on sale for first time to public', *Television and Short-wave World*, 1939, **12**, p. 462
73 Quoted in ref.61, p. 155

Chapter 22
Television in the US, (1935–1941)

Some rivalry, perhaps jealousy, seems to have been engendered by the report of the Television Committee and M-EMI's subsequent, February 1935, choice of the 405-line standard. In the US Sarnoff, President of RCA, in a statement[1] delivered at the May 1935 annual meeting of RCA shareholders presented his opinion as to which country was foremost in television development: 'The results attained by RCA in laboratory experiments go beyond the standards accepted for the inauguration of experimental television service in Europe. We believe we are further advanced scientifically in this field than any other country in the world.'

Sarnoff was in a dilemma. On the one hand he wished to extend RCA's position in high-definition television vis-à-vis that of its competitors, on the other hand he did not want to upset the sales boom in radio receivers which might result if a television service were introduced. In his statement he took some pains to point out that:

> 'The sense of sight which television must eventually add to the body of radiocommunication cannot supplant the service of speech and music which permits any simple event to be simultaneously broadcast to the nation as a whole, and which brings to millions of homes continuous programs of entertainment, information and education.'

For Sarnoff there were several important problems which had to be resolved before a regular public service could be established in the US.

1 'The fact that if the new art of television [were] to make the required technical progress, there [would] be rapid obsolescence of both television transmitters and television receivers.'
2 'The creation of new radio or wire facilities of interconnection before a service on a national basis [could] be rendered.'
3 'Further development through experimentation in the field, of a system of high-definition television which [called] for new radio technique inside and outside the studio, and for the production of home television receivers which

[would] increase the size of the picture and at the same time decrease the price at which the receiver [could] be sold to the public.'

All of this led to one conclusion: ' . . . television [was] not here, nor around the corner.' Still, it was necessary to take the research results of the scientists and engineers out of the laboratory and into the field. For this purpose a plan had to be formulated and further money invested.

Sarnoff described RCA's plan, of field demonstrations, which would cost $1M as follows:

1 'Establish the first modern television transmitting station in the United States, incorporating the highest standards of the art . . .
2 'Manufacture a limited number of television receiving sets. These [would] be placed at strategic points of observation in order that the RCA television system [might] be tested, modified and improved under actual service conditions.
3 'Develop an experimental program service with the necessary studio technique to determine the most acceptable form of television programs.'

The timescale for the project would be from 12 to 15 months to build the experimental television transmitter, to manufacture the observation receivers and to commence the transmission of test programmes. Clearly, Sarnoff was in no hurry to launch a commercial television service which could place a very heavy burden on RCA's financial resources, but he was loth to allow RCA to fall behind its European rivals in technological expertise. That Sarnoff was stimulated to action by the British and German plans to inaugurate regular public television services seems evident from the fact that the forthcoming demonstrations would be the first to be given by RCA since May 1932.

Sarnoff's plan was welcomed by the national press[2,3], although the *New York Times*, in a report headlined 'Million dollar plan brightens television outlook' commented that 'the million [was] looked upon as a "drop in the bucket" in relation to what [would] eventually be spent to put television in the home as a real show'.

This view was to be confirmed in a survey article[4,5]—'Television I: a $13 million "If" '—published in April 1939 in the magazine *Fortune*. The survey suggested that this sum had so far been expended by the five leading television companies, with RCA having contributed between $5 and $10 million and two other firms about $2 million each.

In the United Kingdom, too, the plan to establish the London television station would similarly require a very substantial financial commitment. From the station's inception to the second anniversary in November 1938 of its inaugural broadcast it has been estimated that the British Broadcasting Corporation spent £1 000 000.

RCA's promised television station was opened on 29 June 1936 when some invited guests viewed a programme produced in NBC's facilities at Radio City and transmitted from the Empire State building. Briefly, the facilities[6,7] comprised a completely equipped television studio with three electronic cameras for live programmes, a projection room for televising film, monitoring equipment, a central synchronising generator for producing the line and frame pulses, and video line amplifier and terminal apparatus. From Radio City the video signals could be sent to the Empire State building by either an experimental coaxial cable link or a radio relay circuit operating at 177 MHz. The latter consisted of a transmitter and directional antenna situated on the tenth and 14th floor respectively of the RCA building and a receiver on the 85th floor of the Empire State building. A telephone line was employed for the transmission of the audio signals.

In the Empire State building the video and audio signals were separately amplified and used to modulate 49.75 MHz and 52.00 MHz carriers which were then radiated from a common antenna mounted on the top of the building at a height of 1250′. The line of sight horizon was approximately 45 miles away.

For this system the line standards were 343 lines per picture, 30 pictures per second, 60 frames per second, odd line interlacing, 4:3 aspect ratio and 1.5 MHz video bandwith. This figure was 0.64 of the bandwidth value (2.35 MHz) determined from the usual formula for calculating the video bandwidth, *viz.* $0.5 \times$ (aspect ratio) $\times$ (number of lines)$^2 \times$ (picture frequency). The reduction stemmed from work, undertaken by Engstrom and others at RCA, which showed that equal resolution of the reproduced television image in the vertical and horizontal directions was obtained with a bandwidth only 64 % of that given by the widely accepted formula[8].

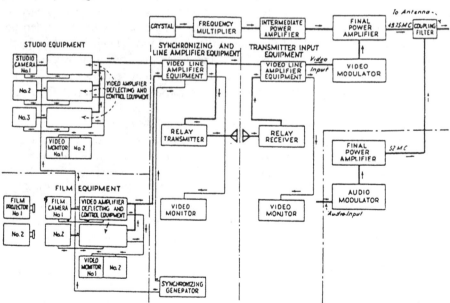

Figure 22.1   Schematic arrangement of the 1937 RCA television system. 441-line images were transmitted from the NBC experimental station in the Empire State Building and successfully received by a selected number of experimental television receivers in the homes of some RCA/NBC engineers

Reproduced by permission of RCA

The reception of the television signals was confined to one hundred receivers distributed in New York City and the surrounding suburban area. Each receiver included 33 valves and a kinescope viewing tube. This was mounted vertically and the 5 × 7 inch greenish hued image was viewed via a front silvered mirror attached to the inside of the cabinet lid.

Although the inauguration of the new system was scheduled for 29 June it was not until 7 July 1936 that the images were considered to be 'worth showing'[9]. A prominent radio industrialist opined: 'It [the television demonstration] was extremely interesting but it is a long way off before it reaches the home. First, it must equal the home movies in quality, and it does not achieve that goal today in clarity or dimension of the pictures, although the images were splendid.'

Among the televised presentations, Major General J G Harbord, chairman of the RCA Board, and D Sarnoff, president, were seen sitting at a desk presenting a review of television's progress, followed by Otto Schairer, vice president of RCA, explaining that although there were no plans to manufacture commercial television sets for 1936, three designs for such sets were being tested. These business matters were succeeded by 20 girls, introduced as the Water Lily Ensemble, who gave a dance, a film featuring the streamliner train, the 'Mercury', and a world of fashion parade with the Bonwit Teller models. Harry Hull, an actor, then delivered a monologue of his role in 'Tobacco Road', Graham McNamee and Ed Wynn gave a comedy sketch, and the programme ended with a film on army manoeuvres.

In all those countries which sought to broadcast high-definition images the question of which standards should be adopted for a high quality national television service had to be resolved. When television was in its infancy the line standards which could be utilised were determined primarily by the bandwidths of the signals which could be easily accommodated in the medium frequency band (0.3 MHz–3.0 MHz). But with the development of radiocommunication systems, electronic circuit techniques, valves, photocells and camera tubes, and the general availability of certain channels in the high frequency (3 MHz–30 MHz) and very high frequency (30 MHz–300 MHz) bands, the issue of line standards was less constrained.

In the United Kingdom the question was soon resolved[10] as only one company, the Marconi-EMI Television Company Ltd in the mid-1930s, offered to the responsible agency (the GPO) an all-electronic high-definition system. M-EMI's choice of 405 lines per picture, 25 pictures per second was not challenged by a superior standard and so the 405-line standard became the national standard. In Germany the 180-line standard predominated television activity until 1936 because no German company in the early 1930s engaged in television research and development on all-electronic television systems of the types which was being undertaken by RCA, Philco, Farnsworth and Marconi-EMI. But in the US the position was more complex. There, the trend towards the utilisation of the h.f. and v.h.f. bands began when it became evident, by 1931, that progress could not be made in the 100 kHz bandwidth channels of the 2000–2950 kHz band.

Authorisation to use some specified v.h.f. bands, *viz.* 43.0–46.0, 48.5–50.3 and 60.0–80.0 MHz, was initially granted to several stations including the Don Lee

Broadcasting System of Los Angeles (W6XAO), the NBC's W2XF and W2XBT in New York, Jenkin's W3XC in Wheaton, Maryland, W1XG in Boston and W8XF in Pontiac, Michigan. In 1933 and 1934 several additional v.h.f. stations were licensed to use the 42.0–56.0 MHz and 60.0–86.0 MHz bands, and in 1936 activity in the 2 MHz band ceased. Subsequently, all v.h.f. stations were required to operate in the 42.0-56.0 MHz and 60.0–86.0 MHz bands[11].

From 1937 frequency allocations were allotted on the basis of a 6 MHz channel bandwidth. Nineteen of these channels were established between 44 MHz and 294 MHz. Later, in 1940, the 44.0–56.0 MHz channel was transferred to frequency modulation sound broadcasting stations and replaced by an extra channel from 60.0–66.0 MHz. The 156.0–162.0 MHz channel was later abolished and the frequency allocation in 1944 comprised 18 channels: from channel 1, 50 to 56 MHz, to channel 18, 288 to 294 MHz.

Since the bandwidth of a television channel depends on the number of lines per picture, the picture rate and the aspect ratio, the bandwidth of 6.0 MHz permitted various line standards. Licencees were free to employ any scanning method but, with the production of images of increasing brightness, and the consequential need to reduce flicker, agreement was reached between 1932 and 1934 that odd line interlacing was the simplest and most satisfactory method of avoiding this disturbing visual defect. The first odd line value to be selected was 343. Later other numbers were chosen. The list of US scanning specifications in chronological order[12] is given in Table 22.1. (The requirement for an odd number of lines and its factorisation into the product of several small integers is explained in Chapter 21.)

The Table shows that the picture rate for the several schemes was either 24 pictures per second or 30 pictures per second. Of the two values the former was attractive for two reasons:

1   the picture rate was the same as that which had been established for motion pictures, which seemed likely to constitute an appreciable part of television programmes;

Table 22.1   *US Scanning specifications*

Date	Organisation	Number of lines	Picture rate/s
1934	RCA	343 (7 × 7 × 7)	24 and 30
1937	Philco	441 (3 × 7 × 3 × 7)	24
1938	Philco	525 (3 × 5 × 5 × 7)	30
1939	Philco	605 (3 × 11 × 11)	24
1939	RMA	441 (3 × 7 × 3 × 7)	30
1940	RCA	507 (3 × 13 × 13)	30
1941	NTSC	525 (3 × 5 × 5 × 7)	30

2   the 24/s rate permitted an increase in picture detail of 25 % $(30 \div 24 = 1.25)$ compared with the 30/s rate, all other parameters remaining the same.

However, this picture rate was incompatible with the frequency, 60 Hz, of the a.c. supply. Experimentation showed that very complete shielding of the cathode-ray display tube and costly filtering of the d.c. supplies, especially in the scanning circuits were necessary to avoid degradation of the scanning raster. This degradation was manifested by vertical and horizontal shifts of the scanning lines from their normal positions.

Effectively, extra detail could be provided but only at extra cost. After argument had taken place for some time on this issue the economic factor prevailed and the picture rate of 30 picture per second was standardised in 1936—a rate which has not been seriously questioned since then.

The resolution of this problem immediately raised another, *viz.* the problem of operating motion picture film (recorded at 24 pictures per second) on a television system working at 30 pictures per second. An increase in film running speed by 25 % would increase the pitch of the reproduced sound and also the speed of motion of the actors and other moving objects in the recorded scene. These effects could not be tolerated. Fortunately, 1/24 is the average of 2/60 and 3/60. Thus, if a film projector could be devised in which a picture was held stationary in the projector for 2/60s, and the next picture was held stationary for 3/60s, the average picture rate would still be 24 pictures per second. Now, in an interlaced television system, each frame is scanned in 1/60s and as each film picture could be scanned for an integer multiple of 1/60s synchronism between the television system and the film projector could be maintained.

In the practical implementation of such a scheme, it was necessary only to devise an intermittent pull-down mechanism which allowed the film to dwell for unequal lengths of time (2/60s and 3/60s) on alternate pictures. Such a system worked very satisfactorily.

If this problem had not been so elegantly solved, either the cost of television receivers in the US would have been substantially increased to permit 24 picture/s operation; or, with the 30 picture/s standard, specially produced motion films for television would have been required.

In the UK the mains supply has a frequency of 50 Hz allowing a television picture rate of 1/25 s, which is so close to the film picture rate of 1/24s that no adverse effects are noticeable.

Movement for an industry-wide consensus on standardisation arose in 1935 after a demonstration by RCA, to the members of the Radio Manufacturers Association (RMA), of the company's 343-line system. In that year the RMA[13] asked its engineering department to determine when it would be appropriate to adopt national television standards. This activity came to the notice of the Federal Communications Commission's chief engineer, T A M Craven, who proposed to the RMA that various branches of the industry should come to an agreement on television performance standards among themselves prior to any Commission action. He stated the FCC opinion that the standards adopted should be such that any television receiver manufactured for general sale to the public must be capable of receiving any broadcast from any station licensed by the FCC. In another communication Craven indicated the necessity of reserving certain regions of the frequency spectrum for television transmissions.

Craven's view stimulated two responses[14]. First, the FCC announced hearings, to commence on 15 June 1936, to determine its long-term policies on the future allocation of frequency channels; and, secondly, the RMA set up two television committees, one on standards and the other on frequency allocations, to prepare a joint report to the hearings.

The work of the RMA's committees was the first major move in television standardisation in the United States. On the worth of the committees' pronouncements an eminent member of the National Television System Committee commented in 1976[14]:

'Forty years later, their judgments display a remarkable understanding of the items to be standardised, albeit that the numbers they initially proposed have survived in only four instances: the channel width, the aspect ratio and the frame and field rates.'

At the 1936 hearings the RMA allocations committee recommended that the FCC establish seven 6 MHz television channels between 42 and 90 MHz and that experimental authorisations be permitted above 120 MHz. The 1936 RMA standards committee proposed[15] 441 lines per picture, 2:1 interlacing to give 30 pictures per second (i.e. 60 frames per second), double sideband negative picture modulation, 2.5 MHz video bandwidth and frequency modulation for the sound signal; the video and audio signal carriers to be spaced c. 3.25 MHz apart with the audio signal carrier being at the higher frequency.

The first demonstration of 441-line television in the US was given by the Philco Radio and Television Corporation on 11 February 1937 to nearly 200 members of the newspaper and technical press at the Germantown Cricket Club in Philadelphia[16]. Philco's 441-line experimental broadcast transmissions had commenced in December 1936 and, as the picture quality steadily improved throughout the tests, Philco felt able in February to show the benefits which would follow from the adoption of the RMA's standards.

Previously, in November 1936, Philco had completed a series of field tests based on a 345-line standard. With the move to 441 lines per picture the company's engineers had had to modify the equipment to accommodate an increase of 65 % in the permitted bandwidth (1.5 MHz to 2.5 MHz) and 29 % more picture elements.

Six receivers were used in the demonstrations[17], each fitted with 12″ cathode-ray tubes displaying images approximately 7.5 × 10 inches in size. The screen material, a mixture of zinc sulphide and other fluorescent compounds, gave a 'black and white image tinged slightly with blue'.

The transmitter of station W3XE operated at 49 MHz, with a peak power output of 4 kW, and excited an antenna mounted 210′ above street level.

Among the televised events were an interview with the manager of the Philadelphia Athletics on the prospects of the forthcoming baseball season, a fashion show staged by Bonwit Teller, 'carefully designed for the delectation of hard-boiled newspaper men' and a variety of televised scenes from the studio, from the roof of the Philco plant and from motion picture film. Of principal technical interest there was a test which compared 345-line and 441-line images. When a pocket watch was placed before the television camera so that its reproduction at the receiver was about life size, the second hand of the watch could not be discerned using 345-line definition, but with the use of 441 lines per picture the second hand was 'clearly visible and could be followed readily'[17]. A contemporary report on the demonstration mentioned that: 'The higher definition produced a smoother, more pleasing picture, in which the line formation was not visible from the ordinary viewing distance'[16]. All of this seemed to vindicate the RMA's opinions.

Interestingly, the chief engineer of Philco at the time of the tests was A F Murray, who had been with Philco's arch rival RCA. He was also the acting chairman of the RMA's committee on television to which the RMA's subcommittees on television standards and frequency allocations reported.

In 1937 the FCC reacted to the findings of its 1936 hearings and established 19 6 MHz channels interspersed with other services in the spectrum from 44 to 294 MHz[13]. On the issue of line standards, the FCC felt that television engineering did not possess the necessary stability needed for governmental sanction.

Undoubtedly, experience of the reception of the high-definition signals radiated from Alexandra Palace, London, contributed to the formulation by the RMA of its standards. The London station was the only station in the world, in 1937, which was operating a regular, public, all-electronic service. Effectively, television administrations and manufacturers of television apparatus in other countries could use the Alexandra Palace transmissions to test their opinions on a wide range of issues without incurring the great expense associated with field testing. Variation of the line standards was not, of course, possible but nevertheless much useful information could be gleaned from careful observation of, for example:

1   the appearances of the images produced by cathode-ray tubes using electrostatic focusing and electrostatic scanning, or electrostatic focusing and magnetic scanning, or magnetic focusing and magnetic scanning;
2   the effect of automobile engine ignition and other sources of interference on synchronisation;
3   the influence of distance, and hence signal strength, on picture quality;
4   the benefits of utilising a directional, rather than an omnidirectional, antenna for reception;
5   the influence of cost of a set, on image brilliance and contrast, picture size and sensitivity;
6   the visual impact of indoor and outdoor programmes and film transmissions;
7   the general quality of the received images, as determined by the line standards, compared with the images produced by home movie projectors.

During the summer of 1937 engineers of the American Hazeltine Service Corporation established a temporary laboratory in England for the purpose of making a survey of television. The survey included observing the transmissions and making measurements of the received signals with special equipment designed for the purpose. Visits were made to Alexandra Palace and to receiver manufacturers, and talks were held with several television engineers. This work led to an important paper[18], by Lewis and Loughren, published in the magazine *Electronics* in October 1937.

The objective of the authors was to determine whether certain aspects of the 'present practice of the proposed television standards in the United States [were] wise'. There were three US standards (on the polarity of the transmission, the transmission of the d.c. or background components and the shape, amplitude and duration of the synchronising pulses), which were the exact reverse of the corresponding standards employed in the UK. Clearly, it was highly desirable that the RMA's standards should be based on sound practical experience.

Lewis and Loughren were most complimentary about British television. 'First let it be said that the British pictures are remarkably good. They are steady; they are brilliant; they have an exceptional amount of detail . . . That [the] British standards constitute a major improvement over present American practice is an inescapable conclusion because television is technically successful and an accomplished fact in England . . . We cannot avoid the fact that the situation in the United States is much less favourable. Unless changes are made in the type of signal which is now being used for experimental transmitters, American receivers will be more expensive, more difficult to service and will give performance inferior to British receivers . . . It will be in steadiness and control that the American picture will suffer . . . In viewing the Alexandra Palace transmissions on a large variety of different receivers there were practically no cases of faulty synchronisation.'

As an indication of the stability of synchronisation of UK sets, Lewis and Loughren noted that reception of the Alexandra Palace transmission at 80 miles (which was beyond the optical horizon) gave an image that 'was not visible except as a hazy movement of light on a grille of noise', yet 'it was easily demonstrated that the grille [raster] was synchronised . . .'. Furthermore, the authors found that even the most excessive automobile static which resulted in 'a snowstorm on the screen' failed to have any disturbing effect on the synchronisation.

In their paper, Lewis and Loughren extolled the virtues of positive modulation and d.c. transmission and observed: 'It would not be putting it too strongly to say that the level of black is a definite foundation upon which the structure of British television is built . . . A final confirmation of the practical nature of the British standard is that the other European countries are adopting its principal features . . . '

Much first-hand knowledge and experience of the British and German television systems was obtained by members of the RMA's committee on television. Among the major industrial firms, apart from the Hazeltine Service Corporation, which sent representatives to Europe to collect data which could be helpful in the formulation of US national standards were the Radio Corporation of America, the Columbia Broadcasting System and Bell Laboratories. Reports on these visits together with details of the latest endeavours being undertaken by organisations in the United States were presented at a meeting, convened by Murray, at the RMA-IRE Fall

Convention held in Rochester, US, in 1937. Among the views expressed were the following[19,20]:

1   Engstrom, of RCA, stated a preference for d.c. transmission as in present British practice, and indicated that 'the "serrated" type [of synchronising pulses] used by RCA and the British Broadcasting Corporation produced better results than any other system tested by RCA, of which there were six or eight which had been examined in the past two or three years'.

2   Lack, of Bell Laboratories, emphasised the advantage and superiority of mechanically scanned film, and reported that 'in Berlin he saw 180-line mechanically scanned film which appeared to have greater detail than pictures of many more lines shown in [the US]'.

3   Goldmark, of CBS, was impressed 'with the sensitivity of [the British] cameras employed and with the strength of the signal'; he considered that the British Scophony mechanical projection system's 'picture was of a very high quality, surpassing all other types in brilliance, definition, gradation and size', and thought that the 441-line image which he had seen in Germany had 'excellent definition, gradation and freedom from geometrical distortion'.

Here, then, were definite opinions on overseas television systems which could be of some assistance to the RMA in the drafting of pragmatic standards.

Further work by the RMA standards committee led, in July 1938, to several additional recommendations to its 1936 standards. D.C. transmission of the brightness of the televised scene was specified—black in the picture was to be represented by a definite carrier level, as in British practice[15], the radiated electromagnetic wave had to be horizontally polarised and details of the line and frame synchronising pulses were delineated—the frame synchronising pulse had to be serrated and include equalising pulses, again as in British practice.

A feature of the RMA's approach was its responsiveness to progress. In June 1938 its 1936 statement on double sideband transmission was replaced by one on vestigial sideband transmission, and the frequency channel needed to accommodate it was rearranged thereby increasing the video bandwidth from 2.5 MHz to 4.0 MHz. This increase permitted an increase in the number of lines per picture from 441 to 560[15] (since the number of lines is proportional to the square root of the bandwidth), but this important variation was not made until March 1941.

These statements were circulated[21] to the RMA's membership and, in the absence of dissent, forwarded in September 1938 to the FCC. However, the FCC again concluded that 'television [was] not ready for standardisation or commercial use by the general public', because of the prospect of rapid obsolescence of equipment, and it therefore decided to continue its policy of not acting on the RMA recommendations.

On one matter the FCC did act decisively. In April 1938 it issued a revised set of 'Rules of practices of procedure', one of which, rule 103.8, limited experimental television licenses to stations actively engaged in research and development in the techniques of television broadcasting[22] (Table 22.2). Stations that had previously been concerned only with programmes and commercialisation were now debarred

Table 22.2   Experimental television broadcasting stations in the US* (August 1938)

Location	Call letters	Licensee	Power
New York, NY	W2XBS	National Broadcasting Company	12 kW
New York, NY	W2XAX	Columbia Broadcasting System	7.5 kW
New York, NY and Camden, NJ	W2XBT	National Broadcasting Company	400 W
Long Island City, NY	W2XDR	Radio Pictures, Inc	1 kW
Camden, NJ	W3XEP	Radio Corporation of America Manufacturing Company	30 kW
Camden, NJ	W1OXX	„ „	50 W
Camden, NJ	W3XAD	„ „ portable	500 W
Philadelphia, Pa	W3XP	Philco Radio and Television Corporation	15 W
Philadelphia, Pa	W3XE	„ „	10 kW
Springfield, Pa	W3XPF	Farnsworth Television Inc.	250 W
Iowa City, Iowa	W9XK	University of Iowa	100 W
Iowa City, Iowa	W9XAI	„ „	100 W
Manhattan, Kansas	W9XAK	Kansas State College of Agriculture and Applied Science	50 W
Kansas City, Mo	W9XAL	First National Television, Inc	300 W
Lafayette, Ind	W9XG	Purdue University	1500 W
Los Angeles	W6XAO	Don Lee Broadcasting System	1 kW
Boston, Mass	W1XG	General Television Corporation	500 W
Jackson, Mich	W8XAN	The Sparks-Withington Co	100 W
Minneapolis, Minn	W9XAT	Dr George W Young	500 W

*Licensed by the Federal Communications Commission

under the rule from transmitting. These stations included for example, W9XAT in Minneapolis, W9XD in Milwaukee and W8XAN in Jackson, Michigan. However, most station owners chose to adopt the RMA standards.

Meanwhile, during the somewhat protracted RMA, FCC and industry discussions on standards, RCA had been conducting television field tests from its Empire State Building experimental station in New York. By the beginning of 1939 these had advanced to the point where the introduction of television to the public, in the areas served by television stations, had been practicable. In addition, the business slump of 1937 which had impeded progress for the first half year of 1938 had declined and the second half of the year had witnessed a substantial improvement in all branches of the radio industry. Sarnoff, the president of RCA, now seemed keen to begin a regular programme service and on 20 October 1938 announced that in April 1939 the RCA manufacturing company would place television receiving sets on the market to coincide with the opening of the New York World's Fair[23]. To support this move the RCA television transmitter, at the top of the Empire State

Building, would broadcast a limited service to the public under the control of the NBC. Such an event almost guaranteed to provide a massive publicity boost to the inauguration of the service and could not be overlooked.

Actually, Sarnoff's 1938 television policy was being forced by several factors and the president of RCA was obliged to amend his previously (1936) reticent position with regard to a public television service to ensure that RCA retained its premier position in the television industry. The first factor concerned the threat to RCA's manufacturing base. In May 1938 Communicating Systems, Inc (later renamed the American Television Corporation), taking advantage of the Empire State Building broadcasts, announced plans[24] to commence selling television sets in the New York and Boston areas where the transmissions from W1XG could be received. The firm had two models in production, with three and five inch screens, and planned to sell these at $150 and $250 respectively. Both models lacked sound reception means but an adaptor could be bought for an extra $15 to $17. Demonstrations of the sets were given in major stores in Manhattan, Brooklyn and the Bronx[25].

Soon afterwards, DuMont also announced[26] that it intended to market television receivers and had arranged with a Madison Square Garden store to have its sets demonstrated. These had 10×8 inch and $8\frac{1}{4} \times 6\frac{1}{4}$ inch screens, and were priced at $650 and $395.

Thus, unless RCA reacted swiftly to these pre-emptive television receiver manufacturing initiatives, its own future production base might be seriously jeopardised.

The second factor concerned the challenges to RCA's television system which were being made by several rival organisations. For many years RCA's position in the US in the field of all-electronic television had been preeminent. The company had developed the world's first electronic camera, the iconoscope, it had conducted extensive field tests of its various systems, it had published many papers on its research activities and it had contributed greatly to the fundamental knowledge necessary to engineer a high-definition system. Now, in 1938, Philco and DuMont were working to develop alternative systems, and CBS in New York and Don Lee in Los Angeles were intent on commissioning their own regular services. RCA could not ignore such moves.

Sarnoff's putative keenness for the start of public television was rather assuaged by remarks made by Otto Schairer[27], vice president of RCA, on the occasion of a large screen (4×6 feet) projection television system demonstration given on 14 February 1939 at the hotel Waldorf-Astoria, New York. There, Schairer sketched the progress of the development of television, but said that it still was the 'equipment of the laboratory' and was not quite ready for commercial application.

Nevertheless, RCA's apprehensions about competitive manufacturing ventures by others were well founded. A few days earlier, on 10 February 1939, the *New York Times* reported[28] the strategy of the Farnsworth Television and Radio Corporation to acquire facilities for manufacturing television apparatus on a commercial scale. Lacking a production capability of its own, the corporation planned to raise new capital (of several million dollars) to buy the business and assets of Capenhart, Inc of Fort Wayne, Indiana, whose plants and equipment would provide the means to produce special radio, television and allied electrical equipment. The report also mentioned that the corporation was contemplating the purchase of the radio plant of the General Household Utilities Company of Marion, Indiana.

These plans to manufacture television receiving sets soon received a setback when on 9 April 1939 it was reported that the FCC 'won't be stampeded blindly into launching [a] television industry'. In a *New York Times* article headed[29]: 'FCC stops gold rush', Craven was quoted as saying:

> 'We want to know about the business end of television. There have been many misgivings on the amount of royalties to be received. Much to the astonishment of the FCC [television] committee, the RMA committee didn't give any consideration to the commission's problems. If the FCC committee accepts the standards offered by the RMA it means almost a monopoly. The standards they propose would put television on a par with the movies in about 1906.
>
> 'If the television development means a limited amount of channels, who, considered on a broad public basis, is entitled to them—the existing broadcasting industry, the moving picture industry or the newspapers? They are all vitally interested.
>
> 'Despite great pressure exerted upon the commission's committee to launch the television industry and still beyond that, pressure brought to bear for the adoption of certain standards, the FCC's watchword is 'caution' in the public interest.
>
> 'What's the necessity of going so fast in this important matter of television?'

To seek the facts the FCC's television committee (comprising commissioners T A M Craven, N S Cass and T H Brown) planned to visit RCA Victor, Philco and Farnsworth in Philadelphia, DuMont in Passaic, New Jersey, NBC, Columbia, AT&T and Armstrong in New York and the General Electric in Schenectady.

Craven's reference to the monopolistic aspect of the RMA standards and the need for circumspection was consonant with the view of the DuMont Laboratories. The Laboratories reasoned that it was desirable to provide some means which would allow a change in the line standards without the necessity of rendering existing apparatus obsolete either at the transmitting or at the receiving ends of a television system. Its view was that a flexible standard, not single values of number of lines and pictures per second, should be favoured. The Laboratories suggested a lower limit of the picture rate of 15 per second and an upper limit of the number of scanning lines in the neighbourhood of 800; such a flexible standard would enable broadcasters and the public to adopt the best compromise at each stage of the art.

In February 1938 the DuMont Laboratories announced[30] that it had developed a system of television transmission which could transmit televised images without the need to send synchronising signals and which utilised one half of the bandwidth of conventional systems.

Since the bandwidth of a television system is directly proportional to the picture rate, then, if the rate is halved, the bandwidth is halved. In the DuMont scheme[31] the rate was reduced from 30 to 15 pictures per second. This reduction naturally led to an increase in flicker and special means had to be introduced to circumvent the effect. The means advanced and demonstrated, by DuMont was to use a 4:1 interlace ratio; i.e. a complete picture comprised four (not two) separately scanned frames, each frame being scanned in 1/60 s.

Such a system could not, it seemed at that time, be implemented using the conventional method of transmitting synchronising pulses because of the severe constraint placed on the timing of the pulses. DuMont's solution was to employ two sweep generators (for the line and frame scans) at the transmitter and to use the outputs from these to modulate two subcarriers which were then combined with the audio and vision signals to modulate the transmitted carrier. In the receiver the two sweep voltages were recovered and applied to the display tube. Thus, exact information on the camera scanning waveforms was sent and no synchronising pulses were necessary. With this novel method DuMont claimed that receivers could be designed which would permit reception to be achieved from numerous stations of differing degrees of definition. This receiver versatility held the promise of diminished obsolescence when higher definition was desired in televised images.

DuMont's proposal[32] was later investigated at length by the National Television System Committee (established in 1940), which concluded that provision for such flexibility would so increase the cost of receivers, and provide so little gain in picture quality, that it was not justified.

Television made its formal debut in the United States on Sunday, 30 April 1939 when the National Broadcasting Company broadcast images of President Roosevelt speaking at the opening of the New York World's Fair[33,34]. The first programme was televised by NBC's mobile television unit using just one camera which was sited c. 50 feet from the President's box. Images were sent by coaxial cable to the control van and the ultra short wave transmitter housed in one wing of the Federal Government Building. From here the images were relayed by a 177 MHz radio link to a relay receiver in the Empire State Building, eight miles from the Fair grounds at Flushing, Long Island. Finally, the television signals were broadcast from station W2XBS using 45.25 MHz and 49.75 MHz carriers for the vision and sound transmissions, respectively.

The programme itself included panoramic shots of the Fair buildings, the parade as it passed the viewing stands, the entrance of the President and Mayor La Guardia and several speakers.

Estimates of the number of sets which received the W2XBS broadcast ranged from 100 to 200 and the number of viewers was possibly 1000.

Of the favourable press reports, the *New York Herald Tribune* noted: 'Amazement and then unbounded enthusiasm marked the start of what will be a semi-weekly feature from now on.' The *New York Times* said: 'Reports from receiving outposts scattered throughout a 50 mile radius of New York indicated that the spectacle by television was highly successful and that a new industry had been launched into the World of Tomorrow.'

The main criticism[35] of television viewers on the fair grounds and at Radio City was that only one camera was used and it was too far away from the speakers thereby causing the images to be too small. 'They also complained of the camera-man's remaining in the same spot for the entire show.'

According to the *New York Times* the use of just one camera led to the British radio officials being 'amazed' at what they called 'the nerve of the Americans'. British practice would have entailed the utilisation of three or four cameras for fading in scenes from different angles to gain variety. 'Then, too, they wondered what would happen if the electric eye burned out at the crucial moment.'

Initially, the broadcast schedule maintained by NBC consisted of a total of 25 hours of television per week. The regular programmes, including plays, variety acts and so on were televised from 8.30 p.m. to 9.30 p.m. on Wednesday and Friday evenings, sports programmes were transmitted on Saturday evenings[35]. Film programmes, to aid the retail industry in demonstrating and installing sets, were timetabled from 11.00 a.m. to 4.00 p.m. on Tuesdays and Fridays and from 4.30 p.m. to 8.00 p.m. on Wednesdays, Fridays and Saturdays. These film programmes consisted of approximately ten minutes of film (mostly of an educational nature) followed by a five minute intermission during which a standard test chart was shown.

Some American television production resources were seen, during a visit in May 1939, by Gerald Cock, the BBC's director of television. His impressions[36] were not particularly favourable:

'In the matter of equipment and general facilities they [were] still well behind us. For example, the NBC [had] one studio in the RCA building, just about half the size of one of the studios at Alexandra Palace—and our own studios [were] too small. The cameras [were] not as light sensitive as ours. Columbia, which [expected] to start a service about now ... [had] taken a huge hall over the main hall of the Grand Central Station in New York, which must be at least ten times the size of one of our studios at Alexandra Palace, but it [had] only two iconoscope cameras and one teleciné apparatus.

' ... I was conscious, too, of the want of a really progressive forward programme policy.

' ... By the way, I have it on reliable authority that the American iconoscope and the British emitron were developed independently but it was left to us to develop the super emitron. American hasn't got one yet.'

Four receiver manufacturers had sets available for purchase within a month of the inauguration of the NBC service. At Macy's store, in New York, the receivers shown in the table opposite were on sale in May 1939.

Cock's comment about 'the want of a really progressive forward programme policy' was tacitly shared by L Espenschied, a senior engineer at Bell Telephone Laboratories. In an internal report he wrote[37]:

'The programs being put out by NBC were distinctly poor in this period [August–September 1939], poor both as productions and in their technical rendition.

Company	Picture tube size (inches)	Price ($)	No. of valves	Remarks
Andrea	5	189.5	16	
DuMont	14	395.0–540.0	22	several models
RCA	5	199.5–295.0		
	9	495.0		
	12	600.0		
Westinghouse	5	199.5		

'During the times I was at home, there was not much to look at, there being generally no progress during the weekends, and the outdoor programs being mainly in the midday hour. The evening hour was generally film. The few live scenes I saw were generally amateurish, poorly lighted, had dark portions and out-of-focus troubles. The films were the usual silly Hollywood stuff . . . the programs had no particular appeal.

'The one program that was most significant of what television might do was a baseball game picked up in Brooklyn . . . One point of view was to the rear and one side of the batter and catcher, showing the action at the plate. Another was of the outfield, taken apparently from the top of the grandstand. The show could have been good, but actually was distinctly mediocre, because of limitations in the pickup technique. Two years ago . . . I saw a corresponding showing of a tennis game at Wembley [UK] that was more clearly reproduced.'

Of the major industrial companies, RCA, GE and Farnsworth supported the RMA's position but, in 1939, Philco, Zenith and DuMont (and soon CBS) strongly dissented. The latter group asserted that the proposed standards were not 'sufficiently flexible to permit certain future technical improvements without unduly jeopardising the initial investment of the public in receivers'. This seemed to be the FCC's opinion also.

Actually, the FCC neither approved nor disapproved of the RMA's standards and accepted that they represented a consensus of expert engineering views. The commission did not believe the standards were objectionable but simply wished to be free to prescribe a better performance for the transmitters which it would licence in the future[38]. It was concerned, too, that its future pronouncements on standards should enable a 'radical reduction' in the price of sets to be accomplished. 'Unless the television receiver of the future is to be within the pocket book capabilities of the average American citizen,' it argued, 'television as a broadcasting service to the general public cannot thrive as a sound business enterprise for any extended period.'

For the FCC the position was simple: 'Considered from the broadcast standpoint . . . television [was] now barely emerging from the first or technical research stage of development.' Therefore the commission counselled 'patience, caution and understanding'. The aim was to develop 'a new and important industry logically and on sound economic principles'.

The FCC's reservations did not deter many aspiring telecasters and by the middle of October 1939 19 licences had been issued in the New York, Philadelphia, Chicago, Washington, Fort Wayne, Cincinnati, Schenectady, Los Angeles and San Francisco areas. A further 23 applications were being processed[39]. However, there was an ominous cloud on the horizon—in the form of colour television—which threatened to change completely the black and white standards of the RMA. Chairman J L Fly, of the FCC, alluded to amendments to come when, in October 1939, he opined: 'I feel that colour in television offers the possibility of a vastly improved system, which gives not merely colour but improvements in terms of definition, contrast and other related factors.' Obviously, the issue of television standards in the US was going to drag on for some considerable time.

The FCC's reluctance to act decisively had an adverse effect on the sale of receivers. By August 1939 only about 800 sets had been sold throughout the US, and another 5000 were occupying space on dealers' shelves[40]. As in the UK, when sales were slow some firms offered discounts in an attempt to stimulate growth. RCA, for example, announced[41] in March 1940 that it planned to reduce prices by approximately 33 % in the expectation that purchases would increase from 2500 to 25 000. GE, DuMont and Andrea reacted similarly.

Three factors militated against a receiver boom[42]:

1   the high costs of the sets;
2   the limited programming schedules of the television broadcasting companies;
3   apprehension among the public that their considerable financial outlays could be wasted as a consequence of changing standards and possibly set obsolescence.

These factors were also apparent during the early months of operation of the London television stations at Alexandra Palace. But, fortunately, in the UK steps were quickly introduced by manufacturers, by the BBC and by the TAC to attract interest in television. The Television Advisory Committee in particular, by recommending in 1938 the adoption of the 405-line standard for three years, removed an important obstacle to future sales.

In the US the FCC's disinclination to set standards had a serious impact on the funding of television programmes. Unlike the position in the UK where programmes were provided by just one organisation, the British Broadcasting Corporation, the programmes in the US required commercial sponsors. But sponsors, for example advertisers, were not willing to invest huge sums to produce attractive programmes which would be seen by the few members of the public who had bought receivers. And the public was not willing to spend large sums on television sets which did not offer attractive programmes and which might after a short time be incapable of receiving a commercial broadcast. There appeared to be a deadlock.

To resolve the dilemma the FCC, in December 1939, without acting on the RMA standards then before it, stated its intention to authorise limited commercial operation of television stations[15]: the funds derived from sponsors would be used primarily for the experimental development of the television programme service.

The FCC scheduled a public hearing for January 1940 to hear all interested parties before proceeding further.

On 29 February 1940 the FCC moved to widen the use of television by adopting new rules[43], to become effective on September 1940, which would permit advertising in connection with programmes for which the production cost was borne by the sponsors. However, there was to be no emphasis on commercial operations at the expense of research. Stations would be allocated into two groups: stations continuing technical investigations might be allowed to utilise more than one channel, but other stations designed to experiment in programme production and technique would be restricted to one channel only. Both classes of stations would remain experimental.

The FCC suggested that receivers should be marketed which were capable of being adjusted to receive any reasonable changes in the number of lines or pictures per second which might be found to be practicable. Larger screen displays were thought to be essential for widespread public acceptance of television and continued experiments in staging and studio aspects of television were deemed to be necessary. 'Actual demonstrations to members of the commission indicate the need for further improvement in the technical quality of television. The evidence reveals a substantial possibility that the art may be on the threshold of significant advance.'

On the vexed subject of full commercialisation the FCC repeated its well known position: 'As soon as the engineering opinion of the industry is prepared to approve any one of the competing systems of broadcasting as the standard system, the Commission will consider the authorisation of full commercialisation.'[15]

The new rules and the more positive approach of the FCC were warmly received by station owners and broadcasters. There was an anticipation that the changes would herald a surge in receiver sales and encourage the construction of new stations. Consequently, plans were formulated by manufacturers to launch sales initiatives. On 20 March 1940 RCA announced[44], via full-page advertisements in some New York newspapers, that television was now ready for the home and that the public could expect more exciting and extensive programming from the NBC. And in the *New York Times* alongside the announcement was another large advertisement for RCA receivers at significantly reduced prices.

RCA's 'forcing the pace' marketing incurred the ire of the FCC and the commission suspended, on 22 March, its recent order authorising limited commercial service. Explaining its sudden decision the FCC stated[45]:

'The current marketing campaign of the Radio Corporation of America is held to be at variance with the intent of the commission's report of February 29 . . . Such action is construed as a disregard of the commission's findings and recommendations for further improvement in the technique and quality of television transmission before sets are widely sold to the public . . . Promotional activities directed to the sale of receivers not only intensifies the danger of these instruments being left in the hands of the public, but may react in the crystallizing of transmission standards at present levels.

Moreover, the possibility of one manufacturer gaining an unfair advantage over competitors may cause them to abandon the further research and experimentation which is in the public interest.'

A further hearing to reappraise the situation was now once again necessary. This was scheduled for 8 April 1940. In addition, following a public outcry against the FCC which was accused of endeavouring to impose an 'alien theory of merchandising' on the United States by attempting to protect consumer interests beyond 'acceptable bounds', there was also in April 1940 a hearing, called by Senator Ernest Lumdean, to consider the FCC's actions and the allegation that it had exceeded its authority by interfering with the freedom of public and private enterprise.

Neither hearing produced an agenda for progress. The Senate hearing ended without any action being taken, and the FCC hearing led to the commission restating its well known opinion that no form of commercial operation should be permitted because of its possible adverse effects on technical experimentation. Once again, the opposition of Philco and DuMont to the RMA's proposals led to an impasse.

This was broken soon afterwards at a meeting between Dr W R G Baker, director of engineering for the RMA, and J L Fly, the chairman of the FCC[14]. Cognisant of the success of Lord Selsdon's Television Advisory Committee, the RMA decided to establish a committee comprising representatives of all pertinent companies and organisations to draft acceptable standards. Both RMA and non RMA members would be invited to participate in the discussions. And so the National Television Standards Committee (NTSC), sponsored by the RMA and supported by the FCC, came into being.

The following companies were each requested to send one representative to the NTSC[46]: Bell Telephone Laboratories, Columbia Broadcasting System, Don Lee Broadcasting System, DuMont Labs Inc, Farnsworth Television and Radio Corporation, General Electric Company, Hazeltine Service Corporation, John V L Hogan, Hughes Tool Company, IRE, Philco Corporation, Radio Corporation of America, Stromberg Carlson Telephone Manufacturing Company, Television Productions and the Zenith Radio Corporation. Eventually, the NTSC had a membership of 168 divided into nine panels each responsible for investigating a particular aspect of standardisation. The nine panels were:

1   systems analysis, chaired by Dr P C Goldmark of CBS;
2   subjective aspects, chaired by Dr A N Goldsmith of the IRE;
3   television spectra, chaired by J E Brown of Zenith Radio Corporation;
4   transmitter power, chaired by Dr E W Engstrom of RCA;
5   transmitter characteristics, chaired by P T Farnsworth of Farnsworth Television and Radio Corporation;
6   transmitter-receiver coordination, chaired by I J Kaer of GE;
7   picture resolution, chaired by D E Harnett of Hazeltine Service Corporation;
8   synchronisation, chaired by T T Goldsmith of DuMont;
9   radiation polarisation, chaired by D B Smith of Philco.

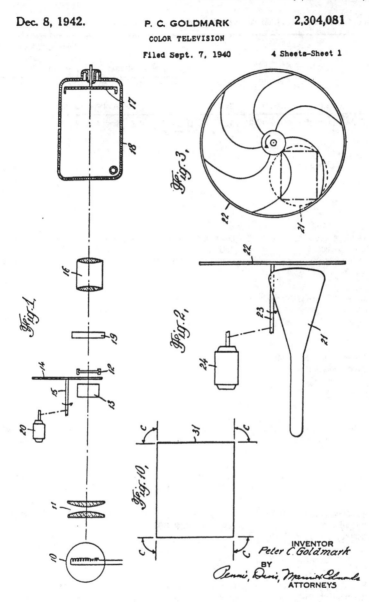

Dec. 8, 1942.　　　　P. C. GOLDMARK　　　　2,304,081

COLOR TELEVISION

Filed Sept. 7, 1940　　　4 Sheets-Sheet 1

INVENTOR
Peter C. Goldmark
BY
ATTORNEYS

Figure 22.2　CBS's colour television system was invented by Dr P C Goldmark. Figs. 1 & 2 of the patent show the layout of the apparatuses at the transmitter and receiver respectively. 14 and 22 are rotating colour filter discs—each containing red, green and blue filters. The transparency 12 is frame scanned by red, green and blue light sequentially. At the receiver the three black and white c.r.t. images corresponding to the transmitted R, G and B image signals are viewed through R, G, and B filters sequentially. The eye-brain combination fuses the three separate images to form a coloured image

Source: *Proc. of the IRE*, **30**, April 1942, pp. 162–182

Table 22.3   Analysis of American and foreign television systems [54]

Number	Designation	Period of operation	When demon-strated	Lines frames fields	As-pect ratio	Type of motion	Per cent of carrier	Description of waveform
				Scanning pattern			Synchronisation system (H = horizontal, V = vertical)	
1	RMA standards	1939–1940	1939–1940	441/ 30–60	4:3	linear	20–25	H: single rect-angular pulse; V: serrated with equalising pulses (see RMA standard M9–211
2	DuMont A (500 kc burst vertical synchronising pulse)	1939–1940	December, 1939 to date	variable (see note A)	4:3	linear	20–25	H: rectangular pulse; V: r–f burst during V blank with H superposed
3	DuMont B (Trans-mission of scan-ning waveforms).	1938–1939	1939	variable	4:3	any (note B)		synchronising inherent in transmission of scanning waveforms
4	Hazeltine (FM for synchronising)	1940	November, 1940	441/ 30–60	4:3	linear	100	any synchron-ising wave shape frequency modulated during H and V blank
5	RCA (507 lines, 495 lines later suggested)	1940	July, 1940	507/ 30–60 495/ 30–60	4:3	linear	20–25	waveform same as RMA
6	RCA (FM for sound)	1940	November, 1940	441/ 30–60	4:3	linear	20–25	RMA M9–211
7	RCA (FM sound quasi-FM for picture)	1940	October, 1940	441/ 30–60	4:3	linear		RMA M9–211
8	RCA (long integ-ration synchron-ising pulse)	1940	October, 1940	441/ 30–60	4:3	linear	20–25	slots in vertical pulse at line freq uency; V pulse = 9 H
9	Philco (525 lines)	1938–1940	1938–1940	525/ 30–60	4:3	linear	over 25	waveform same as RMA M9–211
10	Philco (605 lines)	1938–1940)	1938–1940	605/ 24–48	4:3	linear	over 25	waveform same as RMA M9–211
11	Philco (narrow vertical synchron-ising pulse)	1935–1938	up to 1938	441/ 30–60	4:3	linear	20–25	H: rectangular pulses; V: single narrow rectang-ular pulse, same level

Table 22.3 continued

	Transmitter characteristics						Other characteristics, apparatus employed, advantages claimed by sponsor, etc
Polarity of modulation	Carrier attenuation	D.c. or a.-c. transmission	direction of polarisation	total bandwidth	picture sideband width	carrier separation	
negative	none	d.c.	horizontal	6 Mc	4–4.5 Mc	4.5 Mc	sound pre-emphasised sound and picture carriers of equal power; standards used sometime by RCA, GE, Farnsworth, Philco, CBS, Don Lee, General Television Corporation, Zenith and others
negative	none	d.c.	not specified	6 Mc	4–4.5 Mc	4.5 Mc	note A: designed for flexibility (continuous variability) in line and frame rates, including 15–30 as lower limit
negative (note B)	none (note B)	d.c. (note B)	not specified	6 Mc	4–4.5 Mc	4.5 Mc	note B: receiver automatically follows changes in scanning motion
negative	none	d.c.	not specified	6 Mc	4–4.5 Mc	4.5 Mc	FM modulator required for synchronising; high syncronising amplitude developed; improves picture modulation capability
negative	none	d.c.	horizontal	6 Mc	4–4.5 Mc	4.5 Mc	
negative	none	d.c.	horizontal	6 Mc	4–4.5 Mc	4.5 Mc	FM modulator for sound only; 75 kc maximum deviation
	none	d.c.	horizontal	6 Mc	4–4.5 Mc	4.5 Mc	75 kc maximum deviation for sound; 0.75 Mc maximum deviation for picture; improved transient response reported
negative	none	d.c.	horizontal	6 Mc	4–4.5 Mc	4.5 Mc	V pulse integrated in R–C circuit of long time constant; no equalising pulses used
negative	none	d.c.	vertical	6 Mc	4–4.5 Mc	4.5 Mc	sound carriers staggered
negative	none	d.c.	vertical	6 Mc	4–4.5 Mc	4.5 Mc	greater horizontal definition due to lower frame rate
negative	none	d.c.	vertical	6 Mc	4–4.5 Mc	4.5 Mc	

Table 22.3   continued

12	Farnsworth (Narrow vertical synchronising pulse)	1935–1938	up to 1938	441/30–60	4:3	linear	20–25	H: rectangular pulse; V: single narrow pulse of higher amplitude
13	GE (picture carrier 6db attenuated	1938	1938	441/30–60	4:3	linear	20–25	same as RMA M9–211
14	Kolorama (225 lines)	1935–1940	May, 1939	225/12–24	6:5	linear		by transmission pulse and power line
15	Sound during blanking single carrier system	prior 1940		note C	note C	note C	note C	note C; carrier frequency modulated during horizontal blanking interval with audio frequencies
								American colour
16	CBS (3-colour system no. 3)	1940	August 1940	343/60–120	4:3	linear	note D	adaptable to any synchronising signal that can control filter disks
16a	CBS combination color and black and white	this system same as CBS no. 3, for colour transmissions; For black and white						
17	CBS (3-colour system no. 4)	1940		430/45–180	4:3	linear	note D	note D; quadruple interlace
18	CBS (3-color system no. 5)	1940		550/30–120	4:3	linear	note D	note D; quadruple interlace
19	GE (2 colour system)	1940	November 1940	441/30–60	4:3	linear	20–25	RMA M9–211
								Foreign
20	British (BBC) standard	1936–1939	1936	405/25–50	5:4	linear	30	rectangular H pulses; serrated V pulse; no equalising pulses
21	Scophony (British) (In US also, in 1940)		1937	405/25–50	5:4	linear	30	BBC standard
22	early Baird (British)	1936–1937	1936	240/25	4:3	linear	40	8 per cent line synchronising pulses, single V pulse of 12 lines duration
23	Baird 2-colour (British)		March 1938	120/16.6–100	3:4	linear vertical	40	6 to 1 interlace by nonperiodic pulses

Table 22.3   continued

negative	none	d.c.	not specified	6 Mc	4–4.5 Mc	4.5 Mc	
negative	6 db	d.c.	horizontal	6 Mc	4–4.5 Mc	4.5 Mc	narrower band at transmitter, wider band at receiver
positive	none	d.c.	vertical	100 kc/300 kc	300 kc		mechanical scanners. Transmitter on band at 2000 kc
negative	note C	d.c.	note C	6 Mc or less	4–4.5 Mc		note C: not specified audio frequencies substantially higher than one half line frequency suffer distortion; investigated by RCA, GE, Philco; suggested 1940 by Kallman

television systems

note D	note D	note D	note D	6 Mc	4–4.5 Mc	4.5 Mc	note D: not specified; mechanical filter disks or drums at transmitter and receiver

transmissions, no preference as to specifications is indicated

note D	note D	note D	note D	6 Mc	4–4.5 Mc	4.5 Mc	line scan frequency approximately 19 400 p.p.s
note D	note D	note D	note D	6 Mc	4–4.5 Mc	4.5 Mc	line scan frequency 15 750 p.p.s; Colour coincidences 10 p.s
negative	none	d.c.	horizontal	6 Mc	4–4.5 Mc	4.5 Mc	dichromatic filter disks, synchronising on power line; odd lines always one colour, even lines always other color

systems

positive	none	d.c.	vertical	6 Mc	2.5 Mc	3.5 Mc	double sideband; 2cps/sec tolerance in line frequency for mechanical receivers
positive	none	d.c.	vertical	6 Mc	2.5 Mc	3.5 Mc	mechanical scanners, supersonic light valve; requires BBC tolerance on line synchronising
positive	none	d.c.	vertical	5 Mc	1.5 Mc	3.5 Mc	mechanical scanner at transmitter; live pick up by film only
positive	none	d.c.	vertical	5 Mc	2 Mc	3 Mc	flying spot pick up with rotating colour disks; projected large-screen image

Table 22.3   continued

24	Baird 3-colour (British)	January 1938	120/ 16.6– 100	3:4	linear vertical	40	same as Baird 2-colour system
25	velocity modulation (Br.)	May 1934	60–400 /25	4:3		30	10 per cent line pulses; frame by ratchet circuit from line
26	French (PTT) standards	June 1937	440– 445/ 25–50	5:4	linear	30	
27	Barthelemy (French)	June 1937	450/ 25–50	5:4	linear	17 H 34 V	6 lines paired to form V pulse
28	German standards	1939–1940	441/ 25–50	5:4	linear	30	10 per cent line pulses; single 35 per cent V
29	German proposal	1937–1938	441/ 25–50	5:4	linear	30	burst of 1.1 Mc for 20 per cent line duration as V pulse; square H
30	Italian standards	1939	441/ 21–42	5:4	linear	30	same as German standards
31	Russian standards	1940	441/ 25–50	11.8	linear	20–25	essentially RMA M9–211

The appointment of Dr Baker as the first chairman of the NTSC was announced on 17 July 1940 and the first meeting was held on 31 July 1940. By operating according to parliamentary procedures it was anticipated that an industry-wide technical consensus could be achieved. The panels began their work in September.

Of the issues which divided the industry there were three which were of major importance and required firm resolution:

1   the conflict between fixed and flexible standards;
2   the method of synchronisation;
3   the line standard if a fixed standard were decreed.

These matters had to be clearly delineated by the NTSC. However, just before the nine panels commenced their work, CBS stated, on 29 August 1940[47], that it had perfected a system of colour television that would be ready by 1 January 1941. This was a new factor in the television debate and was one which possibly could challenge all previous notions on television standards.

A private test of CBS's system was given on 28 August to the FCC's chairman, J L Fly. He seems to have been impressed and stated 'that if we can start television off as a colour proposition, instead of a black and white show, it will have a greater acceptance with the public'. CBS claimed that colour television was capable of being accommodated in a 6 MHz channel and that 'existing receivers need not suffer radical changes to adapt them to three colours instead of mere black and white'.

An impressive demonstration of colour television using colour film and slides was mounted on 4 September 1940 by Dr P Goldmark of CBS. A 16 mm film was run

Table 22.3   continued

positive	none	d.c.	vertical	5 Mc	2 Mc	3 Mc	same but three colours laboratory demonstration
positive	none	a.c.	vertical	3 Mc	1.5 Mc	no sound	c.r.t light source on film Amplitude component added
positive		d.c.				4 Mc	standards embrace four systems
positive						4 Mc	Barthélemy interlace; rotating disk synchronising
positive	none	d.c.	vertical	5 Mc	2.0 Mc	2.8 Mc	picture carrier on high side of channel
positive	none	d.c.	vertical	5 Mc	2 Mc	2.8 Mc	
positive	none	d.c.	vertical	5 Mc	2 Mc	2.8 Mc	
negative	none	d.c.	horizontal	6 Mc	4–4.5 Mc	4.5 Mc	sound carrier lower in frequency than picture

at 60 pictures per second, using a continuous motion projector, past a standard Farnsworth image dissector. A six-segment colour filter disc, of 7.5 inch diameter, containing two sets of red, green and blue filters, placed before the image dissector tube, enabled the primary colour components of the film images to be televised. The disc was driven at 1200 r.p.m. (i.e. six filters, RGB RGB, were placed in front of the tube every 1/20 s). Thus, the frame rate was 120 frames per second per primary colour in contrast to the 60 frames per second rate which was currently being discussed by the NTSC.

At the receiver another colour filter disc was used with a 9 inch monchrome (white light) cathode-ray picture tube.

Essentially, Goldmark's system was a frame sequential colour system and was based on work which had been carried out earlier by J L Baird. The number of lines employed was 343, and not 441, to allow use to be made of the higher frame rate. Consequently, the resolution of the colour images was correspondingly poorer. 'Disturbing flicker was evident in bright saturated colours and colour break up appeared at sharp edges. Otherwise it was a convincing demonstration,' noted one observer.

The FCC representatives were quite excited by the test although they were concerned that the method could only work with film. But soon Goldmark[48] was able to adopt a new camera tube, the orthicon, to his colour system, thereby providing a direct pick up capability. This was demonstrated on 12 November 1940.

The introduction of CBS's colour system did not please everyone. RCA, GE, Farnsworth, Philco and DuMont had each invested heavily in all-electronic monochrome television and were highly critical of the reintroduction of mechanical devices into television and the lowering of picture definition. Zenith and Stromberg-Carlson, on the other hand, were just entering the television field and defended the utilisation of colour. CBS itself seemed to be in no hurry to commence television broadcasting. It had bought an experimental television transmitter[49.50,51] in 1938 but had never really pressed its television broadcasting facilities. The CBS radio system was very prosperous and, since CBS was not a manufacturing organisation, unlike RCA, and had no aspiration in manufacturing, it could afford to argue for a delay in the start of commercial television until it had perfected its colour system.

Prior to the arrival of colour television on the television scene the NTSC had had to consider the issue of either fixed standards as propounded by RCA and others or continuously flexible standards as advanced by DuMont. Now there was another possibility, discontinuous flexible standards as advocated by CBS. In this case receivers would be required to receive two different standards, one for monochrome television and one for colour television. Table 22.3 lists the various systems proposed.

On 27 January 1941 the NTSC submitted its report[52] on television standards to the FCC. From the initial meeting on 31 July 1940 to the final meeting in March 1941 the NTSC's 168 committee and panel members had devoted 4000 manhours to meetings, had witnessed 25 demonstrations, on technical matters, and had compiled reports and minutes totalling 600 000 words. The NTSC recommended the following television standards:

1   television channel: 6 MHz wide, with the picture carrier 4.5 MHz lower in frequency than the sound carrier;
2   number of lines: 441;
3   picture frequency: 30 pictures per second with 60 frames per second;
4   aspect ratio: 4 (horizontal): 3 (vertical);
5   vision modulation: amplitude modulation for both picture and synchronising signals, with a decrease in light intensity causing an increase in radiated power;
6   black level: to be represented by a definite carrier level, at 75 % of the peak carrier amplitude;
7   sound modulation: f.m. with a pre-emphasis of 100 μs and a maximum deviation of ± 75 kHz;
8   polarisation: horizontal;
9   synchronising signals: specified.

At its final meeting on 8 March 1941, the NTSC revised its recommendation on the number of lines from 441 to 525 and it recommended the alternative use of f.m. or a.m. for the synchronising signals.

Between the 20 March and 24 March, the FCC conducted hearings on the NTSC report. It was clear that a general industry unity had been accomplished on television standards; the only significant dissenting voice was that of DuMont Laboratories which continued to press its case for flexible standards.

The new rules[53] of the FCC specified that each commercial television station was to broadcast a regular programme schedule for a minimum of 15 hours per week, of which two hours had to be between 2.00 and 11.00 p.m. daily, except Sunday, with at least a one hour programme transmitted on five weekdays between 7.30 and 10.30 p.m. Of the various channels available for television, numbers one to seven inclusive were each permitted to provide a sponsored service, although channels eight to eighteen inclusive were restricted to experimental and television relay development and applications. Above 300 MHz further channels could be utilised for relay and experimental purposes.

And so after approximately five years of rancour and frustration in the industry rules and standards were on hand for commercial television broadcasting. But then, on 27 May 1941, less than one month after the FCC had issued its new rules, President Franklin D Roosevelt declared a state of unlimited national emergency. No raw materials or production capacity could be directed to television. Indeed, RCA's television receiver manufacturing line had ceased to operate for more than a year, and as with other national companies its production capacity was engaged in work for national defence. Nevertheless, notwithstanding these constraints, the official opening of the American television service on a full commercial basis was held on 1 July 1941. Sponsored programmes were now permitted but the only station with paid programmes was NBC. Its inaugural broadcasts were supported by the Bulova Watch company, Sun Oil, Lever Bros and Proctor and Gamble, and included a news broadcast, a television show and a quiz show. Very few saw the television transmissions; one source mentions a figure of c. 5000 sets in the New York area.

For the remainder of 1941 television broadcasting was carried on by only seven of the 22 licensed stations across the US. The seven were: NBC/RCA (W2XBS, New York City), CBS (W2XAX, New York City), DuMont (W2XWV, New York City), Don Lee (W6XAO, Los Angeles), Zenith (W9XZV, Chicago), Philco (W3XE, Philadelphia) and General Electric (W2XB, Schenectady).

The bombing of Pearl Harbour by the Japanese on 7 December 1941 finally led to a cessation of domestic television progress and expansion for the war years. After the entry of the United States into the war, a government order dated 24 February 1942 ruled out the construction of television equipment for civilian use. At the same time the minimum number of hours of transmission required of a licensed station was reduced from 15 to four hours per week.

References

1 Statement of D Sarnoff, president of RCA, delivered at annual meeting of RCA shareholders, New York City, 7 May 1935, post 33/5143, Post Office Records Office, London, UK
2 ANON.: Feature article, *New York Times*, 12 May 1935, **X**, p. 11:1
3 ANON.: 'Television progress in USA', *Television and Short-wave World*, August 1935, pp. 436, 439
4 ANON.: 'Television I: a $13 000 000 "if"', *Fortune*, May 1939, **19**, pp. 52–59, 168, 172, 174, 176, 178, 180, 182
5 ANON.: 'Television II: "Fade in camera one!"', *Fortune*, 1939, Vol. 19, pp. 69–74, 154, 157, 158, 160, 162, 164
6 ANON.: 'RCA describes television system', *Electronics*, January 1937, pp. 8–11, 48
7 ANON.; 'The RCA television system', *Television and Short-wave World*, March 1937, pp. 136–139

8   ANON.: 'Television today—the status quo', *Electronics*, June 1936, pp. 27–30, 53, 54
9   ANON.: 'Television stages first real show', *New York Times*, 8 July 1936, p. 21
10   BURNS, R.W.: 'British television, the formative years' (Peter Peregrinus, London, 1986)
11   FINK, D.G.: 'Television broadcasting practice in America—1927 to 1944', *J. IEE*, September 1945, **92**, pp. 145–164
12   Ref.11, p. 147
13   FINK, D.G.: 'Perspectives on television: the role played by the two NTSCs in preparing television service for the American public', *Proc. IEEE*, September 1976, **64**, (9), pp. 1322–1331
14   Ref.13, p. 1325
15   Ref.13, p. 1326
16   ANON.: 'Philco (USA) television. Results of recent tests on 441-lines', *Television and Short-wave World*, April 1937, p. 206
17   ANON.: 'Philco shows 441-line television', *Electronics*, March 1937, pp. 8–9
18   LEWIS, H.M., and LOUGHREN, A.V.: 'Television in Great Britain', *Electronics*, October 1937, pp. 32–35, 60, 62
19   'Report of the Fall Meeting, Rochester 1937', *Electronics*, December 1937, pp. 11–15, 67–71
20   ANON.: 'Reviewing the video art', *Electronics*, January 1938, pp. 9–11.
21   MURRAY, A.F.: 'RMA completes television standards', *Electronics*, July 1938, pp. 28–29, 55
22   Reports 6, 867, Federal Communications Commission, Washington, US
23   SARNOFF, D.: 'Statement on television', *New York Times*, 3 January 1939, p. 31:3
24   ANON.: 'Communicating Systems, Inc. introduces non-sound picture sets for $125', *New York Times*, 12 May 1938, p. 25:3
25   ANON.: 'Communicating Systems, Inc. plans promotional campaign', *New York Times*, 22 May 1938, **III**, p. 8:6
26   ANON.: 'Communicating Systems, Inc. and A B DuMont Laboratories, Inc. sets to be demonstrated in New York City department stores', *New York Times*, 2 July 1938, p. 39:3
27   ANON.: 'Television images of life-size shown', *New York Times*, 15 February 1939, p. 14:4
28   ANON.: 'Farnsworth plans to raise new capital', *New York Times*, 10 February 1939, p. 36:5
29   ANON.: 'FCC stops the gold rush', *New York Times*, 9 April 1939, section X, p. 12:3
30   ANON.: 'A B DuMont demonstrates new receiving apparatus which would lower costs and permit narrower broadcasting bands', *New York Times*, 6 February 1938, **II**, p. 7:5; 13 February 1938, **X**, p. 12:1
31   ANON.: 'Television without sync signals', *Electronics*, March 1938, pp. 33–34, 68
32   Ref.11, p. 147
33   ANON.: 'America makes a start', *Television and Short-wave World*, June 1939, p. 330
34   ANON.: 'Television for a nation. RCA build 20 transmitters', *Television and Short-wave World*, May 1938, pp. 263–267
35   ANON.: 'Ceremony is carried by television as industry makes its formal bow', *New York Times*, 1 May 1939, p. 8:3
36   COCK, G.: 'My impressions of American television', *Television and Short-wave World*, 1939, **12**, pp. 453–455
37   ESPENSCHIED, L.: 'Television home reception'. Memorandum, 29 September 1939, case 37014, AT&T Archives, Warren
38   ANON.: 'Public verdict awaited', *New York Times*, 28 May 1939, **X**, p. 10:2
39   ANON.: 'Televiews on the air', *New York Times*, 13 October 1939, **IX**, p. 12:1
40   ANON.: 'What's television doing now?', *Business Week*, 12 August 1939, p. 24
41   ANON.: 'Price cut sharply on television sets', *New York Times*, 13 March 1940, p. 34:1
42   KERSTA, N.E.: 'The business side of television', *Electronics*, March 1940, pp. 10–13, 90
43   ANON.; 'FCC moves to widen use of television', *New York Times*, 1 March 1940, p. 13:1
44   ANON.: 'A statement by the RCA', *New York Times*, 20 March 1940, pp. 20–21
45   ANON.: 'FCC suspends limited commercial operation', *New York Times*, 24 March 1940, p. 1:6. Also, 'Television'. Reports of The National Association of Broadcasters, **8**, (15), 12 April 1940, pp. 4163–4168
46   ANON.: 'Television Committee organises', *Electronics*, August 1940, p. 34
47   GOLDMARK, P.C., DYER, J.N., PIORE, E.R., and HOLLYWOOD, J.M.: 'Colour television, part I, *Proc. IRE*, **30**, April 1942, pp. 162–182

48 GOLDMARK, P.C.: 'The maverick inventor' (Saturday Review Press, New York, 1973)
49 ANON.: 'How soon television', *Radio News*, **19**, (6), December 1937, pp.327–328, 361
50 ANON.: 'Columbia's television transmitter', *Television and Short-wave World*, April 1938, p. 197
51 ANON.: 'The television transmitter of the Columbia Broadcasting System', *Television and Short-wave World*, January 1939, pp. 19–21
52 ANON.: 'NTSC proposes television standards', *Electronics*, February 1941, pp. 17–21, 60, 62, 64, 66
53 ANON.: 'Groundwork laid for commercial television', *Electronics*, April 1941, pp. 18–19, 70
54 FINK, D.G.: 'Television standards and practice' (McGraw Hill, New York, 1943)

Chapter 23

The world's first, regular, public, high-definition service, (1936–1939)

Following the inauguration of the London television station at Alexandra Palace and the adoption of the 405-line standard, there were many problems which had to be resolved by the British Broadcasting Corporation (BBC) and the Television Advisory Committee (TAC) before a national (UK) service having a wide popular appeal could be established. Much innovative effort had to be expended by the BBC's producers to determine the type and the form of programmes which would prove attractive to viewers; means had to be devised to enable major outdoor sporting and national events to be televised and the signals relayed to Alexandra Palace; steps had to be taken to stimulate the purchase of sets which were likely to cost several weeks' wages for the average person; the effect of ignition and diathermy interference had to be determined and possible legislation introduced to minimise the detrimental character of spurious radiations; the question of whether to utilise radio links or cable links for transmitting, to the provinces, the video signals generated at the London station had to be most carefully considered since large capital costs would be involved; and the length of time the 405-line standard remained in force before a move to a higher line standard was introduced had to be considered. All these matters fell within the remit of the TAC.

From the opening of the London station in November 1936 an extensive variety of programmes was produced, both in the studios on the premises of the Alexandra Palace, and in the surrounding park. The studio programmes included extracts from West End productions, revues, variety, ballet and illustrated talks and demonstrations, as well as a weekly magazine programme of topical interest called 'Picture Page'. From outside the studio came demonstrations of golf, riding, boxing and other sports. Because of the limited facilities for rehearsal, it was seldom possible, in the early days, to attempt original productions in the studio, but valuable experience was gained in methods of presentation over a very wide field. Transmissions were limited to two hours per day, excluding Sundays.

Some of the programmes which were televised during November and December 1936[1] were:

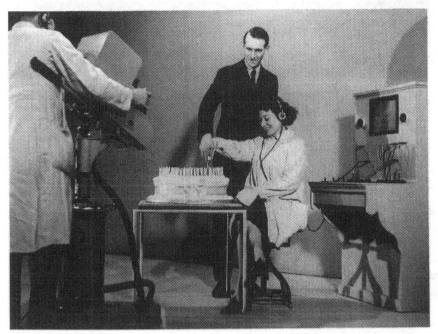

Figure 23.1 *'Picture page'—a weekly magazine programme—celebrating its 100th edition. The producer is Cecil Madden and the announcer is Joan Miller*

Reproduced by permission of the BBC

Studio productions:

'Citizen soldiers of London', a pageant reconstructed from the Lord Mayor's show
'Mr Pickwick', extracts from a new opera by Albert Coates
'The tiger', scenes from the play by Reginald Berkeley
'Murder in the cathedral', scenes from the play by T S Elliot
'Facade', a ballet performed by the Vic-Wells Company
Traditional plays and songs by the children's Theatre Company

Variety:

'Cabaret cartoon', a vaudeville presentation with simultaneous drawings by Harry Rutherford
'Animals all', a programme of animal impersonators from the pantomimes
'Old time music hall', by veterans of variety

Talks:

'Inn signs through the ages', by Montague Weekly
'The modern house', a discussion between John Gloag and Serge Chermayeff
'Ships', a Royal Institution lecture, by Professor G I Taylor
'The pattern of 1936', a review of trade, finance etc., by Professor John Hilton

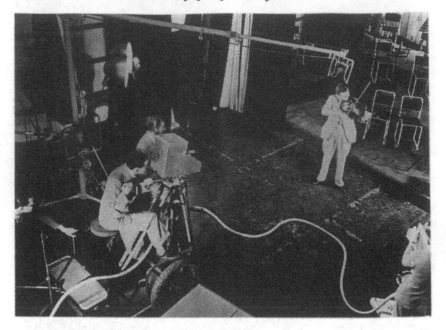

Figure 23.2   Studio at Alexandra Palace showing two emitron cameras in use

Reproduced by permission of the BBC

Outside broadcasts and topical programmes:

A tour of the North London exhibition, at the Alexandra Palace, 'Model theatres',
from the British model theatre and puppet club exhibition
Armistice day programme
Anti-aircraft defence display by the 36th Middlesex anti-aircraft battalion, RE (TA),
and the 61st (11th London) anti-aircraft brigade, RA (TA)
'Diary for 1936', a reconstruction of outstanding items in television programmes

Serial features:

'Picture page', a weekly magazine programme to which the contributors had
included Jim Mollison, Roger Quilter, David Low, and Will Hay
'Starlight', including Bebe Daniels and Ben Lyon, Manuela del Rio, Lisa Minghetti,
Lou Holtz, Sophie Tucker, Noni, Frances Day and George Robey
'London characters'
'Friends from the zoo', animals from the London zoo, introduced by David Seth-Smith

Several of the programmes were considered to be disappointing and the television
correspondent of the *Daily Telegraph* wrote on 24 December[2]:

> 'A visit to a local cinema brings into sharp relief the pathetic triviality of
> the BBC programmes. Television, with its champion cockerels and prize
> fishes seems to offer no comparison as entertainment.'

Specific programmes which the correspondent commented upon included a 15-minute programme showing the stages in the construction of radio transmitter valves, which was given during an afternoon and then repeated in the evening; another programme devoted 15 minutes to a description of the mobile Post Office by an official who retailed a catalogue of 'dreary technicalities'.

Films, apart from newsreels, were all of the semi-educational documentary type made for a full-sized cinema screen, wrote the reporter, and some, for example, 'The land of the Nile' and 'Fisherman's fortune' were each shown on four separate occasions.

Art programmes, too, did not escape his criticism. A number of televised examples of pottery, art and sculpture depended for their appeal on exquisite colouring and were 'meaningless in the black and white of television; others were of interest only to the connoisseur and the dilettante'.

Moreover, vaudeville had not been strongly represented; but champion birds, beasts and fishes were being brought to the studio.

On the credit side 'Picture Page' and the television magazine were felt to be promising and successful features, and 'Starlight', which brought stars to the studio, was another, even though there seemed to be a shortage of front rank artists willing to broadcast for BBC fees.

The *Evening News*[3] on 10 December 1936 mentioned that there was still a need for slicker timing of the television programmes to avoid frequent intervals, and the lighting and make-up problems had not been entirely solved although considerable progress had been made.

Mr J H Thomas, the general manager of A C Cossor Ltd, was reported[4] on 21 December to have said that the television programmes were 'footling; and that unless speedy improvements were effected the programme material would 'not give television a chance'.

Much of this criticism arose because of the limited facilities and money allocated to the service. At Alexandra Palace the staff had to produce shows without a rehearsal studio, and sometimes a performance had to be given with only an hour's rehearsal in the actual transmission studio. Producers, assistant producers and a larger technical staff were required: the BBC was prepared to appoint them but was faced with the problem of finding the money. Gerald Cock, the director of television, was undoubtedly handicapped. Of the £180 000 granted by the BBC and government for television £110 000 had been spent on the station and another £20 000 was being allocated to the purchase of a mobile unit—only £50 000 remained for programmes and maintenance. The cost[3] of the service was being borne by the corporation and until the television service was firmly established there seemed little likelihood of a separate licence fee being approved.

The programmes offered to British viewers in 1937 and subsequently could be classified broadly as studio-produced programmes, televised films and outside broadcasts. Of these the most newsworthy, and the most controversial in regard to their relationships with other interests, were the outside broadcasts (OBs). Ashbridge had had OBs in mind for some time and in December 1935 reported[5] that he had put forward enquiries to both Baird Television and EMI regarding the supply of portable apparatus for this purpose. Possibly he remembered the success which Baird had achieved in 1931 and 1932 with the televising of the Derby, using low-definition equipment. Now there was the prospect of high-definition outside broadcasts with the portable and versatile emitron camera.

Marconi-EMI was unable to undertake the necessary development work at that time because of its heavy involvement with the installation of the Alexandra Palace equipment, but in June 1936 Ashbridge was able to inform the TAC that the company had quoted a figure 'of the order of £14 000' for the supply of an outside broadcast unit[6]. This would comprise three emitron cameras with associated

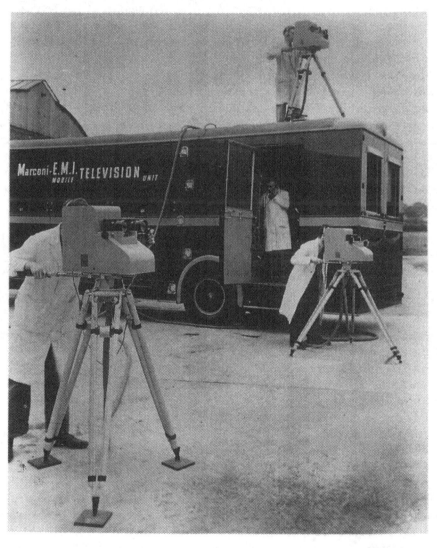

Figure 23.3   One of the outside broadcast mobile television units with three emitron cameras

Reproduced by permission of the Marconi Company Ltd

Figure 23.4   *Interior view of the Marconi EMI outside broadcast television scanning van*

Reproduced by permission of the Marconi Company Ltd

scanning apparatus and a four ton lorry, but excluded sound and power equipment. The quotation and specification mentioned that the cameras would be capable of operating at a maximum distance of 1000 ft from the lorry. On the basis of this information the corporation was given approval to enter into negotiation with the Marconi-EMI company for the acquisition of an OB unit. Later in December the

Figure 23.5   Interior view of the Marconi-EMI mobile camera control unit

Reproduced by permission of Thorn EMI Central Research Laboratories

company tendered a quotation for the supply of apparatus to provide an ultra short wave link between the OB unit and Alexandra Palace, but the price was regarded by Ashbridge as very high. A reduction seems to have been successfully arranged.

Figure 23.6   Photograph of the Marconi-EMI television transmitter and modulator van with mobile antenna (June 1937)

Reproduced by permission of the Marconi Company Ltd

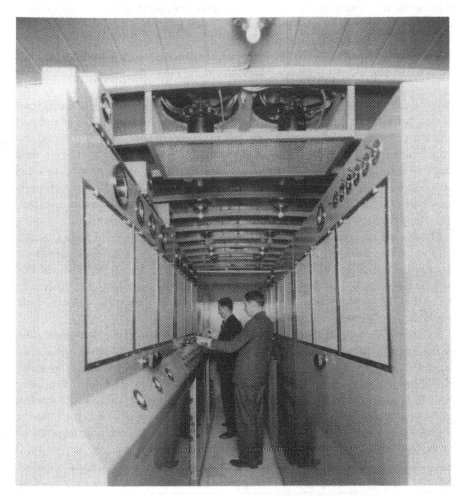

Figure 23.7   Interior view of the transmitter and modulator van (April 1937)

Reproduced by permission of the Marconi Company Ltd

The mobile television unit actually comprised[7] three large vans each about the size of a large coach: one contained the control apparatus and scanning equipment, one housed the power plant and the third was fitted with the VHF radio link transmitter (1 kW). Marconi-EMI's system was entirely self contained and most comprehensive in its specification. The control room unit held two rows of equipment mounted along the sides of the vehicle so as to leave a clear aisle for the engineers operating the controls. Each row comprised six racks each $7\frac{1}{2}$ feet high and $19\frac{1}{2}$ inches wide: the total weight of the vehicle and apparatus was approximately $8\frac{1}{4}$ tons. The engineers were able to view the televised scenes on a monitor fitted into a compartment over the driver's head and if necessary make adjustments by means of controls mounted on the front panels of the equipment racks. In addition to the

vision units, the vehicle also included sufficient faders and amplifiers to complement
the four microphones which were used to pick up local sounds associated with the
scene being televised and the commentary.

The most outstanding broadcast television transmission during 1937 was the televising of the Coronation procession at Apsley Gate, Hyde Park Corner, London, on
its return journey to Buckingham Palace from Westminster Abbey. Three emitron
cameras were used to capture the procession: two were mounted on a special plat-
form at Apsley Gate and were fitted with telephoto lenses to obtain distant and mid-
field shots of the procession and the crowds to the north and south of the gate, and
a third camera was installed on the pavement to the north of the gate to give close-
range views of the royal coach and other important parts of the procession which
passed through the gate. The cameras were connected by about 50 yards of special
cable to the mobile television unit behind the park keeper's lodge, whence the sound
and vision signals were propagated to the London television station.

The outside broadcasts of the Coronation procession represented an outstanding
technical achievement by the Marconi-EMI Television Company. Great credit was
due to Shoenberg and his research team for making the broadcast possible. Well
over 50 000 people saw the scenes televised on Coronation Day and so became
aware that 'television [was] a force to be reckoned with for the provision of entertainment and education in the home'[8].

Two means existed for sending television signals from one place to another, either
by the use of a radio link or by the use of a cable. The first of these options was
dependent upon the allocation of a suitable frequency band. Fortunately this matter
had been considered by the Wireless Telegraphy Board at an informal meeting[9]
held on 13 July 1934. Angwin had represented the Television Committee's interests
on the board as well as those of the Post Office and had proposed that the following
subdivision of the waveband below ten metres should form a working basis during
the preliminary development of the VHF services:

Frequency range (MHz)	Number of channels	Type of service
30– 32	4	television experiments and mobile
32– 35	5	mobile
35– 38	5	fixed and mobile
38– 48	5	television
48– 50	20	broadcasting
50– 56	8	fixed
56– 60	–	amateurs and experimental
60– 85	20	fixed
85– 88	20	broadcasting
88–100	5	television
100–120	12	mobile
120–130	–	amateurs and experimental
130–150	–	fixed
150–300	–	unreserved

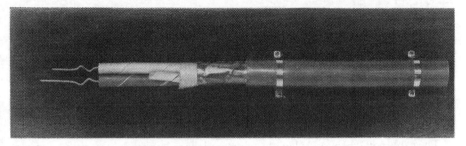

Figure 23.8   Twin-wire balanced television cable

Reproduced by permission of the Science Museum

No recommendations for a subdivision below 1 m were made as there was 'very little data available regarding spacing of channels', although several manufacturers in the UK had produced equipment operating on 17 cm, 26 cm and 65 cm.

Provisionally it was agreed that television broadcasting should be confined to the 38–48 MHz and 88–100 MHz bands. The service could not have unlimited access to the 1 to 10 m band as certain important bands had to be reserved for aircraft services. The civil aviation industry was expanding rapidly and wireless facilities for navigation and communication were essential for its development. In the London area, for example, about 20 aerodromes had to be provided with communication equipment and each one needed at least one separate frequency for the guidance of aircraft in bad weather. Other large towns having airports required similar frequencies. The safety of aircraft was dependent on the efficient operation of a number of very high-frequency beacons which demanded a considerable number of independent channels; no other safety means were as suitable as these.

Confirmation of the allocation of the 38–48 MHz band for television was given by Angwin on 2 October 1935. Later, the band was changed to 40.5–52.5 MHz so as to clear the 30.83–40.5 MHz band which the ICAN had put forward, at a conference held in Paris, for use by the blind landing system[10]. Subsequently the WTB agreed to the BBC's radio link working in the 64–65 ± 2.5 MHz band.

Meanwhile, prior to the opening of the London Television station, EMI had been considering how the signals from the OB cameras could be transmitted to Alexandra Palace. Characteristically Shoenberg and his brilliant research team had realised that OBs would take place, that radio link transmissions might be difficult, that a cable was probably the best solution and that the Post Office had no experience of transmitting television signals[11]. With typical foresight he had initiated a research project to evolve a cable and associated terminal equipment which would send video signals from various locations in London, such as Westminster Abbey, Buckingham Palace, the Houses of Parliament, the Cenotaph and the theatres, to Alexandra Palace. The estimated cost of such a cable, suitable for carrying video signals having frequencies up to 1.5 MHz was £300 per mile: in addition there was the cost of laying the cable in a duct. The total expenditure was thought to be about £900 to £1000 per mile for a cable capable of handling signals up to 2 MHz.

The cable[12], designed by Blumlein and Cork of EMI, was of the twin or balanced pair type, and comprised two 0.08 inch diameter conductors each located centrally within a tube of paper insulation having an external diameter of 0.91 inch. The tubes were twisted together and had copper screening laid around them: the whole structure was encased in a lead sheath. Over the frequency range between 10 Hz and 3 MHz the attenuation varied between 0.036 and 8.0 dB per mile, the velocity of propagation changed from 18 000 to 180 000 miles per second, and the characteristic impedance altered from 3000 to 190 ohms.

The cable could be used in lengths of up to eight miles without the need to employ repeaters, and as the length of the cable route from Broadcasting House to Alexandra Palace was 7.25 miles, the headquarters of the BBC became a convenient repeater station for OB transmissions into Alexandra Palace.

Figure 23.9 shows the cable route in central London. At several locations along the cable, signals could be tapped into it so that interesting OBs could be produced.

Of the factors which generally caused a degradation of image quality the most important was serious electrical interference. The deleterious effect of this was highlighted on 11 November 1937 during the Armistice Day 'two minutes silence' to honour those who had fallen in the First World War. In a report titled 'Television is perfect for two minutes' a *Daily Express* columnist (Jonah Barrington) wrote[13]:

'As cars, buses, lorries outside switched off their engines and came to rest, so did the crackling fade from the sound reception and the spots from the viewing screen, rather as if some unseen smudge had been wiped off a palette. For those two minutes the picture came to us clean, clear and steady—like a photograph. For the first time we were seeing what long-distance television would be like if engineers could solve the interference problem.'

On the broadcast itself Barrington had this to say: 'The transmission touched great heights. Close-ups of the King . . . and of the Prime Minister came through splendidly. We should have liked more.'

The BBC was particularly concerned about this matter since its OB mobile units had to operate frequently in electrically noisy environments, unlike the cameras at Alexandra Palace. The worst disturbances were caused by electrotherapy apparatus in hospitals, but although the Post Office was willing to give advice on remedial measures in a number of cases, the users of such apparatus were unwilling or unable to meet the cost[14]. At the beginning of 1938 the Post Office was powerless to enforce the suppression of interference as it had no legal powers, but in the new Wireless Telegraphy Bill, which it was drafting at that time, provisions empowering the department, in concert with the Electricity Commissioners, to compel owners of interfering apparatus to adopt preventative measures were included.

Prior to the inauguration of the London television station both EMI and Baird Television carried out extended tests on the transmission characteristics (including susceptibility to interference signals) of horizontally polarised and vertically polarised electromagnetic radiations having frequencies in the 6 m to 8 m band. For

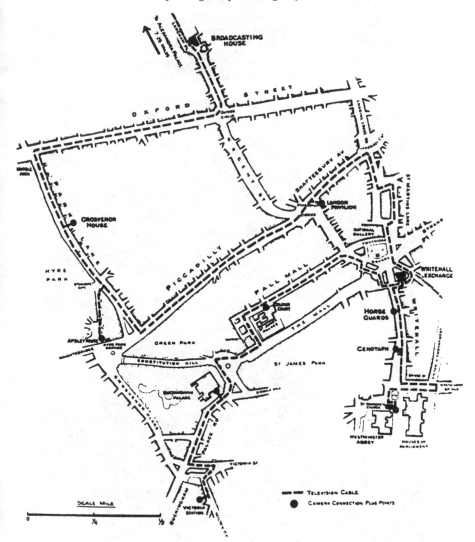

Figure 23.9   *Cable route in London. Camera connection points existed at St Margaret's Church, the Cenotaph, Horse Guards Parade, St James's Palace, Buckingham Palace, Victoria Station, Apsley House, Grosvenor House, the London Pavilion and Broadcasting House*

Reproduced by permission of Thorn EMI Central Research Laboratories

EMI, Blumlein informed[15] the TAC that 'although they had not obtained conclusive experimental data . . . they had found that the increase in noise level [due to automobile ignition] resulting from the use of a vertical aerial as against a horizontal aerial was not appreciable'. He submitted that over flat country a better signal strength was obtained over the ground level when vertically polarised waves were used. Also, he opined that from a practical point of view it was probably much easier

Figure 23.10 HMV model 901, a pre-war television set which sold for 60 guineas

Reproduced by permission of Thorn EMI. Central Research Laboratories

for the average listener to erect a vertical aerial, 'the more so on account of directional effects associated with some types of horizontal aerials'.

In the US Engstrom, as a member of the RMA's television committee, was concerned to ensure that the committee's formulation of its television standards was based on sound experimental testing. He reported[16] in 1937 that studies undertaken by RCA 'revealed definitely that horizontal polarisation [produced] the strongest signal in relation to noise and [was] freer from signal variations'.

Later, in June 1938, Shoenberg told the TAC that 'recent' tests had shown the advantages of utilising horizontally polarised radiations compared to vertically polarised radiations 'both from the points of view of signal strength and a comparative reduction of the effects of interference'[17].

With television systems which employed positive modulation, such as those installed pre-war in the UK, France and Germany, interference signals gave rise to white spots on the television raster: excessive ignition interference, for example, produced a snowstorm image. Similar interference affected television systems based on negative modulation but the effect was to show black spots: these were considered, in the US, to be less objectionable than white spots and so this method of modulation was standardised by the FCC. However, it was found that a 'definitely noticeable loss of line synchronisation could result'[18].

The BBC was, of course, keen, as was the trade, for television broadcasting to prosper, but until it achieved a modicum of popularity as evidenced by increasing sales of sets, the corporation was unlikely to gain much by arguing a case for a separate television licence to cover expenditure on the broadcasts. There were two conflicting aspects to the development of the service: first, from the BBC's point of view, there was insufficient material available for three one-hour periods a day, and hence this limited the broadcasts of television to periods from 3.00 p.m. to 4.00 p.m. and from 9.00 p.m. to 10.00 p.m. on weekdays; and secondly, from the buying public's viewpoint, the price of a good set with a large screen was as much as a small car. At a time when a typical wage[19] was of the order of £2.70 per week, a prospective purchaser had to consider whether television compared favourably, as a form of enjoyment, with, for example, motoring.

Nevertheless, a solution had to be found to the problem of poor sales and it was obvious that the impetus for breaking the deadlock had to be initiated by the Corporation with advice from the TAC. As early as February 1937—only four months after the start of the service—Carpendale had informed the TAC that earnest consideration was being given to the question of providing an additional hour's transmission in the afternoon; but there was the issue of production costs to be reckoned with and the present costs were already causing the Corporation grave concern. Lord Selsdon suggested that the extended programme might consist largely of films to keep costs as low as practical, but Carpendale was somewhat apprehensive about the effect on the reputation of the service if this idea were implemented.

Subsequently, the revised programme timetable was as follows[20]:

23 and 24 August

Three hours' transmission of film with sound announcements from 11.30 a.m. to 12.30 p.m., 3.00 p.m. to 4.00 p.m. and from 9.00 p.m. to 10.00 p.m.

26 August to 4 September

Three hours' transmission from 11.30 a.m. to 12.30 p.m., 4.00 p.m. to 5.00 p.m. and from 9.00 p.m. to 10.00 p.m. (these were to be actuality transmissions)

6 September et seq

10.30 a.m. to 11.00 a.m.	cruciform pattern accompanied by tone
11.00 a.m. to 12.00 noon	television demonstration, film and film magazine with appropriate sound— for trade purposes only
3.00 p.m. to 4.00 p.m.	normal transmission
9.00 p.m. to 10.00 p.m.	normal transmission

The extra hour was not sufficient to stimulate sales and for the quarter ending 30 September the number of receivers sold was about half the number for the previous quarter. Clearly, the position was serious; unless some action could be taken to further sales the service would either die out or it would have to be handed over to private enterprise and be supported by advertisement revenue. There was thus a

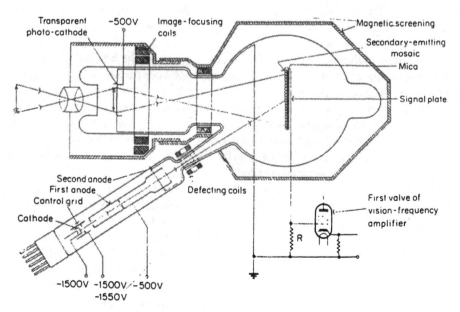

Figure 23.11   The super emitron camera was used for a public broadcast by the BBC, for the first time, on Amistice Day, 1937. In operation the optical image was focused by a lens on to a conducting, transparent photo cathode. The emitted photoelectrons were accelerated and focused, by electric and magnetic fields respectively, on to a secondary electron emitting mosaic. Tests showed that the super emitron was at least ten times as sensitive as the standard emitron. Since the photo cathode could be positioned close to the window of the tube a wide range of lenses could be utilised

Source: *Journal of the IEE*, **84**, 1938, pp. 468–482

necessity for more service hours, particularly in the evenings, and a requirement for a set costing approximately £30.

The BBC in October 1937 planned[21] to present two separate hours of television on Sundays in 1938, and to improve the programmes and their layout by utilising the theatre at Alexandra Palace. It was also intended to develop further the outside broadcasts by the acquisition of an additional mobile unit at a cost of £40 000. The total extra cost of these proposals was estimated to be £100 000 for capital expenditure and £70 000 for running costs. All of this was a move in the right direction but Lord Selsdon felt that greater efforts should be made to extend the evening programmes to be a total of 2½ hours of viewing— 'say, 1½ hours of studio and outside broadcasts and 1 hour of good films'. He understood that old films could be made suitable for television reproduction at a cost of £25 to £30 for 60 000 ft.

Certainly, cost was of central importance in this matter for the expenditure for 1937 was estimated at £65 000 capital and £330 000 running expenses, of which £50 000 was found by the corporation out of 1936 income. Any proposals, therefore, which increased the number of television broadcast hours without incurring a substantial extra cost were of significance to the BBC.

Selsdon's idea was rather negated by the difficulty of obtaining old films at any price. It was thought that about 30 could be procured from Gaumont-British, but this number was much too small to sustain an additional regular period of television. In any case, later enquiries by the BBC to the cinema interests on this matter showed that the film companies were displaying the greatest unwillingness to cooperate, and by January 1938, Graves, the BBC's deputy director general, reported[22]: 'It has proved impossible to obtain English and American feature films. Recently, however it has seemed likely that some good continental films (with English subtitles) may be secured; these can be used as an occasional experiment and afford temporary relief and change from studio programmes.' Again, it was felt that there were not enough of such films to make a special film hour broadcast outside the normal times.

There was another possibility: if the film interests were unwilling to loan or hire full length feature films to the BBC, would they consider allowing trailers to be shown on television?[23] Rosen, of the Radio Manufacturer's Association, put forward this idea as he was under the impression that the television industry in America intended using trailers as a part of their programmes when they started a service. It was thought that such an arrangement would be attractive to the cinema companies and, if so, it might pave the way towards establishing a friendlier relationship between the BBC and the film industry. However, there were difficulties:

1   the present number of viewers was too small to make the proposition attractive to film proprietors—if the potential audience had been substantially greater they might have been induced to provide trailers more suitable for television;
2   the showing of trailers might be thought to infringe the Government's ban on sponsored programmes, although Phillips thought that this arrangement was defensible;
3   some of the trailers were short and poor in quality.

From the film interests point of view the BBC, with its news service, was a potential serious rival with whom it was felt they should not co-operate, at least until their position vis-à-vis the corporation became clearer regarding any deleterious effect on the film industry. Still, the BBC had one likely rejoinder to the cinema companies stance: if it had the power it could ban the cinema reproduction of its (the BBC's) programmes unless the industry could be persuaded to make films available for television.

The use of films would have alleviated the provision of extra television broadcast time in the short term, but in the medium term the BBC had to provide more studio programmes and outside broadcasts.

When the Television Advisory Committee agreed to the adoption of the Marconi-EMI system it was hoped that one of the consequences would be a simplification of receiver design and construction, and hence a reduction in the cost of television sets. Lord Selsdon was very anxious to do everything within his power to develop television and had been rather concerned about the high cost of receivers. He did not have long to wait after the announcement of the single standard for a reduction to take place, for on 9 February 1937 *The Times* reported a decrease in the price of the Marconiphone and HMV television sets—from 120 and 95 guineas to 80 and 60 guineas (a guinea was £1.05), respectively. In addition it was stated[24] that those persons who had bought sets at the original prices would be compensated by the makers.

This sudden drop in prices came about because EMI found that the sales of sets, at the high prices then prevailing, so stagnant that they decided to make an immediate cut. Sterling, the managing director of EMI, hoped that Selsdon and his committee could somehow increase the number of hours during which television was being shown, without thereby incurring very much extra expenditure, so as to stimulate further sales. He told[25] Selsdon that if sales showed good results it was EMI's definite intention to produce a much cheaper television set in the autumn. It was clear that the factors which led to increased sales were lower prices and more broadcasts. The Baird companies had found during the era of low-definition television that the public did not rush to buy receivers when the total number of hours of television per week was low and/or the programmes were arranged at inconvenient times.

Marconiphone's and HMV's initiatives certainly produced immediate results. Osborne, the sales manager of the former company, was able to write to the press controller of the GPO on 19 February 1937 as follows[26]: 'May I say at once that while television sales at the moment show most excellent promise, there still seems to be some doubt in the minds of both trade and public concerning the stability of the present transmission system, and until this is finally removed I feel that it will have a detrimental effect upon future business'. Again, Baird Television had experienced a similar situation during the 30-line definition period. Public reaction to the new invention, although being one of considerable interest, manifested itself initially in a wait and see attitude. Members of the public were loth to pay a high price for a television set which might become redundant in a short time and in effect they required an assurance that the 405-line standards would be in force for some appreciable time. The press announcement of 5 February had stated that the transmission standards would remain unaltered until the end of 1938 but Osborne thought that this could be emphasised. 'The publication of this would, I am sure, react most favourably upon the buying public and would enable the industry as a whole to embark upon its production programme with the feeling of perfect security,' he wrote.

This point was discussed[27] by the TAC on 26 February. Phillips felt that a repetition of the press statement might wait until a press notice was required on some other aspect of the television service. Selsdon agreed but said that in the meantime it might be a good thing if arrangements could be made for a friendly question in the House to be put to the Postmaster General on the point. The secretary of the committee consulted[28] Shoenberg about this, but he and Sterling were of the view that more harm than good would be done by ventilating the matter in this way. On the other hand, they definitely regarded an extension of the television service on weekdays and the opening of a service on Sundays as being of great importance and expressed the hope that a favourable decision on this matter would be reached and publicly announced before Easter.

The need to give the general public a guarantee of several years in respect of the stability of the London station's waveform specification was raised again in October 1937 when the *Wireless and Electrical Trader* in an editorial[29] urged the TAC to look into the matter afresh and let both the industry and the public know where they would be standing in a year's time. For the editor the obvious practical course was to leave the Alexandra Palace transmissions unchanged and to install improved high-definition equipment at the provincial stations when these became available. 'The disadvantages of having the London transmission somewhat out of date for a year or two would be as nothing compared with the need for elaborate adjustment at the public's expense of present day instruments,' the editorial stated.

This statement was discussed at the TAC's meeting in October 1937[21], but not withstanding their discussion on agenda items dealing with sales of sets and the development of the television service, the general feeling of the committee was that no action on the lines indicated in the editorial was required. Their advice seems rather strange when set against the poor sales record and the need for a guarantee to be given to buyers. The committee, which was normally forward looking and perceptive on television matters, was singularly unappreciative of the point raised by the trade journal. However, the RMA understood the position well and, at its meeting with the TAC in December, reaffirmed the view that an official pledge to the public on the issue would assist sales. With this corroboration, the committee agreed to the publication of a press notice on the subject[30]. Initially, the TAC was of the view that the standard should remain unchanged for a period of two years from 1 January 1938, as 'it would be undesirable to specify a longer period in view of the possibility of technical developments', but at the next meeting Sir Frank Smith suggested that the time span should be three years and this was concurred.

Confirmation was given in the reply by the Postmaster General in the House of Commons on 1 February 1938: ' . . . the public may therefore purchase television sets without any fear that they will become obsolete or require substantial alteration for a very considerable time to come.'[31]

This announcement, and the measures which were being implemented to extend and improve the television productions, certainly had a beneficial effect. Whereas 2121 television sets had been sold by the 31 December 1937, a year later the total number of receivers sold was 9315, and by August 1939 the figure was 18 999: television sales were advancing rapidly, only to be stopped by the outbreak of hostilities.

Actually, there was no necessity to alter the transmission standard, for the 405-line system was capable of being developed to give much better pictures than these shown in 1937. The RMA referred to this in December 1937 but the chief engineer of the BBC did not think anything dramatic could be expected in the way of image betterment. Conceivably, he may have been reluctant to spend money on this aspect of the service when so much more could be done with the limited funds at hand.

The RMA's enquiry appears to have been quite valid, for the American Telegraph and Telephone Company had recently given a laboratory demonstration of television transmission along a cable on a standard of 240-lines and Angwin had been informed that the definition realised had been comparable with the results reached over the air by the 405-line transmission. A similar observation had been made by Post Office engineers in 1935 when films had ben televised by the Marconi-EMI and Baird Television systems working on 405-line and 240-line standards respectively. These facts did not show the 405-line system in a poor light, but rather highlighted the improvements which could take place, based on the existing waveform specification, as technical progress in electronics advanced. Both Blumlein and Shoenberg had stated, to the technical subcommittee of the TAC, their belief that the 405-line television demonstrations which they gave in 1936 did not represent the ultimate performance of the chosen standard: as Shoenberg informed the subcommittee, one reason for the choice of 405 lines in 1935 was that it enabled future picture quality to be raised without the requirement for an alteration of system parameters. Blumlein[32] repeated this belief in 1938.

A surprising feature of this subject of line standards concerned the number of lines per picture. When the 405-line system was announced in 1935 there was a considerable outcry against the use of such a high definition by certain sections of the radio and television industry. Now, in April 1938, C O Stanley, a member of the RMA, felt able to tell[33] the chairman of the TAC that opinion in some technical quarters was that 'the present definition was not sufficiently high, and that the question of a higher standard should be considered now, when the number of sets which would need to be adapted was small'. This was rather a volte face for a section of the industry and vindicated Shoenberg's 1935 judgement.

The resolution of the received image was dependent on the receiving equipment: if this was poor, then clearly good pictures could not be reproduced. If, for example, the bandwidth of the various circuits in a receiver was 1 MHz instead of 2.5–3.0 MHz, degradation would result. Both Ashbridge and Angwin[33] thought that 1937 receivers did not take full advantage of the definition offered by the 405-line standard, and the BBC's chief engineer noted that if any improvement was to be made the number of lines per picture would have to be at least doubled. Nevertheless, the point brought up by Stanley was important, for the TAC had to reflect on the standard which should be imposed at the end of the guaranteed period, and associated with this there was the effect of a wider band of frequencies on both free space and cable transmissions to be examined. In retrospect, Ashbridge's view on future line standards—about 811 lines per picture—seems to have been rather pessimistic bearing in mind current British practice.

During the post-1936 period of medium and high-definition television in France, Germany, the UK and the US, various visits were made by senior members of national broadcasting administrations and companies concerned with the development of television to the corresponding organisations overseas. Their objective presumably was to consider and compare the different schemes in operation so that good practices could be adopted and difficulties avoided.

The reports of L Espenschied, an engineer of the Bell Telephone Laboratories, are of some worth since he closely followed the progress of television in the above mentioned countries and was able to write impartial, comparative accounts. On the attributes of the images received in Berlin and London, in July 1937, he wrote[34]:

'The 180-line picture being put out in Berlin is about as good in definition as the 405-line picture of the BBC. Evidently the British . . . are not getting from their pick-up system the definition of which it is theoretically capable, although it looks at the moment better than the RCA result. The Germans employ mechanical scanning, largely from film . . . The picture [seen in Berlin] was good enough to have value for the time being for events of outstanding interest, but would not suffice for continuous consumption of ordinary programmes.

' . . . The British, because of the use of the [emitron] are able to include outdoor pickups in their programmes. I saw a Wimbledon tennis match at the Science Museum . . . The ball could not be followed, but the players could be and, with the running vocal description, it made quite an acceptable show. The way in which the sight and sound supplemented each other was striking. The sight facility added greatly to the vocal account, imperfect though it was.'

Another member of Bell Telephone Laboratories opened, in March 1938[35]:

'The overall picture quality [of British television] seems to be limited less by the location of the receiver than by the technique at Alexandra Palace. The statement has been made by regular viewers that it is possible to determine when a new untrained crew is used at the studio, and this has been borne out even in the writer's experience. There are also, as might be expected, considerable changes from one [emitron] pick up to another. The considerable success of the Armistice Day program was attributed in part to the use of a new and improved [emitron].'

These views supported Shoenberg's and the TAC's contention that improvements in picture quality were possible with the existing 405-line standard. Indeed, steady progress was being effected. When Espenschied visited the London television station in May 1938 he was able to report[36]:

> 'The television image being radiated by the BBC from Alexandra Palace has been improved both technically and in studio technique over the performance last summer, so that it is now not only good enough to have entertainment value but it is to my mind really attractive, at least, it certainly will be with larger screens and enlarged studio facilities now being undertaken.'

Espenschied saw television displays in the viewing rooms at the HMV store in New Bond Street, London, and at Alexandra Palace and observed[37]:

> '[One] picture [was] seen in the viewing room at Alexandra Palace after modulation but before the picture [left] the antenna. The picture was without the spotty distortion seen at the HMV store and was really good; quite brilliant; uniform illumination to edges of picture without appreciable distortion.
>
> 'In the studio a play was being put on of Punch and Judy in life characters. Points noted: The illumination of the scenes being televised seemed to be rather moderate and there was no indication of distress on the part of the actors. Rapid shifting of scenes from one camera to another; movement of the scene from one part of the stage to the other readily followed by the pick up cameras. Supervision of the pickup job, of switching from one camera to another, fade-ins and check on general result, carried out from a control room adjoining the studio and looking out over the studio.'

On television developments in the US, Espenschied witnessed a demonstration of RCA apparatus at a meeting of the New York Electrical Society held on 20 May 1938.[38]

> 'The [RCA] television reproduction I saw was better than that being attained by Rundfunk in Berlin on their 180-line system. This in itself is not saying much, however, because the Germans are not maintaining their 180-line system very well, merely using it to practice programme production while preparing for the 441-line system which is expected to go into service this summer.
>
> 'What I saw in the laboratory of Fernseh and Telefunken was considerably better in size of image, brightness, detail, etc., than the RCA picture.
>
> 'As compared with the BBC, the RCA reproduction is a little poorer than that observed in a viewing room at Alexandra Palace, in these respects: The image as seen on the RCA sets is not as bright and contrasting as those of the London sets; the detail is about the same; but

the uniformity of the field and the depth of focus of the picture, and certainly the smoothness of production of program, are not as good as in London.'

Some of the enhancements in programme merit could be ascribed to the RMA's television development subcommittee which provided much sound advice, to the TAC, on how the BBC's television programmes could be improved. Many of their points seem obvious when set against present day television, but at the time they were made they were obviously necessary. Some of them were[39]:

1   The duplication of programmes in the afternoon and evening of the same day is to be deprecated.
2   The programme transmitted on a Saturday evening should, in general, be of the variety type and consist of a number of short items.
3   In the normal variety programme, the items included are often very similar in character and insufficient variation is introduced, both as regards the artists employed and the type of material used.
4   Transmission of drama have usually been of too ambitious a character, the type of play generally selected has been too morbid and it is felt that short, one-act plays of a more cheerful nature would be preferable. In this connection, it is felt that so large a use of 'horrific' subjects—whether it is in plays or films—is to be deprecated.

Figure 23.12 A super emitron camera with its cover removed

Reproduced by permission of Thorn EMI Central Research Laboratories

5   It is apparent from the transmissions from the studio that in some cases adequate rehearsal has not been possible and the difference between the slickness and perfection of performance which is achieved when sufficient rehearsal has been provided is most marked.

6   It is desired to emphasise the importance of the part played by the announcer. It is felt that the brightness of the television programme is, to a large extent, dependent upon him.

7   The method of displaying television programmes in the *Radio Times* is not such as to give proper prominence to star items.It is suggested that a better use of headlines should be made and, further, that more space should be devoted in the *Radio Times* as a whole to television news.

8   The method of putting out television news is not calculated to secure the widest possible notice in the press. For instance, the announcement that the Cup Final would be televised was first made on the television screen during a television programme, with the result that its news value, as an item of interest to the press, was destroyed at the outset.

9   Any tendency to give information concerning forthcoming television events as a scoop to individual papers is to be deprecated. A positive suggestion is made for a weekly press visit to Alexandra Palace, preferably on Thursday, in order that publicity in the Sunday papers might be secured.

Figure 23.13 The first play to be televised from a theatre was 'When we are married' by J B Priestley. It was broadcast from St. Martin's Theatre, London, on 16 November 1938. Super emitron cameras were used

Reproduced by permission of the BBC

10   As far as possible, in television programmes, all announcements should be accompanied by some picture, either of the announcer or illustrating the announcement being made.

The RMA's perceptive and sensible comments showed that much could be done to enhance programmes, notwithstanding the limited studio space available. In the absence of immediate extra studio space, the BBC could expand its service by making use of more outside broadcast transmissions, and here the corporation had already taken steps to fulfil this. It had ordered an additional mobile unit and planned to transmit a number of attractive OBs from Twickenham, Epsom and several other places. By 25 January 1938 Graves was able to give some details of the BBC's plans in this field[40].

The extra unit was not expected until the summer and, as the existing mobile unit required appreciable time for setting up, testing and subsequent removal to other sites, outside broadcasts could not be given in all afternoon programmes—even assuming good lighting conditions. When the new mobile unit arrived, the BBC planned to provide OBs on Sundays as well as weekdays and to increase their number. The corporation recognised that whatever developments might occur in studio programmes, the future of television would depend very largely on the televising of outside activities. All of this was not to be accomplished without some difficulty for 'considerable opposition had already been experienced and more was anticipated in connection with the televising of important sporting events under commercial auspices'. (Nearly 60 years on the same problem exists.)

The factors which had to be appraised in 1937 undoubtedly had a beneficial effect on the television service; they contributed significantly to the improvement of television productions in 1938. Some appreciation of the progress which the BBC had made, and was making, was given in the comments, on the television programmes for the fortnight ending on the 5 April 1938, offered by the television development subcommittee of the RMA[41]:

1   The programmes transmitted showed, in general, a marked improvement both as regards entertainment value in the items transmitted and the technique of the transmission itself.
2   The crazy programmes 'Nice work' and 'If you can get it' were especially well done and very attractive in character.
3   The transmission of the boat race was another good feature. The idea of putting over a film of the race in a later programme on the same day was particularly happy and indicates, clearly, that there is always a possibility in such topical events as this that something may be obtained which can be repeated by the use of film later on. As a first attempt in this style of transmission the item was a very good one.
4   Transmission of 'Henry IV' was another welcome item and Ernest Milton in the title role was particularly effective. The success of this indicates the wisdom of the policy of cultivating personality in television transmissions.
5   The cabaret transmissions are appreciated, but it is suggested that their attractiveness might be increased from time to time if it were possible to arrange for

a high spot, say, once a month, without detracting from the normal high level.

6   'Picture Page' is considered to be a very popular programme and should be continued with an attempt to improve, as far as possible, upon the material employed in it. It would appear that there is some paucity of good material which it is hoped will be found possible to overcome.

7   It is recommended that during the summer months, once a week for, say, a period of six weeks, outside broadcasts from the zoo would be welcome.

8   Transmissions of comedy films such as Mickey Mouse should be used as much as possible in the afternoon shows, as they are of particular appeal to children.

9   It is again suggested that consideration should be given to the possibility of transmitting in the television programmes items from the 'In town tonight' [sound] broadcast series.

10   Two outside broadcasts of particular interest and merit were the McAvoy fight on 7 April and the soccer international on 9 April. In both cases the arrangements made were altogether adequate and the transmissions resulting were of a very high quality.

11   'Cabaret cruise no.4' on 11 April was also a marked success and its presentation must have done a great deal of good, especially where demonstrations were taking place to prospective customers.

12   The experiment of showing a semieducational film was interesting and it is considered good provided it is put over in such a manner that it stresses the interest rather than the educational aspect.

13   It is felt that the Pepler Masque is a type of entertainment which had such a small appeal that it should not be transmitted very frequently.

From the inception of outside broadcasting on Coronation Day, the BBC's mobile television units enabled a section of the general public to see a wide variety of events in public, social and sporting life—the Wimbledon tennis tournaments, the Lord Mayor's Show, the Cenotaph ceremony on Armistice Day, the Prime Minister alighting from his plane at Heston after his visits to Berchtesgaden and Munich, the Derby, the University Boat Race, the Cup final, Trooping the Colour, the Test Matches from Lord's and the Oval, boxing matches such as those between McAvoy and Harvey, and Boon and Danahar, and so on (Figure 23.14 and 23.15).

An unbiased view of the attractiveness of outside broadcasts was given, in May 1938, by J Royal, the vice president in charge of programmes for the National Broadcasting Company of the US. He saw the Scotland versus England football match, and the Harvey versus McAvoy fight and commented[42]:

'I had never seen that sort of [football] contest before; nevertheless it held my interest. The photographic detail was excellent. We never lost sight of the ball. The trick, of course, is to have enough cameras on the job to follow the fast moving scene . . .

'The lighting effect [of the boxing match] was ideal. The close-ups were extremely interesting as the telephoto lenses glimpsed the infighting . . . The camera men cleverly pointed the lenses into the corners between the rounds. Then, of course, we heard the bell and the smack of the gloves for added reality. The cameras also took close-up shots of celebrities at the ringside and long shots of the arena. I found it most exciting.'

Surprisingly, although the BBC's use of three and four cameras, for fading scenes from different angles to provide variety, was readily apparent from its OBs, the technique was not rapidly adopted in the US. When the NBC televised the opening of the 1939 World's Fair, only one fixed camera was employed and that was positioned too far away from the speakers to give entirely satisfactory images[43].

Several of the pre-war opinions concerning the appropriativeness of attractive events for television seem curious when set against present day practices. Symphony concerts, recitals and operatic performances are now considered to be essential components of a culturally broad-based television service, but in 1938 the telecasting of symphony concerts was not thought, by one senior member of the music establishment, to be worthwhile. Sir Adrain Boult, the chief conductor of the BBC's Symphony Orchestra, could see no practical value in this type of performance[44].

Figure 23.14 Oxford and Cambridge athletic sports at the White City Stadium (1938). The photograph shows a super emitron camera in operation

Reproduced by permission of the BBC

'Television is a medium that calls for sweeping action and movement and no audience of music lovers, no matter how ardent, is interested in a study of still life without the stimulation that comes from physical presence and personal contact in a concert hall or broadcasting studio. No one would care to be a witness at a teleview of a two-hour concert anymore than he would care to witness a motion picture of one.

'The spectators would be bored to distraction in no time, no matter how carefully the cameras were focused. No, televising a full symphony concert, to my mind, would be an impossible and thankless experiment . . .

'[Anyway] you musicians are far too ugly to show up!'

(In 1995 the final concert of the 'Proms' was watched worldwide by 12 million viewers.)

All the above mentioned improvements in the television service were obviously not obtained cheaply. By the end of 1938 the expenditure on the London service was running at the rate of £427 000 per year and the estimate for the year 1939 was £450 000[45]. When the Television Committee had deliberated the question of finance, in 1934, it had recommended that the cost of the television service should, at the outset, be borne by the revenue from the broadcast receiving licence fee (50p per annum). The committee at that time did not favour the immediate introduction

Figure 23.15 Television outside broadcast from Wembley stadium (30 April 1938). Super emitron cameras were used to televise the Football Association's cup final. The photograph shows King George VI arriving at the stadium. One of the cameras can be seen left of centre

Reproduced by permission of the BBC

of a separate licence for television reception but recommended that the question of doing so should be considered later in the light of experience.

Later, when the BBC's new charter and licence came into operation on 1 January 1937, provision was made for the Post Office's share of the licence income (to cover administrative costs) to be fixed at nine percent, and for the Treasury's and the BBC's shares to be 25% and 75% of the balance (the 'net licence revenue'), respectively. The committee on broadcasting (the Ullswater Committee) had suggested that this 75% should be allocated to the BBC 'for purposes other than television' but the Government decided that it should embrace all the corporation's services, including television. There was a possibility of extra funding but this was contingent on the Treasury being satisfied that the BBC's income was insufficient.

During 1937 the BBC financed television out of its 75% share but requested a further share of the revenue to meet increasing television costs. The Treasury was content that an increase was necessary and, by a supplementary estimate presented to Parliament in February 1938, an additional allotment of licence income of eight per cent, amounting to £310 000, was given to the BBC. This was for the cost of the television and foreign language broadcasts incurred during the 15 months to 31 March 1938. Almost the entire amount related to television expenditure.

In November 1937 the Postmaster General, at the request of the Treasury, remitted to the Television Advisory Committee, the 'task of surveying the question of the future development of television in the light of experience so far gained of the service and its cost'.

From Table 23.1 (which shows the actual and estimated revenue and capital expenditures for television for the period 1935 to 1938) it is clear that significant demands were being made on the broadcast receiving licence revenue, even though the number of viewers was relatively small. The fact that this revenue was collected from sound broadcast listeners, the vast majority of whom could not receive television transmissions, was particularly embarrassing.

However, the TAC was convinced that the financing of the London station was not a question merely of providing entertainment for a selected and limited number of persons, or even of providing the basis of a national service, but that it was necessary to enable a new industry concerned with the manufacture of television apparatus to be created. Since Great Britain was the pioneer of television, the TAC advocated the adoption of a bold policy in developing the service so that manufacturers would be placed in a strong position with regard to valuable exports when other countries began to establish their national television networks.

There was a further point which had to be discussed. In paragraph 48 of its report of January 1935, the Television Committee envisaged 'the ultimate establishment of a general television service in this country [the UK]' and the relaying of television productions by land line, or by wireless from one or more main transmitting stations to substations in different parts of the country. It was obvious that the creation of a network of stations to serve all but the most sparsely populated areas would occupy a good many years and require the building of at at least 12 to 15 stations.

The TAC gave much thought to this matter and in December 1938 forwarded to the Postmaster General a report[46] on the expansion of the service. Based on the

Table 23.1   Television expenditure (1935–1938)

Revenue expenditure	1935 (actual)	1936 (actual)	1937 (estimated)	1938 (estimated)
Programmes—excluding staff	76	14 069	89 000	158 000
Engineering—excluding staff	–	8 536	34 000	39 000
Premises, maintenance and overhead charges	13 955	34 300	14 000	44 000
Administration—management charges	2 187	14 544	28 000	39 000
Salaries, wages and pension scheme	548	25 051	63 000	100 500
Income tax	–	8 000	23 000	–
Depreciation	–	7 000	56 000	46 500
	£16 766	£11 500	£307 000	£427 000

Capital expenditure	Gross expenditure	Depreciation provided	Net expenditure outstanding at end of year
1935—actual	24 000	–	24 000
1936—actual	99 000	7 000	92 000
1937—estimated	60 000	56 000	4 000
1938—estimated	70 000	46 500	23 000
	£253 000	£109 500	£143 500

technical experience which had been gained from more than two years working of the London station and the degree of acceptance of television by the general public, a plan of action had been formulated which would lead to the foundation of a semi-national service. The committee foresaw the setting-up of four regional stations, each having a power output of approximately four times the power of the London television station, which would probably be located at or near Birmingham, Moorside Edge near Huddersfield, Westerglen near Falkirk, and Clevedon near Bristol. It was contemplated, not unduly optimistically, to antici-pate that about 25 million people would be within the effective range of one or other of these stations and that the first of them, at Birmingham, would be ready after the end of 1941. Two other stations might be constructed simultaneously and become operational in 1943, and the fourth regional station might follow in late 1944 or early 1945.

Much attention was paid to the financing of the London and proposed regional stations and, in the first instance, it was considered prudent to restrict the regional service to the relaying of programmes initiated in London and thereby avoid the substantial costs of separate regional programmes. Assuming 17 hours per week of television—the 1939 allocation—the committee estimated an annual expenditure of

£900 000 to £1 000 000 (see Appendix 6). Supplementary expenditure 'of substantial amount' would arise if further advancements were desired, such as:

1 the provision of further provincial stations;
2 the extension of programme hours beyond the current weekly figure;
3 the betterment of the quality and scope of programmes above the standard envisioned in the 1938 estimate;
4 the provision of separate programmes from the provincial stations.

Apart from the TAC, provincial traders also wanted an extension of the television service. In a concerted effort a joint committee of dealers and manufacturers (known as the Television Extension Committee) was formed in March 1939 with the objectives of '(bringing) home to the authorities the need for a speed-up in the extension of television into the provinces because of the importance of a new industry at home; because of its export capabilities; and because it (was) so desirable for Britain to maintain the lead which it at present (enjoyed) over all other countries'[47].

On one issue the manufacturers were quite unanimous: the relaying of the Alexandra Palace television productions to the first of the suggested provincial stations, namely Birmingham, should be by radio link rather than by cable. Indeed, so convinced were the manufacturers of the merits of radio links that 'if the Government [proceeded] with the building of a Birmingham television station without delay, [the manufacturers would] be prepared to stand the loss if it should not be a success'[48].

Besides these aspects, the dealers and manufacturers stressed the fact that 'while Alexandra Palace [remained] the only television transmitter in the country large members of the public [would] regard television as being still "experimental". The immediate building of a provincial station would put the seal of certainty on television in London itself as well as in the rest of the country'.

Moreover, the set makers represented that in the absence of some extension they could not accept the risk of expanding their plant to provide for large-scale production: until they could secure economies resulting from such production, prices could not be expected to fall.

As an essential prerequisite to its plan the TAC felt that some augmentation of the London station should be undertaken immediately in order to ensure that the station was an adequate centre for the future national service. It forecast the expenses, for 1939, for the station as £575 000, an increase of £153 000 or 36% over the standard rate of expenditure at the end of 1938. The latter was £158 000 over the actual outgoings for 1937.

All of this overlooked the possible effects of the ominous European political situation. Nevertheless, the TAC's report was submitted to the Treasury for approval of the finance needed to implement the plan. Although the Treasury was fully appreciative of the commercial potentialities inherent in the evolution of television, it opined that the time was not ripe for the allocation of substantial funds to the nascent service. Exceptionally heavy demands were then being imposed on the Exchequer by essential disbursements on rearmament and defence measures generally, and as a consequence the Treasury was unable to authorise, for the year 1939, expenditure of more than the estimated cost, *viz.* £450 000, of running the London service for that year. It suggested that the development of all the likely sources of revenue might provide funds which would facilitate desirable extensions of the service.

The TAC again gave consideration to the cogent points previously discussed. Since it was convinced that there existed 'a danger that if the television service were not quickly extended the infant industry would be killed, and the potential television markets in Empire and other overseas countries lost' it recommended (June 1939):

1   the relaying to the Birmingham station of the London programmes by means of a cable (of the coaxial type), since in the TAC's view it offered important advantages over wireless transmissions;

2   the financing of the station (which was estimated would be c. £50 000 per year for the first three years, rising to £70 000 per year thereafter) by revenue derived from:

    a   the introduction of a special licence, covering both sound and television at a charge of either 50p or £1.00 year, when the number of receiving sets approached say, 50 000,

    b   the imposition of a special charge for large-screen reproduction of television in cinemas,

    c   the withdrawal of the prohibition of sponsored programmes and direct advertising, so far as the television service was concerned.

On point 2*a* it was not known whether the Postmaster General had the legal power to require the taking out of a licence for vision reception: this was an issue on which the law officers of the Crown had been requested to give an opinion.

On point 2*b* the TAC believed that, with the development of large-screen television, cinema proprietors would be eager to give their audience facilities for viewing television broadcasts of events such as the Derby, the Boat Race and football matches. Since television had the advantage, in comparison with an ordinary newsreel, that it showed events at the actual moment of occurrence, it was possible that cinema owners would be willing to pay the BBC for the right to do so, although there were certain 'serious difficulties' which might arise (for example, in relation to copyright and the demand for higher fees for certain outside broadcasts).

Both the Television Committee and the Broadcasting Committee, in 1935, had advised that sponsored programmes should be permitted, at least in the early stages of the television service, but the Government had decided (in Command Paper 5207) not to adopt the recommendation. In this connection the TAC commented as follows[46]:

> 'We are familiar with the arguments against the inclusion of such programmes; and where, as in the case of the sound service, an allocation of upwards of three million pounds a year is available, these arguments may well be considered conclusive. In view, however, of the great difficulty of financing the television service and providing for its extension to

the provinces, we consider that the inclusion of sponsored programmes and even direct advertising in that service would be fully justified.'

(The BBC representatives on the TAC dissociated themselves from this suggestion.) From a technical viewpoint the most interesting, and surprising, aspect of these recommendations was the committee's advice on the need to use a cable, even though the manufacturers had a few months earlier overwhelmingly rejected this form of transmission link. In June 1938 Shoenberg, of EMI, had submitted a paper[17] to the TAC on 'Television relays' and had expounded the reasons why a radio relay link was preferable to a cable link for the propagation of television signals.

1   'A long cable link [while] theoretically possible . . . does not seem to be practically workable with the components at present available.
2   ' . . . with [the] valves at present available, sufficiently stable amplifiers cannot be made to cover the 2.5 MHz band.' (These were required in the repeater units.)
3   'The production of a suitable single sideband signal [was] not known to be feasible yet with a 2.5 MHz band.' (This type of signal was needed for a coaxial cable link.)
4   ' . . . [the] equalisation [of the cable had to] be performed with extreme accuracy using extremely stable components.'

Shoenberg concluded by saying that a 100 mile or more cable link was commercially workable but 'at the present it could only handle a band some 1.0 to 1.5 MHz wide'.

On the other hand, his company had examined the transmission of radio signals in the 60–100 MHz range and it appeared that the London to Birmingham link could be established by three radio hops. The link would comprise a 100 w, 175 MHz transmitter and antenna at Alexandra Palace, and intermediate relay stations sited at elevated positions in the Chiltern Hills and in the Cotswold Hills. Each of these stations would include a receiver and a 100 W transmitter, and the transmitter frequencies would be 185 MHz and 175 MHz, respectively. Soenberg estimated the cost of the complete chain (excluding spares and buildings) as c. £20 000 and thought that it could be installed in about 18 months.

EMI's proposal was examined by the Post Office engineering department. Since several overseas administrations, in addition to the TAC, were, in 1938, concerned with the problem of establishing national television networks, the advantages and disadvantages (as seen by the POED) of cable systems versus radio systems are of some historical significance[49]. A consideration of the various factors serves to explain why in the UK, at least, radio systems were finally adopted in the post war years.

Cable systems
Advantages

1   They were immune from interference if properly screened.
2   The cable could be used for multichannel telephony when not required for television.
3   No frequency-band allocation would be needed.

Disadvantages

1   The initial cost of the cable was considerable: and the cost of the repeater and terminal equipment would be of the same order as the cost of the radio equipment.

2   The electrical characteristics over the band of frequencies necessary were non-uniform.

3   There was a need to frequency translate the television signals upwards for transmission and downwards for reception. (This process would involve complex and precise equipment.)

4   All the apparatus and components used would have to fulfil very exact specifications and would need to be highly stable in use.

Radio systems

Advantages

1   The transmission distortion was much lower than that for cable.

2   Such systems had a lower overall cost as the expense of providing the cable was avoided.

3   The running cost was lower.

4   No frequency translation equipment was required at the terminal stations.

Disadvantages

1   They were liable to interference.

2   Distortion might occur due to multiple reception of a signal (by electromagnetic waves travelling over different paths).

3   Selective fading might arise.

4   The transmission characteristics of the system might be adversely affected by weather conditions.

5   The radio system could not be used for multichannel telephony.

6   Difficulty might be experienced in extending a radio system because of the wide frequency bands necessary.

From these considerations and from some tests the Post Office engineering department (POED) concluded: 'The present position regarding cable transmission [was] that apparatus for translating, amplifying and correcting the signals [was] not yet in a sufficiently advanced state to allow the system to be used for high-definition television.' This view agreed with that of EMI. With regard to radio link systems the POED noted that 'very little practical information' was available: and with regard to EMI's scheme there was a distinct risk that it would be inadequate because of the 'large number of unknown factors present' and the small margin of power.

The department's rather gloomy prognosis for television networking was somewhat assuaged by the knowledge that some satisfactory experience of cable links had been gathered in three countries. In Germany a cable network had been installed to provide a two-way, 180-line, television-telephone service (having a bandwidth of c.

400 kHz) between Berlin, Leipzig, Nuremberg and Munich, and extensions were planned to Hamburg, Frankfurt, Cologne and Vienna. In London the balanced pair, 14 km cable link from the West End to Alexandra Palace was working well, transmitting 405-line video signals over a bandwidth of c. 2.5 MHz. And in the US Bell Telephone Laboratories had transmitted on 9 November 1937 television signals (of 800 kHz bandwidth) along a coaxial cable laid between New York and Philadelphia, a distance of approximately 320 km[50].

The patent for this cable had been filed on 23 May 1929 by Espenschied and Affel[51], and mentioned that one objective was its use as a wideband long-distance transmitting medium for television transmission. Prior to the New York–Philadelphia installation, the cable had entered service on 10 June 1936 when it provided the means for propagating the signals from NBC's studio in New York to the transmitter in the Empire State Building, a distance of 1.5 miles. The intended use of the new cable was, partly, the simultaneous transmission of 240 separate signals, each of 4 kHz bandwidth, of a carrier channel telephony system. Testing of the 960 kHz bandwidth of the cable using one signal demanded the use of a video signal generator. For this purpose a mechanical scanner and motion picture film projector formed the basis of the test equipment.

The six foot diameter steel scanning disc, driven by a ten horsepower motor, incorporated 240 large-aperture lenses, each located at the same distance from the centre of the disc, and rotated at 1440 r.p.m. (i.e. 24 revolutions per second). The lenses focused a light beam to give a square scanning spot on the film of size 0.003×0.003 inches. After passing through the film the light was incident on a photomultiplier and the output from it was applied to a modulator which shifted the band of picture frequencies from 0–800 kHz to 100–900 kHz, the region between 0 and 100 kHz being utilised for the sound and synchronising channels.

The receiver employed a specially designed cathode-ray tube, known as a Davisson tube, and gave an image of 8×10 inches. Although there was some flicker and the brightness of the image was low, it was reported that the picture was of high quality. 'The outstanding characteristics of the image were the crispness of the detail along each line, the sharp demarcation and detail in the shadows, and freedom from phase "hang-over" effects.'

The Post Office engineering department's catalogue of the difficulties associated with television networking did contain one optimistic comment. Given the international standing of Bell Telephone Laboratories the POED had no hesitation in stating that with the large specialist staff which BTL could 'put on to a problem of this kind there [seemed] little doubt that a complete solution [would] be available in the near future'. The department did not have long to wait.

On 21 May 1940 BTL[52] sent 441-line video signals with a bandwidth of 2.7 MHz along the coaxial cable from New York to Philadelphia and back. Demonstrations of such transmissions were subsequently given before the National Television System Committee (NTSC) on 8 November 1940, and before the Institute of Radio Engineers in January 1941. Previously, on 20 May 1939, standard telephone pairs had been used successfully as local transmission lines for the broadcasting of a six-day cycle race at Madison Square Garden, New York. These were the last pre-war demonstrations of television given by Bell Telephone Laboratories. During 1940 television activity ceased, increased effort thereby being made available for defence work.

Returning now to the Television Advisory Committee's proposals, these were submitted to the Treasury with the strong support of the Postmaster General. However, the Treasury was not disposed to sanction the proposed additional expenditure on television on 'a short-term policy'. They asked that the views of the BBC might be sought on certain alternative proposals on the financial provision for both sound and television broadcasting for a term of years. The Treasury considered that this approach would more effectively safeguard the interests of the Exchequer. But before the BBC could respond the Alexandra Palace broadcasts had been suspended (on 1 September 1939).

The closure caused some dismay among the general public and the BBC received many letters asking if it would be possible for the television transmissions to be restored. The corporation's response to these requests was that the television programmes had been withdrawn for reasons affecting national security and the question of resumption could, in wartime conditions, be taken only by the Government.

In Germany television broadcasts continued for several months after the commencement of hostilities and this fact led to questions being raised in the House of Commons in February and May 1940 about the prospect of the television transmissions being resumed. Essentially, the arguments against a resumption were:

1   the frequency bands which had been allocated for television were required for defence purposes;
2   the technicians who were capable of providing a service were needed for war work;
3   the Government had to conserve money and would not be justified in expending resources for a limited audience;
4   manufacturers of television sets were heavily engaged in supplying communications and radar equipment for the war effort.

These cogent points led to one conclusion: The London station had to remain closed for the duration of the war.

Although the post-1940 development of television is beyond the scope of this book, a synoptic view of the spread of television worldwide, to the year 1955, is given in Table 23.3. It is readily seen that just four television systems had reached the standard of excellence required for television broadcasting services. They were all-electronic, high-definition systems (see Table 23.2) based on 405, 525, 625, and 819 lines per picture; (the 625 (USSR) system was a variant of the 625 (CCIR) system). Since the 525-line and 625-line standards are closely similar, and the 405-line and 819-line standards are congruent, it follows that effectively the growth to 1955 of the world's television broadcasting networks has been founded on the US's 525-line, and the UK's 405-line systems. In the main, these stemmed from the endeavours of RCA and EMI. Thus, of the myriad of suggestions which were advanced from c. 1879 for 'seeing by electricity' (see Appendix 2), only those of two well-endowed companies, each having highly skilled and talented research scientists and engineers, operating in an environment conducive to the achievement of progress, succeeded.

Table 23.2 Television system parameters, 1953

system	405	525	625 (CCIR)	625 (USSR)	819
Number of lines per picture	405	525	625	625	819
Video bandwidth (MHz)	3	4	5	6	10.4
Channel width (MHz)	5	6	7	8	14
Sound carrier relative to vision carrier (MHz)	− 3.5	+ 4.5	+ 5.5	+ 6.5	− 11.15
Sound carrier relative to edge of channel (MHz)	+ 2.5	− 0.25	− 0.25	− 0.25	+ 0.10
Interlace	2:1	2:1	2:1	2:1	2:1
Line frequency (Hz)	10 125	15 750	15 625	15 625	20 475
Frame frequency (Hz)	50	60	50	50	50
Picture frequency (Hz)	25	30	25	25	25
Sense of vision modulation	positive	negative	negative	negative	positive
Level of black as % of peak carrier	30	75	75	75	25
Sound modulation	AM	FM	FM	FM	AM

Source: International Telecommunications Union. Quoted in 'Television, a world survey' (UNESCO, 1953) p. 6

What then can be said of the efforts of the many scientists, inventors, innovators and experimentalists who were induced to spend their time and labour seeking after the answer to Job's question: 'Canst thou send lightnings, that they may go, and say unto thee, "Here we are"?': were their efforts in vain? The answer must, of course, be no. As Fontenelle, more than two centuries ago, wrote:

> 'There is an order which regulates our progress. Every science develops after a certain number of preceding sciences have been developed, and only then: it has to await its turn to burst its shell.'

Television is an exemplar of Fontenelle's dictum. The science, and the concomitant technology, of all-electronic, high-definition television had to await the conclusions and the deductions of the enquiries which probed the sciences of photoelectricity, thermionic emission, electromagnetic wave propagation, electron optics, secondary electron emission, radio and electronics. Only then could the technology of television progress on a secure scientific basis. However, although some of the simplistic and empiric notions of the early workers were hopelessly naive and were destined to fail, nevertheless the thoughts and work of many of those who were fascinated by the prospect of distant vision pointed the way forward. By the mid-1930s all was ready for the introduction of high-definition television, the essential fundamentals on which the technology is founded had been determined. The time was opportune for its implementation; it could not have been established before, say, 1930. Nor could its introduction have been undertaken in the mid-1930s without the foresight and the resources of the type provided by EMI and RCA. The whole-

Table 23.3 *Post-war development of television (1955)*

Country; inhabitants per receiver	Date of first post-war transmission; line standard	type of service (initially)	No. of stations by mid-1955 & (January 1953)	Locations	No. of sets at given date & at (January 1953)	Transmissions
Alaska	December 1953 525	commercial	4	2 at Anchorage 2 at Fairbanks	30 000 at 31 March 1955	
Argentina 7056 to 2520	2 November 1951 625 (CCIR)	commercial	3 (1)	1 at Buenos Aires	50 000 at 30 April 1955 (2500 to 7000)	daily
Austria	August 1955	experimental	1	Dobling	500 at 1 August 1955	
Belgium	November 1953 625 (CCIR), 819	experimental	4	Brussels, Antwerp, Liege, Dudelange	48 000 at May 1955	daily
Brazil 751	18 September 1950 525	regular commercial	5 (3)	2 at Sao Paulo 1 at Rio de Janeiro	120 000 at 30 April 1955 (70 000)	
Canada 56	6 September 1952 525	regular	28 (both CBC & private) (2)	too numerous to list	est. 1 780 000 by January 1956	most broadcast c. ten hours daily
Colombia Cuba 55	June 1954 24 October 1950 525	experimental regular (commercial)	1 23 (7)	Bogata Havana, Matanasaz Santa Clara, Cama-guey, Santiago & Havana	10 000 at c. 1955 150 000 at 30 April 1955 (100 000)	daily
Curacao	1955/56	experimental (private station)	1			
Czechoslovakia	June 1954 625 (Sov)	regular	2	Prague, Ostrava	32 000 at January 1956	five days per week

Denmark 7150	October 1951 625 (CCIR)	regular (experimental broadcasts)	1	Copenhagen	at 30 June 1955 (600)	ten hours/ week
Dominican Republic	1 August 1952 525	regular (commercial)	1 (1)	Trujillo	5500 at December 1955 (1200)	
France 704	March 1945 (1935) 819 & 441	regular	11 (permanent & temporary) (2)		183 922 at 4 June 1955 (60 000)	c. 40 hours/ week
GDR	21 November 1952 (1937) 625 (Sov)		9 (1)		between 40 000 & 50 000 (–)	
GFR 8000	25 December 1952 (1937) 626 (CCIR)	regular experimental	(5)		c. 300 000 (6000)	daily
Hawaii	1 December 1952 525	regular (commercial)	3 (2)	Honolulu	200 000 at 31 March 1955 (–)	daily
Italy 7320	September 1949 April 1954 625 (CCIR)	regular (experimental)	1 (RAI) 9	2 at Milan, Torino, Rome, Monte-Penice, Genova-Pontefino, Monte-Sera, Firenze Monte-Pagelia, Monte-Venda	145 000 at May 1955 (5000)	daily 35 hours/ week
Japan 21 000	1 February 1953 525 (April 1940)	regular (NHK) (commercial)	3 2	Nagoya, Tokyo & Osaka Tokyo	115 000 at November 1955 (4000)	daily
Luxembourg	January 1955 819	regular				three to four hours/week
Mexico 578	31 August 1950 525	regular (commercial)	at least 6 (6)		100 000 (50 000)	

Table 23.3 continued

	Date (lines)	Status	No.	Location	Sets	Transmissions
Monaco	January 1955 819			Tele-Monte-Carlo	1200 at 30 April 1955	four hours/day
Morocco	February 1954 819	commercial service suspended in May 1955	2	Casablanca & Rabat	4000 sets sold	daily
Netherlands 2064–1032	October 1951 2 October from 55 625 (CCIR)	experimental regular	(1)		30 000 sets at May 1955 (5000–10 000)	very limited
Norway	14 January 1954	experimental	1	Oslo	12 at 31 December 1954	three times per week, no public transmissions
Phillipines	September 1953	experimental (Bolinao Electronics Corp of Manila)	1			
	end 1954	? (Republic Broadcasting System)	1	Manila	5000 at mid-1955	
Poland	25 October 1952 625 (Sov)	experimental	1 (1)	Warsaw	7000 at end of 1955	four to five hours/day four days per week
Puerto Rico	January 1954 525	television regulated by FCC	2	San Juan, Santurce	60 000 at 30 April 1955	
Saar	June 1954 819	commercial (Telesaar)	1		800 at 30 April 1955	four hours daily in German & French

Country	Start / system	Service type	Stations	Location	Receivers	Frequency
Spain	since 1952	experimental				two or three days/week
Sweden	October 1954 625 (CCIR)	regular (test transmissions)	1	Stockholm	5000 at 28 February 1955	one hour daily
Switzerland	November 1953 625 (CCIR)	experimental (Swiss television service)	1 (1)		8258 at 31 August 1955	almost daily
Thailand	June 1955	government	1	Bangkok	400 at May 1955	daily
USA 7	30 April 1939 525	commercial and non commercial	451 (inc. 17 noncommercial) at November 1955 (139)	too numerous to list	36 000 000 at 30 April 1955 (22 000 000)	
USSR 2400	May 1945 December 1945 625 (Sov)	regular	(3)	Moscow, Leningrad, Kiev, Kharkov, Riga, Odessa, Baku, Tallin	1 000 000 at 30 April 1955 (80 000)	daily
Venezuela	December 1952 625 (CCIR)	commercial and government controlled	3 (1)	Caracas	35 000 at 30 April 1955	
UK 24	7 June 1946 (2 November 1936) 405 30 August 1955 405	regular (non commercial) commercial	5 high power + 5 under construction 1 medium power + 6 low power + 1 under construction 1 + 2 planned (7)	London, nr. Birmingham, nr. Manchester, Cardiff, & 1 in Scotland London + Manchester Birmingham	4 155 989 at January 1955 (2 072 930)	daily, c. 50 hours per week

hearted support of the directors of these companies, the research leadership given by I Schoenberg and A D Blumlein, and by V K Zworykin and E W Engstrom, and the talents and genius of their powerful research and development staffs were essential factors which enabled modern television to be accomplished. The approaches of EMI and RCA to the problem were wholly admirable and commendable: all the more so when the nature of the times (the early 1930s) and of invention are considered. Thomas Sprat in his 'History of the Royal Society' (1667) summed up well the process of invention when he wrote:

> 'Invention is a heroic thing and placed above the reach of a low and vulgar genius. It requires an active, a bold, a nimble, a restless mind: a thousand difficulties must be contemned with which a mean heart would be broken: many attempts must be made to no purpose: much treasure must be scattered without any return: much violence and vigour of thought must attend it: some irregularities and excesses must be granted that could hardly be pardoned by the severe rules of prudence.'

By such invention television has evolved.

References

1   ANON.: 'BBC Annual 1937' (BBC Publication)
2   MARSLAND GANDER, L.: 'Professional touch needed'. *Daily Telegraph*, 24 November 1936
3   ANON.: 'More syncopated dance music', *Evening News*, 10 December 1936
4   ANON.: 'Television to be brighter', *Daily Telegraph*, 21 December 1936
5   Minutes of Sir Noel Ashbridge, remarks made at a Television Advisory Committee, 19 December 1935, Post Office bundle 5536
6   Minutes of the Twenty-eighth meeting of the Television Advisory Committee, 24 June 1936, Post Office Bundle 5536
7   BBC announcement: 'Television in Coronation week', 22 April 1937
8   ANON.: 'The future of television. Coronation success', *Observer*, 23 May 1937
9   Paper 35, Television Advisory Committee, Post Office bundle 5536
10 MILES, W.G.H.: Letter to F W Phillips, 23 January 1936, Post Office bundle post 33/5143, file 7
11 See British patents 452 713 (December 1934), 452 772 (February 1935)
12 COLLARD, J.: 'London's television twin cable links', *POEEJ*, 1937, **30**, Part 3, pp. 215–221
13 BARRINGTON, J.: 'Television is perfect for two minutes', *Daily Express*, 12 November 1937
14 Minutes of the forty-eighth meeting of the Television Advisory Committee, 14 January 1938, Post Office bundle 5536
15 Technical Subcommittee, minutes of the fourth meeting, 26 February 1935, minute post 33/5533
16 ANON.: 'Reviewing the video art', *Electronics*, January 1938, pp. 9–11
17 SHOENBERG, I.: 'Relay of television by radio links',15 June 1938, post 33/5536, file 52, Post Office Records Office
18 ANON.: 'Television in the field', *Electronics*, June 1939, pp. 13–15, 90
19 'Annual abstract of statistics' no. 85, 1937–1947 (HMSO, London, 1948)
20 Paper 48, Television Advisory Committee, Post Office bundle 5536

21   Minutes of the forty-fourth meeting of the Television Advisory Committee, 5 October 1937, Post Office bundle 5536
22   Minutes of the forty-ninth meeting of the Television Advisory Committee, 25 January 1938, Post Office bundle 5536
23   Minutes of the forty-sixth meeting of the Television Advisory Committee, 7 December 1937, Post Office bundle 5536
24   ANON.: 'Television receivers–reduction in prices', *The Times*, 9 February 1937
25   SELSDON: Letter to N Ashbridge, 10 February 1937, Post Office bundle 5536
26   OSBORNE, G.M.: Letter to J H Brebner, press controller, GPO, 19 February 1937, Post Office bundle 5536
27   Minutes of the fortieth meeting of the Television Advisory Committee, 26 February 1937, Post Office bundle 5536
28   VARLEY ROBERTS, J.: Memorandum to F W Phillips, 5 March 1937, Post Office bundle 5536
29   Editorial: 'Television anniversary', *Wireless and Electrical Trader*, 2 October 1937
30   RMA proposed statement to the press. Post Office bundle 5536
31   Paper 55, Television Advisory Committee, Post Office bundle 5536
32   BLUMLEIN, A.D.: 'The transmitted waveform', *J. IEE*, 1938, **83**, pp. 758–766
33   Minutes of the fiftieth meeting of the Television Advisory Committee, 29 April 1938, Post Office bundle 5536
34   ESPENSCHIED, L.: 'Notes on television developments in Europe, 21 July 1937, 37127–1, AT&T Archives, Warren, New Jersey
35   GILMAN, G.W.: Memorandum on discussion of television in Mr R Bown's office, 11 March 1938, 37127-1, AT&T Archives,Warren, New Jersey
36   ESPENSCHIED, L.: 'Technical developments abroad', 26 May 1938, 37127-1, AT&T Archives, Warren, New Jersey
37   ESPENSCHIED, L.: 'Television in London', 20 May 1938, 37127-3, AT&T Archives, Warren, New Jersey
38   ESPENSCHIED, L.: 'RCA Television demonstration', 25 May 1938, 37127-2, AT&T Archives, Warren, New Jersey
39   Paper 54, Television Advisory Committee, Post Office bundle 5536
40   Minutes of the forty-ninth meeting of the Television Advisory Committee, 25 January 1938, Post Office bundle 5536
41   Papers 54A and 54B, Television Advisory Committee, Post Office bundle 5536
42   ROYAL, J.: 'On a holiday overseas', *New York Times*, 29 May 1938, **X**, 10:1
43   COCK, G.: 'My impressions of American television', *Television and Short-wave World*, August 1939, pp. 453–455
44   BOULT, Sir Adrian.: 'Music versus cameras', *New York Times*, 29 May 1938, **X**, 10:6,7
45   PHILLIPS, F.W.: 'Television'. Paper, 11 April 1940, Post 33/5533, file 19 Post Office Records Office
46   C.A.T.: 'Television'. A paper, 1 June 1943, minute P327, file 5, Post Office Records Office
47   ANON.: 'Provincial traders urge television speed-up', *The Wireless and Electrical Trader*, 8 April 1939, **58**, (764), p.45
48   ANON.: 'Television manufacturers explain their offer', *The Wireless and Electrical Trader*, April 1939, p. 127
49   'Memorandum on the transmission of television signals over cable and radio links'. Post Office Engineering Department, 6 July 1938, Post 33/5536, file 52, Post Office Records Office
50   ANON.: 'Bell Labs test coaxial cable', *Electronics*, December 1937, pp. 18–19
51   ESPENSCHIED, L., and AFFEL, H.A.: US patent 1835 031, 9 December 1931
52   ANON.: 'Principal Bell System dates in television', November 1946, AT&T Archives, Warren, New Jersey

Appendix 1

The charge storage principle

The essential feature of the charge storage principle is that, during part at least of a picture frame period, the photoelectric emission due to the incident light falling on a mosaic of photoelectric cells is stored as a charge on the mosaic, each of the cells of which is associated with a capacitor. The capacitors are then discharged in sequence by a switching mechanism, such as a beam of electrons: the resulting train of electrical pulses constitutes the picture signal.

The principle[1] can easily be illustrated by comparing the magnitude of the output of an iconoscope with the output obtained from a photocell and Nipkow disc unit, Figure 16.10. From the Figure, the voltage developed across R is $R(F/n)S$ where F is the luminous flux corresponding to the total image, S is the sensitivity of the photocell, n is the number of picture elements and R is the input resistance.

The time constant CR of the input circuit (C being the capacitance to ground of the photoelectric element and associated circuits) must clearly be equal to or less than $1/(Nn)$, the time taken to scan a picture element, (N being the number of picture frames per second). Hence, substituting for R, the voltage across R becomes $FS/(NCn^2)$.

Now the charge Q on one picture element of the iconoscope mosaic is equal to $(F/n)St$, where t, the time during which light is incident on the element, is $1/N$. Therefore, the output voltage V, which is equal to Q/C' (C' being the total input capacitance to ground of the iconoscope and associated circuits), is determined by V equal to $FS/(nNC')$.

From these two expressions for the output voltages, the ratio of the outputs is nC/C'; or n, if C' is equal to C. Hence, for a picture mosaic having 70 000 elements, the net theoretical gain of the iconoscope compared with the conventual Nipkow disc system is 70 000 times.

By 1933 Zworykin had achieved an increase in gain of about ten per cent of n, giving a net gain of several thousand times.

The principle of charge storage is one of the most important in television science. It is a principle which is inherent in all television cameras of the iconoscope/ emitron, super emitron and orthicon types, but not in Farnsworth's image dissector tube.

The principle according to some writers seems to be implicit in Campbell Swinton's description[2] of his camera tube. As stated in Chapter 6, he wrote: 'It is further to be noted that, as each of the metallic cubes in the screen acts as an independent photoelectric cell, and is only called upon to act once in a tenth of a second, the arrangement has obvious advantages over other arrangements that have been suggested in which a single photoelectric cell is called upon to produce the many thousands of separate impulses that are required to be transmitted through the line wire per second, a condition which no known form of photoelectric cell will admit of. Again it may be pointed out that sluggishness on the part of the metallic cubes . . . in acting photoelectrically in no wise interferes with the correct transmission and reproduction of the image, provided all portions of the image are at rest.'

Professor J D McGee considered[3] that, although the storage principle is not stated explicitly in this or any other paragraph of Campell Swinton's writings, it is difficult to escape the conclusion that it is this principle which he had in mind in the paragraph quoted above. 'This is borne out by the specific and separate reference to the sluggishness of the photoelectric effect he opined.

Professor F Schroter[4] was also in agreement with this view, but Dr D Gabor[5] could not accept it. 'There is no doubt that he [Campbell Swinton] had only photo conductivity in mind, otherwise he would not have referred to the sluggishness of the cells. This effect is eliminated by the principle of using a great number of cells and scanning them in succession. But as regards storage, a photoconducting cell is a rather different matter from a photoelectric layer backed by an insulator. The photocell integrates over any length of time. On the other hand, the best a photoconducting cell can do is to reach an equilibrium with the illumination, and this equilibrium is only delayed by the sluggishness due to transient polarisation. Miller and Strange have found certain charge storage effects, similar to those exhibited by every imperfect dielectric, but these are not due to the illumination but to charges produced by the scanning beam. These produce a "memory" effect, which is, however, pronounced not in the bright but in the dark parts of the picture. The illumination itself produces no surface charges.'

Gabor's objections seem to have been founded on the premise that Campbell Swinton had in mind only photoconductive mosaics. But he specifically mentioned that the metallic cubes which composed his mosaic would be 'made of some metal such as rubidium . . . ' Consequently, his mosaic was a photoemissive mosaic and not a photoconducting mosaic.

Dr H G Lubszinski[6] also did not agree with McGee's conclusions. ' . . . I feel certain that at the time he could not [possibly] have realised the importance of storage.'

The controversy has arisen because Campbell Swinton's description of his camera tube makes no reference to a signal plate or a means of charge storage. It is most likely that, in 1911, he did not appreciate the significance of charge storage as a means of improving the signal-to-noise ratio of the picture signal generating system, since the concept of signal-to-noise ratio was unknown at that time and electronic amplifiers had not yet been invented. However, it was obviously desirable in the pre-electronic era for inventors to devise picture signal producing devices which would give as large a signal as possible. This is what Campbell Swinton sought to achieve.

If a single scanning cell is used in a television camera the cell, ideally, must respond to light changes in $1/(Nn)$ of a second. For N equal to ten frames per second and n equal to 10 000 (corresponding to a square image of 100 lines) the response time of the cell must be less than 10 µs. On the other hand, if a scanned mosaic of n cells is utilised each cell has a time of 100 000 µs to react to the changes of incident light flux provided that each cell is connected to an appropriate circuit storage element. Campbell Swinton's account clearly shows that he partially understood these facts and was concerned to maximise the output signal from his proposed camera tube.

In 1911 it was known that the response of certain photosensitive materials, particularly selenium, was not instantaneous when subjected to rapidly changing light fluxes. Sale had determined the transient light characteristic of selenium in 1873 and other experimenters had shown that as the frequency of an intermittent light source, incident on a bar of selenium, increased, so the change in resistance decreased; that is, the selenium bar was sluggish in its response to light flux changes.

This response is analogous to the charging and discharging of a capacitor (a circuit storage element) via a resistor. It is possible that Campbell Swinton reasoned by analogy from these well known (in 1911) observations that selenium possessed some form of inherent storage capacity when exposed to light. Surely then, (he, possibly, argued) just as the charge of a capacitor increases as the charging time increases, so the photoconductive/photoelectric effect would likewise increase with increasing exposure of a cell to the incident light; i.e. a cell exposed for 10^{-1}s would give a much greater output than a cell exposed for 10^{-5}s. Unfortunately, Campbell Swinton, who was not professionally educated as an engineer, failed to realise the circuit connections necessary for the practical implementation of this argument, even though he was thinking along the correct lines.

References

1 ZWORYKIN, V.K.: 'Television with cathode rays' *Journal IEE*, 1933, **73**, pp. 437–451
2 SWINTON, A.A.C.: Presidential address, *J. Rontgen Society*, January 1912, **8**, 30, pp. 1–5
3 McGEE, J.D., and LUBSZYNSKI, H.G.: 'EMI cathode ray television transmission tubes', *J. IEE*, 1938, **84**, pp. 468–482
4 SCHRÖTER, F.: 'Handbuch der Bildtelegraphie und des Fernsehens' (Julius Springer, Berlin, 1932) p. 61
5 Ref. 3, p. 476
6 LUBSZYNSKI, H.G.: 'Some early developments of television camera tubes at EMI Research Laboratories' *IEE Conf. Publ.*, 271, 1986, p. 60

Appendix 2
Some patents on scanning 1889–1933

No.	Patent no. (Journal)	Patent date country	Patent holder/ inventor	Type of scanner
1	30 105	6/1/1884 (G)	Nipkow	aperture disc, single spiral
2	(Lumière Électrique)	1889	Weiller	mirror drum
3	5031	24/2/1897	Szczepanik	vibrating mirrors (electrical)
4	(Lumière Électrique)	1898	Brillouin	two lens discs
5	26 586	6/12/1904	Belin	vibrating mirrors (mechanical)
6	29 428	31/12/1904	Ribbe	aperture band
7	27 570	13/12/1907	Rosing	mirror polyhedron
8	7219	1/4/1908	Adamian	aperture disc, single spiral
9	(Nature)	1908	Campbell Swinton	electron scanned photo matrix
10	30 188	24/12/1908 (UK) 24/12 1908 (Dn)	Anderson	aperture disc, spiral segments and aperture band
11	32 220	24/1.1910 (Sw)	Ekstrom	rotating oblique mirror
12	5486	4/3/1911	Rosing	mirror polyhedra with luminescent screen
13	15/270	25/6/1914	Hart	lens drum
14	161 706	19/1/1920 6/8/1920	Baden-Powell	mirror drum with plane or concave mirrors
15	200 643	16/5/1922	Wade	mirror polyhedron
16	185 463	4/5/1921	Whiston	aperture band
17	209 049	16/11/1923 (UK) 27/12/ 1922 (Fr)	Belin	oscillating mirrors (electrical)

18	209 406	18/12/1923 (F)	Mihaly	mirror galvanometer (in conjunction with phonic motor)
		2/1/1923 (H)		
19	218 766	18/4/1923	Stephenson and Walton	(*i*) mirror polyhedra (*ii*) mirror polyhedron and disc (*iii*) two aperture discs
20	222 597	12/7/1923	Robb	vibrating mirrors (electrical)
21	222 604	26/7/1923	Baird	commutated lamp bank
22	225 860	4/12/1924 (UK)	Nisco	aperture-drum
		4/12/1923 (I)		
23	228 961	7/9/1923	Western Electric	mirror galvanometer, CRO
24	230 576	29/12/1923	Baird	Discs, lenses in spiral
25	236 464	8/1/1924	Sensicle	mirror polhedra (2)
26	240 463	23/9/1925 (UK)	BTH(ass. by	aperture-endless band (Tr), aperture-disc, single spiral and multispirals with slotted drum
		23/9/1924 (US)	Hoxie)	
27	252 387	20/5/1926 (UK)	Metro-Vick	vibrating mirrors (electrical using piezo electric crystals and electromagnet)
		25/5/1925 (US)	(ass. by Whitten)	
28	253 957	1/1/1925	Baird	(*i*) Lens disc with slotted member disc, lenses in spiral (*ii*) two lens discs, circle of lenses with rocking mirror
29	255 057	3/7/1926 UK)	Zworykin	CRO use of mosaic colour screens
		13/7/1925 (US)		
30	261 195	30/11/1925	Belin	vibrating mirrors (mechanical)
31	263 005	23/3/1926	Baxter	aperture discs, one with single spiral and one with radial slots
32	264 174	7/1/1927 (UK)	Telefunken	mirror galvanometer
		8/1/1926 (G)		
33	265 640	6/8/1925	Baird	circle of lenses
34	266 564	1/9/1925	Baird	disc, lenses in spiral
35	266 591	1/9/25	Baird	aperture disc. multi spiral
36	269 658	20/1/1926	Baird	lens disc, aperture disc prisms (two)
37	271 131	9/2/1926	Clay	aperture disc, single and slotted (Tr) CRO aperture band
38	279 067	4/10/1927 (UK)	Westinghouse	rotary prism and rotating disc with concave mirrors
		15/10/1926 (US)		
39	279 457	19/10/1927 (UK)		B.T.H. (Ass. of Alexanderson mirror drum with lens, lens drum)
		19/10/1926 (US)		
40	280 630	17/8/1926	Hall	two slotted discs, one with radial slots and one with curved or radial slots
41	287 643	24/12/1926	Rtcheouloff	magnetically moved photo-element or fluorescent element

42	288 680	4/1/1927	Rtcheouloff	magnetically moved photo-element or fluorescent element
43	289 307	15/10/1926	Baird	two aperture/lens discs, many arrangements
44	290 245	11/5/1927 (US)	Zworykin	rotary prisms
45	291 365	30/5/1928 (UK) 30/5/1927 (F)	Thurm and Gaisenband	mirror polyhedron and drum
47	291 786	26/4/1928 (UK) 10/6/1927 (G)	Telefunken	spiral of cells or lamps on disc (R)
48	293 308	2/7/1927 (US)	ERP (ass. of Ives)	aperture disc single spiral
49	293 474	3/3/1927	Dawson and Milner	moving arc associated with stationary apertured screen
50	294 257	20/7/1928 (UK) 21/7/1927 (F)	Thurm	aperture band
51	294 267	21/1/1927	Baird	aperture disc, multi spiral
52	295 653 (void)	16/8/1927	Szenyovszky and Schulzsche AG	(*i*) two drums (*ii*) two discs
53	297 078	19/3/1928 (UK) 14/9/1927 (US)	ERP (ass. of Hartley and Ives)	aperture disc, single spiral, commutated lamp bank (R) (intermediate film)
54	297 147	16/6/1927	Valensi	two discs
55	297 152	17/6/1927	Beatty	aperture discs, pairs of spirals
56	298 245	20/7/1928 (D)	Siemens	vibrating mirrors (mechanical)
57	298 255	8/6/1927	Aspden	(*i*) spiral of cells/lamps on disc (*ii*) aperture disc, multi spiral
58	299 076	20/6/1927	Baird	lens disc, circle of lenses
59	302 187	11/6/1927	Baird	commutated bank of lamps
60	302 228	28/11/1928 (UK) 12/12/1927 (D)	Telefunken	aperture disc, multiturn spiral
61	302 240	6/4/1927 (US)	ERP (ass. of Ives and Gray)	aperture disc, single spiral
62	303 771	5/7/1927	Baird	two aperture discs
63	304 730	25/1/1928	Strange	aperture drum
64	305 079	26/9/1927	Rowe and Rowe	two discs or bands
65	306 961	27/2/1929 (UK) 29/2/1928 (US)	Thuau and Antranikian	circle of lenses (with or without slotted mask)
66	308 277	20/3/1928 (US)	BTH (ass.of Kell)	aperture dixc, spiral segments
67	309 965	19/10/1927	Baird	aperture disc, lens disc.
68	314 591	4/1/1928	Baird	aperture disc, spiral segments
69	315 308	31/5/308 (US)	Ives	(*i*) aperture disc, single spiral (*ii*) aperture disc, spiral segments
70	315 362 (void)	12/7/1928	Tihanyi	electron scanned photo matrix

71	315 417	9/7/1929 (UK) 13/7/1928 (US)	Nyquist	aperture discs, single spiral and spiral segments
72	361 85	19/7/1929 (UK) 24/7/1928 (US)	Mertz (ass. to ERP)	aperture disc, spiral segments
73	318 331	22/6/1928	Roberts	electron phalanx (Tr)
74	318 565	1/6/1928	Ass. T and T (ass. of Pollak)	electron scanned photo matrix
75	319 307	20/6/1928	Baird	aperture disc, spiral segments (Tr), single spiral (R)
76	319 454	2/8/1928	De Wet	slotted disc
77	320 999	28/8/1928	Shinton	mirror drum and slotted disc
78	321 389	5/6/1928	Baird	aperture disc, spiral segments
79	321 441	11/7/1928	Baird	aperture disc, spiral segments
80	321 935	9/8/1928	Sharma	luminescent screen (R)
81	320 993	24/8/1928	Sharma	electron scanned composite plate
82	322 822	11/7/1928	Baird	spiral of cells, lamps on disc
83	324 904	4/10/1928	Baird	commutated contacts (R)
84	325 362	24/1/1929	Aspden	aperture disc, spiral segments
85	325 790	24/10/1928	Todd	lens disc with slotted member
86	326 251	10/10/1928	Baird	lensed disc with slotted member
87	328 286	25/10/1928	Walton	Fixed prisms and moving element
88	328 616	31/1/1929	Baird	drum with concave mirrors
89	332 284	18/4/1929	Western Electric	aperture disc, single spiral
90	332 559	22/4/1929	Philips	aperture disc, spiral segments
91	332 971	9/5/1929	Mather	aperture-endless band
92	334 234	29/5/1929	Codelli	mirror oscilloscope
93	335 638	2/7/1929	Western Electric	aperture disc, single spiral
94	336 267	13/7/1928 (US)	Nyquist	aperture disc, multiple spiral
95	341 353	10/4/1930 (UK 15/4/1929 (US)	BTH (ass. of Kell)	lens disc
96	343 084	13/11/1929	HMV and Bowman Manifold	lens disc, circle of lenses
97	346 456	29/10/1929	Ehrenhaft	modified lens disc
98	347 435	29/3/1930 (UK) 1/4/1929 (US)	MWT (ass. of Smith)	mirror drums and fixed mirrors
99	347 969	10/7/1930 (UK) 14/8/1929 (US)	BTH (ass. of Bentley)	two discs or bands disc drum
100	348 139	5/2/1930	Fleming	scintillating disc
101	348 211	14/2/1930	Baird	commutated lamp bank
102	348 260	1/3/1930 (UK) 13/3/1929 (US)	Telefunken	slotted disc
103	348 414	30/5/1930 (UK)	Telefunken	mirror drum

		8/6/1929 (US)		
104	348 638	7/2/1930	Baird	disc, lenses in spiral mirror drum
105	349 773	27/5/1930 (UK) 27/6/1929 (US)	Smith	aperture disc, single spiral
106	350 512	13/2/1930	Gardiner and Wilson	aperture disc
107	350 926	13/2/1930	Kolster-Brandes	two piezoelectric crystals
108	351 972	1/4/1930 (UK) 1/4/1929 (US)	MWT (ass. of Cioffari)	mirror drum and fixed mirrors
109	352 693	19/6/1930 (UK) 19/6/1929 (D)	Skaupy	prisms moving or stressed
110	353 471	24/4/1930 (UK) 25/5/1929 (US)	ERP (ass. of Horton)	aperture disc, multispiral
111	354 572	28/8/1930 (UK) 25/9/1929 (F)	Comp-a-Gaz	aperture disc, spiral segments
112	355 795	24/5/1930	Maloney	aperture disc, multiple spiral
113	355 890	7/7/1930 (UK) 5/7/1929 (US)	Zworykin (Ass. to Westinghouse)	mirror galvanometer (Tr) CRO (R)
114	356 760	11/3/1930	Von Bronk	commutated cell vibrating mirrors, electrical
115	356 880	21/7/1930 (UK) 19/7/1929 (US)	MWT (ass. of Smith)	aperture drum
116	357 687	21/7/1930 (UK) 19/7/1929 (US)	MWT(ass. T A Smith)	aperture or lens disc, multiturn spiral
117	358 050	2/7/1930 (UK) 9/7/1929 (US)	BTH (ass. of Moore)	rotatable mirror
118	358 087	1/7/1930 (UK) 1/7/1929 (F)	Loiseau	crystal, moving zone of compression
119	358 183	7/8/1930 (UK) 8/5/1929 (US)	Communication	moving arc
120	358 411	16/3/1931 (UK) 25/3/1930 (G)	Hatzinger	mirror screw
121	358 433	2/7/1930	Maloney	aperture disc, single spiral aperture drum
122	358 916	14/7/1930	Wilson	scintillating drum interior mirrored drum
123	358 920	14/7/1930	Triggs	miror polyhedron
124	359 981	30/7/1930	Baird	commutated lamp bank
125	360 850	7/8/1930 (UK) 31/5/1930 (US)	Communication	moving arc
126	360 942	6/8/1930	Baird	mirror drum, mirror drum with lens
127	362 144	21/11/1930	Browne	lens band
128	362 950	12/9/1930	Gretton and Haskell	two slotted drums

129	365 241	12/11/1930 25/8/1931	Baird	lens disc
130	365 632	20/10/1930	Aspden	aperture drum
131	366 045	24/7/1930 (UK) 24/7/1929 (US)	Fries	mirror polyhedron and rocking mirror
132	366 392	20/5/1931 (UK) 24/5/1930 (F)	Vorobieff	rotary prisms
133	366 453 (void)	31/7/1930 (UK) 23/7/1929 (Switz)	Buol	lens drum
134	366 477	29/9/1930 (UK) 30/9/1929 (US)	Kell	cathode-ray (receiver)
135	368 069	9/3/1931 (UK) 8/3/1930 (G)	Rosenfelder	mirror drum with curved mirrors
136	368 262	24/11/1930 (UK) 22/11/1929 (US)	BTH (ass. by L J Hartley)	aperture drum, single spiral (Tr) Aperture disc and rotating prism (R), (disc has circle of holes)
137	368 309	27/8/1930	Farnsworth	electron phalanx (Tr)
138	373 288	26/2/1931	MWT and Dowsett	aperture drum
139	373 539	4/11/1931 (UK) 4/11/1930 (F)	Comp-a-Gaz	mirror drum and fixed mirrors
140	373 540	10/11/1931 (UK) 26/11/1930 (F)	Comp-a-Gaz	oscillating mirror with static mirrors
141	374 094	6/2/1931	Marconi and Dowsett	mirror polyhedron and drum
142	374 114	4/3/1931 8/5/1931	Baird	aperture disc, multiple spiral
143	374 391	30/1/1932 (UK)	Karolus	mirror drum with curved mirrors
144	374 564	9/4/1931	Baird	mirror polyhedron and fixed mirror
145	375 385	23/3/1931	MWT and Dowsett	aperture disc/drum
146	375 589	27/9/1931 (UK) 7/8/1930 (US)	Cawley	aperture disc
147	377 187	13/1/1931	Todd	mirror polyhedron and lens circle
148	377 622	30/3/1932 (UK) 13/4/1931 (US)	Zworykin (ass. to MWT)	electron scanned fluorescent screen
149	378 079	6/5/1931	MWT and Dowsett and Levin	mirror drum and fixed mirrors
150	378 948	20/5/1931	MWT and Dowsett, Levin	mirror polyhedron and drum

			and Walker	
151	379 303	20/2/1931	Walton	fixed prism and moving element
152	379 317	20/2/1931	Walton	fixed prisms and moving element
153	380 419	21/6/1932 (UK) 27/6/1931 (G)	TKD	mirror screw
154	380 839	24/6/1931 (UK) 28/6/1930 (US)	ERP	aperture disc/drum
155	381 898	30/5/1932	Baird	mirror drum and slotted disc
156	384 463	18/1/1932	Koeppe	aperture band
157	384 947	30/7/1932 (UK) 5/8/1931 (G)	TKD	mirror screw
158	386 183	14/4/1932 (UK) 24/4/1931 (US)	TV Laboratories	moving arc
159	387 087	2/12/1931 (UK) 22/12/1930 (G)	Ardenne	electron scanned fluorescent screen
160	387 206	26/5/1932 (UK) 30/5/1931 (G)	Siemens	electron scanned fluorescent colour screen
161	387 536	29/3/1932 (UK) 27/3/1931 (G)	Ardenne	cathode ray
162	388 071	19/8/1931 (UK) 30/12/1930 (US)	Communication	moving arc
163	388 422	25/8/1931 (UK) 26/8/1930 (US)	ERP	disc and drum
164	389 751	27/9/1932	Jackson	slotted cylinders
165	390 158	28/9/1931 (UK) 4/10/1930 (US)	ERP	aperture disc
166	392 730	29/4/1932	Baird	mirror drum and slotted disc
167	393 657	12/12/1931 (UK) 13/12/1930 (US)	ERP	circle (s) of lenses, disc drum, two disc or bands
168	394 446	18/3/1932 (UK) 14/4/1931 (G)	Von Laczay	mirror wreath with rotating mirror
169	394 597	30/12/1932 (UK) 6/1/1932 (US)	Ballard (ass. to MWT)	electron scanned photo matrix (Tr)
170	394 904	23/1/1933 (UK) 2/1/1932 (F)	Comp-a-Gaz	mirror drum and fixed mirrors
171	395 985	20/2/1933 (UK) 20/2/1932 (G)	Telefunken	cathode ray
172	397 648	5/7/1932	Cooper	aperture drum with slits, drum with strip mirrors (slits/strips at different angles)
173	399 108	28/3/1933 (UK) 28/3/1932 (US)	Peck	scintillating disc

175	402 069	11/3/1932	Hazell and Dent	bank of cells and bank of neon lamps
176	405 977	19/6/1931 (G) 16/6/1932 (UK)	Von Ardenne	electron scanned screen
177	406 709	2/9/1932	MWT and Dowsett and Walker	mirror drum and reflectors
178	411 489	8/10/1932	MWT and Dowsett and Kemp	aperture/or lensed drums one or two pairs used
179	413 894	13/9/1933	RCA	iconoscope (Tr), CRT(R), apertured disc (Tr) CRT(R)
180	414 730	9/2/1933	EMI and Browne	mirror polyhedron and slit
181	415 036	14/2/1933	Baird	lens disc and slotted disc
182	418 527	25/5/1933	Baird	mirror drum and sloted disc
183	418 759	2/6/1933	Baird	mirror polyhedrons (three or more)
184	419 120	27/11/1933 (UK) 26/11/1932 (G)	IMK sydicate	static mirror drum and rotating mirror
185	420 391	19/7/1933 (UK) 19/7/1932 (US)	MWT (ass. of Ballard)	CRT (interlaced scanning)

Characteristics of mechanical optical scanning systems

	Scanning sytem	Light output expression in lumens	Light output in lumens =lux for 1m² picture			B	D	t	p	q	k
			120 lines	240 lines	405 lines						
1	Aperture disc	$\pi^3 BD^n t/4N^2 k^2 = A$	0.7	0.045		80 000	100	0.1	1		2
2	Mirror drum	$32\pi^4 BD^2 t/(Nn)^2 = M$	0.004	0.000065		80 000	100	0.1	1		
3	Double reflection system	$M.4$	0.016	0.000028		80 000	100	0.1	1		
4	Double mirror drum	$M.p^4/2$	1.6	0.018		80 000	50	0.1	4		
5	Mihaly-Traub	$M.2q^2$	0.08	0.0012		80 000	10	0.1	1	30	
6	Combination of 5 and 2	$M.q^2 p^4$	3.2	0.36		80 000	10	0.1	1	10	
7	Split beam system	$M.q^2$	0.004	0.000065		80 000	10	0.1	1	10	
8	Combination of 7 and 2	$M.q^2 p^4/2$	0.025	0.00042		80 000	10	0.1	4	10	
9	Mirror screw virtual image	$\pi Bt/N$	0.3	0.075		2000		0.1	1		
10	Mirror screw real image	$M.\pi/n$	0.002	0.000063		80 000	$r = 5$	0.1			2
11	Lens disc	A	0.7	0.045		80 000	100	0.1			2
12	Combination of 11 and 2	$A.p^2/2$	0.9	0.05		80 000	50	0.1	4		2
13	Multiplying Lens drum	$A.2p^2$	2.8	0.18		80 000	50	0.1	4		2
14	Concave lens disc and 2	$A.p^2/2$	3.6	0.24		80 000	100	0.1	4		2
15	Aperture disc film transmitter	$A.p^2/4$	0.15	0.01	0.0015	20 000	20	0.1	4		2
16	Scophony film transmitter	$A.4p^2$	1.40	0.10	0.018	20 000	20	0.1	4		2
17	Mihaly-Traub film transmitter	$M.2q^2 p^4$	10.4	0.9	0.05	20 000	10	0.1	4	10	2

In all the above cases the picture ratio $K = 4/3$, t is the transmission loss ratio.

B = brightness, of a source of light
D = effective diameter of scanning member
K = picture ratio, length/width
N = number of picture elements, $N = Kn^2$
k = aperture ratio or f/number of lens
n = number of lines
p = number of revolutions of scanning member to effect one complete picture scan
q = number of ring, or complementary, mirrors
t = transmission loss factor

Taken from: Myers, L.M.: 'Television optics' (Pitman, London, 1936)

Appendix 4

'Brief survey of the present television situation'

(*The Gramophone Company Ltd, 27 October 1930*)

Prior to engaging in television development, The Gramophone Company carried out a survey of the patents which related to television, and made assessments of the systems which were then being engineered. The Company's conclusions influenced its future television policy and led to its work on zone television. Extracts from the 'Brief survey of the present television situation' are given at length since the unpublished survey is unique and has some historical importance. The document highlights the 1930 patent position, the limitations of the various existing low-definition television systems and the need for a 'picture of at least 150 lines':

'With the possible exception of the cathode ray system, the whole television field has been almost completely explored and [sic] patents covering the essential details have long since expired. This is shown in the following list:

1	Transmitting and receiving apparatus		
	a	Nipkow disc	1884
	b	lens wheel	1891
	c	cylinder with a spiral of holes or lenses, probably about	1895
	d	Weiller's wheel	1889
	e	mirror oscillograph	1897
	f	cathode-ray tubes	1906
2	Light sensitive cells		
	a	selenium cell	1880
	b	photocell	1907
3	Modulated light sources		
	a	Kerr cell	1884
	b	filament lamps	1875

c	cathode-ray tubes	1906
d	light valve	1907
e	glow lamp with flat plate	1908
f	glow lamps with commutator (about)	1924

4 Synchronising devices
  a   use of a.c. mains
  b   transmission of signals on a separate line or carrier $\Big\}$ patent freee
  c   transmission of synchronising impulses along the same channel as the
      picture signals

The only possibilities in this list for development work to give a strong patent position are:

1 Cathode-ray methods, which are being developed extensively by RCA.
2 Improvements in modulated light sources.
3 Development of methods other than cathode ray which will produce a fine grain picture and which probably will utilise more than one channel.

This will be clear from the following summary of systems which are now in use or which have been reported in the press.'

(*The survey continues with short descriptions of the systems of Baird Television Company, Mihaly, Telefunken, Jenkins Television Corporation, The Short Wave and Television Company, and RCA.*)

'RCA
Under this heading must now be included the work carried out by the GEC and Westinghouse. Their reports indicate that they consider the cathode-ray method of reception to be the one of immediate importance, since it is likely to be most profitable from the point of view of a sound patent position, and is the most probable method for home use in the immediate future.

We have had no detailed reports up to date but the system uses a 80 line picture, 12 pictures per second, size of picture approximately five inches square.

It is reported that RCA expect to have this system in a commercial form by the end of the present year.

The advantages of this system are:

1 Simplicity of operation.
2 Absence of mechanically moving parts.
3 A greater scope for further development.

Its disadvantages are:

1 The present picture is small.
2 High voltages are required.
3 Probable short life of cathode tubes.
4 Greenish colour of the picture and the difficulty of maintaining a uniformly illuminated screen.'

(The survey continues with a mention of Zworykin's work on cathode ray tubes, and a description of GEC's work under the supervision of Dr Alexanderson.)

'The GEC has also been experimenting with a multichannel system which comprises a lens disc scanner in conjunction with seven photocells. The receiving device is a mirror wheel with seven Kerr cells as the modulated light sources.

This system gave a picture of 168 lines, and bears a very close resemblance to the five-channel television now under construction in this Department.

In the above systems, it will be noted that the patent position of Baird, Telefunken, Mihaly, Jenkins and the Short Wave & Television Co is weak, that the transmissions carried out by them are of a purely experimental nature, and that it appears unlikely their system[1] with the possible exception of Telefunken [which used a Kerr cell], will be developed into a commercial possibility.

All the systems have been designed to fit into the normal broadcast wavelengths with a maximum picture impulse frequency of about 5 kilo cycles. It appears unlikely that they will be capable of further improvement on account of the limitations of the picture "grain", by the present broadcasting conditions.

With regard to picture grain, there appears to be no accepted view as to the minimum number of elements required to build up a picture which will have a definitive "interest" value.

In the case of the Baird system, the number of picture elements is 1800—$12\frac{1}{2}$ pictures per second are transmitted, and a maximum modulation frequency of approximately 10 000 is required. This maximum is not necessary for the production of reasonably good pictures and Baird achieves tolerable definition with rather less than half the maximum frequency range.

In the case of the RCA cathode-ray method with 6400 elements transmitted at the rate of 12 times per second, a maximum modulation frequency of 38 400 is required and a good definition is probably achieved with about 12 000.

In the case of Dr Alexanderson's 168 line picture, the maximum modulation frequency will be of the order of 100 kilocycles.

Our own system gives approximately the same definition as Dr Alexanderson's, but it is designed to test the question of picture grain since the number of lines per picture may be varied from 60 upwards.

Pending a decision on the minimum number of elements necessary for the production of a good picture it is clear that an 80-line picture at least, will be required, but in the future something of the order of 150 lines will be required.

If the above estimate of the number of picture elements is conceded it is clear that television on the normal broadcast band of wavelengths is impossible, and that it will be necessary to employ wavelengths of less than 150 metres, and ultimately of ultra short wavelengths.'

(From this survey the writer, G E Condliffe, opined that 'the vast majority' of the Baird patents were useless with the 'solitary exception' of 320 639 which claimed broadly the use of the same channel for synchronising and picture impulses.)

[1] These systems used a neon lamp; Telefunken also had a Kerr cell system

'With regard to their system it is significant to note that similar apparatus is used by Mihaly, Telefunken, Jenkins and the Short Wave & Television Company. The latter two companies are almost equivalent to the Baird Company in regard to their patent situation and to their future prospects.

Before television can become an established fact, a system must be produced which will project a picture of at least 150 lines i.e. roughly 15 000 elements; further, the picture must be projected on to a screen which has an area of not less than 6" × 6", and preferably one foot square.

The Baird system[2] is incapable of modification to produce these results, even if the broadcasting conditions allowed it.'

G E Condliffe
Advanced development division

C O Browne's diary (04.09.1933–17.04.1935)
C O Browne was one of EMI's outstanding engineers. At the Gramophone Company he had engineered the 150-line zone television system which the company had exhibited at the January 1931 Physical and Optical Societies Exhibition, London. Later, following the merger of HMV and the Columbia Graphophone Company in 1931, Browne had contined to work on television problems. He was much involved in the development of EMI's high-definition system.

Browne's diary, which is the only extant diary which records details of EMI's technical progress, covers the period September 1933 to April 1935. In it brief details of the research and development effort and of the visits made by personages having an interest in the advancement of television have been noted.

His entry for 10 March 1935 refers to iconoscope no. 83 (sic). The statement lends weight to the opinions of McGee, Shoenberg, Preston, Broadway and Lubszynski that EMI's development of the iconoscope (which EMI called the emitron) was independent of RCA's development of the iconoscope. Otherwise it would not have been necessary for EMI to have engaged in such an extended experimental (and costly) programme to fabricate such a tube.

The diary records the following visits and demonstrations:

Date	Visitor	Organisastion
04.09.1933	D Sarnoff	RCA
17.11.1933	Ramsay Macdonald, Lord Marks, L Sterling, A Clark	H M Government, EMI
08.12.1933	directors	EMI
22.12.1933	directors	EMI
12.01.1934	N Ashbridge, H L Kirke	BBC

[2] Condliffe here alludes to Baird's neon lamp system

23.01.1934	H L Kirke,	BBC
	D C Birkinshaw	

(25.01.1934 Browne started work with McGee on emitron)

26.01.1934	Vanderville	?
20.03.1934	J F W Reith	BBC

(This was the 21st demonstration of EMI's system)

27.04.1934	Taylor	Taylor, Taylor, Hobson Ltd
11.05.1934	representatives	GPO

(First demonstration of direct television)

01.06.1934	members	Television Committee
05.06.1934	representatives	Telefunken
11.06.1934	A Clark, O F Brown,	EMI, Television Committee
	R A Watson Watt	Radio Research Station
13.06.1934	members	Television Committee
14.06.1934	representatives	Admirally
11.09.1934	H M Dowsett	Marconi Wireless Telegraph
10.10.1934	Sir Stafford Cripps	H M Government
29.10.1934	Dr V K Zworykin	RCA
02.11.1934	Vanderville	?
30.11.1934	Prince of Wales	Royalty
14.12.1934	members	Television Committee
11.01.1935	O F Brown, Sir Frank	Television Committee
	Smith	
14.01.1935	D C Birkinshaw	BBC
22.01.1935	Lord Invergordon	
24.01.1935	staff from head office	EMI

(The report of the Television Committee was published 31.01.1935,)

12.02.1935	'Two lots of vistitors'	?

(06.02.1935, first mention of 405 lines is given)

13.02.1935	representatives	GPO
14.02.1935	representatives	Telefunken
18.02.1935	Lord Selsdon	TAC

(The diary mentions 'Good demonstration of street scenes, studio, view from Labs.')

25.02.1935	N Ashbridge, Sir Frank	BBC and TAC
	Smith, O F Brown and	
	T C MacNamara	
26.02.1935	R M Ellis	RMA
27.02.1935	A J Gill	GPO
11.03.1935	Vanderbilt	?
23.03.1935	Messrs Reddie &	patent agents
	Grose	
27.03.1935	Vanderbilt &	?
	Frenchmen	
12.04.1935	members	TAC
17.04.1935	N E Davis	MWT

Appendix 5

Comparison of intensity and velocity modulation*

	Intensity modulation	Velocity modulation
Scanning	signal storage tube or flying spot available for direct vision	signal storage for direct vision impracticable
Transmission efficiency	almost all of straight portion of transmitter characteristic can be used for useful picture signals	owing to implicit synchronisation only a small fraction of the transmitter characteristic can be used for useful picture signals. Requires 20 to 100 times the transmitter power necessry for amplitude modulation
Frequency bandwidth and Resolution	equal definition at all intensities	low definition in dark parts. Requires greater frequency band to give the same performance as amplitude modulation
Background brightness component	transmitted	not transmitted
Transmitter equipment	high quality amplifiers feasible	high quality amplifiers may be found impracticable
Receiver equipment	hard cathode-ray tube practically necessary, mechanical scanning can be used	reduced peak brilliancy of cathode-ray tube required; soft tube can be used

*BLUMLEIN, A.D.: 'Velocity modulation'. 24 November 1934, post 33/4682, part 1, file 3

Appendix 6

Estimated cost of a television service from London and four regional stations

Revenue expenditure	London station (3 studios, including present theatre)	Four regional stations	Total
Programmes–artists, orchestras, etc.	151 000	-	151 000
Cable link	6 000	104 000	110 000
Engineering–power and maintenance	58 000	40 000	98 000
Premises–maintenance and minor alterations	33 000	4 000	37 000
major alterations	7 000		7 000
Salaries and pension scheme	146 000	32 000	178 000
Depreciation	92 000	85 000	177 000
Overheads	50 000	27 000	77 000
Income tax	17 000	25 000	42 000
	£560 000	£317 000	£877 000
			say £900 000

Capital expenditure			
Improvements to plant	85 000		85 000
Theatre equipment	90 000		90 000
Land, premises and plant	-	512 000	512 000
	£175 000	£512 000	£687 000

Bibliography

ABRAMSON, A.: 'Electronic motion pictures' (University of California Press, Berkley, 1955)
ABRAMSON, A.: 'The history of television, 1880 to 1941' (McFarland, Jefferson, 1987)
ABRAMSON, A,: 'Zworykin, pioneer of television' (University of Illinois Press, Urbana, 1995)
AISBERG, E., and ASCHEN, R.: 'Theorie et practique de la television' (Chiron, Paris, 1932)
ARCHER, G.L.: 'History of radio to 1926' (American Historical Society, New York, 1938)
ARCHER, G.L.: 'Big business and radio' (American Historical, New York, 1939)
BAIRD, J.L.: 'Sermons, soap and television' (Royal Television Society, London, 1988)
BAIRD, M.: 'Television Baird' (HAUM, Cape Town, 1973)
BAKER,T.T.: 'Wireless pictures and television' (Constable, London, 1926)
BAKER,T.T.: 'The telegraphic transmission of photographs' (Constable, London,1910)
BAKER,W.J.: 'A history of the Marconi Company' (Methuen, London,1970)
BARNOUW, E.: 'A tower in Babel' (Oxford University Press, 1966)
BARNOUW, E.: 'The golden web' (Oxford University Press, New York, 1968)
BENSON, T.W.: 'Fundamentals of television' (Mancall, New York, 1930)
BLAKE, G.G.: 'History of radio telegraphy and telephony' (Radio Press, London, 1926)
BRIGGS, A.: 'The golden age of wireless' (Oxford University Press, London, 1965)
BRIDGEWATER, T.H.: 'A A Campbell Swinton' (Royal Television Society, London, 1982)
BRUCE, R.V.: 'Bell' (Gallancz, London, 1973)
BRUCH, W.: 'Die Fernseh story' (Franckh'sche Verlagshandlung, Stuttgart, 1969)
BURNS, R.W.: 'British television, the formative years' (Peter Peregrinus, London, 1986)
CAMM, F.J.: 'Newnes television and short-wave handbook' (Newnes, London, 1934)
CAMM, F.J.: 'Newnes television manual' (Newnes, London, 1942)
CERAM, C.W.: 'Archaeology of the cinema' (Harcourt, Brace and World, New York, 1965)
CHAPPLE, H.J.B.: 'Popular television' (Pitman, London, 1935)
COCKING, W.T.: 'Television receiving equipment' (Iliffe, London, 1940)
CRAWLEY, C.G.: 'From telegraphy to television', (Warne, London, 1931)
DE FOREST, L.: 'Television, today and tomorrow' (The Dial Press, New York, 1942)
DINSDALE, A.: 'Television' (Pitman, London, 1926)
DINSDALE, A.: 'Television' (Television Press, London, 1928)
DUNLAP, O.E.: 'The outlook for television' (Harper, New York, 1932)
DUNLAP, O.E.: 'The future of television' (Harper, New York, 1947)
DUNLAP, O.E.: 'Dunlap's radio and television almanac' (Harper, New York, 1951)
ECKHARDT, G.H.: 'Electronic television' (Goodheart-Wilcox, Chicago, 1936)
EDER, J.M.: 'History of photography' (Columbia UP, New York, 1945)
EDDY, W.C.: 'Television–the eyes of tomorrow' (Prentice-Hall, New York, 1945)
EVERSON, G.: 'The story of television, the life of Philo T Farnsworth' (Norton, New York, 1949)
FAGEN, M.D. (Ed.): 'A history of engineering and science in the Bell System (Bell Telephone Laboratories, 1975)

FELIX, E.H.: 'Television, its methods and uses' (McGraw-Hill, New York, 1939)
FINK, D.G.: 'Principles of television engineering' (McGraw-Hill, New York, 1940)
FINK, D.G.: 'Television standards and practice' (RMA-NTSC, New York, 1943)
FRIEDEL, W.: 'Elektrisches Fernesehen: Fernkinematographie and Bildfernübertragung' (Meusser, Berlin, 1925)
FUCHS, G.: 'Die Bildtelegraphie', (Siemens, Berlin, 1926)
GARRATT, G.R.M., and PARR, G.: 'Television' (HMSO, London, 1937), a Science Museum booklet
GEDDES, K.: 'Broadcasting in Britain: 1922–1972' (HMSO, London, 1972), a Science Museum booklet
GORHAM, M.: 'Television. Medium of the future' (Marshall, London, 1949)
GORHAM, M.: 'Broadcasting and television since 1900' (Dakers, London, 1952)
GOROKHOV, P.K.: 'Boris L'vovich Rosing' (Hayka, Moscow, 1964)
HALLORAN, A.H.: 'Television with cathode rays' (Pacific Radio, San Francisco, 1936)
HATSCHEK, P.: 'Electron optics' (American Photographic, Boston, 1948)
HEMARDINQUER, P.: 'La télévision et ses progrès' (Dunod, Paris, 1933)
HEMARDINQUER, P.: 'Technique et pratique de la télévision' (Dunod, Paris, 1948)
HENDRICKS, G.: 'Eadweard Muybridge, the father of the motion picture' (Grossman, New York, 1975)
HUBBELL, R.: 'Four thousand years of television' (Harrap, London, 1946)
JENKINS, C.F.: 'Animated pictures' (McQueen, Washington, DC, 1898)
JENKINS, C.F.: 'Vision by radio, radio photographs, radio photograms' (Jenkins Laboratories, Washington, DC, 1925)
JENKINS,C.F.: 'Radiomovies, radiovision, television' (Jenkins Laboratories, Washington, DC, 1929)
KORN, A., and GLATZEL, B.: 'Handbuch der Phototelegraphie und Telautographie' (Nemnich, Leipzig, 1911)
LARNER, E.T.: 'Practical television' (Ernest Benn, New York, 1928)
LEE, R.E.: 'Television: the revolution' (Essential Books, New York, 1944)
LIESEGANG, R. (Ed.): 'Beitrage zum Problem des elektrischen Fernsehens' (Dusseldorf, 1891)
LOHR, L.R.: 'Television broadcasting' (McGraw Hill, New York, 1940)
LYONS, E.: 'David Sarnoff' (Harper & Row, New York, 1966)
McGOWAN, K.: "Behind the screen' (Dell, New York, 1967)
MacLAURIN, W.R.: 'Invention and innovation in the radio industry' (Macmillan, New York, 1949)
MADDOX, B.: 'Beyond Babel. New Directions in communications' (Andre Deutsch, London, 1972)
MALOFF, I.G., and EPSTEIN, D.W.: 'Electron optics in television' (McGraw Hill, New York, 1938)
MARTIN, M.J.: 'Wireless transmission of photgraphs' (The Wireless Press, London, 1916)
MARTIN, M.J.: 'The electrical transmission of photographs' (Pitman, London, 1921)
Mc ARTHUR, T., and WADDELL, P.: 'The secret life of John Logie Baird' (Hutchinson, London, 1986)
MESNEY, R. : 'Television et transmission des images' (Armand Golin, Paris, 1933)
MOSELEY, S.A.: 'John Baird. The romance and tragedy of the pioneer of television' (Odhams Press, London, 1952)
MOSELEY, S.A., and CHAPPLE, H.J.B.: 'Television, today and tomorrow' (Pitman, London, 1931)
MYERS, L.M.: 'Television optics' (Pitman, London, 1936)
NORMAN, B.: 'Here's looking at you' (BBC and RTS, London, 1984)
OGDEN,W.B.: 'The television business' (Ronald Press, New York, 1961)
PARR, G.: 'The cathode-ray tube and its applications' (Chapman and Hall, London, 1943)
POHL, R.: 'Die elektrische Fernubertragung von Bildern' (Braunschweig, 1910)
QUIGLEY, M.: 'Magic shadows' (Georgetown UP, Washington, 1948)
REYNER, J.H.: 'Television theory and practice' (Chapman and Hall, London, 1934)
REYNER, J.H.: 'Cathode ray oscillographs' (Pitman, London, 1939)
RICHARDS, V. : 'From crystal to television' (Black, London, 1928)
ROBINSON, D.: 'The history of world cinema' (Stein and Day, New York, 1973)
ROBINSON, E.H.: 'Televiewing' (Selwyn & Blount, London, 1935)

ROSS, G.: 'Television jubilee. The story of 25 years of BBC television' (Allen, London, 1961)
ROWLAND, J.H.: 'The television man' (Lutterworth Press, London, 1966)
SHELDON, H.H., and GRISEWOOD, E.N.: 'Television: present methods of picture transmission' (Library Press, London, 1930)
SHIERS, G. (Ed.): 'Technical development of television' (Arno, New York, 1977)
SINGLETON, T.: 'The story of Scophony' (Royal Television Society, London, 1988)
STURMEY, S.G.: 'The economic development of radio' (Duckworth, London, 1958)
SCHROTER, F.: 'Handbuch der Bildtelegraphie und des Fernsehens' (Springer, Berlin, 1932)
SCHROTER, F.: 'Fernsehen' (Springer, Berlin, 1937)
TILTMAN, R.F.: 'Baird of television' (Seeley Service, London, 1933)
UDELSON, J.H.: 'The great television race: a history of the American television industry (1925–1941)' (University of Alabama Press, c.1982)
VON ARDENNE, M.: 'Funkemfangs Technik' (Rothgiesser & Diesing, Berlin, 1934)
VON ARDENNE, M.: 'Television reception' (Chapman and Hall, London, 1936)
VON ARDENNE, M.: 'Cathode ray tubes' (Pitman, London, 1939)
VON MIHALY, D.: 'Das elektrische Fernesehen und das Telehor' (Krayn, Berlin, 1923)
WALDROP, F.C. and BORKIN, J.: 'Television: a struggle for power' (William Morrow, New York, 1938)
WEST, A.G.D. (Ed.): 'Television today. Practice and principles explained' (Newnes, London, 1935)
WILSON, J.C.: 'Television engineering' (Pitman, London, 1937)
ZWORYKIN, V.K., and MORTON, G.A.: 'Television. The electronics of image transmission' (Wiley, New York, 1940)
ZWORYKIN, V.K., and WILSON, E.D.: 'Photocells and their applications' (Wiley, New York, 1930)

Index